John Senior

International Textbooks in Electrical Engineering

Consulting Editor

ROBERT F. LAMBERT
Professor of Electrical Engineering
University of Minnesota

an introduction to

Random Signals and Communication Theory

an introduction to

Random Signals and Communication Theory

B. P. LATHI

Electrical Engineering Department
Bradley University

INTERNATIONAL TEXTBOOK COMPANY
Scranton, Pennsylvania

 Printed in the United States of America by The Haddon Craftsmen, Inc., Scranton, Pennsylvania. Library of Congress Catalog Card Number: 68-16126.

For
Anjali

Preface

Among the many changes that have taken place in the undergraduate Electrical Engineering curriculum during the past few years, the introduction of probabilistic concepts and the analysis of random signals are significant. The statistical approach is essential to the understanding of modern communication theory. Presently there exist a number of excellent books which treat these topics. These are, however, primarily addressed to the graduate students. There is a need for an introductory text on random signals and modern communication theory. The present work was undertaken with this purpose in mind. Although primarily intended for seniors, it could be also used by first-year graduate students. The requisite background in statistical concepts and random processes is developed in the earlier chapters. The later chapters deal with analog and digital communication, signal detection, and information theory.

In presenting the subject at an introductory level, a teacher faces the challenge of having both to generate and to sustain the interest of the student. This can usually be done by introducing the theoretical development along with physical interpretations and examples. In consonance with this view, the approach to the subject in this book is physical rather than axiomatic. Wherever possible, theoretical results are interpreted heuristically, supported by carefully chosen examples and illustrations. Logical motivation is provided for introducing new concepts. As a result, the development of the subject matter follows naturally. This is in contrast to the axiomatic approach, where the definitions of the important concepts appear as arbitrary postulates.

I am grateful to Professors I. M. Jacobs, T. Kailath, J. H. Park, Jr., D. J. Sakrison, and T. L. Stewart for their useful comments and to Professor R. R. Boorstyn who read most of the manuscript and made many valuable suggestions. Thanks are due to Mr. J. Pichamuthu for his painstaking editorial assistance and to Professors Philip Weinberg, Head of the Electrical Engineering Department at Bradley, and R. L. Gonzales for making available departmental assistance. I wish to express my appreciation to Mrs. Patricia Pettit, Mrs. Evelyn Kahrs, and Mrs. Evabeth Stone for their efficient and careful typing. Finally I wish to acknowledge the patience and understanding of my wife Rajani during the preparation of the book.

B. P. LATHI

Peoria, Illinois
June, 1968

Note to the Instructor

This book can be used as a text for two areas:

i) The analysis of random signals (signal analysis in general)

ii) Introduction to communication theory

The first two chapters serve as a self-contained introduction to signal analysis and the theory of probability. Chapters 3 and 4 deal with the concepts, analysis, transmission, and filtering of random signals. Chapter 5 treats the transmission of analog data; this includes classical modulation (both amplitude modulation and angle modulation) and digital methods (PCM) of transmitting analog data. Transmission of digital data (using M-ary signaling) is discussed in Chapter 6 (the study is limited to optimum coherent reception). Chapter 7 introduces the reader to important concepts of information theory. The performance of various communication systems are compared in the light of Shannon's results.

The book can serve either as an introduction to more advanced study or as a terminal course in communication theory.

Contents

1 Signal Analysis . . . 1

2 Elements of Probability Theory . . . 83

5 Analog Data Communication: Modulation . . . 323

6 Digital Data Communication: Signal Detection . . . 382

7 Introduction to Information Theory . . . 428

1

Signal Analysis

Meaningful communication is possible only when there is some randomness or uncertainty involved in the messages. This can be easily seen from the fact that a statement of certainty conveys no information. Thus a message "man is mortal" conveys very little or no information at all. But a statement from a weatherman that on a certain day in January, the temperature in Minneapolis reached 120°F certainly conveys a lot of information. This is because of the high uncertainty (low probability) of this message. Evidently uncertainty (or randomness) is the essence of communication. In addition, the channel disturbances (noise) are random in nature. Indeed, as will be shown later that if the channel disturbance were not random, communication would cease to be a problem. For these reasons, the study of random signals is fundamental in understanding the process of communication.

Since we are concerned here with communication by means of signals, the first chapter will be devoted to signal analysis. In Chapter 1, we shall restrict ourselves to nonrandom signals and later in Chapter 3, shall extend the analysis to random signals.

In this chapter we shall briefly review certain important topics in signal analysis. These concepts will then be extended to random signals in later chapters.

1-1. ANALOGY BETWEEN VECTORS AND SIGNALS

A problem is better understood or better remembered if it can be associated with some familiar phenomenon. Therefore we always search for analogies while studying a new problem. In the study of abstract problems, analogies are very helpful, particularly if the problem can be shown to be analogous to a concrete phenomenon. It is then easy to gain some insight into the new problem from the knowledge of the analogous phenomenon. Fortunately, there is a perfect analogy between vectors and signals which leads to a better understanding of signal analysis. We shall now briefly review the properties of vectors.

Vectors. A vector is specified by magnitude and direction. We shall denote all vectors by boldface type and their magnitudes by lightface type, for example, **A** is a certain vector with magnitude A. Consider two vectors $\mathbf{v}_1$ and $\mathbf{v}_2$ as shown in Fig. 1-1. Let the component of $\mathbf{v}_1$ along $\mathbf{v}_2$ be given by $C_{12}\mathbf{v}_2$. How do we interpret physically the component of one vector along the other vector? Geometrically the component of a

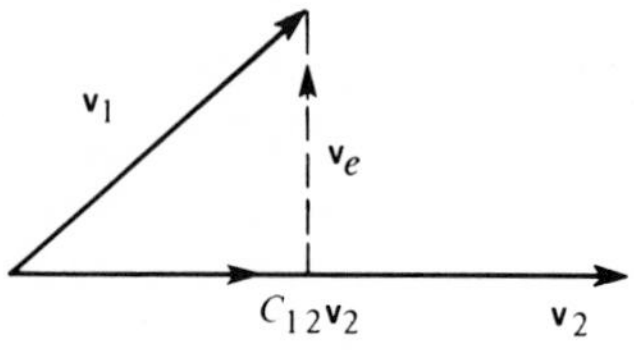

FIG. 1-1.

vector $\mathbf{v}_1$ along the vector $\mathbf{v}_2$ is obtained by drawing a perpendicular from the end of $\mathbf{v}_1$ on the vector $\mathbf{v}_2$, as shown in Fig. 1-1. The vector $\mathbf{v}_1$ can now be expressed in terms of vector $\mathbf{v}_2$.

$$\mathbf{v}_1 = C_{12}\mathbf{v}_2 + \mathbf{v}_e \tag{1-1a}$$

However, this is not the only way of expressing vector $\mathbf{v}_1$ in terms of vector $\mathbf{v}_2$. Figure 1-2 illustrates two of the infinite alternate possibilities. Thus in Fig. 1-2a

$$\mathbf{v}_1 = C_1\mathbf{v}_2 + \mathbf{v}_{e1} \tag{1-1b}$$

and in Fig. 1-2b,

$$\mathbf{v}_1 = C_2\mathbf{v}_2 + \mathbf{v}_{e2} \tag{1-1c}$$

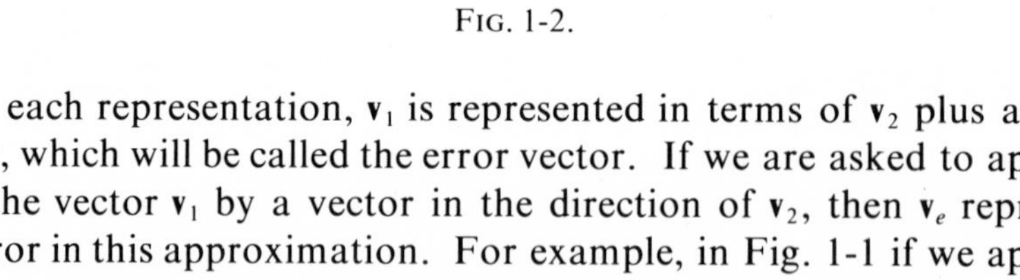

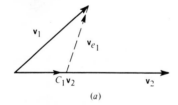

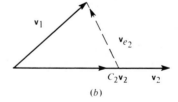

FIG. 1-2.

In each representation, $\mathbf{v}_1$ is represented in terms of $\mathbf{v}_2$ plus another vector, which will be called the error vector. If we are asked to approximate the vector $\mathbf{v}_1$ by a vector in the direction of $\mathbf{v}_2$, then $\mathbf{v}_e$ represents the error in this approximation. For example, in Fig. 1-1 if we approximate $\mathbf{v}_1$ by $C_{12}\mathbf{v}_2$, then the error in the approximation is $\mathbf{v}_e$. If $\mathbf{v}_1$ is approximated by $C_1\mathbf{v}_2$ as in Fig. 1-2a, then the error is given by $\mathbf{v}_{e1}$, and so on. What is so unique about the representation in Fig. 1-1? It is immediately evident from the geometry of these figures that the error vector is smallest in Fig. 1-1. We can now formulate a quantitative definition of a component of a vector along another vector. The component of a vector $\mathbf{v}_1$ along the vector $\mathbf{v}_2$ is given by $C_{12}\mathbf{v}_2$, where C_{12} is chosen such that the error vector is minimum.

Let us now interpret physically the component of one vector along another. It is clear that the larger the component of a vector along the other vector, the more closely do the two vectors resemble each other in their directions, and the smaller is the error vector. If the component of a vector $\mathbf{v}_1$ along $\mathbf{v}_2$ is $C_{12}\mathbf{v}_2$, then the magnitude of C_{12} is an indication of the similarity of the two vectors. If C_{12} is zero, then the vector has no component along the other vector, and hence the two vectors are mutually perpendicular. Such vectors are known as *orthogonal vectors.* If the vectors are orthogonal, then the parameter C_{12} is zero.

For convenience, we define the dot product of two vectors **A** and **B** as

$$\mathbf{A} \cdot \mathbf{B} = AB \cos \theta$$

where θ is the angle between vectors **A** and **B**. It follows from the definition that

$$\mathbf{A} \cdot \mathbf{B} = \mathbf{B} \cdot \mathbf{A}$$

According to this notation,

$$\text{Component of } \mathbf{A} \text{ along } \mathbf{B} = A \cos \theta = \frac{\mathbf{A} \cdot \mathbf{B}}{B}$$

and

$$\text{Component of } \mathbf{B} \text{ along } \mathbf{A} = B \cos \theta = \frac{\mathbf{A} \cdot \mathbf{B}}{A}$$

Similarly,

$$\text{Component of } \mathbf{v}_1 \text{ along } \mathbf{v}_2 = \frac{\mathbf{v}_1 \cdot \mathbf{v}_2}{v_2} = C_{12} v_2$$

Therefore

$$C_{12} = \frac{\mathbf{v}_1 \cdot \mathbf{v}_2}{v_2^2} = \frac{\mathbf{v}_1 \cdot \mathbf{v}_2}{\mathbf{v}_2 \cdot \mathbf{v}_2} \tag{1-2}$$

Note that if $\mathbf{v}_1$ and $\mathbf{v}_2$ are orthogonal, then

$$\mathbf{v}_1 \cdot \mathbf{v}_2 = 0 \tag{1-3}$$

and

$$C_{12} = 0$$

Signals. The concept of vector comparison and orthogonality can be extended to signals.[1] Let us consider two signals, $f_1(t)$ and $f_2(t)$.

[1]We shall often use the terms signals and functions interchangeably. A signal is a function of time. However, there is one difference between signals and functions. A function $f(t)$ can be a multivalued function of variable t. But the physical signal is always a single-valued function of t. Hence, whenever we use a term function, it will be understood that it is a single-valued function of the independent variable.

Suppose we want to approximate $f_1(t)$ in terms of $f_2(t)$ over a certain interval $(a < t < b)$ as follows.

$$f_1(t) \simeq C_{12} f_2(t) \qquad \text{for } a < t < b \tag{1-4}$$

How shall we choose C_{12} in order to achieve the best approximation? Obviously, we must find C_{12} such that the error between the actual function and the approximated function is minimum over the interval $(a < t < b)$. Let us define an error function $f_e(t)$ as

$$f_e(t) = f_1(t) - C_{12} f_2(t) \tag{1-5}$$

One possible criterion for minimizing the error $f_e(t)$ over the interval a to b is to minimize the average value of $f_e(t)$ over this interval; that is, to minimize

$$\frac{1}{(b-a)} \int_a^b [f_1(t) - C_{12} f_2(t)]\, dt$$

However, this criterion is inadequate because there can be large positive and negative errors present that may cancel each other in the process of averaging and give the false indication that the error is zero. For example, if we approximate a function sin t with a null function $f(t) = 0$ over an interval 0 to 2π, the average error will be zero, indicating wrongly that sin t can be approximated to zero over the interval 0 to 2π without any error. This situation can be corrected if we choose to minimize the average (or the mean) of the square of the error instead of the error itself. Let us designate the average of $f_e^2(t)$ by ϵ.

$$\epsilon = \frac{1}{(b-a)} \int_a^b f_e^2(t)\, dt = \frac{1}{(b-a)} \int_a^b [f_1(t) - C_{12} f_2(t)]^2\, dt \tag{1-6}$$

To find the value of C_{12} that will minimize ϵ, we must have

$$\frac{d\epsilon}{dC_{12}} = 0 \tag{1-7}$$

That is,

$$\frac{d}{dC_{12}} \left[\frac{1}{(b-a)} \int_a^b [f_1(t) - C_{12} f_2(t)]^2\, dt \right] = 0 \tag{1-8}$$

Changing the order of integration and differentiation, we get

$$\frac{1}{b-a} \left[-2 \int_a^b f_1(t) f_2(t)\, dt + 2C_{12} \int_a^b f_2^2(t)\, dt \right] = 0 \tag{1-9}$$

Hence

$$C_{12} = \frac{\displaystyle\int_a^b f_1(t) f_2(t)\, dt}{\displaystyle\int_a^b f_2^2(t)\, dt} \tag{1-10}$$

Observe the similarity between Eqs. 1-10 and 1-2, which expresses C_{12} for vectors.

By analogy with vectors, we say that $f_1(t)$ has a component of waveform $f_2(t)$, and this component has a magnitude C_{12}. If C_{12} vanishes, then the signal $f_1(t)$ contains no component of signal $f_2(t)$ and we say that the two functions are orthogonal over the interval (a, b). It therefore follows that the two functions $f_1(t)$ and $f_2(t)$ are orthogonal over the interval (a, b) if

$$\int_a^b f_1(t)\, f_2(t)\, dt = 0 \tag{1-11}$$

Observe the similarity between Eq. 1-11 derived for orthogonal functions and Eq. 1-3 derived for orthogonal vectors.

We can easily show that the functions $\sin n\omega_0 t$ and $\sin m\omega_0 t$ are orthogonal over any interval $(t_0, t_0 + 2\pi/\omega_0)$ for integral values of n and $m (n \neq m)$. Consider the integral I:

$$\begin{aligned} I &= \int_{t_0}^{t_0+2\pi/\omega_0} \sin n\omega_0 t \sin m\omega_0 t\, dt \\ &= \int_{t_0}^{t_0+2\pi/\omega_0} \frac{1}{2}[\cos (n - m)\,\omega_0 t - \cos (n + m)\,\omega_0 t]\, dt \\ &= \frac{1}{2\omega_0}\left[\frac{1}{(n - m)} \sin (n - m)\,\omega_0 t - \frac{1}{(n + m)} \sin (n + m)\,\omega_0 t\right]_{t_0}^{t_0+2\pi/\omega_0} \end{aligned}$$

Since n and m are integers, $(n - m)$ and $(n + m)$ are also integers. In such a case the integral I is zero. Hence the two functions are orthogonal. Similarly, for integral values of m and n it can be shown that $\sin n\omega_0 t$ and $\cos m\omega_0 t$ are orthogonal functions and $\cos n\omega_0 t$, $\cos m\omega_0 t$ are also mutually orthogonal (for $m \neq n$).

Example 1-1. A rectangular function $f(t)$ (Fig. 1-3) is defined by

$$f(t) = \begin{cases} 1, & (0 < t < \pi) \\ -1, & (\pi < t < 2\pi) \end{cases}$$

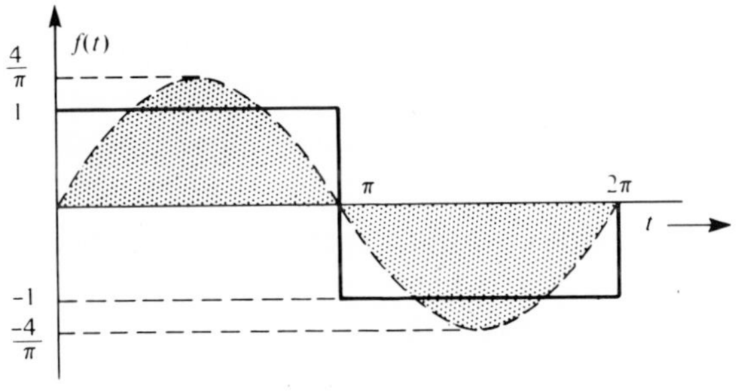

FIG. 1-3.

Approximate this function by a waveform sin t over the interval $(0, 2\pi)$ such that the mean-square error is minimum.

Solution. The function $f(t)$ will be approximated over the interval $(0, 2\pi)$, as

$$f(t) \simeq C_{12} \sin t$$

We shall find the optimum value of C_{12} which will minimize the mean-square error in this approximation. According to Eq. 1-10 to minimize the mean-square error:

$$C_{12} = \frac{\int_0^{2\pi} f(t) \sin t \, dt}{\int_0^{2\pi} \sin^2 t \, dt}$$

$$= \frac{1}{\pi}\left[\int_0^{\pi} \sin t \, dt + \int_{\pi}^{2\pi} - \sin t \, dt\right.$$

$$= \frac{4}{\pi}$$

Thus

$$f(t) \simeq \frac{4}{\pi} \sin t$$

represents the best approximation of $f(t)$ by a function sin t which will minimize the mean-square error.

By analogy with vectors, we may say that the rectangular function $f(t)$ shown in Fig. 1-3 has a component of function sin t and the magnitude of this component is $4/\pi$.

What is the significance of orthogonality of two functions? In the case of vectors, orthogonality implies that one vector has no component along the other. Similarly, a function does not contain any component of the form of the function which is orthogonal to it. If we try to approximate a function by its orthogonal function, the error will be larger than the original function itself, and it is better to approximate a function with a null function $f(t) = 0$ rather than with a function orthogonal to it. Hence the optimum value of $C_{12} = 0$ in such a case.

Orthogonal Vector Space. The analogy between vectors and signals may be extended further. Let us now consider a three-dimensional vector space described by rectangular coordinates, as shown in Fig. 1-4. We shall designate a vector of unit length along the x axis by $\mathbf{a}_x$. Similarly, unit vectors along the y and z axes will be designated by $\mathbf{a}_y$ and $\mathbf{a}_z$, respectively. Since the magnitude of vectors $\mathbf{a}_x$, $\mathbf{a}_y$, and $\mathbf{a}_z$ is unity, it follows that for any general vector $\mathbf{F}$

$$\text{Component of } \mathbf{F} \text{ along the } x \text{ axis} = \mathbf{F} \cdot \mathbf{a}_x$$

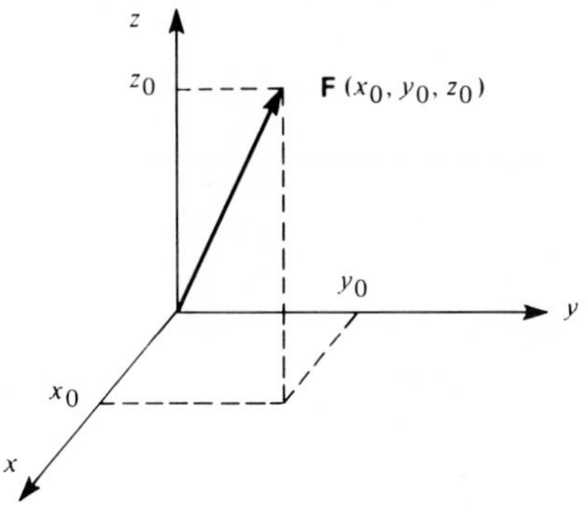

FIG. 1-4.

$$\text{Component of } \mathbf{F} \text{ along the } y \text{ axis} = \mathbf{F} \cdot \mathbf{a}_y$$
$$\text{Component of } \mathbf{F} \text{ along the } z \text{ axis} = \mathbf{F} \cdot \mathbf{a}_z$$

A vector **F** drawn from the origin to a general point (x_0, y_0, z_0) in space has components x_0, y_0, and z_0 along the x, y, and z axes, respectively. We can express this vector **F** in terms of its components along the three mutually perpendicular axes:

$$\mathbf{F} = x_0\mathbf{a}_x + y_0\mathbf{a}_y + z_0\mathbf{a}_z$$

Any vector in this space can be expressed in terms of the three vectors $\mathbf{a}_x$, $\mathbf{a}_y$, and $\mathbf{a}_z$.

Since the three vectors $\mathbf{a}_x$, $\mathbf{a}_y$, and $\mathbf{a}_z$ are mutually perpendicular, it follows that

$$\mathbf{a}_x \cdot \mathbf{a}_y = \mathbf{a}_y \cdot \mathbf{a}_z = \mathbf{a}_z \cdot \mathbf{a}_x = 0 \tag{1-12}$$

and since these are unit vectors

$$\mathbf{a}_x \cdot \mathbf{a}_x = \mathbf{a}_y \cdot \mathbf{a}_y = \mathbf{a}_z \cdot \mathbf{a}_z = 1$$

The properties of the three vectors, as expressed by Eq. 1-12, can be succinctly expressed by

$$\mathbf{a}_m \cdot \mathbf{a}_n = \begin{cases} 0 & m \neq n \\ 1 & m = n \end{cases} \tag{1-13}$$

where m and n can assume any symbol x, y, and z.

Now we make an important observation. If the coordinate system has only two axes, x and y, then the system is inadequate to express a general vector **F** in terms of the components along these axes. This system can only express two components of vector **F**. Therefore, to express any general vector **F** in terms of its coordinate components the system of coordinates must be complete. In this case there must be three coordinate axes.

A single straight line represents a one-dimensional space; a single plane represents a two-dimensional space; and our universe, in general, has three dimensions of space. We may extend our concepts as developed here to a general n-dimensional space. Such a physical space, of course, does not exist in nature. Nevertheless, there are many analogous problems that may be viewed as n-dimensional problems. For example, a linear equation in n independent variables may be viewed as a vector expressed in terms of its components along n mutually perpendicular coordinates.

If unit vectors along these n mutually perpendicular coordinates are designated as $\mathbf{x}_1, \mathbf{x}_2, \ldots, \mathbf{x}_n$ and a general vector $\mathbf{F}$ in this n dimensional space has components $C_1, C_2, \ldots, C_n$ respectively along these n coordinates, then

$$\mathbf{F} = C_1\mathbf{x}_1 + C_2\mathbf{x}_2 + C_3\mathbf{x}_3 + \cdots + C_n\mathbf{x}_n \tag{1-14}$$

All the vectors $\mathbf{x}_1, \mathbf{x}_2, \ldots, \mathbf{x}_n$ are mutually orthogonal, and the set must be complete in order for any general vector $\mathbf{F}$ to be represented by Eq. 1-14. The condition of orthogonality implies that the dot product of any two vectors $\mathbf{x}_n$ and $\mathbf{x}_m$ must be zero, and being unit vectors the dot product of any vector with itself must be unity. This is the direct extension of Eq. 1-13 and can be expressed as

$$\mathbf{x}_m \cdot \mathbf{x}_n = \begin{cases} 0 & m \neq n \\ 1 & m = n \end{cases} \tag{1-15}$$

The constants $C_1, C_2, C_3, \ldots, C_n$ in Eq. 1-14 represent the magnitudes of the components of $\mathbf{F}$ along the vectors $\mathbf{x}_1, \mathbf{x}_2, \mathbf{x}_3, \ldots, \mathbf{x}_n$ respectively. It follows that

$$C_r = \mathbf{F} \cdot \mathbf{x}_r \tag{1-16}$$

This result can also be obtained by taking the dot product of both sides in Eq. 1-14 with vector $\mathbf{x}_r$. We have

$$\mathbf{F} \cdot \mathbf{x}_r = C_1\mathbf{x}_1 \cdot \mathbf{x}_r + C_2\mathbf{x}_2 \cdot \mathbf{x}_r + \cdots + C_r\mathbf{x}_r \cdot \mathbf{x}_r + \cdots + C_n\mathbf{x}_n \cdot \mathbf{x}_r \tag{1-17}$$

From Eq. 1-15 it follows that all the terms of the form $C_j\mathbf{x}_j \cdot \mathbf{x}_r (j \neq r)$ on the right-hand side of Eq. 1-17 are zero. Therefore

$$\mathbf{F} \cdot \mathbf{x}_r = C_r\mathbf{x}_r \cdot \mathbf{x}_r = C_r \tag{1-18}$$

We call the set of vectors $(\mathbf{x}_1, \mathbf{x}_2, \ldots, \mathbf{x}_n)$ an orthogonal vector space. In general, the product $\mathbf{x}_m \cdot \mathbf{x}_n$ can be some constant k_m instead of unity. When k_m is unity, the set is called normalized orthogonal set, or *orthonormal* vector space. Therefore, in general, for orthogonal vector space $\{\mathbf{x}_r\} \cdots (r = 1, 2, \ldots, n)$ we have

$$\mathbf{x}_m \cdot \mathbf{x}_n = \begin{cases} 0 & m \neq n \\ k_m & m = n \end{cases} \tag{1-19}$$

For an orthogonal vector space, Eq. 1-18 is modified to

$$\mathbf{F} \cdot \mathbf{x}_r = C_r \mathbf{x}_r \cdot \mathbf{x}_r = C_r k_r$$

and

$$C_r = \frac{\mathbf{F} \cdot \mathbf{x}_r}{k_r}$$

We shall now summarize the results of our discussion. For an orthogonal vector space $\{\mathbf{x}_r\} \cdots (r = 1, 2, \ldots)$,

$$\mathbf{x}_m \cdot \mathbf{x}_n = \begin{cases} 0 & m \neq n \\ k_m & m = n \end{cases} \tag{1-20}$$

If this vector space is complete, then any vector **F** can be expressed as

$$\mathbf{F} = C_1 \mathbf{x}_1 + C_2 \mathbf{x}_2 + \cdots + C_r \mathbf{x}_r + \cdots \tag{1-21}$$

where

$$C_r = \frac{\mathbf{F} \cdot \mathbf{x}_r}{k_r} = \frac{\mathbf{F} \cdot \mathbf{x}_r}{\mathbf{x}_r \cdot \mathbf{x}_r} \tag{1-22}$$

Orthogonal Signal Space. We shall now apply certain concepts of vector space to gain some intuition about signal analysis. We have seen that any vector can be expressed as a sum of its components along n mutually orthogonal vectors, provided these vectors formed a complete set of coordinate system. We therefore suspect that it may be possible to express any function $f(t)$ as a sum of its components along a set of mutually orthogonal functions if these functions form a complete set. We shall now show that this indeed is the case.

Approximation of a Function by a Set of Mutually Orthogonal Functions. Let us consider a set of n functions $g_1(t), g_2(t), \ldots, g_n(t)$ which are orthogonal to one another over an interval a to b; that is,

$$\int_a^b g_j(t) g_k(t) = 0 \qquad j \neq k \tag{1-23a}$$

and let

$$\int_a^b g_j^2(t)\, dt = K_j \tag{1-23b}$$

Let an arbitrary function $f(t)$ be approximated over an interval (a, b) by a linear combination of these n mutually orthogonal functions.

$$f(t) \simeq C_1 g_1(t) + C_2 g_2(t) + \cdots + C_k g_k(t) + \cdots + C_n g_n(t) \tag{1-24}$$

$$= \sum_{r=1}^{n} C_r g_r(t) \tag{1-25}$$

For the best approximation we must find the proper values of constants $C_1, C_2, \ldots, C_n$ such that ϵ, the mean-square error $f_e(t)$, is minimized.

By definition,

$$f_e(t) = f(t) - \sum_{r=1}^{n} C_r g_r(t)$$

and

$$\epsilon = \frac{1}{b-a} \int_a^b \left[f(t) - \sum_{r=1}^{n} C_r g_r(t) \right]^2 dt \tag{1-26}$$

It is evident from Eq. 1-26 that ϵ is a function of $C_1, C_2, \ldots, C_n$ and to minimize ϵ, we must have

$$\frac{\partial \epsilon}{\partial C_1} = \frac{\partial \epsilon}{\partial C_2} = \cdots = \frac{\partial \epsilon}{\partial C_j} = \cdots = \frac{\partial \epsilon}{\partial C_n} = 0 \tag{1-27}$$

Let us consider the equation

$$\frac{\partial \epsilon}{\partial C_j} = 0 \tag{1-28}$$

Since $(b - a)$ is constant, Eq. 1-28 may be expressed as

$$\frac{\partial}{\partial C_j} \int_a^b \left[f(t) - \sum_{r=1}^{n} C_r g_r(t) \right]^2 dt = 0 \tag{1-29}$$

When we expand the integrand, we note that all the terms arising due to the cross-product of the orthogonal functions are zero by virtue of orthogonality; that is, all the terms of the form $\int g_j(t) g_k(t)\, dt$ are zero as expressed in Eq. 1-23 (for $j \neq k$). Similarly, the derivative with respect to C_j of all the terms that do not contain C_j are zero. This leaves only two nonzero terms in Eq. 1-29 as follows:

$$\frac{\partial}{\partial C_j} \int_a^b [-2C_j f(t) g_j(t) + C_j^2 g_j^2(t)]\, dt = 0 \tag{1-30}$$

Changing the order of differentiation and integration in Eq. 1-30, we get

$$2 \int_a^b f(t) g_j(t)\, dt = 2C_j \int_a^b g_j^2(t)\, dt \tag{1-31}$$

Therefore

$$C_j = \frac{\int_a^b f(t) g_j(t)\, dt}{\int_a^b g_j^2(t)\, dt} \tag{1-32}$$

$$= \frac{1}{K_j} \int_a^b f(t) g_j(t)\, dt \tag{1-33}$$

We may summarize this result as follows. Given a set of n functions $g_1(t), g_2(t), \ldots, g_n(t)$ mutually orthogonal over the interval (a, b), it is possible to approximate an arbitrary function $f(t)$ over this interval by a linear combination of these n functions.

$$f(t) \simeq C_1 g_1(t) + C_2 g_2(t) + \cdots + C_n g_n(t)$$

$$= \sum_{r=1}^{n} C_r g_r(t) \tag{1-34}$$

For the best approximation, that is, the one that will minimize the mean of the square error over the interval, we must choose the coefficients $C_1, C_2, \ldots, C_n$, etc., as given by Eq. 1-32.

Evaluation of Mean-Square Error. Let us now find the value of ϵ when optimum values of coefficients $C_1, C_2, \ldots, C_n$ are chosen according to Eq. 1-32. By definition,

$$\epsilon = \frac{1}{b-a} \int_a^b \left[f(t) - \sum_{r=1}^{n} C_r g_r(t) \right]^2 dt \tag{1-35}$$

$$= \frac{1}{b-a} \int_a^b f^2(t)\,dt + \sum_{r=1}^{n} C_r^2 \cdot \int_a^b g_r^2(t)\,dt$$

$$- 2 \sum_{r=1}^{n} C_r \int_a^b f(t) g_r(t)\,dt \tag{1-36}$$

But from Eq. 1-33 it follows that

$$\int_a^b f(t) g_r(t)\,dt = C_r \int_a^b g_r^2(t)\,dt = C_r K_r \tag{1-37}$$

Substituting Eq. 1-37 in Eq. 1-36, we get

$$\epsilon = \frac{1}{(b-a)} \int_a^b f^2(t)\,dt + \sum_{r=1}^{n} C_r^2 K_r - 2 \sum_{r=1}^{n} C_r^2 K_r$$

$$= \frac{1}{(b-a)} \int_a^b f^2(t)\,dt - \sum_{r=1}^{n} C_r^2 K_r \tag{1-38}$$

$$= \frac{1}{(b-a)} \int_a^b f^2(t)\,dt - (C_1^2 K_1 + C_2^2 K_2 + \cdots + C_n^2 K_n) \tag{1-39}$$

One can therefore evaluate the mean-square error by using Eq. 1-39.

Representation of a Function by a Closed or a Complete Set of Mutually Orthogonal Functions. It is evident from Eq. 1-39 that if we increase n, that is, if we approximate $f(t)$ by a larger number of orthogonal functions, the error will become smaller. But by its very definition, ϵ is a positive quantity; hence in the limit as the number of terms is made infinity, the sum $\sum_{r=1}^{\infty} C_r^2 K_r$ may converge to the integral

$$\int_a^b f^2(t)\,dt$$

and then $\epsilon \longrightarrow 0$. Thus

$$\int_a^b f^2(t)\,dt \rightarrow \sum_{r=1}^{\infty} C_r^2 K_r \tag{1-40}$$

Under these conditions $f(t)$ is represented by the infinite series:

$$f(t) = C_1 g_1(t) + C_2 g_2(t) + \cdots + C_r g_r(t) + \cdots \tag{1-41}$$

The infinite series on the right-hand side of Eq. 1-40 thus converges to $f(t)$ such that the mean square of the error approaches zero. The series is said to *converge in the mean.*

A set of functions $g_1(t), g_2(t), \ldots, g_r(t)$ mutually orthogonal over the interval (a, b) is said to be a complete or a closed set if there exists no function $x(t)$ for which it is true that

$$\int_a^b x(t)\,g_k(t)\,dt = 0 \qquad \text{for } k = 1, 2, \ldots$$

If a function $x(t)$ could be found such that the above integral is zero, then obviously $x(t)$ is orthogonal to each member of the set $g_r(t)$ and consequently is itself a member of the set. Evidently the set cannot be complete without $x(t)$ being its member.

Let us now summarize the results of this discussion. For a set $\{g_r(t)\}$, $(r = 1, 2, \ldots)$ mutually orthogonal over the interval (a, b),

$$\int_a^b g_m(t)\,g_n(t)\,dt = \begin{cases} 0 & \text{if } m \neq n \\ K_m & \text{if } m = n \end{cases} \tag{1-42}$$

If this set is complete, then a function $f(t)$, can be expressed as

$$f(t) = C_1 g_1(t) + C_2 g_2(t) + \cdots + C_r g_r(t) + \cdots \tag{1-43}$$

where

$$C_r = \frac{\int_a^b f(t)\,g_r(t)\,dt}{K_r} = \frac{\int_a^b f(t)\,g_r(t)\,dt}{\int_a^b g_r^2(t)\,dt} \tag{1-44}$$

Comparison of Eqs. 1-42 to 1-44 with Eqs. 1-20 to 1-22 brings out forcefully the analogy between vectors and signals. Any vector can be expressed as a sum of its components along n mutually orthogonal vectors, provided these vectors form a complete set. Similarly, a function $f(t)$ can be expressed as a sum of its components along mutually orthogonal functions, provided these functions form a closed or a complete set.

Equation 1-43 shows that $f(t)$ contains a component of signal $g_r(t)$, and this component has a magnitude C_r. Representation of $f(t)$ by a set of infinite mutually orthogonal functions is called a *generalized Fourier series* representation of $f(t)$.

Orthogonality in Complex Functions. In the above discussion, we have considered only real orthogonal functions of real variable. If $f_1(t)$ and $f_2(t)$ are complex functions of real variable t, then it can be shown that $f_1(t)$ can be approximated by $C_{12}f_2(t)$ over an interval (a, b):

$$f_1(t) \simeq C_{12}f_2(t)$$

The optimum value of C_{12} to minimize the mean-square error magnitude is given (Ref. 1-1) by

$$C_{12} = \frac{\int_a^b f_1(t)f_2^*(t)\,dt}{\int_a^b f_2(t)f_2^*(t)\,dt} \tag{1-45}$$

where $f_2^*(t)$ is a complex conjugate of $f_2(t)$.

It is evident from Eq. 1-45 that two complex functions $f_1(t)$ and $f_2(t)$ are orthogonal over the interval (a, b) if

$$\int_a^b f_1(t)f_2^*(t)\,dt = \int_a^b f_1^*(t)f_2(t)\,dt = 0 \tag{1-46}$$

For a set of complex functions $\{g_r(t)\}$, $(r = 1, 2, \ldots)$ mutually orthogonal over the interval (a, b):

$$\int_a^b g_m(t)g_n^*(t)\,dt = \begin{cases} 0 & \text{if } m \neq n \\ K_m & \text{if } m = n \end{cases} \tag{1-47}$$

if this set of functions is complete, then a function $f(t)$ can be expressed as

$$f(t) = C_1g_1(t) + C_2g_2(t) + \cdots + C_rg_r(t) + \cdots \tag{1-48}$$

where

$$C_r = \frac{1}{K_r}\int_a^b f(t)g_r^*(t)\,dt \tag{1-49}$$

If the set of functions is real, then $g_r^*(t) = g_r(t)$, and all the results for complex functions reduce to those obtained for real functions in Eqs. 1-42 to 1-44.

1-2. SOME EXAMPLES OF ORTHOGONAL FUNCTIONS

Representation of a function over a certain interval by a linear combination of mutually orthogonal functions is called Fourier series representation of a function. There exist, however, a large number of sets of orthogonal functions, and hence a given function may be expressed in terms of different sets of orthogonal functions. In vector space this is analogous to the representation of a given vector in different sets of coordinate systems. Each set of orthogonal functions corresponds to a coordinate system. Some of the examples of sets of orthogonal functions are trigonometric functions, exponential functions, Legendre polynomials, Jacobi polynomials, etc. Bessel functions also form a special kind of orthogonal functions.

Trigonometric Fourier Series. We have already shown that functions $\sin \omega_0 t$, $\sin 2\omega_0 t$, etc., form an orthogonal set over any interval $(t_0, t_0 + 2\pi/\omega_0)$. This set, however, is not complete. This is evident from the fact a function $\cos n\omega_0 t$ is orthogonal to $\sin m\omega_0 t$ over the same interval. Hence, to complete the set, we must include cosine as well as sine functions. It can be shown that the composite set of functions consisting of a set $\cos n\omega_0 t$ and $\sin n\omega_0 t$ for $(n = 0, 1, 2, \ldots)$ forms a complete orthogonal set. Note that for $n = 0$, $\sin n\omega_0 t$ is zero, but $\cos n\omega_0 t = 1$. Thus we have a complete orthogonal set represented by functions 1, $\cos \omega_0 t$, $\cos 2\omega_0 t, \ldots, \cos n\omega_0 t, \ldots$; $\sin \omega_0 t$, $\sin 2\omega_0 t$, $\ldots, \sin n\omega_0 t, \ldots$, etc. It therefore follows that a function $f(t)$ can be represented in terms of these functions over any interval $(t_0, t_0 + 2\pi/\omega_0)$. Thus

$$f(t) = a_0 + \sum_{n=1}^{\infty} a_n \cos n\omega_0 t + b_n \sin n\omega_0 t \qquad (t_0 < t < t_0 + T) \quad T = 2\pi/\omega_0 \tag{1-50}$$

Equation 1-50 is the trigonometric Fourier series representation of $f(t)$ over an interval $(t_0, t_0 + T)$. The various constants a_n and b_n are given by Eq. 1-44. Therefore

$$a_0 = \frac{1}{T} \int_{t_0}^{t_0+T} f(t)\, dt \tag{1-51a}$$

$$a_n = \frac{2}{T} \int_{t_0}^{t_0+T} f(t) \cos n\omega_0 t\, dt \tag{1-51b}$$

$$b_n = \frac{2}{T} \int_{t_0}^{t_0+T} f(t) \sin n\omega_0 t\, dt \tag{1-51c}$$

The constant term a_0 in the series is given by Eq. 1-51a. It is evident that a_0 is the average value of $f(t)$ over the interval $(t_0,\ t_0 + T)$. Thus, a_0 is the d-c component of $f(t)$ over this interval.

Exponential Fourier Series. It can be easily shown that a set of exponential functions $\{e^{jn\omega_0 t}\}$, $(n = 0, 1, 2, \dots)$ is orthogonal over an interval $(t_0,\ t_0 + 2\pi/\omega_0)$ for any value of t_0. Note that this is a set of complex functions. We can demonstrate the orthogonality of this set by considering the integral

$$I = \int_{t_0}^{t_0+2\pi/\omega_0} (e^{jn\omega_0 t})(e^{jm\omega_0 t})^* \, dt = \int_{t_0}^{t_0+2\pi/\omega_0} e^{jn\omega_0 t} e^{-jm\omega_0 t} \, dt \qquad (1\text{-}52)$$

If $n = m$, the integral I is given by

$$I = \int_{t_0}^{t_0+2\pi/\omega_0} dt = \frac{2\pi}{\omega_0}$$

If $n \neq m$, the integral in Eq. 3-64 is given by

$$I = \frac{1}{j(n - m)\omega_0} e^{j(n-m)\omega_0 t} \Big|_{t_0}^{t_0+2\pi/\omega_0}$$

$$= \frac{1}{j(n - m)\omega_0} e^{j(n-m)\omega_0 t}[e^{j2\pi(n-m)} - 1]$$

Since both n and m are integers, $e^{j2\pi(n-m)}$ is equal to unity and hence the integral is zero:

$$I = 0$$

Thus

$$\int_{t_0}^{t_0+2\pi/\omega_0} e^{jn\omega_0 t}(e^{jm\omega_0 t})^* \, dt = \begin{cases} \dfrac{2\pi}{\omega_0} & m = n \\ 0 & m \neq n \end{cases} \qquad (1\text{-}53)$$

As before, let

$$\frac{2\pi}{\omega_0} = T$$

It is evident from Eq. 1-53 that the set of functions

$$\{e^{jn\omega_0 t}\}, \qquad (n = 0, \pm 1, \pm 2, \dots)$$

is orthogonal over the interval $(t_0,\ t_0 + T)$, where $T = 2\pi/\omega_0$. Further, it can be shown that this is a complete set. It is therefore possible to represent a function $f(t)$ by a linear combination of exponential functions over an interval $(t_0,\ t_0 + T)$:

$$f(t) = \sum_{n=-\infty}^{\infty} F_n e^{jn\omega_0 t} \qquad (t_0 < t < t_0 + T) \qquad (1\text{-}54)$$

where $\omega_0 = 2\pi/T$ and the summation in Eq. 1-54 is for integral values of n from $-\infty$ to ∞, including zero. Representation of $f(t)$ by exponential series, as shown in Eq. 1-54, is known as *exponential Fourier series* representation of $f(t)$ over the interval $(t_0,\ t_0 + T)$. The various coefficients in this series can be evaluated by using Eq. 1-49:

$$F_n = \frac{\displaystyle\int_{t_0}^{t_0+T} f(t)(e^{jn\omega_0 t})^*\,dt}{\displaystyle\int_{t_0}^{t_0+T} e^{jn\omega_0 t}(e^{jn\omega_0 t})^*\,dt}$$

$$= \frac{1}{T}\int_{t_0}^{t_0+T} f(t)\,e^{-jn\omega_0 t}\,dt \qquad (1\text{-}55)$$

It should be noted that the trigonometric and the exponential Fourier series are not two different types of series but two different ways of expressing the same series. The coefficients of one series can be obtained from those of the other. This can be seen from Eqs. 1-51 and 1-55. From these equations, it follows that

$$\begin{aligned} a_0 &= F_0 \\ a_n &= F_n + F_{-n} \\ b_n &= j(F_n - F_{-n}) \end{aligned} \qquad (1\text{-}56)$$

and

$$F_n = \frac{1}{2}(a_n - jb_n)$$

1-3. REPRESENTATION OF A PERIODIC FUNCTION BY THE FOURIER SERIES OVER THE ENTIRE INTERVAL $(-\infty < t < \infty)$

Thus far we have been able to represent a given function $f(t)$ by the Fourier series over a finite interval $(t_0,\ t_0 + T)$. Outside this interval the function $f(t)$ and the corresponding Fourier series need not be equal. If, however, the function $f(t)$ happens to be periodic, it can be shown that the series representation applies to the entire interval $(-\infty,\ \infty)$. This can be easily shown by considering some function $f(t)$ and its exponential Fourier series representation over an interval $(t_0,\ t_0 + T)$.

$$f(t) = \sum_{n=-\infty}^{\infty} F_n e^{jn\omega_0 t} \qquad (1\text{-}57)$$

The equality holds over the interval $(t_0 < t < t_0 + T)$. The two sides of the equation need not be equal outside the interval. It is easy to see, however, that the right-hand side of the Eq. 1-57 is periodic with period

$T = \omega_0/2\pi$. This follows from the fact that

$$e^{jn\omega_0 t} = e^{jn\omega_0(t+T)}; \quad T = \frac{2\pi}{\omega_0}$$

It is therefore obvious that if $f(t)$ is periodic with period T, then the equality 1-57 will hold over the entire interval $(-\infty, \infty)$. Thus for a periodic function $f(t)$,

$$f(t) = \sum_{n=-\infty}^{\infty} F_n e^{jn\omega_0 t} \qquad (-\infty < t < \infty)$$

where

$$F_n = \frac{1}{T}\int_{t_0}^{t_0+T} f(t)\, e^{-jn\omega_0 t}\, dt \tag{1-58}$$

Note that the choice of t_0 is immaterial.

1-4. REPRESENTATION OF AN ARBITRARY FUNCTION OVER THE ENTIRE INTERVAL $(-\infty, \infty)$; THE FOURIER TRANSFORM

We have seen that a periodic signal can be represented by a Fourier series. This is equivalent to representing a signal by its frequency components, $\omega_0, 2\omega_0, \ldots$, etc., where $\omega_0 = 2\pi/T$. We now wish to represent a nonperiodic function by its frequency components. This problem can be handled by constructing a periodic function of period T, so that $f(t)$ represents the first cycle of this periodic waveform. In the limit, we let the period T become very large, such that the periodic function tends to have only one cycle in the interval $(-\infty < t < \infty)$ and is identical to $f(t)$. The fundamental frequency $\omega_0 = 2\pi/T$ becomes smaller as T is made larger and the frequency spectrum becomes denser. In the limit as $T \to \infty$, the spectrum becomes continuous. The signal $f(t)$ is now represented as a continuous sum (integral) of various exponential components (Ref. 1-2):

$$f(t) = \frac{1}{2\pi}\int_{-\infty}^{\infty} F(\omega)\, e^{j\omega t}\, d\omega \tag{1-59a}$$

where

$$F(\omega) = \int_{-\infty}^{\infty} f(t)\, e^{-j\omega t}\, dt \tag{1-59b}$$

The pair of equations 1-59 is known as the Fourier transform pair, where Eq. 1-59b represents the direct Fourier transform of $f(t)$ and Eq. 1-59a represents the inverse Fourier transform of $F(\omega)$. This pair is also expressed symbolically as

$$\mathfrak{F}[f(t)] = F(\omega)$$

$$\mathfrak{F}^{-1}[F(\omega)] = f(t)$$

Existence of the Fourier Transform. It is evident from the definition of the Fourier transform that $F(\omega)$ will exist if

$$\int_{\infty}^{\infty} |f(t)|\, dt \qquad < \infty$$

This condition is sufficient for the existence of the Fourier transform. It is, however, not a necessary condition. We shall see later that some functions like sin $\omega_0 t$, cos $\omega_0 t$, $u(t)$ that are not absolutely integrable do have Fourier transforms.

Time-Domain and Frequency-Domain Representation of a Signal. The Fourier transform is a tool that resolves a given signal into its frequency components. The function $F(\omega)$ is the direct Fourier transform of $f(t)$ and represents amplitudes of various frequency components. Therefore, $F(\omega)$ is the frequency-domain representation of $f(t)$. Time-domain representation specifies a function at each instant of time, whereas frequency-domain representation specifies the relative amplitudes of the frequency components of the function. Either representation uniquely specifies the function. However, the function $F(\omega)$ is complex in general, and needs two plots for its graphical representation:

$$F(\omega) = |F(\omega)|\, e^{j\theta(\omega)}$$

Thus $F(\omega)$ may be represented by a magnitude plot $|F(\omega)|$ and a phase plot $\theta(\omega)$. In many cases, $F(\omega)$ is either real or imaginary and only one plot is necessary. $F(\omega)$, however, is in general a complex function of ω, and we shall now show that for a real function $f(t)$,

$$F^*(\omega) = F(-\omega)$$

We have

$$F(\omega) = \int_{-\infty}^{\infty} f(t)\, e^{-j\omega t}\, dt \qquad (1\text{-}60a)$$

Therefore

$$F(-\omega) = \int_{-\infty}^{\infty} f(t)\, e^{j\omega t}\, dt \qquad (1\text{-}60b)$$

From Eqs. 1-60 it follows that if $f(t)$ is a real function of t, then

$$F^*(\omega) = F(-\omega) \qquad (1\text{-}61a)$$

Thus, if

$$F(\omega) = |F(\omega)|\, e^{j\theta(\omega)}$$

then

$$F(-\omega) = |F(\omega)|\, e^{-j\theta(\omega)} \qquad (1\text{-}61b)$$

It is evident from this equation that the magnitude function $|F(\omega)|$ is an even function of ω, and the phase function $\theta(\omega)$ is an odd function of ω.

1-5. FOURIER TRANSFORM OF SOME USEFUL FUNCTION

1. Single-Sided Exponential Signal $e^{-at}u(t)$.

$$f(t) = e^{-at}u(t)$$

$$\begin{aligned} F(\omega) &= \int_{-\infty}^{\infty} e^{-at}u(t)\,e^{-j\omega t}\,dt \\ &= \int_{0}^{\infty} e^{-(a+j\omega)t}\,dt \qquad (1\text{-}62) \\ &= \frac{1}{a+j\omega} \\ &= \frac{1}{\sqrt{a^2+\omega^2}}\,e^{-j\tan^{-1}\left(\frac{\omega}{a}\right)} \end{aligned}$$

Here

$$|F(\omega)| = \frac{1}{\sqrt{a^2+\omega^2}} \quad \text{and} \quad \theta(\omega) = -\tan^{-1}\left(\frac{\omega}{a}\right)$$

The magnitude spectrum $|F(\omega)|$ and the phase spectrum $\theta(\omega)$ are shown in Fig. 1-5b.

Note that the integral 1-62 converges only for $a > 0$. For $a < 0$, the Fourier transform does not exist.

2. Double-Sided Exponential Signal $e^{-a|t|}$.

$$f(t) = e^{-a|t|}$$

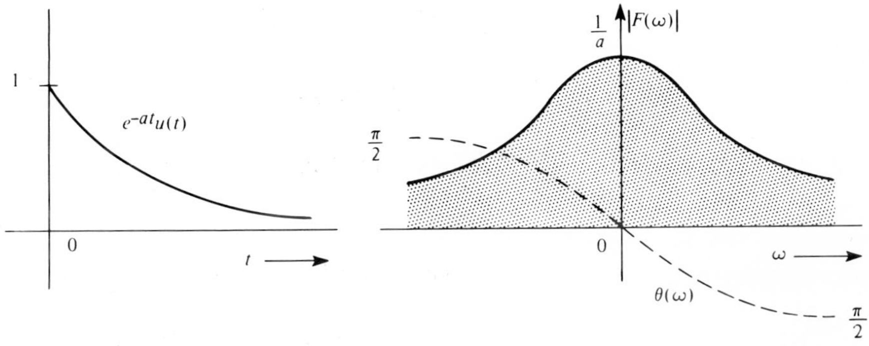

FIG. 1-5.

and

$$
\begin{aligned}
F(\omega) &= \int_{-\infty}^{\infty} e^{-a|t|} e^{-j\omega t}\, dt \\
&= \int_{-\infty}^{0} e^{(a-j\omega)t}\, dt + \int_{0}^{\infty} e^{-(a+j\omega)t}\, dt \\
&= \frac{2a}{a^2 + \omega^2} \qquad (1\text{-}63)
\end{aligned}
$$

Note that the phase spectrum $\theta(\omega) = 0$ in this case. The magnitude spectrum is sketched in Fig. 1-6b.

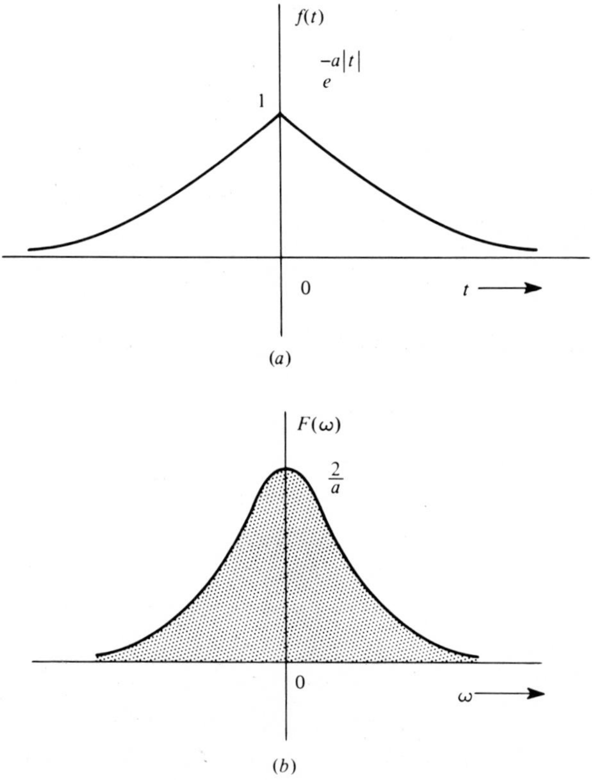

FIG. 1-6.

3. A Gate Function $G_\tau(t)$. A gate function $G_\tau(t)$ is a rectangular pulse of unit height and width τ. The function shown in Fig. 1-7a is obviously $AG_\tau(t)$. For this function

$$F(\omega) = \int_{-\tau/2}^{\tau/2} A e^{-j\omega t}\, dt$$

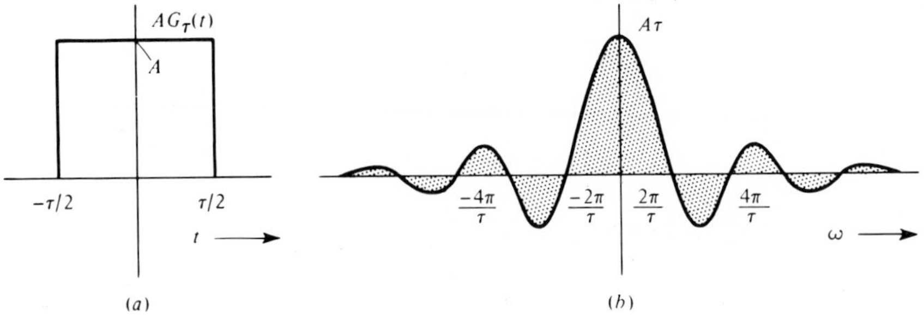

FIG. 1-7.

$$= \frac{A}{j\omega}\left(e^{j\omega\tau/2} - e^{-j\omega\tau/2}\right)$$

$$= A\tau \frac{\sin(\omega\tau/2)}{(\omega\tau/2)}$$

Here we encounter the function of the form $(\sin x)/x$. This function plays a very important role in signal theory and is known as the sampling function abbreviated by $\mathrm{Sa}(x)$:

$$\mathrm{Sa}(x) = \frac{\sin x}{x}$$

The sampling function is sketched in Fig. 1-8. Thus for a gate function $AG_\tau(t)$,

$$F(\omega) = A\tau\,\mathrm{Sa}\left(\frac{\omega\tau}{2}\right)$$

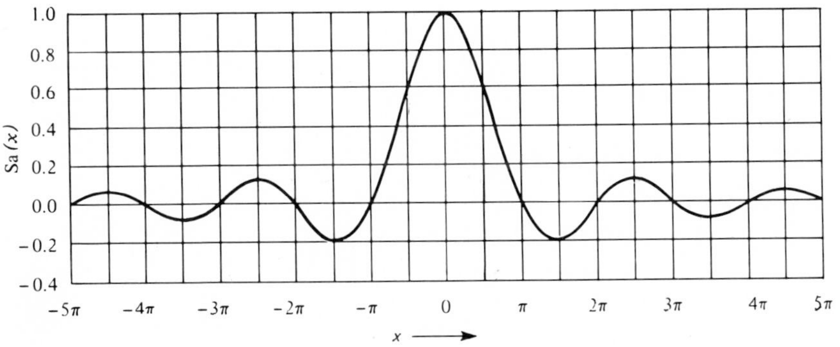

FIG. 1-8.

1-6. THE UNIT IMPULSE FUNCTION

The unit impulse function was originally defined by P. A. M. Dirac by the equations

$$\int_{-\infty}^{\infty} \delta(t)\,dt = 1 \tag{1-64}$$

and

$$\delta(t) = 0 \qquad t \neq 0$$

Thus the impulse function $\delta(t)$ is zero everywhere except at the origin and it has a unit area. Intuitively this function may be conceived as a narrow pulse with a large height and having a unit area. In the limit as the width of the pulse is made zero, its height increases in such a way that the area of the pulse always remains unity. This function is obviously not a true function in a mathematical sense where a function is defined for every value of t. This function has been rigorously justified by a theory of distribution by Schwartz (Ref. 1-3). A version of this theory (generalized functions) was given by Temple (Ref. 1-4). In this approach, the impulse function is defined as a sequence of regular functions and all operations on impulse function are viewed as the operation on this sequence. One may define the impulse function accordingly by a number of sequences as shown in Fig. 1-9. Figure 1-10 shows the

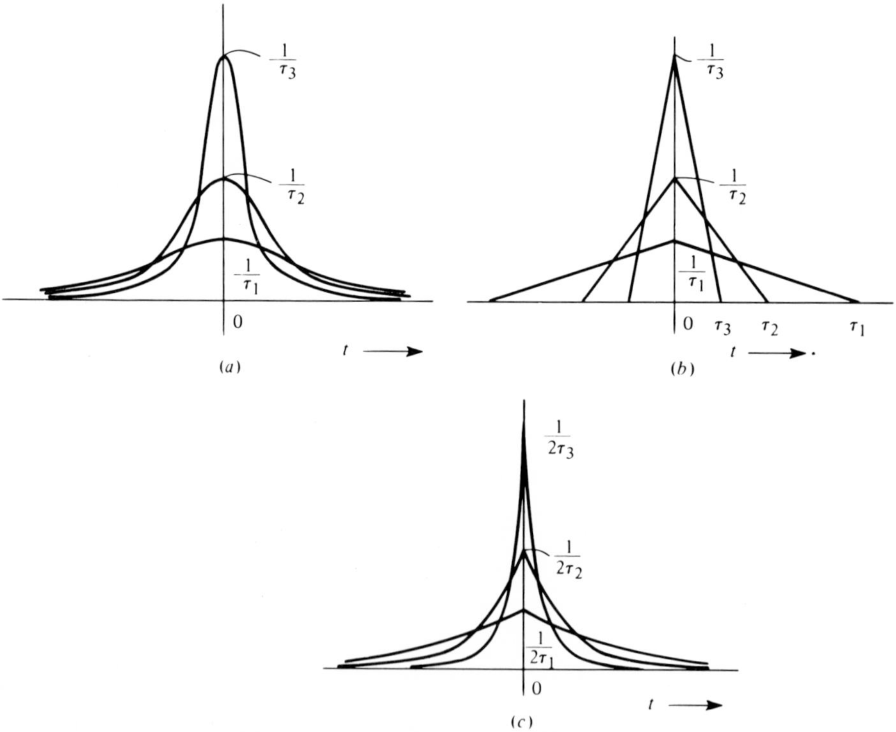

FIG. 1-9. (a) Gaussian pulse sequence. (b) Triangular pulse sequence. (c) Two-sided exponential pulse sequence.

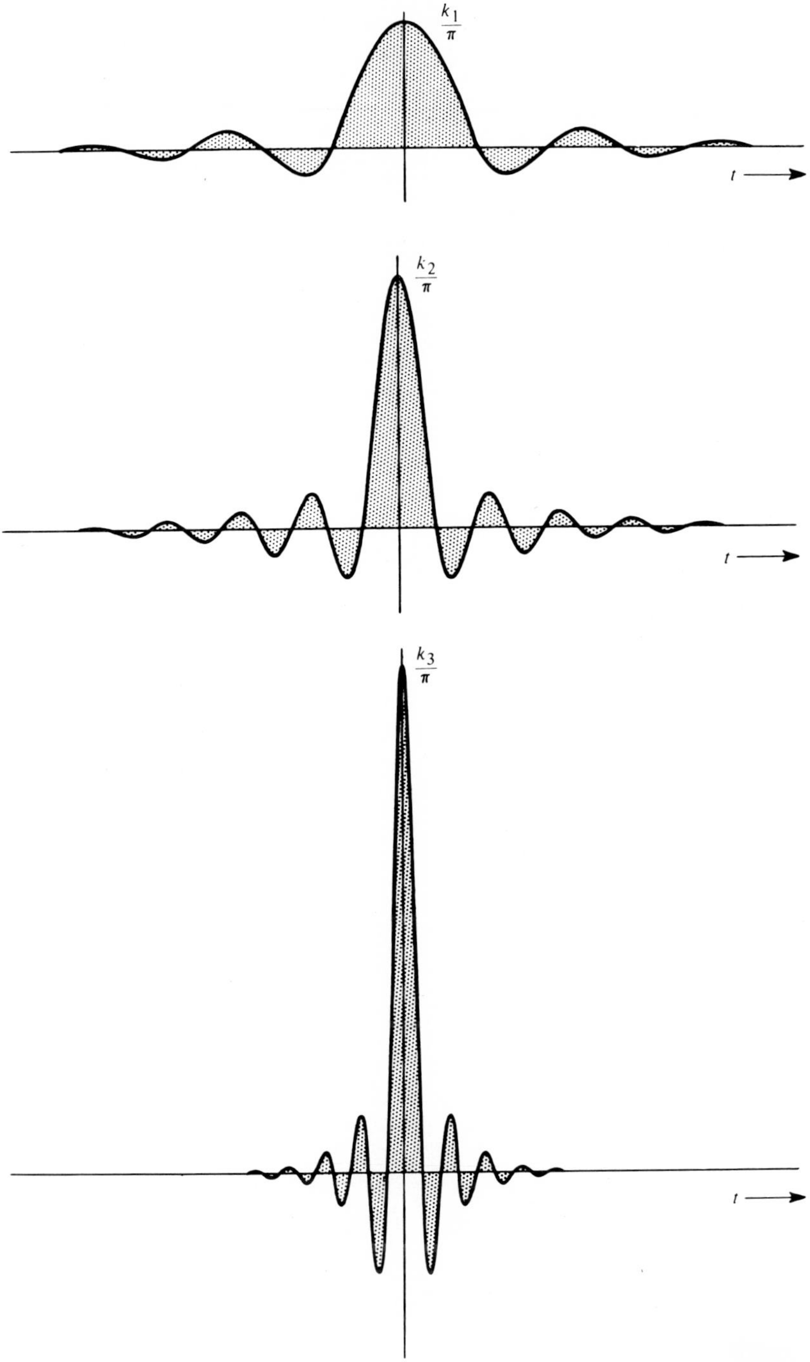

FIG. 1-10. The impulse function as a limit of the sequence of a sampling function.

impulse function as a limit of sampling functions. All these sequences satisfy Eq. 1-64.

1. *Gaussian Pulse*:

$$\delta(t) = \lim_{\tau \to 0} \frac{1}{\tau} e^{(-\pi t^2 / \tau^2)}$$

2. *Triangular Pulse*:

$$\delta(t) = \lim_{\tau \to 0} \frac{1}{\tau} \left[1 - \frac{|t|}{\tau}\right] \cdots |t| < \tau$$

3. *Exponential Pulse*:

$$\delta(t) = \lim_{\tau \to 0} \frac{1}{2\tau} e^{(-|t|/\tau)}$$

4. *Sampling Function.* It can be shown that

$$\int_{-\infty}^{\infty} \frac{k}{\pi} \operatorname{Sa}(kt)\, dt = 1 \tag{1-65}$$

As k is made larger and larger, the amplitude of the function $(k/\pi)\operatorname{Sa}(kt)$ becomes larger. The function oscillates faster and decays inversely proportional to t. In the limit as $k \to \infty$, the function concentrates at the origin and the net area under the curve is unity as seen from Eq. 1-65. Hence the sampling function $(k/\pi)\operatorname{Sa}(kt)$ becomes an impulse function in the limit as $k \to \infty$.

$$\lim_{k \to \infty} \frac{k}{\pi} \operatorname{Sa}(kt) = \delta(t) \tag{1-66}$$

5. *Sampling Square Function*:

$$\delta(t) = \lim_{k \to \infty} \frac{k}{\pi} \operatorname{Sa}^2(kt) \tag{1-67}$$

This follows from the argument similar to those employed for the sampling function and the fact that

$$\int_{-\infty}^{\infty} \frac{k}{\pi} \operatorname{Sa}^2(kx)\, dx = 1$$

The entire area of an impulse function is concentrated at $t = 0$, and we may write

$$\int_{-\infty}^{\infty} \delta(t)\, dt = \int_{0^-}^{0^+} \delta(t)\, dt = 1 \tag{1-68}$$

Since $\delta(t) = 0$ everywhere except at $t = 0$, we have

$$\begin{aligned} \int_{-\infty}^{\infty} f(t)\,\delta(t)\, dt &= f(0) \int_{-\infty}^{\infty} \delta(t)\, dt \\ &= f(0) \end{aligned} \tag{1-69a}$$

It also follows that

$$\int_{-\infty}^{\infty} f(t)\,\delta(t-\tau)\,dt = f(\tau) \tag{1-69b}$$

Equation 1-69 represents the *sampling property* (or *sifting property*) of the impulse function. We have shown here that Eq. 1-69 follows from the definition of the impulse function in Eq. 1-64. Actually, in the rigorous approach, impulse function is defined by Eq. 1-69. This way we define the impulse function by its integral property (Eq. 1-69b) rather than as a function of time defined for every t. It can be easily shown that the definition 1-64 does not specify a unique function (Ref. 1-5). For a rigorous approach Eq. 1-69 is taken as the starting point or the definition of the impulse function.

The Fourier Transform of an Impulse Function. The Fourier transform of a unit impulse function $\delta(t)$ is given by

$$\mathfrak{F}[\delta(t)] = \int_{-\infty}^{\infty} \delta(t)\,e^{-j\omega t}\,dt$$

From the sampling property of an impulse function expressed in Eq. 1-69, it is evident that the integral on the right-hand side is unity. Hence

$$\mathfrak{F}[\delta(t)] = 1 \tag{1-70}$$

Therefore the Fourier transform of a unit impulse function is unity (Fig. 1-11).

It is thus evident that an impulse function has a uniform spectral density over the entire frequency interval. In other words, an impulse function contains all frequency components with the same relative amplitudes. From Eq. 1-70 it follows that

$$\mathfrak{F}^{-1}[1] = \frac{1}{2\pi}\int_{-\infty}^{\infty} e^{j\omega t}\,d\omega = \delta(t)$$

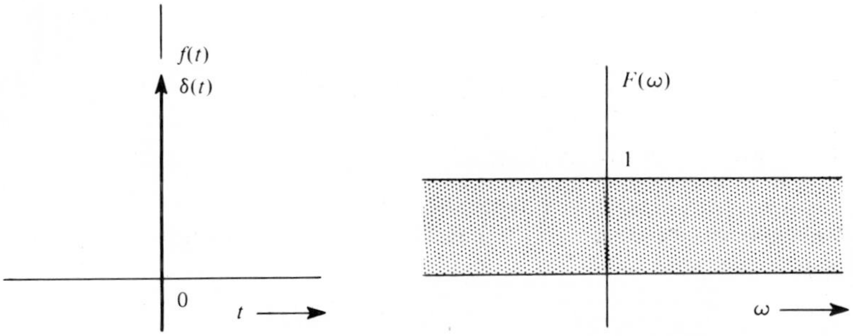

FIG. 1-11. An impulse function and its transform.

Hence we have

$$\int_{-\infty}^{\infty} e^{j\omega x}\,d\omega = 2\pi\,\delta(x) \tag{1-71}$$

Fourier Transform of a Constant. Let us now find the Fourier transform of a function $f(t) = A$. By definition

$$\mathfrak{F}[A] = \int_{-\infty}^{\infty} Ae^{-j\omega t}\,dt$$

$$= A\int_{-\infty}^{\infty} e^{j\omega x}\,dx$$

From Eq. 1-71 it now follows that

$$\mathfrak{F}[A] = 2\pi A\,\delta(\omega) \tag{1-72}$$

Thus when $f(t)$ equals a constant it contains only a frequency component $\omega = 0$ (Fig. 1-12). This is a logical result, since a constant function is a d-c signal ($\omega = 0$), and does not have any other frequency components.

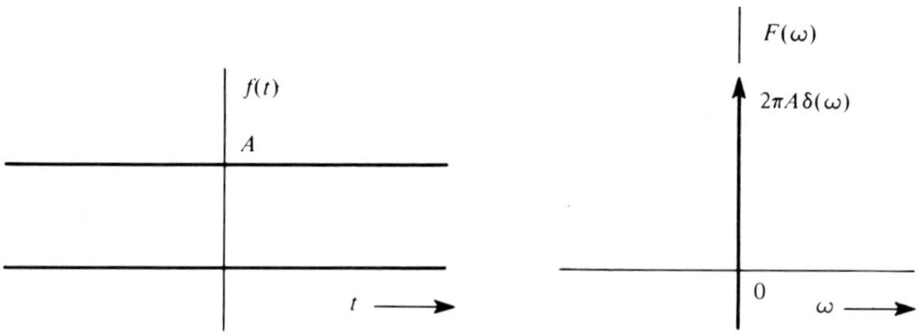

FIG. 1-12. A constant and its transform.

This result (Eq. 1-72) can also be derived alternatively by realizing that

$$A = \lim_{\tau\to\infty} AG_\tau(t)$$

Hence

$$\mathfrak{F}[A] = \mathfrak{F}\left[\lim_{\tau\to\infty} AG_\tau(t)\right]$$

$$= \lim_{\tau\to\infty} A\tau\,\mathrm{Sa}\left(\frac{\omega\tau}{2}\right)$$

Now use of Eq. 1-66 leads to Eq. 1-72.

Transform of sgn(t). The signum function (Fig. 1-13) sgn(t) is defined as

$$\mathrm{sgn}(t) = \begin{cases} 1 & t > 0 \\ -1 & t < 0 \end{cases} \tag{1-73}$$

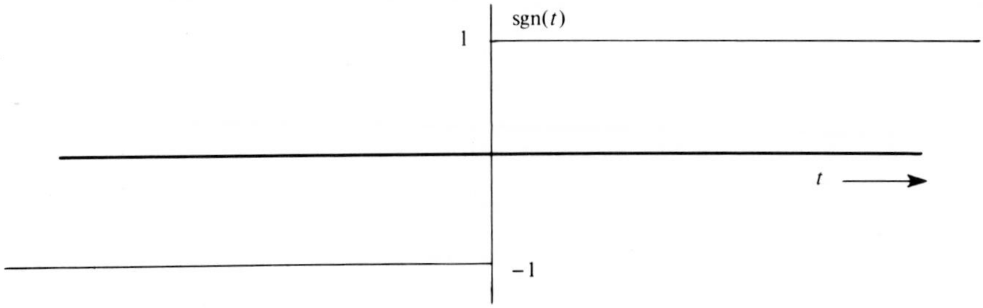

FIG. 1-13.

This function can be expressed as

$$\operatorname{sgn}(t) = \lim_{a \to 0} [e^{-at}u(t) + e^{at}u(-t)]$$

and

$$\begin{aligned} F(\omega) &= \lim_{a \to 0} \left[\int_0^{\infty} e^{-(a+j\omega)t}\, dt + \int_{-\infty}^{0} e^{(a-j\omega)t}\, dt \right] \\ &= \lim_{a \to 0} \frac{-2j\omega}{a^2 + \omega^2} \\ &= \frac{2}{j\omega} \end{aligned} \tag{1-74}$$

Transform of Unit Step Function $u(t)$.

$$u(t) = \frac{1}{2} + \frac{1}{2} \operatorname{sgn}(t)$$

Using results in Eqs. 1-72 and 1-74 we get[2]

$$\mathfrak{F}[u(t)] = \pi\delta(\omega) + \frac{1}{j\omega} \tag{1-75}$$

Transform of an Exponential $e^{j\omega_0 t}$.

$$\begin{aligned} \mathfrak{F}[e^{j\omega_0 t}] &= \int_{-\infty}^{\infty} e^{j\omega_0 t} e^{-j\omega t}\, dt \\ &= \int_{-\infty}^{\infty} e^{j(\omega_0 - \omega)t}\, dt \end{aligned}$$

[2]Here we are using linearity property of Fourier transform. This property states that if

$$\mathfrak{F}[f_1(t)] = F_1(\omega), \qquad \mathfrak{F}[f_2(t)] = F_2(\omega),$$

then

$$\mathfrak{F}[f_1(t) + f_2(t)] = F_1(\omega) + F_2(\omega)$$

See Eq. 1-83.

Let $t = -y$ and $\omega - \omega_0 = x$. This yields

$$\mathfrak{F}[e^{j\omega_0 t}] = \int_{-\infty}^{\infty} e^{jxy}\, dy$$

From Eq. 1-71 it now follows that

$$\begin{aligned}\mathfrak{F}[e^{j\omega_0 t}] &= 2\pi\delta(x) \\ &= 2\pi\delta(\omega - \omega_0)\end{aligned} \tag{1-76a}$$

Similarly it follows that

$$\mathfrak{F}[e^{-j\omega_0 t}] = 2\pi\delta(\omega + \omega_0) \tag{1-76b}$$

The Fourier transform of $e^{j\omega_0 t}$ is therefore a single impulse of strength 2π at $\omega = \omega_0$. Note that the signal $e^{j\omega_0 t}$ is not a real function of time, and hence it has a spectrum which exists at $\omega = \omega_0$ alone. We have shown previously that for any real function of time, the spectral density function $F(\omega)$ satisfies (see Eq. 1-61):

$$F^*(\omega) = F(-\omega)$$

and

$$|F(\omega)| = |F(-\omega)|$$

Transform of cos $\omega_0 t$ and sin $\omega_0 t$.

$$\begin{aligned}\mathfrak{F}[\cos \omega_0 t] &= \mathfrak{F}\left[\frac{1}{2}(e^{j\omega_0 t} + e^{-j\omega_0 t})\right] \\ &= \frac{1}{2}\{\mathfrak{F}[e^{j\omega_0 t}] + \mathfrak{F}[e^{-j\omega_0 t}]\} \\ &= \pi[\delta(\omega - \omega_0) + \delta(\omega + \omega_0)]\end{aligned} \tag{1-77a}$$

Similarly it can be readily shown that

$$\mathfrak{F}[\sin \omega_0 t] = j\pi[\delta(\omega + \omega_0) - \delta(\omega - \omega_0)] \tag{1-77b}$$

The transform of sin $\omega_0 t$ can be expressed alternatively by noting that $\pm j = e^{\pm j\pi/2}$.

$$\mathfrak{F}[\sin \omega_0 t] = \pi[\delta(\omega + \omega_0)e^{j\pi/2} + \delta(\omega - \omega_0)e^{-j\pi/2}] \tag{1-77c}$$

It is evident from Eq. 1-77a and 1-77c, that the Fourier spectrum of cosine and sine waves consists of two impulses $\pm\omega_0$ (Fig. 1-14). This is the expected result since these signals do not contain components of any other frequencies but ω_0. Note the phase difference of $\pi/2$ in the spectra of sine and cosine signals. This is also an expected result.

1-7. THE FOURIER TRANSFORM OF A PERIODIC FUNCTION

We have considered a Fourier transform as a limiting case of the Fourier series by letting the period of a periodic function become in-

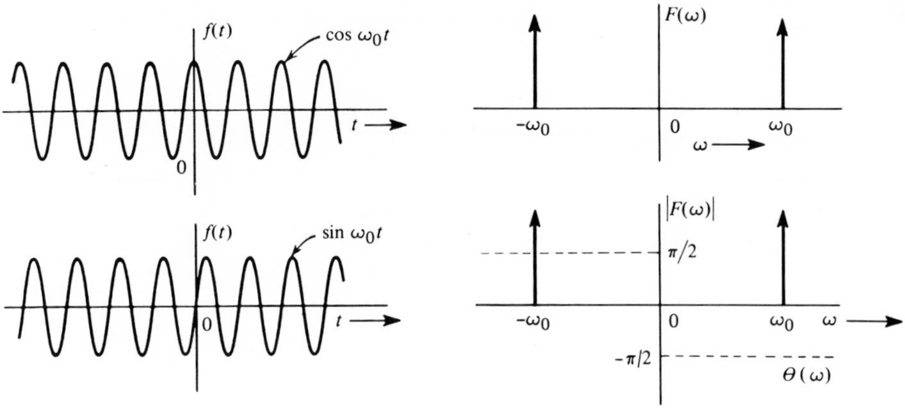

FIG. 1-14.

finite. We shall now proceed in the opposite direction and show that the Fourier series is just a limiting case of the Fourier transform. This point of view is very useful, since it permits a unified treatment of both the periodic and the nonperiodic functions.

We can express a periodic function by its Fourier series. The Fourier transform of a periodic function is then the sum of Fourier transforms of its individual components. We can express a periodic function $f(t)$ with period T as

$$f(t) = \sum_{n=-\infty}^{\infty} F_n e^{jn\omega_0 t} \qquad \omega_0 = \frac{2\pi}{T}$$

Taking the Fourier transforms of both sides, we have

$$\mathfrak{F}[f(t)] = \mathfrak{F} \sum_{n=-\infty}^{\infty} F_n e^{jn\omega_0 t}$$

$$= \sum_{n=-\infty}^{\infty} F_n \mathfrak{F}(e^{jn\omega_0 t})$$

Substituting the transform of $e^{jn\omega_0 t}$ from Eq. 1-76a, we get

$$\mathfrak{F}[f(t)] = 2\pi \sum_{n=-\infty}^{\infty} F_n \delta(\omega - n\omega_0) \tag{1-78}$$

This is a significant result. Relation 1-78 states that the spectral density function or the Fourier transform of a periodic signal consists of impulses located at the harmonic frequencies of the signal and that the strength of each impulse is the same as 2π times the value of the corresponding coefficient in the exponential Fourier series. The sequence of equidistant impulses is just a limiting form of continuous density function. The result should, of course, be no surprise, since we know that a periodic function contains components only of discrete harmonic frequencies.

Example 1-2. We shall find a Fourier transform of a sequence of equidistant impulses of unit strength and separated by T seconds as shown in Fig. 1-15. This function is very important in sampling theory, and hence it is convenient to denote this function by a special symbol $\delta_T(t)$. Thus

$$\begin{aligned}\delta_T(t) &= \delta(t) + \delta(t - T) + \delta(t - 2T) + \delta(t - nT) + \cdots \\ &\quad + \delta(t + T) + \delta(t + 2T) + \cdots + \delta(t + nT) + \cdots \\ &= \sum_{n=-\infty}^{\infty} \delta(t - nT) \qquad (1\text{-}79)\end{aligned}$$

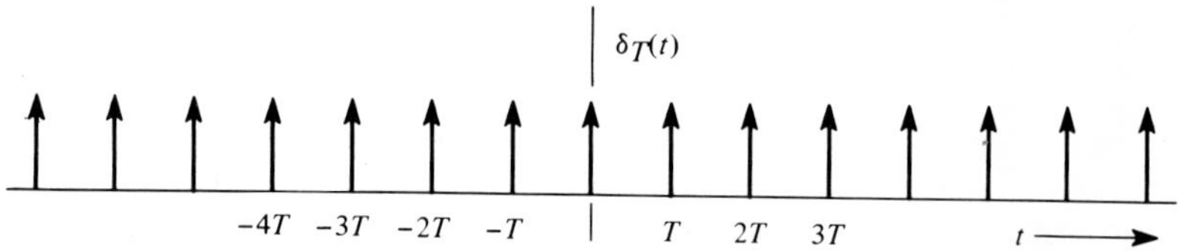

FIG. 1-15. The sequence of a uniform equidistant impulse function.

This is obviously a periodic function with period T. We shall first find the Fourier series for this function.

$$\delta_T(t) = \sum_{n=-\infty}^{\infty} F_n e^{jn\omega_0 t}$$

where

$$F_n = \frac{1}{T} \int_{-T/2}^{T/2} \delta_T(t)\, e^{-jn\omega_0 t}\, dt$$

Function $\delta_T(t)$ in the interval $(-T/2, T/2)$ is simply $\delta(t)$. Hence

$$F_n = \frac{1}{T} \int_{-T/2}^{T/2} \delta(t)\, e^{-jn\omega_0 t}\, dt$$

From the sampling property of an impulse function as expressed in Eq. 1-69, the above equation reduces to

$$F_n = \frac{1}{T}$$

Consequently, F_n is a constant $(1/T)$. It therefore follows that

$$\delta_T(t) = \frac{1}{T} \sum_{n=-\infty}^{\infty} e^{jn\omega_0 t}$$

To find the Fourier transform of $\delta_T(t)$, we use Eq. 1-78. Since in this case $F_n = 1/T$, it is evident that

$$\mathfrak{F}[\delta_T(t)] = 2\pi \sum_{n=-\infty}^{\infty} \frac{1}{T} \delta(\omega - n\omega_0)$$

$$= \frac{2\pi}{T} \sum_{n=-\infty}^{\infty} \delta(\omega - n\omega_0)$$

$$= \omega_0 \sum_{n=-\infty}^{\infty} \delta(\omega - n\omega_0)$$

$$= \omega_0 \delta_{\omega_0}(\omega) \tag{1-80}$$

Relation 1-80 is very significant. It states that the Fourier transform of a unit-impulse train of period T is also a train of impulses of strength ω_0 and separated by ω_0 radians ($\omega_0 = 2\pi/T$). Therefore, the impulse-train function is its own transform. It is evident that as the period of the impulses increases, the frequency spectrum becomes denser.

Transforms of some useful functions are listed in Table 1-1.

TABLE 1-1
TABLE OF FOURIER TRANSFORMS

$f(t)$	$F(\omega)$
1. $e^{-at}u(t)$	$1/(a + j\omega)$
2. $te^{-at}u(t)$	$1/(a + j\omega)^2$
3. $\lvert t \rvert$	$-2/\omega^2$
4. $\delta(t)$	1
5. 1	$2\pi\delta(\omega)$
6. $u(t)$	$\pi\delta(\omega) + 1/j\omega$
7. $\cos \omega_0 t\, u(t)$	$\frac{\pi}{2}[\delta(\omega - \omega_0) + \delta(\omega + \omega_0)] + \frac{j\omega}{\omega_0^2 - \omega^2}$
8. $\sin \omega_0 t\, u(t)$	$\frac{\pi}{2j}[\delta(\omega - \omega_0) - \delta(\omega + \omega_0)] + \frac{\omega_0}{\omega_0^2 - \omega^2}$
9. $\cos \omega_0 t$	$\pi[\delta(\omega - \omega_0) + \delta(\omega + \omega_0)]$
10. $\sin \omega_0 t$	$j\pi[\delta(\omega + \omega_0) - \delta(\omega - \omega_0)]$
11. $e^{-at} \sin \omega_0 t\, u(t)$	$\frac{\omega_0}{(a + j\omega)^2 + \omega_0^2}$
12. $\frac{\mathrm{w}}{2\pi} \mathrm{Sa} \frac{(\mathrm{w}t)}{2}$	$G_{\mathrm{w}}(\omega)$
13. $G_\tau(t)$	$\tau \mathrm{Sa}\left(\frac{\omega\tau}{2}\right)$
14. $\left.\begin{matrix} 1 - \frac{\lvert t \rvert}{\tau} \cdots \lvert t \rvert < \tau \\ 0 \qquad \cdots \lvert t \rvert > \tau \end{matrix}\right\}$	$\tau\left[\mathrm{Sa}\left(\frac{\omega\tau}{2}\right)\right]^2$
15. $e^{-a\lvert t \rvert}$	$\frac{2a}{a^2 + \omega^2}$
16. $\frac{1}{\sigma\sqrt{2\pi}} e^{-t^2/2\sigma^2}$	$e^{-\sigma^2\omega^2/2}$
17. $\delta_T(t)$	$\omega_0 \delta_{\omega_0}(\omega) \quad \left(\omega_0 = \frac{2\pi}{T}\right)$

1-8. SOME PROPERTIES OF THE FOURIER TRANSFORM

The Fourier transform is a tool for expressing a function in terms of its exponential components of various frequencies. It has already been pointed out that the Fourier transform of a function is just another way of specifying the function. We therefore have two descriptions of the same function: the time-domain and frequency-domain descriptions. It is very illuminating to study the effect in one domain caused by certain operations over the function in the other domain. We may ask, for example: If a function is differentiated in the time domain, how is the spectrum of the derivative function related to the spectrum of the function itself? What happens to the spectrum of a function if the function is shifted in the time domain? We shall now seek to evaluate the effects on one domain caused by certain important operations on the function in the other domain.

It is important to point out at this stage that there is a certain amount of symmetry in the equations defining the two domains. This can be easily seen from the equations defining the Fourier transform.

$$F(\omega) = \int_{-\infty}^{\infty} f(t)\, e^{-j\omega t}\, dt$$

and

$$f(t) = \frac{1}{2\pi} \int_{-\infty}^{\infty} F(\omega)\, e^{j\omega t}\, d\omega \tag{1-81}$$

We therefore should expect this symmetry to be reflected in the properties. For example, we expect that the effect on the frequency domain due to differentiation in the time domain should be similar to the effect on the time domain due to differentiation in the frequency domain. We shall see that this indeed is the case.

For convenience, the correspondence between the two domains will be denoted by a double arrow. Thus the notation

$$f(t) \leftrightarrow F(\omega)$$

denotes that $F(\omega)$ is the direct Fourier transform of $f(t)$ and that $f(t)$ is the inverse Fourier transform of $F(\omega)$ related by Eq. 1-81.

Symmetry Property. If

$$f(t) \leftrightarrow F(\omega)$$

then

$$F(t) \leftrightarrow 2\pi f(-\omega) \tag{1-82}$$

Proof. From Eq. 1-81 it follows that

$$2\pi f(-t) = \int_{-\infty}^{\infty} F(\omega)\, e^{-j\omega t}\, d\omega$$

Since ω is a dummy variable in the above integral it may be replaced by another variable x. Therefore

$$2\pi f(-t) = \int_{-\infty}^{\infty} F(x) e^{-jxt}\, dx$$

Hence

$$2\pi f(-\omega) = \int_{-\infty}^{\infty} F(x) e^{-jx\omega}\, dx$$

$$= \mathfrak{F}[F(t)]$$

Hence

$$F(t) \leftrightarrow 2\pi f(-\omega)$$

This property is illustrated in Fig. 1-16.

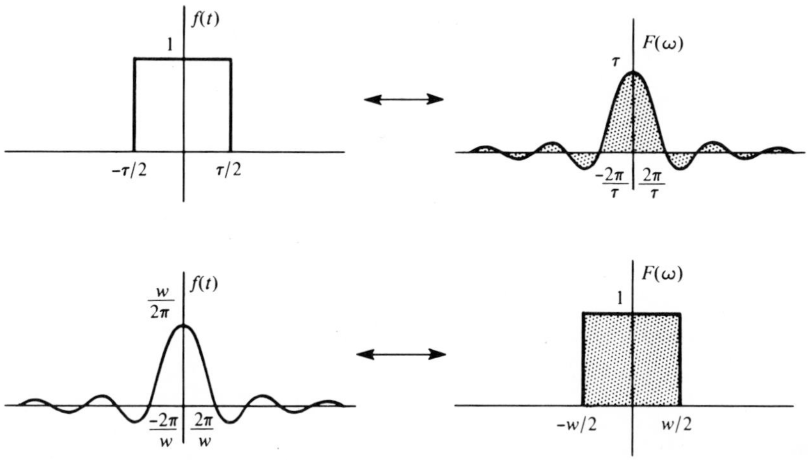

FIG. 1-16. Symmetry property of the Fourier transform.

It can be easily seen that the Fourier transform of a gate function is a sampling function, whereas the Fourier transform of a sampling function is a gate function. The symmetry property holds for all, even $f(t)$. If $f(t)$ is not an even function, then the symmetry is not so perfect; nevertheless there is some measure of symmetry, as indicated by Eq. 1-82.

Linearity Property. If

$$f_1(t) \leftrightarrow F_1(\omega)$$

$$f_2(t) \leftrightarrow F_2(\omega)$$

then, for any arbitrary constants a_1 and a_2,

$$a_1 f_1(t) + a_2 f_2(t) \leftrightarrow a_1 F_1(\omega) + a_2 F_2(\omega) \tag{1-83}$$

The proof is trivial. The above is also valid for finite sums:

$$a_1 f_1(t) + a_2 f_2(t) + \cdots + a_n f_n(t) \leftrightarrow a_1 F_1(\omega) + a_2 F_2(\omega) + \cdots + a_n F_n(\omega)$$

Scaling Property. If

$$f(t) \leftrightarrow F(\omega)$$

then for a real constant a,

$$f(at) \leftrightarrow \frac{1}{|a|} F\left(\frac{\omega}{a}\right) \tag{1-84}$$

Proof. By definition,

$$\mathfrak{F}[f(at)] = \int_{-\infty}^{\infty} f(at)\, e^{-j\omega t}\, dt$$

Let $x = at$. Then, for positive constant a,

$$\mathfrak{F}[f(at)] = \frac{1}{a} \int_{-\infty}^{\infty} f(x)\, e^{(-j\omega/a)x}\, dx$$

$$= \frac{1}{a} F\left(\frac{\omega}{a}\right)$$

Hence

$$f(at) \leftrightarrow \frac{1}{a} F\left(\frac{\omega}{a}\right)$$

Similarly, it can be shown that if $a < 0$,

$$f(at) \leftrightarrow \frac{-1}{a} F\left(\frac{\omega}{a}\right)$$

Consequently it follows that

$$f(at) \leftrightarrow \frac{1}{|a|} F\left(\frac{\omega}{a}\right)$$

Significance of the Scaling Property. The function $f(at)$ represents function $f(t)$ compressed in the time scale by a factor of a. Similarly, a function $F(\omega/a)$ represents a function $F(\omega)$ expanded in the frequency scale by the same factor a. The scaling property therefore states that compression in the time domain is equivalent to expansion in the frequency domain and vice versa. This result is also obvious intuitively, since compression in the time scale by a factor a means that the function is varying rapidly by the same factor, and hence the frequencies of its components will be increased by the factor a. We therefore expect its frequency spectrum to be expanded by the factor a in the frequency scale. Similarly, if a function is expanded in the time scale, it varies slowly, and hence the frequencies of its components are lowered. Thus the frequency spectrum is compressed. As an example, consider the signal $\cos \omega_0 t$. This signal has frequency components at $\pm\omega_0$. The signal $\cos 2\omega_0 t$ represents compression of $\cos \omega_0 t$ by a factor of two, and its frequency components lie at $\pm 2\omega_0$. It is therefore evident that the frequency spectrum has been expanded by a factor of two. The effect of scaling is demonstrated in Fig. 1-17.

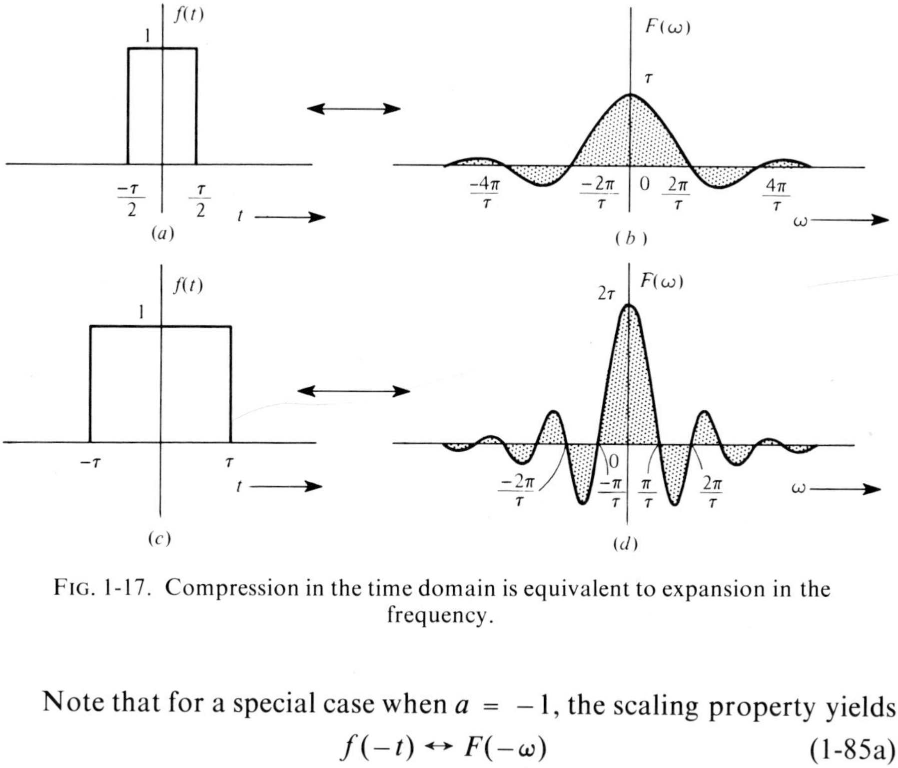

FIG. 1-17. Compression in the time domain is equivalent to expansion in the frequency.

Note that for a special case when $a = -1$, the scaling property yields

$$f(-t) \leftrightarrow F(-\omega) \tag{1-85a}$$

Using this equation and applying the symmetry property (Eq. 1-82) to Eqs. 1-74 and 1-75 the reader can easily show that

$$\frac{j}{\pi t} \leftrightarrow \operatorname{sgn}(\omega) \tag{1-85b}$$

and

$$\left[\frac{1}{2}\,\delta(t) + \frac{j}{2\pi t}\right] \leftrightarrow u(\omega) \tag{1-85c}$$

Time-Shifting Property. If

$$f(t) \leftrightarrow F(\omega)$$

then

$$f(t - t_0) \leftrightarrow F(\omega)\, e^{-j\omega t_0} \tag{1-86}$$

Proof.

$$\mathfrak{F}[f(t - t_0)] = \int_{-\infty}^{\infty} f(t - t_0)\, e^{-j\omega t}\, dt$$

Let

$$t - t_0 = x$$

then

$$\mathfrak{F}[f(t - t_0)] = \int_{-\infty}^{\infty} f(x)\, e^{-j\omega(x + t_0)}\, dx$$

$$= F(\omega)\, e^{-j\omega t_0}$$

This theorem states that if a function is shifted in the time domain by t_0 seconds, then its magnitude spectrum $F(\omega)$ remains unchanged, but the phase spectrum is changed by an amount $-\omega t_0$. This result is also obvious intuitively, since the shifting of a function in the time domain really does not change the frequency components of the signal but each component is shifted by an amount t_0. A shift of time t_0 for a component of frequency ω is equivalent to a phase shift of $-\omega t_0$.

We may state that a shift of t_0 in the time domain is equivalent to multiplication by $e^{-j\omega t_0}$ in the frequency domain.

Frequency-Shifting Property. If

$$f(t) \leftrightarrow F(\omega)$$

then

$$f(t)\, e^{j\omega_0 t} \leftrightarrow F(\omega - \omega_0) \tag{1-87}$$

Proof.

$$\mathfrak{F}[f(t)\, e^{j\omega_0 t}] = \int_{-\infty}^{\infty} f(t)\, e^{-j\omega_0 t}\, e^{-j\omega t}\, dt$$

$$= \int_{-\infty}^{\infty} f(t)\, e^{-j(\omega - \omega_0)t}\, dt$$

$$= F(\omega - \omega_0)$$

The theorem states that a shift of ω_0 in the frequency domain is equivalent to multiplication by $e^{j\omega_0 t}$ in the time domain. Note the dual nature between the time-shift and the frequency-shift properties. It is evident that multiplication by a factor $e^{j\omega_0 t}$ translates the whole frequency spectrum $F(\omega)$ by an amount ω_0. Hence this property is also known as the *frequency-translation theorem*.

In communication systems it is often desirable to translate the frequency spectrum. This is usually accomplished by multiplying a signal $f(t)$ by a sinusoidal signal. This process is known as *modulation*. Since a sinusoidal signal of frequency ω_0 can be expressed as the sum of exponentials, it is evident that multiplication of a signal $f(t)$ by a sinusoidal signal (modulation) will translate the whole frequency spectrum. This can be easily shown by observing the identity

$$f(t)\cos\omega_0 t = \frac{1}{2}[f(t)e^{j\omega_0 t} + f(t)e^{-j\omega_0 t}]$$

Using the frequency-shift theorem, it therefore follows that if

$$f(t) \leftrightarrow F(\omega)$$

then

$$f(t)\cos\omega_0 t \leftrightarrow \frac{1}{2}[F(\omega+\omega_0) + F(\omega-\omega_0)] \tag{1-88a}$$

Similarly, it can be shown that

$$f(t)\sin\omega_0 t \leftrightarrow \frac{j}{2}[F(\omega+\omega_0) - F(\omega-\omega_0)] \tag{1-88b}$$

Thus the process of modulation translates the frequency spectrum by the amount $\pm\omega_0$. This is a very useful result in communication theory. An example of frequency translation caused by modulation is shown in Fig. 1-18. This result is also known as the *modulation theorem.*

Time Differentiation and Integration. If

$$f(t) \leftrightarrow F(\omega)$$

then[3]

$$\frac{df}{dt} \leftrightarrow (j\omega)F(\omega) \tag{1-89a}$$

and

$$\int_{-\infty}^{t} f(\tau)\,d\tau \leftrightarrow \frac{1}{j\omega}F(\omega) \tag{1-89b}$$

provided[4] that $F(\omega)/\omega$ is bounded at $\omega = 0$. This is equivalent to saying that $F(0) = 0$ or

$$\int_{-\infty}^{\infty} f(t)\,dt = 0$$

Proof.

$$f(t) = \frac{1}{2\pi}\int_{-\infty}^{\infty} F(\omega)e^{j\omega t}\,d\omega$$

Therefore

$$\frac{df}{dt} = \frac{1}{2\pi}\frac{d}{dt}\int_{-\infty}^{\infty} F(\omega)e^{j\omega t}\,d\omega$$

[3]Equation 1-89a does not guarantee the existence of the transform of df/dt. It merely says that if that transform exists it is given by $j\omega F(\omega)$.

[4]If this condition is not satisfied, then Eq. 1-89b is modified. See Prob. 22 at the end of the chapter.

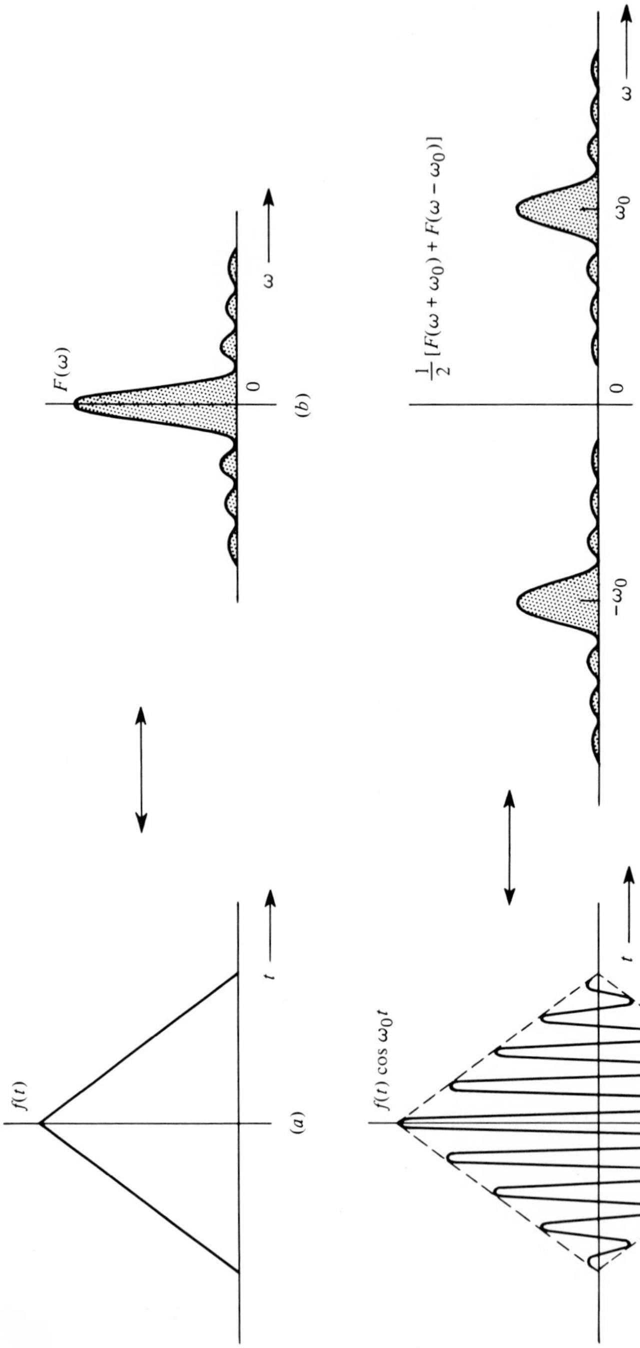

FIG. 1-18. Effect of modulation on the frequency spectrum.

Changing the order of differentiation and integration, we get

$$\frac{df}{dt} = \frac{1}{2\pi} \int_{-\infty}^{\infty} j\omega F(\omega)\, e^{j\omega t}\, d\omega$$

It is now evident from the above equation that

$$\frac{df}{dt} \leftrightarrow j\omega F(\omega)$$

Now consider the function

$$\phi(t) = \int_{-\infty}^{t} f(\tau)\, d\tau$$

Then

$$\frac{d\phi(t)}{dt} = f(t)$$

Hence, if

$$\phi(t) \leftrightarrow \Phi(\omega)$$

then

$$f(t) \leftrightarrow j\omega \Phi(\omega)$$

That is,

$$F(\omega) = j\omega \Phi(\omega)$$

Therefore

$$\Phi(\omega) = \frac{1}{j\omega} F(\omega)$$

and thus

$$\int_{-\infty}^{t} f(\tau)\, d\tau \leftrightarrow \frac{1}{j\omega} F(\omega)$$

Note that this result is valid only if $\Phi(\omega)$ exists, that is, if $\phi(t)$ is absolutely integrable. This is possible only if

$$\lim_{t \to \infty} \phi(t) = 0$$

That is,

$$\int_{-\infty}^{\infty} f(t)\, dt = 0$$

This is equivalent to the condition that $F(0) = 0$, since

$$\int_{-\infty}^{\infty} f(t)\, dt = F(\omega)\big|_{\omega=0}$$

The time-differentiation and time-integration theorems as expressed in Eqs. 1-89a and 1-89b are also obvious intuitively. The Fourier transform actually expresses a function $f(t)$ in terms of a continuous sum of exponential functions of the form $e^{j\omega t}$. The derivative of $f(t)$ is therefore equal to the continuous sum of the derivatives of the individual exponential components. But the derivative of an exponential function $e^{j\omega t}$ is equal to $j\omega e^{j\omega t}$. Therefore, the process of differentiation of $f(t)$ is equivalent to multiplication by $j\omega$ of each exponential component. Hence

$$\frac{df}{dt} \leftrightarrow j\omega F(\omega)$$

A similar argument applies to integration.

We conclude that differentiation in the time domain is equivalent to multiplication by $j\omega$ in the frequency domain, and that integration in the time domain is equivalent to division by $j\omega$ in the frequency domain. The time-differentiation theorem proves convenient in deriving the Fourier transform of some piecewise continuous functions. (See problems 14 and 15 at the end of the chapter.)

Frequency Differentiation. If

$$f(t) \leftrightarrow F(\omega)$$

then

$$-jtf(t) \leftrightarrow \frac{dF}{d\omega} \tag{1-90}$$

This theorem can be proved along the same lines as the time-differentiation theorem. From this theorem, we conclude that differentiation in the frequency domain is equivalent to multiplication by $-jt$ in the time domain.

The Convolution Theorem. The convolution theorem is perhaps one of the most powerful tools in frequency analysis. It permits the easy derivation of many important results, and will be used often in this text.

Given two functions $f_1(t)$ and $f_2(t)$, we form the integral

$$f(t) = \int_{-\infty}^{\infty} f_1(\tau) f_2(t - \tau)\, d\tau \tag{1-91}$$

This integral defines the convolution of functions $f_1(t)$ and $f_2(t)$. The convolution integral (Eq. 1-91) is also expressed symbolically as

$$f(t) = f_1(t) * f_2(t)$$

Again, as usual, we have two theorems—time convolution and frequency convolution.

Time-Convolution Theorem. If

$$f_1(t) \leftrightarrow F_1(\omega)$$

and

$$f_2(t) \leftrightarrow F_2(\omega)$$

then

$$\int_{-\infty}^{\infty} f_1(\tau) f_2(t - \tau)\, d\tau \leftrightarrow F_1(\omega) F_2(\omega)$$

That is,

$$f_1(t) * f_2(t) \leftrightarrow F_1(\omega) F_2(\omega) \tag{1-92}$$

Proof.

$$\mathfrak{F}[f_1(t) * f_2(t)] = \int_{-\infty}^{\infty} e^{-j\omega t} \left[\int_{-\infty}^{\infty} f_1(\tau) f_2(t - \tau)\, d\tau\right] dt$$

$$= \int_{-\infty}^{\infty} f_1(\tau) \left[\int_{-\infty}^{\infty} e^{-j\omega t} f_2(t - \tau)\, dt\right] d\tau$$

From the time-shifting property (Eq. 1-86), it is evident that the integral inside the bracket, on the right-hand side, is equal to $F_2(\omega) e^{-j\omega\tau}$. Hence

$$\mathfrak{F}[f_1(t) * f_2(t)] = \int_{-\infty}^{\infty} f_1(\tau) e^{-j\omega\tau} F_2(\omega)\, d\tau$$

$$= F_1(\omega) F_2(\omega)$$

Frequency Convolution. If

$$f_1(t) \leftrightarrow F_1(\omega)$$

and

$$f_2(t) \leftrightarrow F_2(\omega)$$

then

$$f_1(t) f_2(t) \leftrightarrow \frac{1}{2\pi} \int_{-\infty}^{\infty} F_1(u) F_2(\omega - u)\, du$$

TABLE 1-2

Operation	$f(t)$	$F(\omega)$
1. Scaling	$f(at)$	$\frac{1}{\lvert a \rvert} F\left(\frac{\omega}{a}\right)$
2. Time shifting	$f(t - t_0)$	$F(\omega) e^{-j\omega t_0}$
3. Frequency shifting	$f(t) e^{j\omega_0 t}$	$F(\omega - \omega_0)$
4. Time differentiation	$\frac{d^n f}{dt^n}$	$(j\omega)^n F(\omega)$
5. Frequency differentiation	$(-jt)^n f(t)$	$\frac{d^n F}{d\omega^n}$
6. Time integration	$\int_{-\infty}^{t} f(\tau)\, d\tau$	$\frac{1}{(j\omega)} F(\omega)$
7. Time convolution	$f_1(t) * f_2(t)$	$F_1(\omega) F_2(\omega)$
8. Frequency convolution	$f_1(t) f_2(t)$	$\frac{1}{2\pi} [F_1(\omega) * F_2(\omega)]$

That is,

$$f_1(t)f_2(t) \leftrightarrow \frac{1}{2\pi}[F_1(\omega) * F_2(\omega)] \tag{1-93}$$

This theorem can be proved in exactly the same way as the time-convolution theorem because of the symmetry in the direct and inverse Fourier transform.

We therefore conclude that the convolution of two functions in the time domain is equivalent to multiplication of their spectra in the frequency domain and that multiplication of two functions in the time domain is equivalent to convolution of their spectra in the frequency domain.

Table 1-2 shows some of the important properties of the Fourier transform. Note the duality between the time and the frequency domains.

1-9. SOME CONVOLUTION RELATIONSHIPS

The symbolic representation of convolution suggests that convolution is a special kind of multiplication. Indeed, it is possible to write the laws of convolution algebra along lines that are similar to those for ordinary multiplication.

Commutative Law.

$$f_1(t) * f_2(t) = f_2(t) * f_1(t) \tag{1-94a}$$

This relationship can be easily proved as follows:

$$f_1(t) * f_2(t) = \int_{-\infty}^{\infty} f_1(\tau)f_2(t-\tau)\,d\tau$$

Changing the variable τ to $t - x$, we get

$$\begin{aligned} f_1(t) * f_2(t) &= \int_{-\infty}^{\infty} f_2(x)f_1(t-x)\,dx \\ &= f_2(t) * f_1(t) \end{aligned}$$

Distributive Law.

$$f_1(t) * [f_2(t) + f_3(t)] = f_1(t) * f_2(t) + f_1(t) * f_3(t) \tag{1-94b}$$

The proof is trivial.

Associative Law.

$$f_1(t) * [f_2(t) * f_3(t)] = [f_1(t) * f_2(t)] * f_3(t) \tag{1-94c}$$

This law follows from the convolution theorem and from the fact that

$$F_1(\omega)[F_2(\omega)F_3(\omega)] = [F_1(\omega)F_2(\omega)]F_3(\omega)$$

1-10. GRAPHICAL INTERPRETATION OF CONVOLUTION

The graphical interpretation of convolution is very useful in system analysis as well as in communication theory. It permits one to grasp visually the results of many abstract relationships. This is particularly true in communication theory. For the sake of illustration, let us consider $f_1(t)$ and $f_2(t)$ as rectangular and triangular pulses as shown in Fig. 1-19a. We shall find the convolution $f_1(t) * f_2(t)$ graphically. By definition,

$$f_1(t) * f_2(t) = \int_{-\infty}^{\infty} f_1(\tau) f_2(t - \tau)\, d\tau \tag{1-95}$$

The independent variable in the convolution integral is τ (Eq. 1-95). The functions $f_1(\tau)$ and $f_2(-\tau)$ are shown in Fig. 1-19b. Note that $f_2(-\tau)$ is obtained by folding $f_2(\tau)$ about the vertical axis passing through the origin. The term $f_2(t - \tau)$ represents the function $f_2(-\tau)$ shifted by t seconds along the positive τ axis. Figure 1-19c shows $f_2(t_1 - \tau)$. The value of the convolution integral at $t = t_1$ is given by the integral in Eq. 1-95 evaluated at $t = t_1$. This is clearly the area under the product curve of $f_1(\tau)$ and $f_2(t_1 - \tau)$. This area is shown shaded in Fig. 1-19d. The value of $f_1(t) * f_2(t)$ at $t = t_1$ is equal to this shaded area and is plotted in Fig. 1-19f. We choose different values of t, shift the function $f_2(-\tau)$ accordingly, and find the area under the new product curve. These areas represent the value of the convolution function $f_1(t) * f_2(t)$ at the respective values of t. The plot of the area under the product curve as a function of t represents the desired convolution function $f_1(t) * f_2(t)$.

The graphical mechanism of convolution can be appreciated by visualizing the function $f_2(-\tau)$ as a rigid frame which is being progressed along the τ axis by t_1 seconds. The function represented by this frame is multiplied by $f_1(\tau)$, and the area under the product curve is the value of the convolution function at $t = t_1$. Therefore, to find the value of $f_1(t) * f_2(t)$ at any time, say $t = t_0$, we displace the rigid frame representing $f_2(-\tau)$ by t_0 seconds along the τ axis and multiply this function with $f_1(\tau)$. The area under the product curve is the desired value of $f_1(t) * f_2(t)$ at $t = t_0$. To find the function $f_1(t) * f_2(t)$, we progress the frame successively by different amounts and find the areas of the product curve at various positions. The plot of the area as a function of displacement of the frame represents the required convolution function $f_1(t) * f_2(t)$.

Convolution of a Function with a Unit Impulse. The convolution of a function $f(t)$ with a unit impulse $\delta(t)$ yields the function $f(t)$ itself.

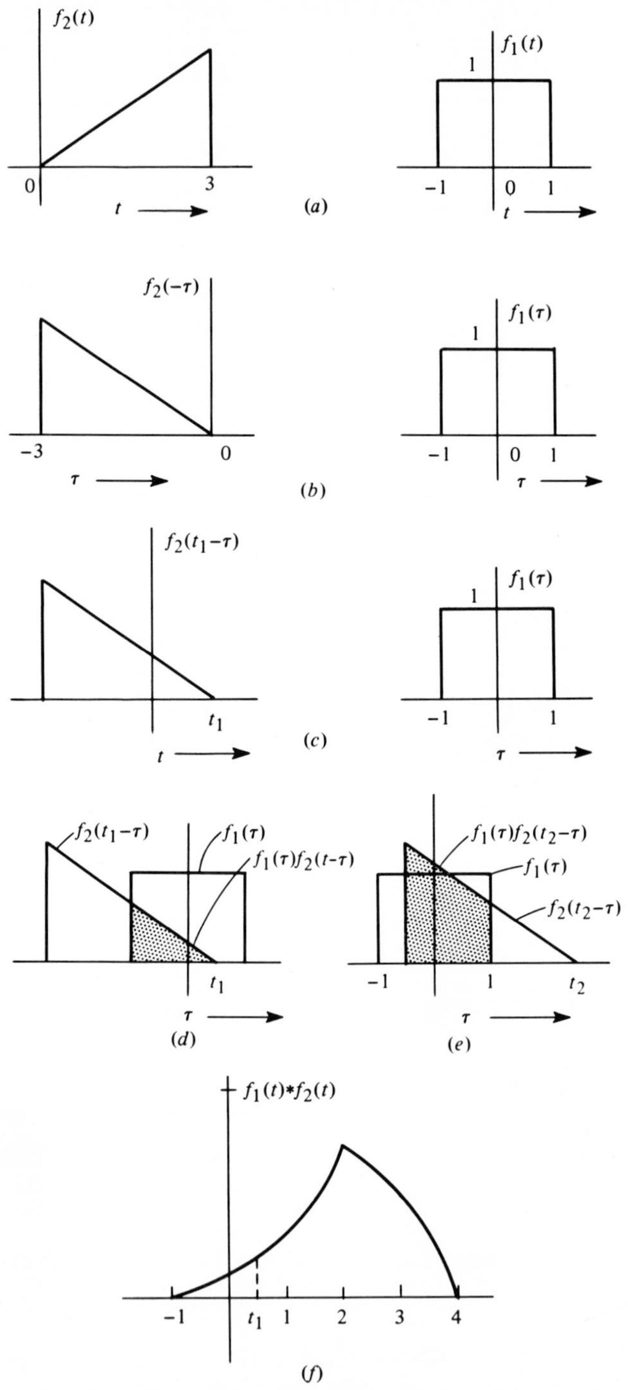

$f_2(t)$
0
3
t
(a)
$f_1(t)$
1
−1
0
1
t
$f_2(-\tau)$
−3
0
τ
(b)
$f_1(\tau)$
1
−1
0
1
τ
$f_2(t_1-\tau)$
t_1
t
(c)
$f_1(\tau)$
1
−1
1
τ
$f_2(t_1-\tau)$
$f_1(\tau)$
$f_1(\tau)f_2(t-\tau)$
t_1
τ
(d)
$f_1(\tau)f_2(t_2-\tau)$
$f_1(\tau)$
$f_2(t_2-\tau)$
−1
1
t_2
τ
(e)
$f_1(t)*f_2(t)$
−1
t_1
1
2
3
4
(f)

FIG. 1-19.

This can be easily proved by using the sampling property 1-69b.

$$f(t) * \delta(t) = \int_{-\infty}^{\infty} f(\tau)\,\delta(t - \tau)\,d\tau$$

$$= f(t) \tag{1-96}$$

This result is also evident graphically. The reader may verify this result by convolving $f(t)$ with $\delta(t)$ graphically. A simple extension of Eq. 1-96 yields

$$f(t) * \delta(t - T) = f(t - T)$$

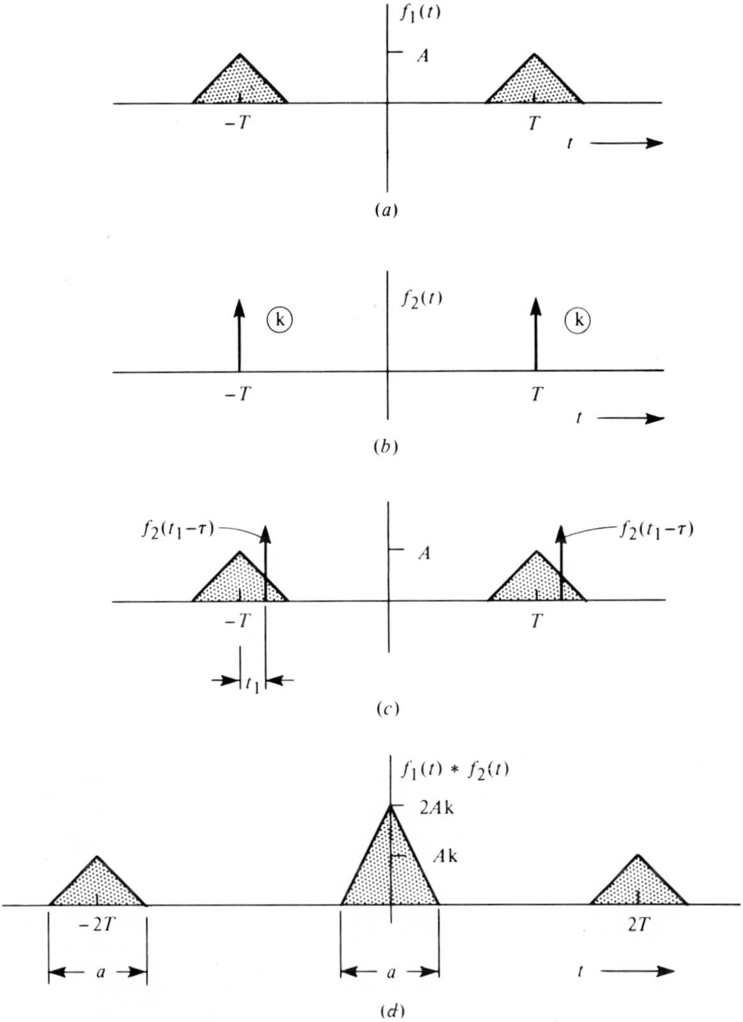

FIG. 1-20.

$$f(t - t_1) * \delta(t - t_2) = f(t - t_1 - t_2) \tag{1-97}$$
$$\delta(t - t_1) * \delta(t - t_2) = \delta(t - t_1 - t_2)$$

Example 1-3. Find graphically the convolution of $f_1(t)$ (Fig. 1-20a) with a pair of impulses of strength k each, as shown in Fig. 1-20b.

Solution. Following the procedure of graphical convolution described earlier, we fold back $f_2(\tau)$ about the vertical axis to obtain $f_2(-\tau)$. Since $f_2(\tau)$ is an even function of τ, $f_2(-\tau) = f_2(\tau)$. The convolution of $f_1(\tau)$ with $f_2(\tau)$ thus reduces to convolution of $f_1(\tau)$ with two impulses. From the property of an impulse function to reproduce the function by convolution (Eq. 1-96), it can be easily seen that each impulse produces a triangular pulse of height Ak at the origin ($t = 0$). Hence the net height of the triangular pulse is $2Ak$ at the origin. As the function $f_2(t - \tau)$ is moved farther in the positive direction the impulse originally located at $-T$ encounters the triangular pulse at $\tau = T$, and reproduces the triangular pulse of height Ak at $t = 2T$. Similarly, the impulse originally located at T reproduces the triangular pulse of height Ak at $t = -2T$.

1-11. SAMPLING THEOREM

The sampling theorem is very significant in the communication theory. As before, there is a sampling theorem in time domain and its dual in frequency domain.

Sampling Theorem (Time Domain). This theorem states that:

A band-limited signal which has no spectral components above the frequency B Hertz is uniquely determined by its samples taken at uniform intervals less than $1/2B$ seconds apart.

This theorem is known as the uniform sampling theorem since it pertains to the specification of a given signal by its samples at uniform intervals of $1/2B$ seconds.[5]

The sampling theorem can be easily proved with the help of the frequency convolution theorem. Consider a band-limited signal $f(t)$ which has no spectral components above B Hz (Fig. 1-21). Suppose we multiply the function $f(t)$ by a uniform train of impulse function $\delta_T(t)$. The

[5]This theorem is actually a special case of the general sampling theorem which states:

If a signal is band-limited and if the time interval is divided into equal parts forming subintervals such that each subdivision comprises an interval T seconds long where T is less than half the period of the highest significant frequency (B) component of the signal; and if one instantaneous sample is taken from each subinterval in any manner; then a knowledge of the instantaneous magnitude of each sample plus a knowledge of the instants within each subinterval at which the sample is taken contains all the information of the original signal.

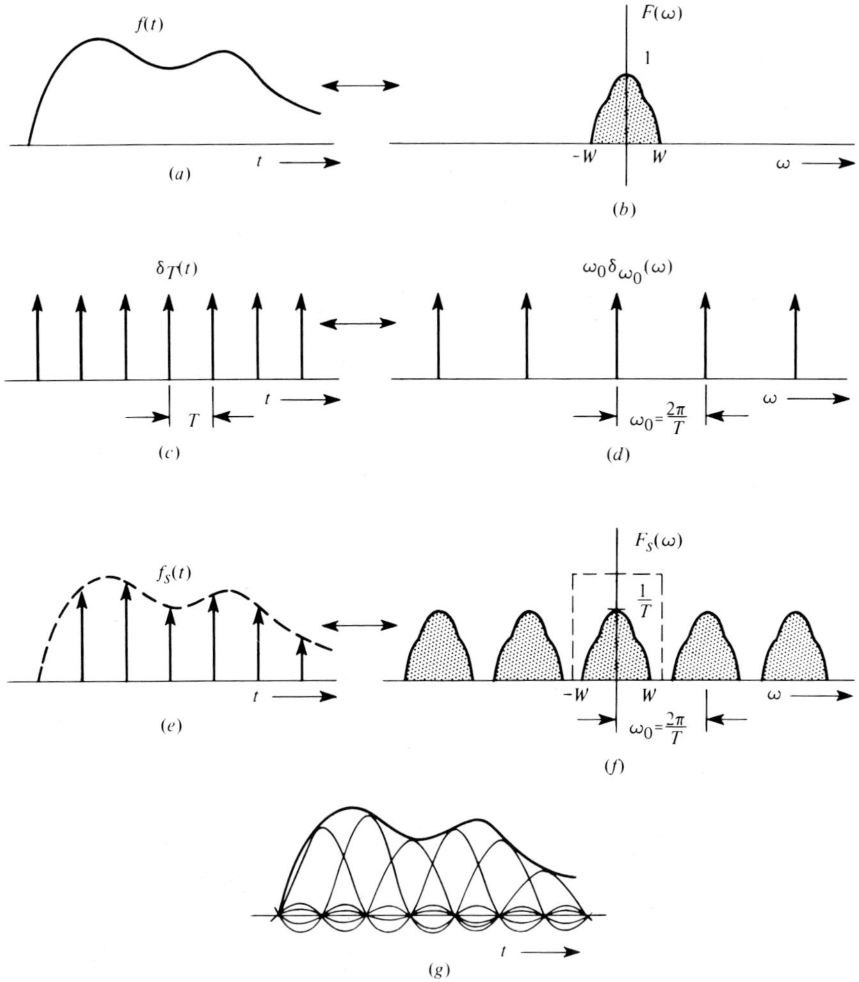

FIG. 1-21.

product function is a sequence of impulses located at regular intervals of T seconds and having strengths equal to the values of $f(t)$ at the corresponding instants. The product $f(t)\,\delta_T(t)$ indeed represents (Fig. 1-21e) the function $f(t)$ sampled at a uniform interval of T seconds. We shall denote this sampled function by $f_s(t)$.

$$f_s(t) = f(t)\,\delta_T(t) = \sum_{n=-\infty}^{\infty} f(nT)\,\delta(t - nT) \tag{1-98}$$

The frequency spectrum of $f(t)$ is $F(\omega)$, and from Eq. 1-80, we have

$$\delta_T(t) \leftrightarrow \omega_0\delta_{\omega_0}(\omega) \qquad \omega_0 = \frac{2\pi}{T}$$

The Fourier transform of $f(t)\,\delta_T(t)$ will, according to the frequency-convolution theorem, be given by (Eq. 1-93)

$$2\pi f_s(t) \leftrightarrow F(\omega) * \omega_0 \delta_{\omega_0}(\omega)$$

Substituting $\omega_0 = 2\pi/T$, we get

$$f_s(t) \leftrightarrow \frac{1}{T} F(\omega) * \delta_{\omega_0}(\omega) \tag{1-99}$$

The convolution of $F(\omega)$ and $\delta_{\omega_0}(\omega)$ is shown in Fig. 1-21f. It is obvious that the convolution of $F(\omega)$ with $\delta_{\omega_0}(\omega)$ will reproduce $F(\omega)$ every ω_0 radians periodically. Note that $F(\omega)$ will repeat periodically without overlap as long as

$$\omega_0 \geq 2W \qquad (W = 2\pi B)$$

That is,

$$\frac{2\pi}{T} \geq 2(2\pi B)$$

or

$$T \leq \frac{1}{2B}$$

Therefore, as long as we sample $f(t)$ at regular intervals less than $1/2B$ seconds apart, $F_s(\omega)$, the spectral density function of $f_s(t)$, will be a periodic replica of $F(\omega)$ and therefore contains all the information of $f(t)$. We can easily recover $F(\omega)$ from $F_s(\omega)$ by allowing the sampled signal to pass through a low-pass filter which will allow only frequency components below B Hz and attentuate all the higher-frequency components. It is therefore evident that the sampled function $f_s(t)$ contains all the information of $f(t)$. The ideal filter characteristic required to recover $f(t)$ from $f_s(t)$ is shown dotted in Fig. 1-21f.

Note that if the sampling interval T becomes greater than $1/2B$, then the convolution of $F(\omega)$ with $\delta_{\omega_0}(\omega)$ yields $F(\omega)$ repeating periodically. But now there is an overlap between successive cycles and $F(\omega)$ cannot be recovered from $F_s(\omega)$. Therefore, if the sampling interval T is made too large, the information is partly lost, and the function $f(t)$ cannot be recovered from the sampling function. The conclusion is quite logical since it is reasonable to expect that the information will be lost if the sampling is too slow. The maximum interval of sampling $T = 1/2B$ is also called the *Nyquist interval*.

Most of the signals in practice closely approximate the band-limited signals. It should be stated here that strictly speaking a band-limited signal does not exist in the real world. It can be shown that if a signal exists over a finite interval of time it contains the components of all frequencies (reference 1-6). However, for all signals in practice the spectral density functions diminish at higher frequencies. Most of the energy is carried by components lying within a certain frequency interval, and for all practical purposes a signal may be considered to be band-limited. The error introduced by ignoring high-frequency components is negligible.

The proof of the sampling theorem given above is essentially a graphical proof. We can obtain the same result analytically. From Eq. 1-99, we have

$$\begin{aligned} F_s(\omega) &= \frac{1}{T} F(\omega) * \delta_{\omega_0}(\omega) \\ &= \frac{1}{T} F(\omega) * \sum_{n=-\infty}^{\infty} \delta(\omega - n\omega_0), \qquad n = 0, \pm 1, \pm 2, \ldots \\ &= \frac{1}{T} \sum_{n=-\infty}^{\infty} F(\omega - n\omega_0) \end{aligned} \tag{1-100}$$

The right-hand side of Eq. 1-100 represents $F(\omega)$ repeating itself periodically every ω_0 rps. Obviously if $\omega_0 \geq 2W$ (Fig. 1-21), all the information in $F(\omega)$ is preserved.

Recovering $f(t)$ from its Samples. It is evident from Fig. 1-21f (or Eq. 1-100) that

$$F(\omega) = TF_s(\omega)\, G_{\omega_0}(\omega) \tag{1-101}$$

The application of the time-convolution theorem to Eq. 1-101 yields

$$\begin{aligned} f(t) &= Tf_s(t) * \frac{\omega_0}{2\pi} \mathrm{Sa}\left(\frac{\omega_0 t}{2}\right) \\ &= f_s(t) * \mathrm{Sa}\left(\frac{\omega_0 t}{2}\right) \\ &= \sum_n f(nT)\,\delta(t - nT) * \mathrm{Sa}\left(\frac{\omega_0 t}{2}\right) \\ &= \sum_n f(nT)\,\mathrm{Sa}\left(\frac{\omega_0}{2}(t - nT)\right) \qquad \text{(1-102a)} \\ &= \sum_n f(nT)\,\mathrm{Sa}\left(\frac{\omega_0 t}{2} - n\pi\right) \qquad \text{(1-102b)} \end{aligned}$$

When the sampling is performed at the Nyquist rate, $\omega_0 = 2W$, and

$$\begin{aligned} f(t) &= \sum_n f(nT)\,\mathrm{Sa}[W(t - nT)] \qquad \text{(1-103a)} \\ &= \sum_n f(nT)\,\mathrm{Sa}(Wt - n\pi) \qquad \text{(1-103b)} \end{aligned}$$

Equation 1-103a shows the method of reconstructing $f(t)$ from its samples. Each sample is multiplied by a sampling function and all the resulting waveforms are added to obtain $f(t)$. The graphical process of reconstruction is shown in Fig. 1-21g. Note that the sampling function $\mathrm{Sa}(Wt - n\pi)$ is zero at all sampling instants except the nth.

We have proved the sampling theorem for a band-limited signal whose spectrum is centered at $f = 0$. It can be shown that the theorem holds even if the signal has a bandpass spectrum of bandwidth $2B$ Hz and centered at some frequency f_0.

Sampling Theorem (Frequency Domain). The sampling theorem in time domain has a dual which states: A time-limited signal that is zero for $|t| > T$ is uniquely determined by the samples of its frequency spectrum at uniform intervals less than $1/2T$ Hz apart (or π/T rps apart).

The proof of this theorem is similar to that of in time domain. The dual of Eq. 1-103 is given by

$$F(\omega) = \sum_{n=-\infty}^{\infty} F\left(\frac{n\pi}{T}\right) \mathrm{Sa}(\omega T - n\pi) \qquad (1\text{-}104)$$

The sampling theorem has a deep significance in communication theory. It allows us to replace a continuous band-limited signal by a discrete sequence of its samples without the loss of any information. The information content of a continuous band-limited signal is thus equivalent to discrete pieces of information. Since the sampling principle specifies the least number of discrete values necessary to reproduce a continuous signal, the problem of transmitting such a signal is reduced to that of transmitting a finite number of values. Such discrete information may be transmitted by a group of pulses whose amplitudes may be varied (pulse-amplitude modulation) according to sample values. Other forms of modulation include pulse-position modulation, where the position of pulse is varied, the pulse-width modulation (variation of pulse width) of pulse-code modulation.

1-12. TRANSMISSION OF SIGNALS THROUGH LINEAR SYSTEMS

The sampling property of the impulse function allows us to represent a signal $f(t)$ as

$$f(t) = f(t) * \delta(t) = \int_{-\infty}^{\infty} f(\tau)\,\delta(t - \tau)\,d\tau \qquad (1\text{-}105)$$

This representation can be viewed as an expression of $f(t)$ in terms of impulse components, where $f(t)$ is expressed as a continuous sum (integral) of impulse function.

If a linear time invariant system has a unit impulse response $h(t)$ (Fig. 1-22), then it follows that the response $r(t)$ to signal $f(t)$ will be given by

$$r(t) = \int_{-\infty}^{\infty} f(\tau) h(t - \tau) \, d\tau \tag{1-106}$$

This follows from Eq. 1-105 and the principle of superposition. Thus

$$r(t) = f(t) * h(t)$$

and

$$R(\omega) = F(\omega) H(\omega) \tag{1-107}$$

where

$$h(t) \leftrightarrow H(\omega) \qquad \text{and} \qquad r(t) \leftrightarrow R(\omega)$$

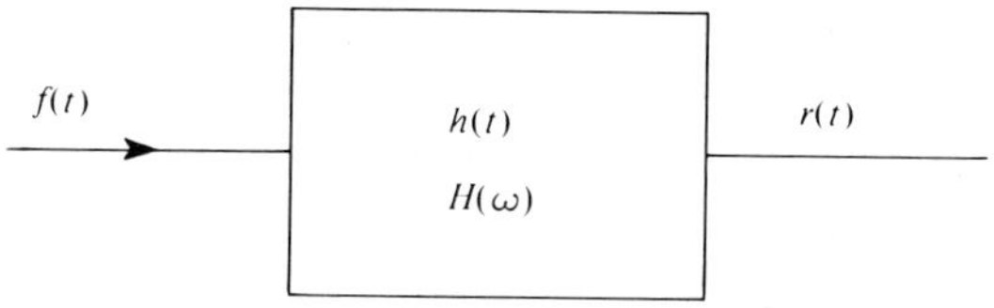

FIG. 1-22.

The unit-impulse response $h(t)$ is the response of the system to a unit impulse applied at $t = 0$. Obviously for any physical system $h(t)$ cannot appear before $t = 0$. Hence $h(t) = 0$ for $t < 0$. Such signals are called *causal* signals. Thus, for all physical realizable systems $h(t)$ is causal. We can, however, postulate systems for which $h(t)$ is not causal. Such systems are of course physically unrealizable.

The causality of $h(t)$ is the criterion of physical realizability in time domain. In the frequency domain this criterion implies that a necessary and sufficient condition for a magnitude function $|H(\omega)|$ to be physically realizable is that

$$\int_{-\infty}^{\infty} \frac{|\ln |H(\omega)||}{1 + \omega^2} \, d\omega < \infty$$

This is the Paley-Wiener criterion of physical realizability (Refs. 1-5, 1-7). This magnitude function $|H(\omega)|$ must, however, be square integrable for the Paley-Wiener criterion to be valid.[6]

A system whose magnitude function $H(\omega)$ violates the Paley-Wiener criterion, has a noncasual-impulse response and is consequently physically unrealizable.

[6]If the magnitude function $|H(\omega)|$ satisfied Paley-Wiener criterion, it does not follow that the system is physically realizable. It merely says that a suitable phase function $\theta(\omega)$ may be associated with $|H(\omega)|$ so that the resulting transfer function is physically realizable.

1-13. THE ENERGY-DENSITY SPECTRUM

A useful parameter of a signal $f(t)$ is its normalized energy. We define the normalized energy (or simply the energy) E of a signal $f(t)$ as the energy dissipated by a voltage $f(t)$ applied across a 1-ohm resistor (or by a current $f(t)$ passing through a 1-ohm resistor). Thus[7]

$$E = \int_{-\infty}^{\infty} f^2(t)\,dt \tag{1-108a}$$

The concept of signal energy is meaningful only if the integral in Eq. 1-108a is finite. The signals for which E is finite are known as *energy signals* or *pulse signals*. In case of some signals, e.g., periodic signals the integral in Eq. 1-108a is obviously infinite and the concept of energy of a signal is meaningless. In such cases, we consider average power of the signal instead of its energy. Such signals are called *power signals* and will be discussed in a later section.

If $F(\omega)$ is the Fourier transform of $f(t)$,

$$f(t) = \frac{1}{2\pi}\int_{-\infty}^{\infty} F(\omega)\,e^{j\omega t}\,d\omega$$

and the energy E of $f(t)$ is given by

$$E = \int_{-\infty}^{\infty} f^2(t)\,dt = \int_{-\infty}^{\infty} f(t)\left[\frac{1}{2\pi}\int_{-\infty}^{\infty} F(\omega)\,e^{j\omega t}\,d\omega\right]dt$$

Interchanging the order of integration on the right-hand side, we get

$$E = \int_{-\infty}^{\infty} f^2(t)\,dt = \frac{1}{2\pi}\int_{-\infty}^{\infty} F(\omega)\left[\int_{-\infty}^{\infty} f(t)\,e^{j\omega t}\,dt\right]d\omega$$

The inner integral on the right-hand side is obviously $F(-\omega)$. Hence we have

$$\int_{-\infty}^{\infty} f^2(t)\,dt = \frac{1}{2\pi}\int_{-\infty}^{\infty} F(\omega)\,F(-\omega)\,d\omega$$

We have already proved (see Eq. 1-61) that for a real $f(t)$,

$$F^*(\omega) = F(-\omega)$$

Hence

$$F(\omega)\,F(-\omega) = |\,F(\omega)\,|^2 \tag{1-109}$$

[7]For real signals the energy E is defined by Eq. 1-108a. If, however, the signal $f(t)$ is a complex function of variable t, then the energy E dissipated by voltage $f(t)$ applied across a 1 ohm resistor is given by

$$E = \int_{-\infty}^{\infty} |f^2(t)|\,dt = \int_{-\infty}^{\infty} f(t) f^*(t)\,dt \tag{1-108b}$$

This is the definition of energy E for complex signals.

and

$$\int_{-\infty}^{\infty} f^2(t)\,dt = \frac{1}{2\pi}\int_{-\infty}^{\infty} |F(\omega)|^2\,d\omega \tag{1-110a}$$

$$= \int_{-\infty}^{\infty} |F(\omega)|^2\,df \tag{1-110b}$$

This equation[8] states that the energy of a signal is given by the area under $|F(\omega)|^2$ curve (integrated with respect to the frequency variable $f = \omega/2\pi$).

Eq. 1-110b has interesting interpretation. Let us consider a signal $f(t)$ applied to an ideal bandpass filter, whose transfer function $H(\omega)$ is shown in Fig. 1-23. This filter suppresses all frequencies except a narrow-band $\Delta\omega$ centered at frequency ω_0. If $R(\omega)$ is the Fourier transform of the response $r(t)$ of this filter, then

$$R(\omega) = F(\omega)H(\omega)$$

and the energy E_0, of the output $r(t)$ is given by Eq. 1-110a:

$$E_0 = \frac{1}{2\pi}\int_{-\infty}^{\infty} |F(\omega)H(\omega)|^2\,d\omega$$

Since $H(\omega) = 0$ everywhere except over a narrow-band $\Delta\omega$ where it has a unity magnitude, we have for $\Delta\omega \rightarrow 0$

$$E_0 = 2\frac{1}{2\pi}|F(\omega_0)|^2\,\Delta\omega$$

$$= 2|F(\omega_0)|^2\,\Delta f$$

The energy of the output signal is thus $2|F(\omega_0)|^2\,\Delta f$. As can be seen from Fig. 1-23, only the frequency components of $f(t)$ which lie in the narrow-band $\Delta\omega$ are transmitted and the remaining frequency compo-

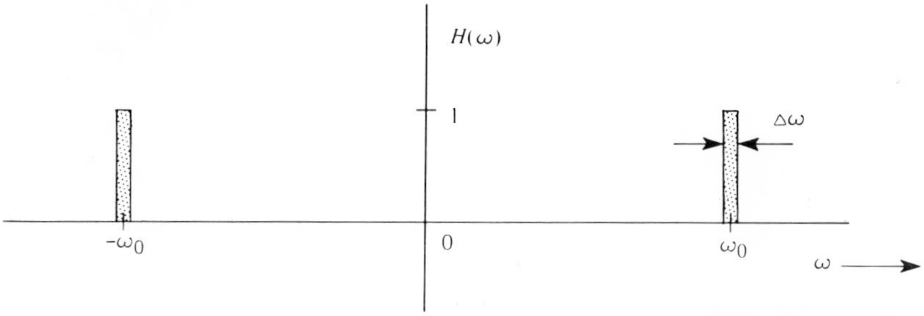

FIG. 1-23.

[8]Equation 1-110, which corresponds to Parseval's theorem (Eq. 1-40) for non-periodic signals, is called Parseval's theorem or Plancherel's theorem.

nents are suppressed. Obviously $2\,|F(\omega_0)|^2\,\Delta f$ represents the contribution to the energy of $f(t)$ by frequency components of $f(t)$ lying in the narrow-band Δf centered at ω_0. Hence $|F(\omega)|^2$ is the energy per unit bandwidth (in Hertz) contributed by frequency components in the narrow-band Δf centered around frequency $f = 2\pi/\omega$. Note that we consider positive as well as negative frequencies. The contribution of positive band is $|F(\omega)|^2\,\Delta f$ and there is an equal contribution of the negative band (because $|F(\omega)|^2$ is an even function of ω). This is clearly illustrated by Fig. 1-24. Thus $|F(\omega)|^2$ is the energy-density spectrum, since it represents the energy per unit bandwidth (in Hertz).

We now define the energy-density spectrum $\Phi_f(\omega)$ as[9]

$$\Phi_f(\omega) = |F(\omega)|^2 \tag{1-111}$$

The energy density provides us with a relative contribution of energy by various frequency components.

The total energy E is given by

$$E = \frac{1}{2\pi}\int_{-\infty}^{\infty} \Phi_f(\omega)\,d\omega \tag{1-112a}$$

$$= \int_{-\infty}^{\infty} \Phi_f(\omega)\,df \tag{1-112b}$$

Since $\Phi_f(\omega)$ is an even function of ω, we may express these equations as

$$E = \frac{1}{\pi}\int_{0}^{\infty} \Phi_f(\omega)\,d\omega \tag{1-113a}$$

$$= 2\int_{0}^{\infty} \Phi_f(\omega)\,df \tag{1-113b}$$

Thus instead of integrating the energy-density spectrum from $-\infty$ to ∞, it is necessary to integrate it only from 0 to ∞ and multiply by 2. Equation 1-111 defines bilateral energy density, since the energy is obtained by integrating over negative as well as positive frequencies. The unilateral energy density is defined as $2\,|F(\omega)|^2$, since it is necessary to integrate this density only over positive frequencies. Throughout this book we shall use bilateral energy density.

[9]In literature $\Phi_f(\omega)$ is defined alternatively as

$$\Phi_f(\omega) = 2\,|F(\omega)|^2$$

This is the unilateral energy-density spectrum. In this case the energy of positive and negative frequencies is lumped together. The total energy E is found by integrating $\Phi_f(\omega)$ over 0 to ∞ instead of $-\infty$ to ∞. Energy density is also defined as $1/\pi\,|F(\omega)|^2$. In this case the energy density is defined as the energy per radian bandwidth.

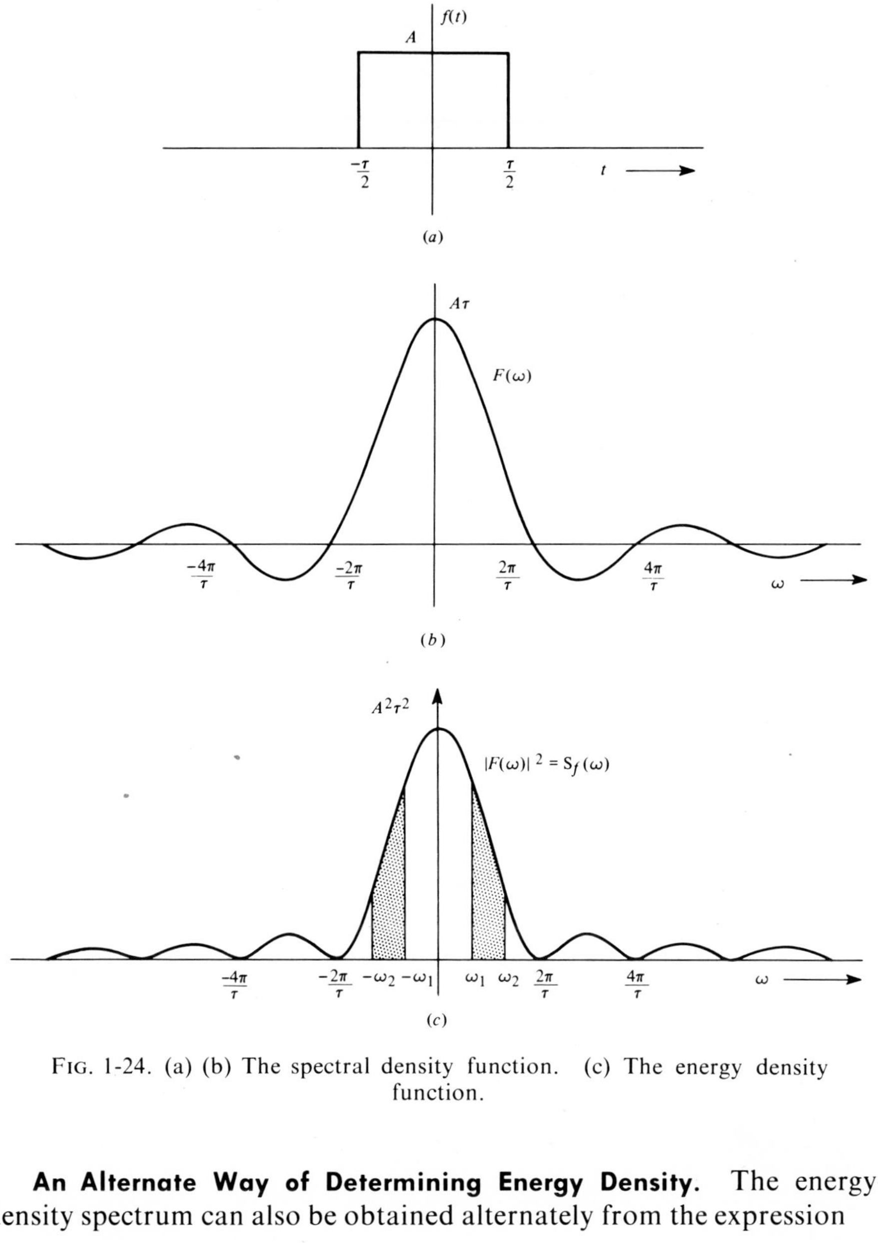

FIG. 1-24. (a) (b) The spectral density function. (c) The energy density function.

An Alternate Way of Determining Energy Density. The energy-density spectrum can also be obtained alternately from the expression

$$\phi_f(\tau) = \int_{-\infty}^{\infty} f(t) f(t + \tau)\, dt \tag{1-114}$$

A change of variable t to $t - \tau$ in Eq. 1-114 yields

$$\int_{-\infty}^{\infty} f(t) f(t + \tau)\, dt = \int_{-\infty}^{\infty} f(t) f(t - \tau)\, dt \tag{1-115}$$

It is evident from Eq. 1-115 that the function $\phi_f(\tau)$ is an even function of τ. We shall now show that the Fourier transform of $\phi_f(\tau)$ is equal to $|F(\omega)|^2$, that is

$$\phi_f(\tau) \leftrightarrow \Phi_f(\omega)$$

This can be shown by observing that $\phi_f(\tau)$ has a similar nature to that of $f(\tau) * f(\tau)$. Indeed $\phi_f(\tau)$ is obtained by multiplying $f(t)$, by $f(t)$ shifted by τ seconds, and finding the area under the product curve. This is exactly what is done in convolution except that $f(t)$ is inverted about the vertical axis before shifting. It is, therefore, evident that $\phi_f(\tau)$ can be found by convolving $f(\tau)$ with $f(-\tau)$. Hence

$$f(\tau) * f(-\tau) = \int_{-\infty}^{\infty} f(t) f(t + \tau)\, dt = \int_{-\infty}^{\infty} f(t) f(t - \tau)\, dt \quad (1\text{-}116a)$$

Since

$$f(\tau) \leftrightarrow F(\omega)$$

and from Eq. 1-85a, we have

$$f(-\tau) \leftrightarrow F(-\omega)$$

It follows that[10]

$$\phi_f(\tau) = f(\tau) * f(-\tau) \leftrightarrow F(\omega) F(-\omega) = |F(\omega)|^2 \quad (1\text{-}116b)$$

This result can also be obtained directly. It can be easily shown that

$$\begin{aligned}
\mathfrak{F}\left[\int_{-\infty}^{\infty} f(t) f(t + \tau)\, dt\right] &= \int_{-\infty}^{\infty} e^{-j\omega t}\, d\tau \int_{-\infty}^{\infty} f(t) f(t + \tau)\, dt \\
&= \int_{-\infty}^{\infty} f(t)\, dt \int_{-\infty}^{\infty} f(t + \tau)\, e^{-j\omega t}\, d\tau \\
&= \int_{-\infty}^{\infty} f(t)\, dt \int_{-\infty}^{\infty} f(x)\, e^{-j\omega(x - t)}\, dx \\
&= \int_{-\infty}^{\infty} f(t)\, e^{j\omega t}\, dt \int_{-\infty}^{\infty} f(x)\, e^{-j\omega x}\, dx \\
&= F(\omega) F(-\omega) \\
&= |F(\omega)|^2
\end{aligned}$$

Thus we have

$$\phi_f(\tau) \leftrightarrow |F(\omega)|^2$$

[10]If $f(t)$ is a complex function of time, then we define

$$\phi_f(\tau) = \int_{-\infty}^{\infty} f^*(t)\, f(t + \tau)\, dt = \int_{-\infty}^{\infty} f(t)\, f^*(t - \tau)\, dt$$

The reader can easily show that in this case also

$$\phi_f(\tau) \leftrightarrow F(\omega)\, F^*(\omega) = |F(\omega)|^2$$

or

$$\phi_f(\tau) \leftrightarrow \Phi_f(\omega)$$

That is,

$$\phi_f(\tau) \leftrightarrow \Phi_f(\omega)$$

where $\Phi_f(\omega)$ is the energy-density spectrum of $f(t)$.

For linear systems, a simple relationship exists between the energy-density spectra of the input signal $f(t)$ and the response $r(t)$. If $H(\omega)$ is the transfer function, then

$$R(\omega) = H(\omega)\,F(\omega) \tag{1-117}$$

The energy-density spectrum of the driving function is $|F(\omega)|^2$ and that of the response is $|R(\omega)|^2$. It is evident from Eq. 1-117 that

$$|R(\omega)|^2 = |H(\omega)|^2\,|F(\omega)|^2 \tag{1-118}$$

Hence the energy-density spectrum of the response is given by the energy-density spectrum of the driving function multiplied by $|H(\omega)|^2$.

1-14. THE POWER-DENSITY SPECTRUM

It was mentioned earlier, that some signals, e.g., a periodic signal, or an idealized noise signal in the interval $(-\infty, \infty)$, have infinite energy. In such cases a meaningful parameter is the average power of a signal $f(t)$, which is defined as the average power dissipated by a voltage $f(t)$ applied across a 1-ohm resistor (or by a current $f(t)$ passing through a 1-ohm resistor). Thus the average power P of a signal $f(t)$ is by definition[11]

$$P = \lim_{T\to\infty} \frac{1}{T} \int_{-T/2}^{T/2} f^2(t)\,dt \tag{1-119a}$$

Note that the average power P defined above is also the mean-square value of $f(t)$. Henceforth for convenience we shall indicate the time average of any quantity by a wavy overline. For example, the time average of a signal $f(t)$ will be denoted by $\widetilde{f(t)}$.

$$\widetilde{f(t)} = \lim_{T\to\infty} \frac{1}{T} \int_{-T/2}^{T/2} f(t)\,dt$$

Similarly, the mean-square value of $f(t)$, being a time average of $f^2(t)$, will be denoted by $\widetilde{f^2(t)}$. Thus

[11]For a complex signal $f(t)$, the average power dissipated by a voltage $f(t)$ applied across a 1-ohm resistor is given by

$$P = \lim_{T\to\infty} \frac{1}{T} \int_{-T/2}^{T/2} |f(t)|^2\,dt = \lim_{T\to\infty} \frac{1}{T} \int_{-T/2}^{T/2} f(t)\,f^*(t)\,dt \tag{1-119b}$$

This is the definition of power P for complex signals.

$$\widetilde{f^2(t)} = \lim_{T\to\infty} \frac{1}{T} \int_{-T/2}^{T/2} f^2(t)\,dt$$

In general,

$$\widetilde{\phi[f(t)]} = \lim_{T\to\infty} \frac{1}{T} \int_{-T/2}^{T/2} \phi[f(t)]\,dt$$

Since the average power P by definition (Eq. 1-119a) is the mean-square value of $f(t)$, we have

$$P = \widetilde{f^2(t)} = \lim_{T\to\infty} \frac{1}{T} \int_{-T/2}^{T/2} f^2(t)\,dt$$

We can now proceed along the same lines as those used for obtaining energy density.

Let us form a new function $f_T(t)$ by truncating $f(t)$ outside the interval $|t| > T/2$. The truncated function (Fig. 1-25) can be expressed as $f(t)$ multiplied by a gate function $G_T(t)$.

$$f_T(t) = \begin{cases} f(t) & |t| < T/2 \\ 0 & \text{Otherwise} \end{cases}$$

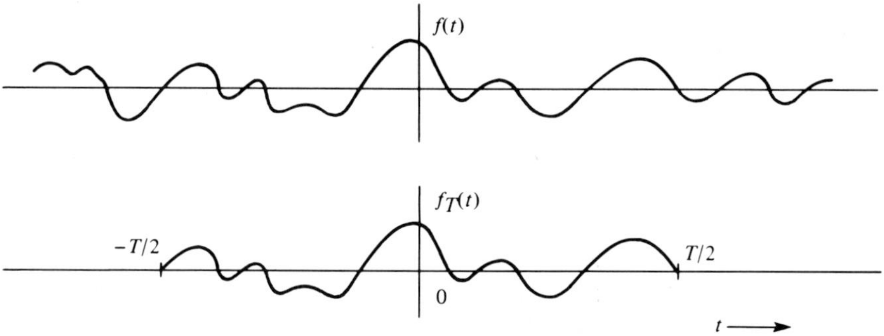

FIG. 1-25.

As long as T is finite, $f_T(t)$ has finite energy. Let

$$f_T(t) \leftrightarrow F_T(\omega)$$

Then the energy E_T of $f_T(t)$ is given by

$$E_T = \int_{-\infty}^{\infty} f_T^2(t)\,dt = \int_{-\infty}^{\infty} |F_T(\omega)|^2\,df$$

But

$$\int_{-\infty}^{\infty} f_T^2(t)\,dt = \int_{-T/2}^{T/2} f^2(t)\,dt$$

Hence

$$\frac{1}{T}\int_{-T/2}^{T/2} f^2(t)\,dt = \int_{-\infty}^{\infty} \frac{|F_T(\omega)|^2}{T}\,df$$

and the average power P is given by

$$P = \lim_{T\to\infty}\frac{1}{T}\int_{-T/2}^{T/2} f^2(t)\,dt = \int_{-\infty}^{\infty} \lim_{T\to\infty}\frac{|F_T(\omega)|^2}{T}\,df$$

As T increases, the energy of $f_T(t)$ also increases. Thus $|F_T(t)|^2$ increases with T. In the limit as $T \to \infty$, the quantity $|F_T(\omega)|^2/T$ may approach a limit. Assuming that such a limit exists, we define $\mathcal{S}_f(\omega)$, the power-density spectrum of $f(t)$ as

$$\mathcal{S}_f(\omega) = \lim_{T\to\infty}\frac{|F_T(\omega)|^2}{T} \tag{1-120}$$

Hence

$$P = \widetilde{f^2(t)} = \lim_{T\to\infty}\frac{1}{T}\int_{-T/2}^{T/2} f^2(t)\,dt = \int_{-\infty}^{\infty} \mathcal{S}_f(\omega)\,df \tag{1-121}$$

$\mathcal{S}_f(\omega)$ is called the *average power-density spectrum*[12] or simply the *power-density spectrum* (also *spectral-power density*). It is evident from the definition 1-120, that a power-density spectrum is a real and even function of ω.

[12]If the signal $f(t)$ is complex, then the average power dissipated by voltage $f(t)$ applied across 1 Ω resistor is given by

$$\begin{aligned}
P &= \lim_{T\to\infty}\frac{1}{T}\int_{-T/2}^{T/2} |f^2(t)|\,dt \\
&= \lim_{T\to\infty}\frac{1}{T}\int_{-T/2}^{T/2} f(t)\,f^*(t)\,dt \\
&= \lim_{T\to\infty}\frac{1}{T}\int_{-\infty}^{\infty} f_T(t)\,f_T^*(t)\,dt \\
&= \lim_{T\to\infty}\frac{1}{T}\int_{-\infty}^{\infty} f_T^*(t)\,dt\int_{-\infty}^{\infty}\frac{1}{2\pi}F_T(\omega)\,e^{j\omega t}\,d\omega \\
&= \lim_{T\to\infty}\frac{1}{T}\int_{-\infty}^{\infty} F_T(\omega)\,F_T^*(\omega)\,d\omega
\end{aligned}$$

Hence

$$\mathcal{S}_f(\omega) = \lim_{T\to\infty}\frac{F_T(\omega)\,F_T^*(\omega)}{T} = \lim_{T\to\infty}\frac{|F_T(\omega)|^2}{T}$$

Note that in general for a complex signal, $F_T^*(\omega) \neq F_T(-\omega)$.

Input and Response Power Densities in a Linear System. Let us apply a power signal $f(t)$ at the input of a linear system with transfer function $H(\omega)$ and let the output signal be $r(t)$. We shall express this fact symbolically as

$$f(t) \rightarrow r(t)$$

According to our notation $f_T(t)$ and $r_T(t)$ represents $f(t)$ and $r(t)$ respectively truncated beyond $|t| > T/2$. Let us now apply at the input the truncated signal $f_T(t)$. This is the same as applying the signal $f(t)$ over the interval $|t| < T/2$ and no signal beyond this interval. The response in general will not be $r_T(t)$. The response will extend beyond $t = T/2$. However, since the input is zero for $t > T/2$, for a stable system, the response for $t > T/2$, must decay with time. In the limit as $T \rightarrow \infty$, this contribution (beyond $t = T/2$) will be of no significance when viewed in the proper perspective of a signal of infinite time duration.[13] Hence in the limit as $T \rightarrow \infty$, the response to $f_T(t)$ may be considered to be $r_T(t)$ without much error.

$$\lim_{T \rightarrow \infty} f_T(t) \rightarrow r_T(t)$$

and

$$\lim_{T \rightarrow \infty} R_T(\omega) = H(\omega) F_T(\omega)$$

Also by definition $\mathcal{S}_r(\omega)$, the power-density spectrum of the output signal $r(t)$ is given by

$$\begin{aligned} \mathcal{S}_r(\omega) &= \lim_{T \rightarrow \infty} \frac{1}{T} |R_T(\omega)|^2 \\ &= \lim_{T \rightarrow \infty} \frac{1}{T} |H(\omega) F_T(\omega)|^2 \\ &= |H(\omega)|^2 \lim_{T \rightarrow \infty} \frac{1}{T} |F_T(\omega)|^2 \\ &= |H(\omega)|^2 \mathcal{S}_f(\omega) \end{aligned} \tag{1-122}$$

It is therefore evident that the output-signal power density is given by $|H(\omega)|^2$ times the power density of the input signal.

The mean-square value of a signal is given by $1/2\pi$ times the area under its power-density spectrum. Hence the mean-square value of the response $r(t)$ is given by

$$\overline{r^2(t)} = \frac{1}{2\pi} \int_{-\infty}^{\infty} |H(\omega)|^2 \mathcal{S}_f(\omega)\, d\omega$$

[13]Similar argument applies for $t < -T/2$. The response for $t > -T/2$ due to the input signal $f(t)$ applied before $t = -T/2$ will be insignificant in the perspective as $T \rightarrow \infty$.

Note that $\widetilde{r^2(t)}$ the mean-square value of $r(t)$ is by definition the power of $r(t)$.

Interpretation of Power-Density Spectrum. The power-density spectrum has a physical interpretation very similar to that of energy density. Let us consider a power signal $f(t)$ applied at the input of an ideal bandpass filter whose transfer function $H(\omega)$ is shown in Fig. 1-23. This filter suppresses all frequencies except a narrow-band $\Delta\omega(\Delta\omega \to 0)$ centered at frequency ω_0. The power-density spectrum of the output signal will be given by $\mathcal{S}_f(\omega_0)$ over the narrow-band $\Delta\omega$ centered at $\pm\omega_0$ as shown in Fig. 1-26.

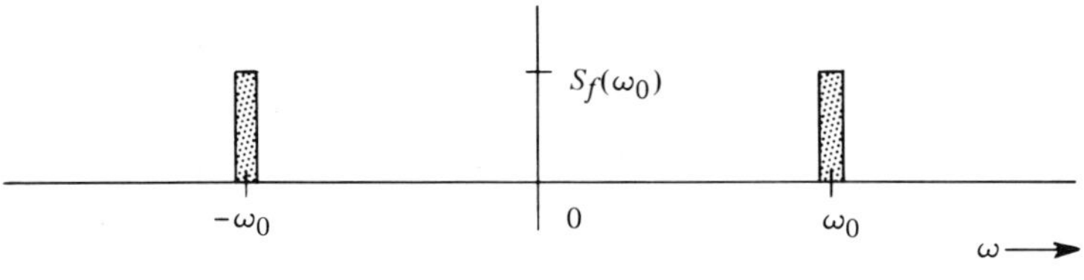

FIG. 1-26.

The power P_0 of the output signal according to Eq. 1-121 is given by

$$P_0 = 2\mathcal{S}_f(\omega_0)\,\Delta f$$

It can be seen that this is the power contribution of frequency components of $f(t)$ lying in the band Δf centered at ω_0. As usual we have negative and positive frequency components. We therefore attribute $\mathcal{S}_f(\omega_0)\,\Delta f$ contribution to positive frequencies and the equal amount to negative frequencies. Thus $\mathcal{S}_f(\omega)$ is the power per unit bandwidth (Hertz) contributed by frequency components centered at frequency ω.

Power-Density Spectrum of a Periodic Signal. Consider a periodic signal $f(t)$ and its Fourier series representation

$$f(t) = \sum_{n=-\infty}^{\infty} F_n e^{jn\omega_0 t}$$

$F(\omega)$, the Fourier transform of $f(t)$ is given by Eq. 1-78 as

$$F(\omega) = 2\pi \sum_{n=-\infty}^{\infty} F_n\,\delta(\omega - n\omega_0)$$

The truncated function $f_T(t)$ can be obtained by multiplying $f(t)$ by a gate function

$$f_T(t) = G_T(t)f(t)$$

Using the frequency-convolution theorem, we get

$$F_T(\omega) = \frac{1}{2\pi}\left[T\,\mathrm{Sa}\left(\frac{\omega T}{2}\right) * F(\omega)\right]$$

$$= T\,\mathrm{Sa}\left(\frac{\omega T}{2}\right) * \sum_{n=-\infty}^{\infty} F_n \delta(\omega - n\omega_0)$$

$$= T \sum_{n=-\infty}^{\infty} F_n\left[\mathrm{Sa}\left(\frac{\omega T}{2}\right) * \delta(\omega - n\omega_0)\right]$$

$$= T \sum_{n=-\infty}^{\infty} F_n\,\mathrm{Sa}\left[\frac{(\omega - n\omega_0)T}{2}\right]$$

Therefore

$$\lim_{T\to\infty} \frac{|F_T(\omega)|^2}{T} = \lim_{T\to\infty} T \sum_{n=-\infty}^{\infty} |F_n|^2\,\mathrm{Sa}^2\left[\frac{(\omega - n\omega_0)T}{2}\right] \tag{1-123}$$

Note that as $T \to \infty$, the function $\mathrm{Sa}^2[(\omega - n\omega_0)T/2]$ tends to be concentrated at $\omega = n\omega_0$. Hence the expression for $|F_T(\omega)|^2$ in Eq. 1-123 does not have any cross-product terms since each component exists where all other components are zero.

Use of Eq. 1-67 in Eq. 1-123 now yields

$$\mathcal{S}_f(\omega) = \lim_{T\to\infty} \frac{|F_T(\omega)|^2}{T} = 2\pi \sum_{n=-\infty}^{\infty} |F_n|^2 \delta(\omega - n\omega_0) \tag{1-124}$$

For a sinusoidal signal,

$$f(t) = A\cos(\omega_0 t + \theta)$$

$$|F_1| = |F_{-1}| = \frac{A}{2}$$

and

$$|F_n| = 0 \qquad n \neq \pm 1$$

Hence from Eq. 1-124, we have

$$\mathcal{S}_f(\omega) = \frac{\pi A^2}{2}[\delta(\omega + \omega_0) + \delta(\omega - \omega_0)]$$

Note that 2π times the area under the power-density spectrum $\mathcal{S}_f(\omega)$ is $A^2/2$. This is the expected result, since the time mean square value of $A\cos(\omega_0 t + \theta)$ is $A^2/2$.

Alternate Way of Obtaining Power-Density Spectrum. For finite energy signals, it was shown that the energy-density spectrum may be obtained from the Fourier transform of a function

$$\int_{-\infty}^{\infty} f(t)f(t+\tau)\,dt \quad \text{or} \quad \int_{-\infty}^{\infty} f(t)f(t-\tau)\,dt$$

A similar situation exists with regards to the power-density spectrum of a nonfinite energy signal. We shall now show that the power-density spectrum is given by the Fourier transform of $\mathcal{R}_f(\tau)$ defined as

$$\mathcal{R}_f(\tau) = \lim_{T\to\infty} \frac{1}{T} \int_{-T/2}^{T/2} f(t)f(t+\tau)\,dt \tag{1-125a}$$

Changing the variable $t + \tau$ to t in this equation, we obtain

$$\mathcal{R}_f(\tau) = \lim_{T\to\infty} \frac{1}{T} \int_{-T/2}^{T/2} f(t)f(t-\tau)\,dt \tag{1-125b}$$

From Eqs. 1-125a and 1-125b it follows that $\mathcal{R}_f(\tau)$ is an even function of τ. Consider a truncated signal $f_T(t)$. We observe that

$$\lim_{T\to\infty} \int_{-T/2}^{T/2} f(t)f(t+\tau)\,dt = \int_{-\infty}^{\infty} f_T(t)f_T(t+\tau)\,dt \tag{1-126a}$$

Now if $f_T(t) \leftrightarrow F_T(\omega)$, then on lines similar to those used to derive Eq. 1-116, we get

$$\mathcal{F}\left[\int_{-\infty}^{\infty} f_T(t)f_T(t+\tau)\,dt\right] = F_T(\omega)F_T(-\omega) = |F_T(\omega)|^2 \tag{1-126b}$$

and hence from Eqs. 1-126a and 1-126b, we obtain

$$\lim_{T\to\infty} \frac{1}{T} \int_{-T/2}^{T/2} f(t)f(t+\tau)\,d\tau \leftrightarrow \lim_{T\to\infty} \frac{|F_T(\omega)|^2}{T} = \mathcal{S}_f(\omega)$$

It therefore follows that

$$\mathcal{F}[\mathcal{R}_f(\tau)] = \mathcal{S}_f(\omega) \tag{1-127}$$

The function $\mathcal{R}_f(\tau)$ is called the *time-autocorrelation function* of $f(t)$. It is obvious that $\mathcal{R}_f(\tau)$ is the time average of $f(t)f(t+\tau)$. Hence[14]

$$\mathcal{R}_f(\tau) = \widetilde{f(t)f(t+\tau)} \tag{1-128a}$$

[14]The reader can easily show that for complex signals, the power-density spectrum $\mathcal{S}_f(\omega)$ is given by the Fourier transform of $\mathcal{R}_f(\tau)$ defined as

$$\begin{aligned}\mathcal{R}_f(\tau) &= \widetilde{f^*(t)f(t+\tau)} = \widetilde{f(t)f^*(t-\tau)} \qquad (1\text{-}128\text{b})\\ &= \lim_{T\to\infty} \frac{1}{T} \int_{-T/2}^{T/2} f(t)f^*(t+\tau)\,dt\\ &= \lim_{T\to\infty} \frac{1}{T} \int_{-T/2}^{T/2} f(t)f^*(t-\tau)\,dt\end{aligned}$$

The definition of the time-autocorrelation function as given in Eq. 1-128b is more general. Eq. 1-128a is a special case of this when $f(t)$ is a real signal.

1-15. GEOMETRICAL REPRESENTATION OF SIGNALS: THE SIGNAL SPACE

The analogy between signals and vectors has been observed in the earlier sections. This at once suggests the possibility of a geometrical representation of signals.

Consider n-dimensional vector space with unit vectors $\mathbf{\Phi}_1, \mathbf{\Phi}_2, \ldots, \mathbf{\Phi}_n$ along the coordinate axis. These vectors which form the coordinate system are called the *basis vectors*. A set of n basis vectors is not unique. One can choose any n independent vectors in this space as the basis vectors. The basis vectors may or may not be mutually orthogonal. In our discussion, however, we shall assume the set of basis vectors to be mutually orthogonal and each normalized to a unit length. Such a set is called an *orthonormal set*.

Once the n basis vectors are chosen, any vector $\mathbf{A}$ in this space can be specified by n numbers $(a_1, a_2, \ldots, a_n)$ which represent the magnitude of components of $\mathbf{A}$ along the n basis vectors respectively. The vector $\mathbf{A}$ can be expressed as

$$\mathbf{A} = \sum_{j=1}^{n} a_j \mathbf{\Phi}_j$$

where, for normalized basis (see Eq. 1-16),

$$a_j = \mathbf{A} \cdot \mathbf{\Phi}_j$$

If the vectors are not orthonormal they can be easily made so by dividing the vector by its magnitude (length). It should be stressed here that the basis vectors must form a complete set of vectors.

The vector $\mathbf{A}$ can be specified by an n-tuple $(a_1, a_2, \ldots, a_n)$ represented geometrically by a point $(a_1, a_2, \ldots, a_n)$ in the coordinate system formed by the basis vectors. We have an analogous situation for signals. From the discussion in Sec. 1-1, if n signals $\phi_1(t), \phi_2(t), \ldots, \phi_n(t)$, where n may be finite or infinite, are orthonormal basis signals, then a signal $x(t)$ in this space can be expressed as

$$x(t) = \sum_j x_j \phi_j(t) \tag{1-129a}$$

where, for normalized basis signals,[15]

$$x_j = \int_{-\infty}^{\infty} x(t)\, \phi_j(t)\, dt \tag{1-129b}$$

[15]In case the basis signals are complex,

$$x_j = \int_{-\infty}^{\infty} x(t)\, \phi_j^*(t)\, dt$$

See Eq. 1-45.

If the basis signals are not orthonormal, they can be easily made so by multiplying by a normalizing constant. The basis signals $\{\phi_j(t)\}$ must form a complete set of signals for the space.

Once the basis signals $\{\phi_j(t)\}$ are specified, we can represent the signal $x(t)$ by an n-tuple $(x_1, x_2, \ldots, x_n)$. Alternately we may represent this signal geometrically by a point $(x_1, x_2, \ldots, x_n)$ in an n-dimensional space. We can now associate a vector $\mathbf{x}(x_1, x_2, \ldots, x_n)$ with signal $x(t)$. Note that the basis signal $\phi_1(t)$ is represented by the corresponding basis vector $\mathbf{\Phi}_1(1, 0, 0, \ldots, 0)$ and $\phi_2(t)$ is represented by $\mathbf{\Phi}_2(0, 1, 0, \ldots, 0)$, and so on.

Representation of signals by a set of orthonormal signals is not new to the reader. The Fourier series representation in the earlier section is one such example. Another useful example which we have studied earlier is representation of a band-limited signal in terms of its samples and sampling functions. From Eq. 1-103b we have

$$x(t) = \sum_{n=-\infty}^{\infty} x(nT_s)\,\mathrm{Sa}(Wt - n\pi) \qquad \begin{aligned} T_s &= 1/2B \\ W &= 2\pi B \end{aligned} \tag{1-130}$$

where B is the highest frequency contained in $x(t)$ and $x(nT_s)$ is the nth sample of the signal $x(t)$ sampled every T_s seconds. The sampling functions appearing in Eq. 1-130 are orthogonal functions, since

$$\int_{-\infty}^{\infty} \mathrm{Sa}(Wt - n\pi)\,\mathrm{Sa}(Wt - k\pi)\,dt = \begin{cases} 0 & n \neq k \\ 1/2B & n = k \end{cases} \tag{1-131}$$

Note that $\mathrm{Sa}(Wt - n\pi)$ is an orthogonal set but is not orthonormal. We can normalize the set by multiplying each member by $1/\sqrt{2B}$. Thus a set $\{(1/\sqrt{2B})\,\mathrm{Sa}(Wt - k\pi)\}$, $k = 0, \pm 1, \pm 2, \ldots$, is an orthonormal set. If we let

$$\phi_k(t) = \frac{1}{\sqrt{2B}}\,\mathrm{Sa}(Wt - k\pi) \tag{1-132a}$$

then

$$x(t) = \sum_{k=-\infty}^{\infty} x_k \phi_k(t) \tag{1-132b}$$

where

$$x_k = \sqrt{2B}\,x(kT_s) \tag{1-132c}$$

Thus a signal $x(t)$ can be represented by a point in an orthogonal vector space. The coordinates of the point are given by

$$(\ldots, x_{-k}, \ldots, x_{-2}, x_{-1}, x_0, x_1, x_2, \ldots, x_k, \ldots)$$

where x_k is given by Eq. 1-132c. This is obviously an infinite dimensional space.

The Scalar Product. In a certain signal space, let $x(t)$ and $y(t)$ be two signals represented by vectors $\mathbf{x}(x_1, x_2, \ldots, x_n)$ and $\mathbf{y}(y_1, y_2, \ldots, y_n)$. If $\{\phi_k(t)\}$ are the orthonormal basis signals, then

$$x(t) = \sum_i x_i \phi_i(t)$$

$$y(t) = \sum_j y_j \phi_j(t)$$

Hence

$$\int_{-\infty}^{\infty} x(t)\, y(t)\, dt = \int_{-\infty}^{\infty} \left[\sum_i x_i \phi_i(t)\right]\left[\sum_j y_j \phi_j(t)\right] dt$$

The basis signals are orthonormal. Hence

$$\int_{-\infty}^{\infty} \phi_i(t)\, \phi_j(t)\, dt = \begin{cases} 1 & \text{for } i = j \\ 0 & \text{for } i \neq j \end{cases}$$

Use of this result now yields

$$\int_{-\infty}^{\infty} x(t)\, y(t)\, dt = \sum_k x_k y_k \tag{1-133}$$

However, the right-hand side of Eq. 1-133 is by definition the scalar product of vectors **x** and **y**:

$$\mathbf{x} \cdot \mathbf{y} = \sum_k x_k y_k$$

Hence we have

$$\mathbf{x} \cdot \mathbf{y} = \int_{-\infty}^{\infty} x(t)\, y(t)\, dt \tag{1-134}$$

If $x(t)$ and $y(t)$ are mutually orthogonal, then it follows from Eq. 1-134, that the corresponding vectors **x** and **y** are also orthogonal. *We conclude that the integral of the product of two signals is equal to the scalar product of the corresponding vectors.*

Energy of a Signal. For a signal $x(t)$, the energy E is given by

$$E = \int_{-\infty}^{\infty} x^2(t)\, dt$$

It follows from Eq. 1-134 that

$$E = \mathbf{x} \cdot \mathbf{x} = |\mathbf{x}|^2 \tag{1-135}$$

where $|\mathbf{x}|^2$ is the square of the length of the vector **x**. Hence the signal energy is given by the square of the length of the corresponding vector.

The Dimensionality of a Signal Space. The dimensionality of a vector space is defined as the minimum number of specifications required to describe a vector in this space. The dimensionality of a space is also equal to the maximum number of independent vectors in the space. This can be easily seen from the definition of independent vector set in a space. A set of vectors is said to be independent if none of the vectors in this set can be represented in terms of the remaining vectors. Thus if $\mathbf{x}_1, \mathbf{x}_2, \ldots, \mathbf{x}_n$ is a set of independent vectors, then it is impossible to find constants $a_1, a_2, \ldots, a_n$, not all zero, such that

$$a_1\mathbf{x}_1 + a_2\mathbf{x}_2 + \cdots + a_n\mathbf{x}_n = 0$$

If the space has maximum of n independent vectors, then we must be able to express any other vector $\mathbf{y}$ in this space as a linear combination of these n vectors. If not, then $\mathbf{y}$ becomes a member of the independent vector set. But this is not possible on the assumption that there are a maximum of n independent vectors. Thus any vector in this space can be specified by n numbers and hence is an n-dimensional space. It is obvious from this discussion, that the dimensionality of a space can be at most as large as the total number of vectors in the space.

If we consider a subset of vectors, in a given space it is possible to reduce the dimensionality. Thus in a three-dimensional space, if we consider all vectors lying in one plane the dimensionality of this particular subset can be reduced to two. If we consider all the vectors lying along a straight line, the dimensionality can be reduced to one.

In general, the vector space representing signals has infinite dimensions. This can be seen from the Fourier Series representation or a sampling-function representation of a signal (Eq. 1-130). Any other basis signals would also lead to infinite dimensions. In practice, however, the higher dimensions may be ignored because they often have negligible contributions. Thus in a Fourier series representation, first few terms may describe the signal with fair accuracy. The contribution of the higher harmonics being negligible can be ignored. Thus one can reduce the dimensionality of a signal set to a finite number.

Consider the case of signals which are time limited to T secs., that is $f(t) = 0$ for $|t| < T/2$. Such signals therefore exist only in the finite time interval of duration T and are zero outside this interval. It can be shown that such signals (time-limited) cannot be band-limited. In general a signal cannot be simultaneously time-limited and band-limited. If a signal is time-limited, its spectrum exists in the entire frequency interval $(-\infty, \infty)$. On the other hand, if a signal has a band-limited spectrum, it must exist in the entire time interval $(-\infty, \infty)$. This follows from Paley-Wiener criterion (a band-limited signal violates Paley-Wiener criterion).

Thus if a signal is time-limited its bandwidth $B = \infty$. The minimum sampling interval (Nyquist interval) is obviously zero and we need infinite samples to specify such a signal completely. The signal therefore has infinite dimensions. However, for all physical signals, the spectrum magnitude decays with frequency and the high-frequency components beyond some frequency B Hz have negligible contribution. In such a case the signal may be considered essentially band-limited to B Hz. We may choose B as high as we wish to reduce the error in this assumption. Since the Nyquist interval is $1/2B$ sec., we need a total of $2TB$ samples to specify the signal time limited to T seconds. Hence the signal has $2TB$ dimensions. An estimate of the error in this assumption is given by Landau and Pollak (Ref. 1-8). A brief discussion of their results is given in the appendix at the end of the chapter.

In certain problems of communication, we deal with a finite number of waveforms. If there are a total of m messages represented by m waveforms, then obviously the dimensions of this signal space can be at most m. Consider the case of binary waveforms, one of which is shown in Fig. 1-27a. This waveform is composed of 10 binary pulses, each of which can assume 2 values ($\pm A$). We can therefore construct a total of $2^{10} = 1{,}024$ waveforms by different combination of the pulses. Thus $m = 1{,}024$ but the dimensionality of this set of signals is only 10. It can be easily seen that this set of signals can be described by 10 orthonormal functions shown in Fig. 1-27b:

$$\phi_j(t) = u(t - j + 1) - u(t - j)$$

Each of the 1,024 signals can be expressed as

$$s_i(t) = A \sum_{j=1}^{10} a_{ij} \phi_j(t)$$

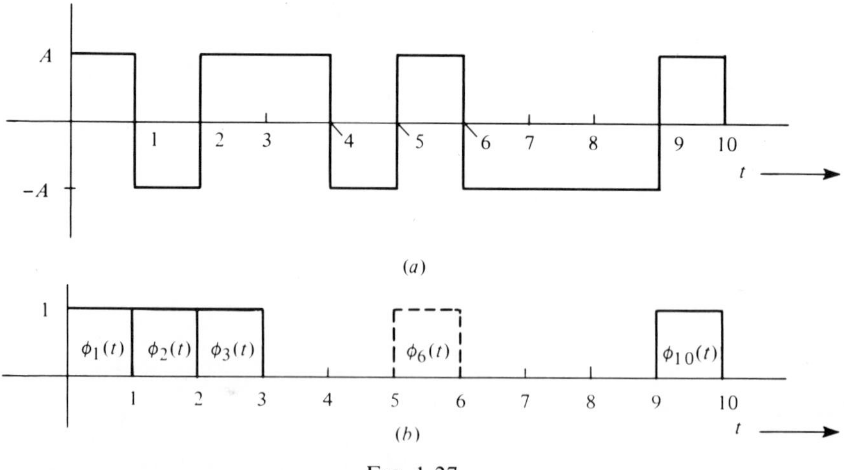

FIG. 1-27.

where a_{ij} is either $+1$ or -1. Note that in this case there is no approximation involved in determining the dimensionality of the signal space. As long as we restrict to the signals to the form in Fig. 1-27a, the dimensionality is exactly 10. This is considerably less than $2TB$. The reduction in dimensionality is the result of restricting to a certain subset of signals in a general signal space. This is analogous to the situation in a three-dimensional vector space where the dimensionality of a subset of vectors restricted to a plane is reduced to two.

Figure 1-28 shows four signals and their geometrical representation (using the basis signals in Fig. 1-27). Note that these signals are two-dimensional. Observe that waveforms $f_1(t)$ and $f_4(t)$ are orthogonals, since

$$\int_{-\infty}^{\infty} f_1(t)\, f_4(t)\, dt = 0$$

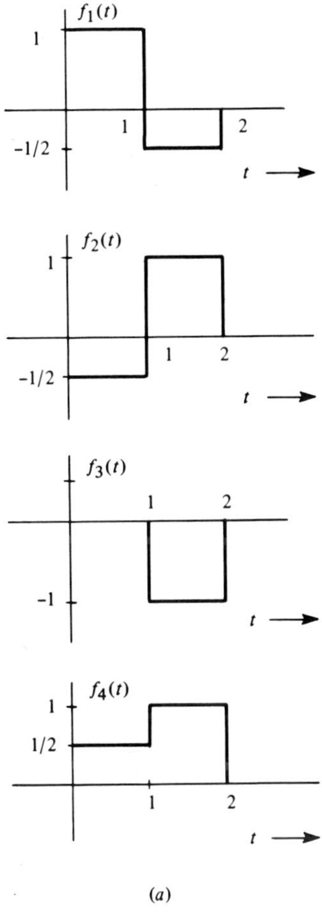

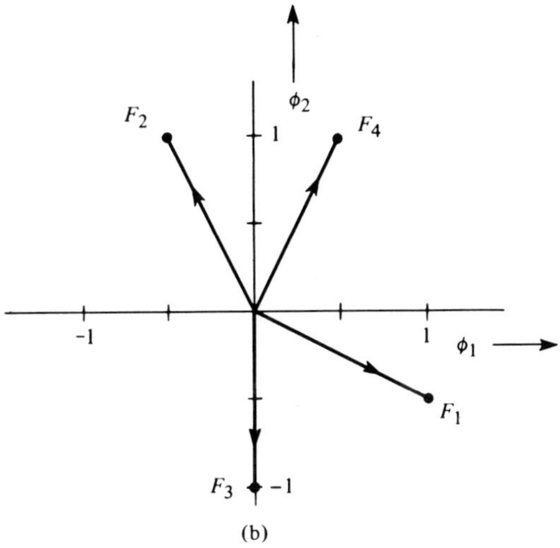

FIG. 1-28.

In Fig. 1-28b the vectors F_1 and F_4 representing these signals are also geometrically orthogonal.

Each point in the signal space in Fig. 1-28b corresponds to some waveform.

1-16. GRAM-SCHMIDT ORTHOGONALIZATION OF A VECTOR SET

We have defined the dimensionality of a vector space to be equal to the maximum number of independent vectors in the space. Thus in an n-dimensional space, there can be no more than n vectors that are independent. Alternatively it is always possible to find a set of n vectors which are independent. Once such a set is chosen, any vector in this space can be expressed in terms of (as a linear combination of) the vectors in this set. This set forms what we commonly refer to as a basis set which forms the coordinate system. This set of n independent vectors is by no means unique. The reader is familiar with this fact in the physical space of three dimensions where one can find an infinite number of independent sets of three vectors. This is obvious from the fact that we have infinite possible coordinate systems. The orthogonal set, however, is of special interest since it is easier to deal with compared to nonorthogonal sets. If we are given a set of n independent vectors, it is possible to obtain from this set another set of n independent vectors which are orthogonal. This is done by *Gram-Schmidt process* of orthogonalization.

In order to get a physical insight into this procedure, we shall consider a simple case of two-dimensional space. Let $\mathbf{x}_1, \mathbf{x}_2$ be the two independent vectors in a two-dimensional space (Fig. 1-29). We wish to generate a new set of two orthogonal vectors from $\mathbf{x}_1$ and $\mathbf{x}_2$. The new orthogonal vectors will be denoted by $\mathbf{y}_1$ and $\mathbf{y}_2$. For convenience we shall choose

$$\mathbf{y}_1 = \mathbf{x}_1 \tag{1-136a}$$

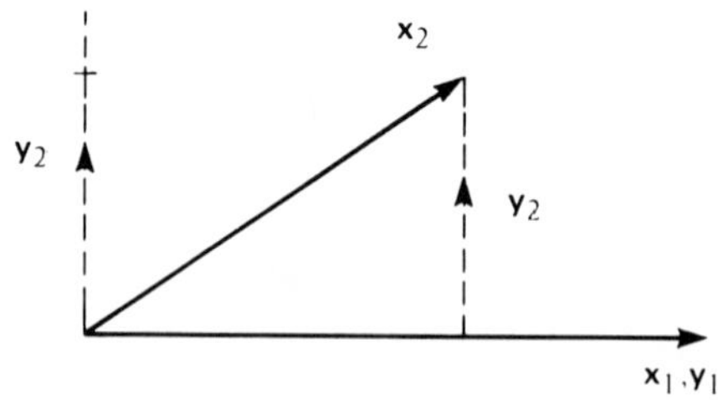

FIG. 1-29.

$\mathbf{y}_2$ is orthogonal to $\mathbf{y}_1$ (and $\mathbf{x}_1$) as shown in Fig. 1-29. We now wish to express $\mathbf{y}_2$ in terms of $\mathbf{x}_1$ and $\mathbf{x}_2$. This can be easily done by observing the fact that the vector $\mathbf{x}_2$ is equal to the sum of vector $\mathbf{y}_2$ and the projection of $\mathbf{x}_2$ upon $\mathbf{x}_1$. Thus

$$\mathbf{x}_2 = \mathbf{y}_2 + C_{12}\mathbf{x}_1 = \mathbf{y}_2 + C_{12}\mathbf{y}_1 \tag{1-136b}$$

where C_{12} is given by (see Eq. 1-2)

$$C_{12} = \frac{\mathbf{x}_1 \cdot \mathbf{x}_2}{|\mathbf{x}_1|^2}$$

But since (by Eq. 1-136a) $\mathbf{y}_1 = \mathbf{x}_1$,

$$C_{12} = \frac{\mathbf{y}_1 \cdot \mathbf{x}_2}{|\mathbf{y}_1|^2}$$

Hence

$$\begin{aligned}\mathbf{y}_2 &= \mathbf{x}_2 - C_{12}\mathbf{y}_1 \\ &= \mathbf{x}_2 - \frac{\mathbf{y}_1 \cdot \mathbf{x}_2}{|\mathbf{y}_1|^2}\mathbf{y}_1\end{aligned} \tag{1-136c}$$

Equations 1-136a and 1-136c yield the desired orthogonal set. Note that this set is not unique. There are infinite possible orthogonal vector sets $(\mathbf{y}_1, \mathbf{y}_2)$ that can be generated from $(\mathbf{x}_1, \mathbf{x}_2)$. In our derivation we could as well have started with $\mathbf{y}_1 = \mathbf{x}_2$ instead of $\mathbf{y}_1 = \mathbf{x}_1$. This starting point would have yielded an entirely different set. In general,

$$\begin{aligned}\mathbf{y}_1 &= a_{11}\mathbf{x}_1 + a_{12}\mathbf{x}_2 \\ \mathbf{y}_2 &= a_{21}\mathbf{x}_1 + a_{22}\mathbf{x}_2\end{aligned} \tag{1-137a}$$

and

$$\mathbf{y}_1 \cdot \mathbf{y}_2 = 0 \tag{1-137b}$$

Substituting Eq. 1-137a in Eq. 1-137b, we get one equation in 4 unknown coefficients a_{11}, a_{12}, a_{21}, and a_{22}. Hence there are infinite possible solutions. One can choose any three of the coefficients arbitrarily and the remaining fourth coefficient will be determined from Eqs. 1-137a and 1-137b.

The reader can easily extend these results to a three-dimensional case. If vectors $\mathbf{x}_1, \mathbf{x}_2, \mathbf{x}_3$ form an independent set in this space, then it can be easily shown that one possible orthogonal vector set $\mathbf{y}_1$, $\mathbf{y}_2$, $\mathbf{y}_3$ in this space is given by

$$\begin{aligned}\mathbf{y}_1 &= \mathbf{x}_1 \\ \mathbf{y}_2 &= \mathbf{x}_2 - \frac{\mathbf{y}_1 \cdot \mathbf{x}_2}{|\mathbf{y}_1|^2}\mathbf{y}_1 \\ \mathbf{y}_3 &= \mathbf{x}_3 - \frac{\mathbf{y}_1 \cdot \mathbf{x}_3}{|\mathbf{y}_1|^2}\mathbf{y}_1 - \frac{\mathbf{y}_2 \cdot \mathbf{x}_3}{|\mathbf{y}_2^2|}\mathbf{y}_2\end{aligned}$$

These results can be easily extended to n-dimensional space. In general if we are given n independent vectors $\mathbf{x}_1, \mathbf{x}_2, \ldots, \mathbf{x}_n$, then proceeding along similar lines one can obtain an orthogonal set $\mathbf{y}_1, \mathbf{y}_2, \ldots, \mathbf{y}_n$ where

$$\mathbf{y}_1 = \mathbf{x}_1$$

and

$$\mathbf{y}_j = \mathbf{x}_j - \sum_{k=1}^{j-1} \frac{\mathbf{y}_k \cdot \mathbf{x}_j}{|\mathbf{y}_k|^2} \mathbf{y}_k \qquad j = 2, 3, \ldots, n \tag{1-138}$$

Note that this is one of the infinitely many orthogonal sets that can be formed from the set $\mathbf{x}_1, \mathbf{x}_2, \ldots, \mathbf{x}_n$. Moreover this set is not an orthonormal set. The orthonormal set $\hat{\mathbf{y}}_1, \hat{\mathbf{y}}_2, \ldots, \hat{\mathbf{y}}_n$ can be obtained by normalizing the lengths of the respective vectors.

$$\hat{\mathbf{y}}_k = \frac{\mathbf{y}_k}{|\mathbf{y}_k|}$$

We can apply these concepts to signal space, since there is one to one correspondence between the signals and vectors. If we have n independent signals $x_1(t), x_2(t), \ldots, x_n(t)$, we can form a set of n orthogonal signals $y_1(t), y_2(t), \ldots, y_n(t)$ where

$$y_1(t) = x_1(t)$$

$$y_j(t) = x_j(t) - \sum_{k=1}^{j-1} C_{kj}\, y_k(t) \qquad j = 2, 3, \ldots, n \tag{1-139}$$

where

$$C_{kj} = \frac{\int y_k(t)\, x_j(t)\, dt}{\int y_k^2(t)\, dt} \tag{1-140}$$

Note that this is one of the infinitely many possible orthogonal sets that can be formed from the set $x_1(t), x_2(t), \ldots, x_n(t)$. The set can be normalized by dividing each signal $y_j(t)$ by its energy.

Example 1-4. The exponential signals

$$g_1(t) = e^{-pt}u(t)$$
$$g_2(t) = e^{-2pt}u(t)$$
$$\vdots$$
$$g_n(t) = e^{-npt}u(t)$$

form an independent set of signals in n-dimensional space where n may be any integer. This set, however, is not orthogonal. We can use Gram-Schmidt process to obtain an orthogonal set for this space. If $y_1(t), y_2(t), \ldots, y_n(t)$ is the desired orthogonal basis set, we choose

$$y_1(t) = g_1(t) = e^{-pt}u(t)$$

From Eqs. 1-139 and 1-140 we have

$$y_2(t) = x_2(t) - C_{12}\, y_1(t)$$

where

$$C_{12} = \frac{\int_{-\infty}^{\infty} y_1(t)\, x_2(t)\, dt}{\int_{-\infty}^{\infty} y_1^2(t)\, dt}$$

$$= \frac{\int_0^{\infty} e^{-pt} e^{-2pt}\, dt}{\int_0^{\infty} e^{-2pt}\, dt}$$

$$= \frac{2}{3}$$

Hence

$$y_2(t) = \left(e^{-2pt} - \tfrac{2}{3} e^{-pt}\right) u(t)$$

Similarly we can proceed and find the remaining functions $y_3(t), \ldots, y_n(t)$, etc. The reader can verify that all this represents a mutually orthogonal set.

APPENDIX 1-1. DIMENSIONALITY OF SIGNALS (REF. 1-8)

According to Landau and Pollak, if (a) $x(t)$ is time limited to T seconds $[x(t) = 0$ for $|t| < T/2]$, (b) the energy of $x(t)$ that falls outside the bandwidth B Hz is at most k times the total signal energy E—that is,

$$\int_{-B}^{B} |X(\omega)|^2 df > E(1 - k)$$

where E, the signal energy, is given by

$$E = \int_{-\infty}^{\infty} |X(\omega)|^2 df = \int_{-\infty}^{\infty} x^2(t)\, dt$$

then there exists a particular set of orthonormal functions $\{\phi_j(t)\}$; $j = 0, 1, 2, \ldots$ such that the energy of error signal in the approximation

$$x(t) \simeq \sum_{j=1}^{L-1} x_j \phi_j(t)$$

is less than $12kE$, that is,

$$\int_{-\infty}^{\infty} \left[x(t) - \sum_{j=1}^{L-1} x_j \phi_j(t)\right]^2 dt < 12kE$$

where

$$L = \text{largest integer} < 2TB + 1$$

It can be observed that the error-signal energy is less than $12kE$ and can be made as small as desired by reducing k. This can be accomplished by increasing B. As $B \to \infty$, $k \to 0$.

REFERENCES

1. S. Mason and H. Zimmerman, *Electronic Circuits, Signals and Systems*, Wiley, New York, 1960.

2. B. P. Lathi, *Signals, Systems and Communication*, Wiley, New York, 1965.

3. L. Schwartz, *Theorie des Distributions*, Vols. I and II, Hermann et Cie., Paris, 1950 and 1951.

4. M. J. Lighthill, *Fourier Analysis and Generalized Functions*, Cambridge U. P., New York, 1959.

5. A. Papoulis, *Fourier Integral and its Applications*, McGraw-Hill, New York, 1962.

6. J. M. Wozencraft and I. M. Jacobs, *Principles of Communication Engineering*, Wiley, New York, 1965.

7. R. E. A. C. Paley and N. Wiener, Fourier Transforms in the Complex Domain, *American Math. Soc. Colloquium Publication 19*, New York, 1934.

8. H. L. Landau and H. O. Pollak, "Prolate Spheroidal Wave Functions, Fourier Analysis, and Uncertainty: III: The Dimensions of Space of Essentially Time- and Band-Limited Signals," *Bell Systems Tech. J.*, *41* (July 1962), pp. 1295–1336.

PROBLEMS

1-1. Show that if the two signals $f_1(t)$ and $f_2(t)$ are orthogonal over an interval (t_1, t_2), then the energy of the signal $[f_1(t) + f_2(t)]$ is equal to the sum of the energies of $f_1(t)$ and $f_2(t)$. The energy of a signal $f(t)$ over the interval (t_1, t_2) is defined as

$$\text{Energy} = \int_{t_1}^{t_2} f^2(t)\, dt$$

Extend this result to n number of mutually orthogonal signals.

1-2. The rectangular function $f(t)$ in Fig. 1-3 is approximated by the signal $(4/\pi) \sin t$. Show that the error function

$$f_e(t) = f(t) - \frac{4}{\pi} \sin t$$

is orthogonal to the function $\sin t$ over the interval $(0, 2\pi)$. (Can you give the qualitative reason for this?) Hence, show that the energy of $f(t)$ is the sum of energies of $f_e(t)$ and $(4/\pi) \sin t$.

1-3. Prove that if $f_1(t)$ and $f_2(t)$ are complex functions of a real variable t, then the component of $f_2(t)$ contained in $f_1(t)$ over the interval (t_1, t_2) is given by

$$C_{12} = \frac{\displaystyle\int_{t_2}^{t_2} f_1(t) f_2^*(t)\, dt}{\displaystyle\int_{t_1}^{t_2} f_2(t) f_2^*(t)\, dt}$$

The component is defined in the usual sense to minimize the magnitude of the mean-square error. Now, show that the signals $[f_1(t) - C_{12} f_2(t)]$ and $f_2(t)$ are mutually orthogonal.

1-4. Find the component of a waveform $\sin \omega_2 t$ contained in another waveform $\sin \omega_1 t$ over the interval $(-T, T)$ for all real values of ω_1 and $\omega_2 (\omega_1 \neq \omega_2)$. How does this component vary with T? Show that as T is made infinite, the component vanishes. Show that the above result holds for any pair of the functions $\sin \omega_1 t$, $\sin \omega_2 t$, $\cos \omega_1 t$, $\cos \omega_2 t$.

1-5. The two periodic functions $f_1(t)$ and $f_2(t)$ with zero d-c components have arbitrary waveforms with periods T and $\sqrt{2}T$, respectively. Show that the components in $f_1(t)$ of waveform $f_2(t)$ is zero over the interval $(-\infty < t < \infty)$. Show that the above result is true for any two periodic functions if the ratio of their periods is an irrational number, and provided that either $f_1(t)$ or $f_2(t)$ or both have a zero-average value (zero d-c component).

1-6. Show that an arbitrary function $f(t)$ can always be expressed as a sum of an even function $f_e(t)$ and an off function $f_o(t)$.

$$f(t) = f_e(t) + f_o(t)$$

Hence, find the even and odd components of the functions $u(t)$, $e^{-at}u(t)$, and e^{jt}. (*Hint*: $f(t) = \frac{1}{2}[f(t) + f(-t)] + \frac{1}{2}[f(t) - f(-t)]$.)

1-7. Show that the Fourier transform of $f(t)$ may also be expressed as

$$\mathfrak{F}[f(t)] = F(\omega) = \int_{-\infty}^{\infty} f(t) \cos \omega t - j \int_{-\infty}^{\infty} f(t) \sin \omega t \, dt$$

Further, show that if $f(t)$ is an even function of t, then

$$F(\omega) = 2 \int_0^{\infty} f(t) \cos \omega t \, dt$$

and if $f(t)$ is an odd function of t, then

$$F(\omega) = -2j \int_0^{\infty} f(t) \sin \omega t \, dt$$

1-8. A function $f(t)$ can be expressed as a sum of even function and odd function (see Prob. 1-6).

$$f(t) = f_e(t) + f_o(t)$$

where $f_e(t)$ is an even function of t and $f_o(t)$ is an odd function of t. (a) Show that if $F(\omega)$ is the Fourier transform of a real signal $f(t)$, then $\mathrm{Re}[F(\omega)]$ is the Fourier transform of $f_e(t)$ and $j\,\mathrm{Im}[F(\omega)]$ is the Fourier transform of $f_o(t)$. (b) Show that if $f(t)$ is complex

$$f(t) = f_r(t) + jf_i(t)$$

and if $F(\omega)$ is the Fourier transform of $f(t)$, then

$$\mathfrak{F}[f_r(t)] = \tfrac{1}{2}[F(\omega) + F^*(-\omega)]$$

$$\mathfrak{F}[f_i(t)] = \frac{1}{2j}[F(\omega) - F^*(-\omega)]$$

Hint: $f^*(t) = f_r(t) - jf_i(t)$ and $\mathfrak{F}[f^*(t)] = F^*(-\omega)$.

1-9. Find the Fourier transforms of functions $f(t)$ shown in the accompanying illustration.

1-10. Using the sampling property of the impulse function, evaluate the following integrals

$$\int_{-\infty}^{\infty} \delta(t - 2) \sin t \, dt$$

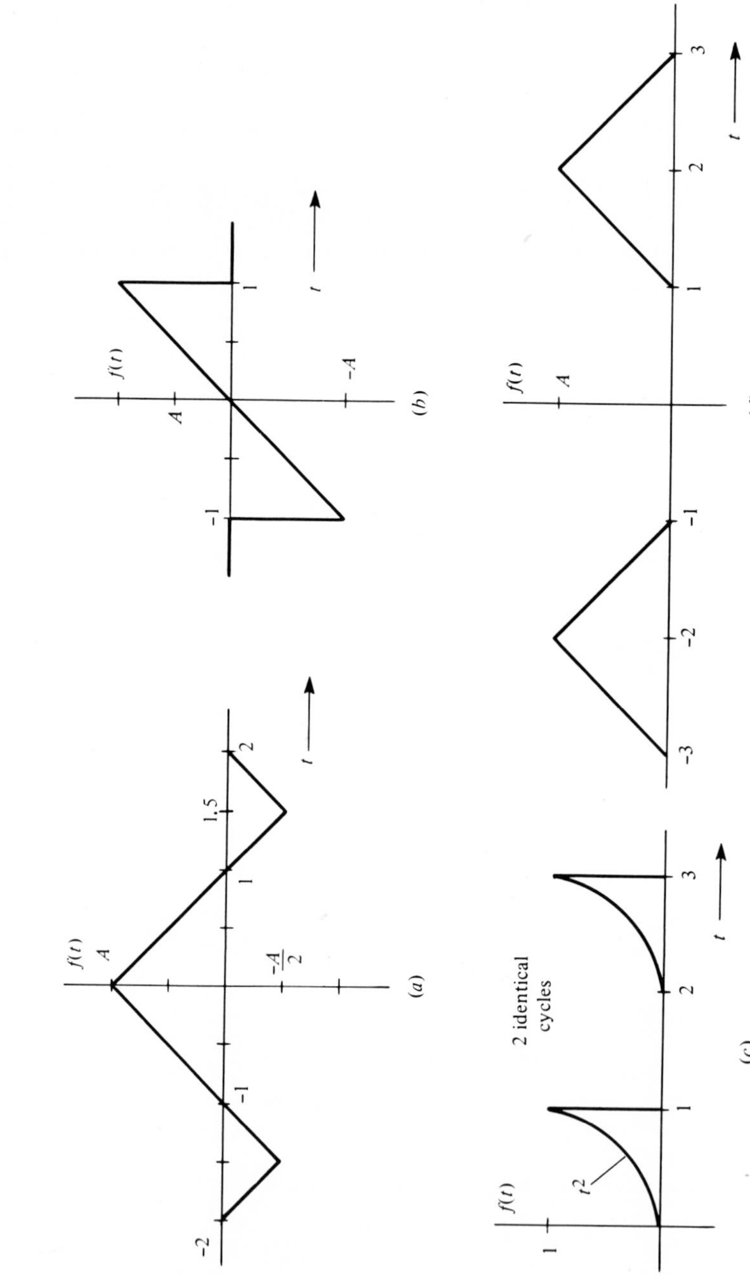

f(t)
A
$-\frac{A}{2}$
-2
-1
1
1.5
2
t
(a)
f(t)
A
-A
-1
1
t
(b)
f(t)
1
t^2
2 identical cycles
1
2
3
t
(c)
f(t)
A
-3
-2
-1
1
2
3
t
(d)

FIG. P. 1-9.

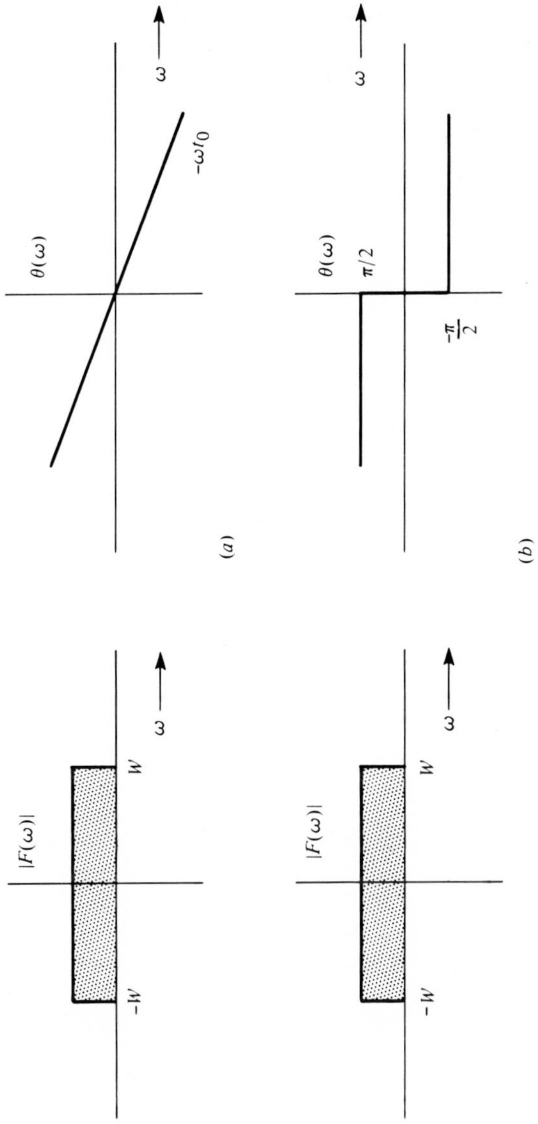

FIG. P. 1-11.

$$\int_{-\infty}^{\infty} \delta(2t + 3) e^{-t}\, dt$$

$$\int_{-\infty}^{\infty} \delta(1 - t)(t^3 + 4)\, dt$$

1-11. Determine the functions $f(t)$ whose Fourier transforms are shown in the accompanying figure.

1-12. Find the function $f(t)$ whose Fourier transforms are as shown in the figure. (*Hint*: Use the modulation theorem.)

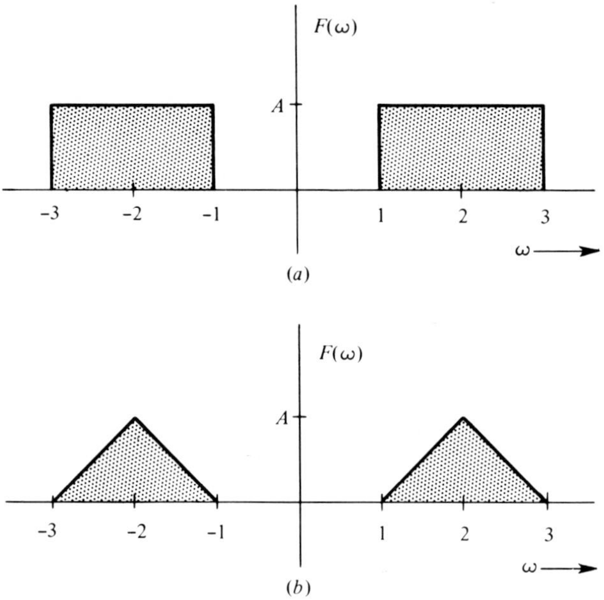

FIG. P. 1-12.

1-13. Find the Fourier transform of the functions shown in the figure using the modulation theorem. Sketch the frequency spectrum in each case.

1-14. Find the Fourier transforms of functions $f(t)$ in Prob. 1-9 using frequency-differentiation property, the time-shifting property and the Table 1-1.

1-15. Find the Fourier transform of a function shown in the accompanying figure by: (a) Straightforward integration; (b) using only the time-integration property and transform Table 1-1; (c) using only the time-differentiation property, the time-shifting property, and transforming Table 1-1; (d) using only the time-shifting property and transform Table 1-1.

1-16. Determine the Fourier transforms of the following: (a) (i) $\cos \omega_0 t\, u(t)$, (ii) $\sin \omega_0 t\, u(t)$. (b) If $f(t) \leftrightarrow F(\omega)$, determine the Fourier transforms of the following: (i) $tf(2t)$, (ii) $(t - 2)f(-2t)$, (iii) $t(df/dt)$, (iv) $f(1 - t)$, (v) $(1 - t)f(1 - t)$.

1-17. The nth moment m_n of a function $f(t)$ is defined by

$$m_n = \int_{-\infty}^{\infty} t^n f(t)\, dt$$

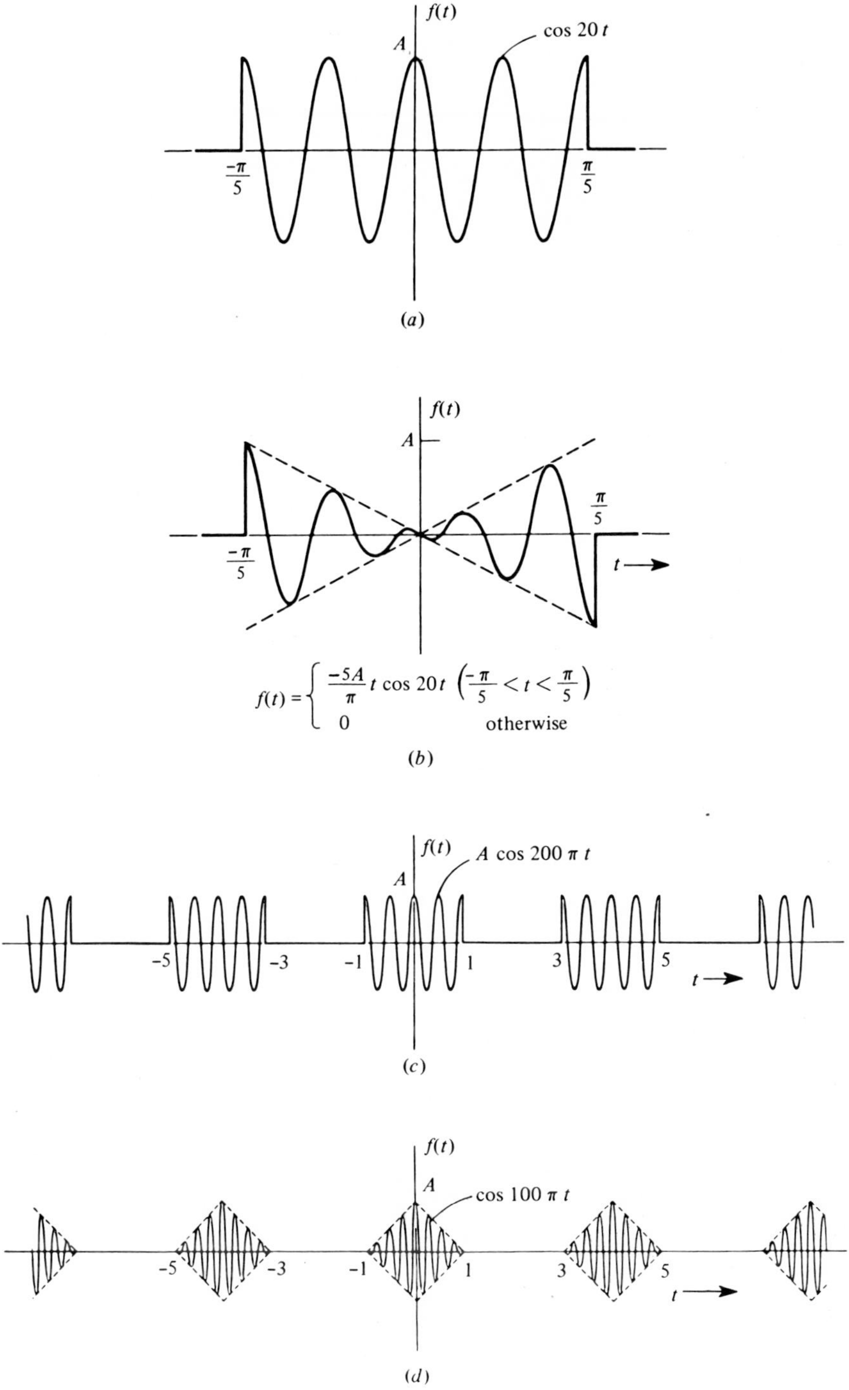
f(t)
cos 20 t
A
−π/5
π/5
(a)
f(t)
A
π/5
−π/5
t →
$f(t) = \begin{cases} \dfrac{-5A}{\pi} t \cos 20t & \left(\dfrac{-\pi}{5} < t < \dfrac{\pi}{5}\right) \\ 0 & \text{otherwise} \end{cases}$
(b)
f(t)
A cos 200 π t
A
−5
−3
−1
1
3
5
t →
(c)
f(t)
A
cos 100 π t
−5
−3
−1
1
3
5
t →
(d)

FIG. P. 1-13.

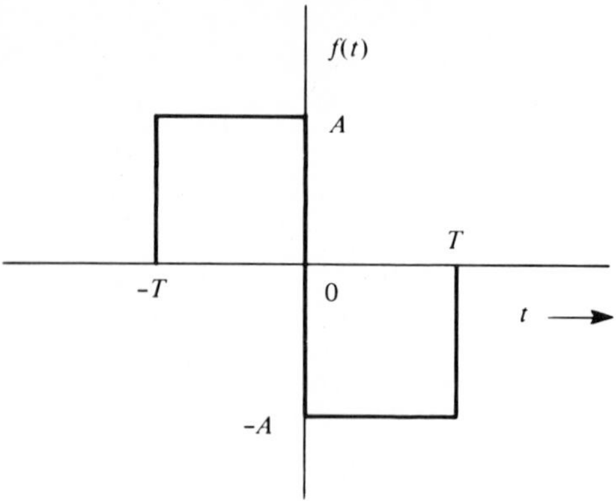

FIG. P. 1-15.

Using the frequency-differentiation theorem, show that

$$m_n = (j)^n \frac{d^n F(0)}{d\omega^n}$$

Using this result, show that Taylor's series expansion of $F(\omega)$ can be expressed as

$$F(\omega) = m_0 - jm_1\omega - \frac{m_2\omega^2}{2!} + \frac{jm_3\omega^3}{3!} + \frac{m_4\omega^4}{4!} + \cdots = \sum_{n=0}^{\infty} (-j)^n m_n \frac{\omega^n}{n!}$$

Determine the various moments of a gate function and, using the above equation, find its Fourier transform.

1-18. Evaluate the following convolution integrals: (a) $u(t) * u(t)$, (b) $u(t) * e^{-t}u(t)$, (c) $e^{-t}u(t) * e^{-2t}u(t)$, (d) $u(t) * tu(t)$, (e) $e^{-t}u(t) * tu(t)$, (f) $e^{-2t}u(t) * e^{-t}$. Verify your results for parts (b) through (f) by using Fourier transforms.

1-19. If $f(t)$ is a continuous signal band-limited to W rps, then show that

$$\frac{k}{\pi}\,[f(t) * \mathrm{Sa}(kt)] = f(t) \qquad \text{for } k \geq W$$

1-20. Evaluate the inverse Fourier transform of $\mathrm{Sa}^2(Wt)$ by using time-convolution theorem. Evaluate the convolution integral graphically.

1-21. Determine the minimum sampling rate and the Nyquist interval for the following signals: (a) $\mathrm{Sa}(100t)$, (b) $\mathrm{Sa}^2(100t)$, (c) $\mathrm{Sa}(100t) + \mathrm{Sa}(50t)$, (d) $\mathrm{Sa}(100t) + \mathrm{Sa}^2(60t)$.

1-22. The time-integration theorem (Eq. 1-89b) holds only if

$$\int_{-\infty}^{\infty} f(t)\,dt = 0$$

If this condition is not satisfied, then show that

$$\int_{-\infty}^{t} f(\tau)\,d\tau \leftrightarrow \pi F(\omega)\,\delta(\omega) + \frac{1}{j\omega} F(\omega)$$

Hint: Express $\int_{-\infty}^{t} f(\tau)\,d\tau$ as a convolution of $f(t)$ and $u(t)$.

1-23. A signal $f(t) = 2e^{-t}u(t)$ is passed through an ideal lowpass filter with cutoff frequency 1 rps. Find the energy-density spectrum of the output of the filter. Determine the energy of the input signal and the output signal.

1-24. Find the power (mean-square value) and sketch the power-density spectrum of each of the following signals: (a) $A \cos (2000\pi t) + B \sin (200\pi t)$, (b) $[A + \sin (200 \cdot \pi t)] \cos (2000\pi t)$, (c) $A \cos (200\pi t) \cos (2000\pi t)$, (d) $A \sin (200\pi t) \cos (2000\pi t)$, (e) $A \sin (300\pi t) \cos (2000\pi t)$, (f) $A \sin^2 (200\pi t) \cos (2000\pi t)$.

1-25. A periodic signal $f(t)$ shown in part (a) of the accompanying figure is transmitted through a system with transfer function $H(\omega)$ shown in part (b). For each of the three different values of $T (T = 2\pi/3, \pi/3$, and $\pi/6)$ find the power and sketch the power-density spectrum of the output signal.

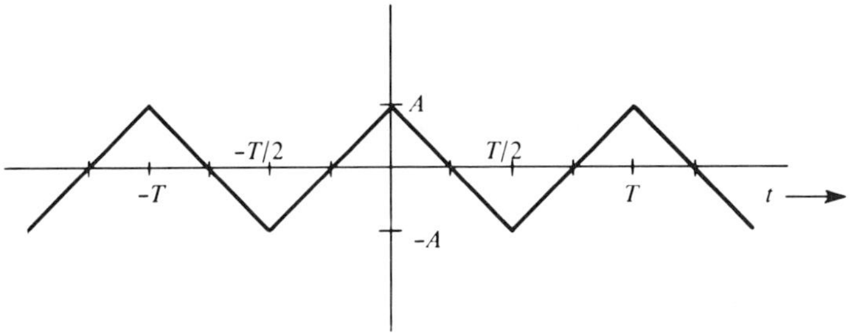

(*a*)

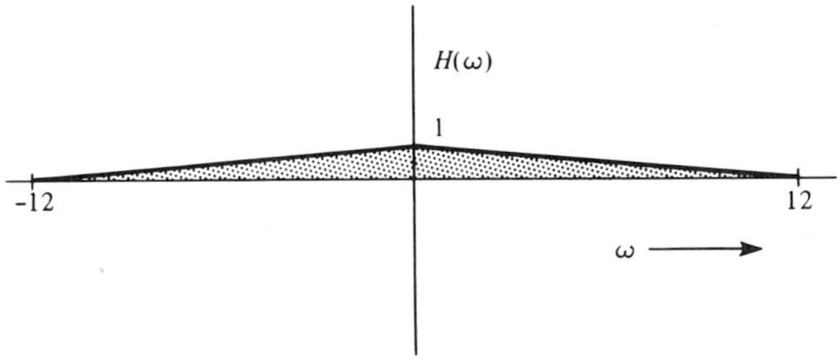

(*b*)

FIG. P. 1-25.

1-26. Find the mean-square value of the output voltage $v_o(t)$ of an R-C network shown in the illustration if the input voltage has a power-density spectrum $S_i(\omega)$

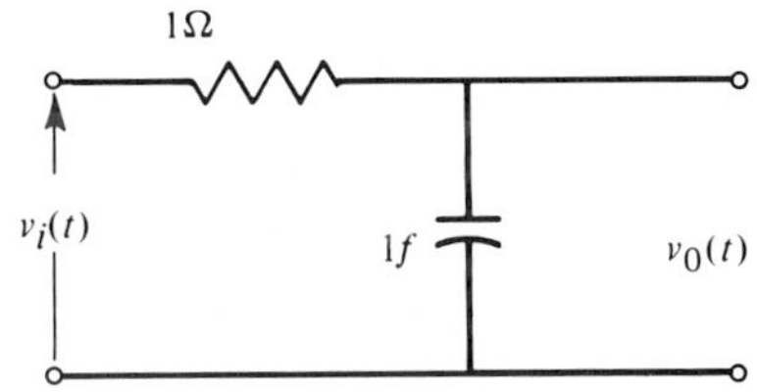

FIG. P. 1-26.

given by (a) $\mathcal{S}_i(\omega) = K$, (b) $\mathcal{S}_i(\omega) = G_2(\omega)$ (gate function with cutoff at $\omega = 1$), (c) $\mathcal{S}_i(\omega) = \pi[\delta(\omega + 1) + \delta(\omega - 1)]$. In each case calculate the power (the mean-square value) of the input signal.

1-27. The basis signals of a three-dimensional signal space are given by

$$\phi_1(t) = p(t), \quad \phi_2(t) = p(t - T) \quad \text{and} \quad \phi_3(t) = p(t - 2T)$$

where

$$p(t) = \frac{1}{\sqrt{T}}[u(t) - u(t - T)]$$

Sketch the waveforms of signals which are located at (1, 1, 0), (2, −1, 1), (3, 2, −1/2), and (−1/2, −1, 1) in this space.

1-28. The basis signals of a five-dimensional signal space are given by

$$\phi_k(t) = p[t - (k - 1)T] \qquad k = 1, 2, 3, 4, 5$$

where

$$p(t) = \frac{1}{\sqrt{T}}[u(t) - u(t - T)]$$

(a) Sketch the waveforms of signals located at (−1, 2, 3, 1, 4), (2, 1, −4, 4, 2), (3, −2, 3, 4, 1) and (−2, 4, 2, 2, 0) in this space. (b) Point out the pairs of signals which are orthogonal.

1-29. For the signal space in Prob. 1-27, find another orthonormal basis set by using Gram-Schmidt process.

1-30. A signal set consists of the following three signals:

$$(2\cos\omega_0 t + \cos 2\omega_0 t), \quad \cos 2\omega_0 t \quad \text{and} \quad (\cos\omega_0 t + 3\cos 3\omega_0 t)$$

These signals exist over the interval $(0, 2\pi/\omega_0)$ and are zero outside this interval. Find two different sets of orthonormal basis signals for these signals.

2

Elements of Probability Theory

2-1. INTRODUCTION

So far we have studied various types of signals such as periodic and nonperiodic signals. These signals were such that their values at all instants were known either from their graphical or analytical description. Such signals which can be specified analytically or graphically are called *deterministic* signals. There is an important class of signals, such as noise signals, about which we have only partial specifications. The variations of these signals are extremely complex. The physical events which generate such a signal may be difficult to predict. Consider for example the outcome of rolling a die. It is difficult to predict this outcome because it depends upon many factors such as the exact shape of the die, the nature of its edges and surface of the die and of the surface on which it is thrown, the distribution of its mass, the initial condition in which it is thrown, the amount of force and its complex direction imparted by the player, etc. It can be seen that some of these factors defy analytical description. Evidently the exact outcome of such a process cannot be predicted. Such processes may, however, possess some kind of regularity—that is, they may be characterized by certain averages. In the above example, we may say that on the average each of the six possible outcomes will appear 1/6 of the times. These processes are therefore random mechanisms. The behavior of random mechanisms can, therefore, be predicted only on the average basis and there is an ignorance or uncertainty regarding its complete behavior.

In communication theory random signals assume a very important role for a number of reasons. On every communication channel there is an unavoidable random-noise signal which contaminates messages. Moreover, as we shall see in Chapter 7, a message can transmit information only if it is unpredictable. The amount of information is proportional to the uncertainty about the signal. The plausibility of this statement can be seen from the fact that if a message is deterministic, it is known completely and, therefore, reception of such a message conveys no additional information. In statistical communication theory, both the messages and the noise are treated as random signals which can be described by their statistical properties.

In this chapter we shall review the basic concepts in the theory of probability which is the basis for describing random signals.

2-2. RELATIVE-FREQUENCY APPROACH TO PROBABILITY

The theory of probability is a branch of applied mathematics dealing with the effects of chance. The throw of a die discussed in Sec. 2-1 illustrates the effect of chance. The outcome of that experiment depends upon the combination of many factors which are unpredictable. We can, however, predict the average behavior for a large number of experiments. Intuitively, we know that any of the six outcomes is equally likely or equally probable (if the die is fair). Thus the idea of chance is related to likelihood or probability. Note that the concept of comparing the likelihood of events follow naturally. For example, in the month of January in Minneapolis cold temperatures are more likely than warm. In tossing a fair coin, we are certain that the outcome of a head or a tail is equally likely or equally probable. This may also be interpreted to mean that if a coin is tossed a large number of times, the head and the tail will appear about equally. This example provides us a clue to define probability of a certain event quantitatively.

In an experiment repeated N times, if the event A occurs m times, than $P(A)$, the probability of event A is defined as

$$P(A) = \lim_{N \to \infty} \frac{m}{N} \tag{2-1}$$

This definition of probability is known as the *relative-frequency definition* as it defines the probability by relative frequency of occurrence of the event. It is also known as the empirical definition. Note that in defining $P(A)$ in Eq. 2-1 it is implicitly assumed that the limit as $N \to \infty$ exists. It also follows from this definition that the probability of an event is a nonnegative quantity (less than unity):

$$P(A) \geq 0 \tag{2-2}$$

Mutually Exclusive Events. A set of events is said to be mutually exclusive or incompatible (also called disjoint) if occurrence of any event in the set precludes the possibility of simultaneous occurrence of any other event in the set.

Consider for example the following events in an experiment of tossing two dice.

A_1: the total number of points on the dice to be 10.

A_2: the total number of points on the dice to be 11.
A_3: at least one of the numbers be 6.

We observe that if event A_1 occurs, then it is impossible for event A_2 also to occur at the same time. Hence occurrence of A_1 automatically precludes the possibility of occurrence of A_2, and vice versa. Hence A_1 and A_2 are mutually exclusive. On the other hand, if the event A_1 occurs it is also possible for the event A_3 to occur at the same time. If, for example, the numbers of the dice are 6 and 4, then it is obvious that both events A_1 and A_3 have occurred at the same time. Hence A_1 and A_3 are not mutually exclusive. Similarly it can be seen that A_2 and A_3 are not mutually exclusive events.

Total Probability. In an experiment a particular event can take place in a number of ways. Consider for example the tossing of two dice, where the desired event is "the sum of the points appearing on the dice to be 10." This event can materialize as a result of any of the three combinations, (6,4), (5,5), (4,6).

Let us start with a simple case where a given event materializes as a result any one of the two events A and B. Such an event is denoted by $A + B$. Thus the event $A + B$ occurs, when either the event A or the event B materializes. If the experiment is performed N times and m_1 and m_2 are the number of outcomes favorable to A and B respectively, then the number of trials favorable to the event $A + B$ are $m_1 + m_2$ provided the events A and B are mutually exclusive. Then by definition

$$P(A + B) = \lim_{N \to \infty} \frac{m_1 + m_2}{N}$$

$$= P(A) + P(B) \tag{2-3}$$

Note that Eq. 2-3 holds if and only if A and B are mutually exclusive. If this condition is not satisfied the events A and B can occur simultaneously and the outcomes of some trials are favorable to both A and B. Hence the number of trials favorable to the event $A + B$ will be less than $m_1 + m_2$ and we have

$$P(A + B) \leq P(A) + P(B) \tag{2-4}$$

The equality sign holds only if A and B are mutually exclusive.

The reader can extend the results in Eq. 2-3 to any number of events.

$$P(A_1 + A_2 + \cdots + A_n) = P(A_1) + P(A_2) + \cdots + P(A_n)$$

if $A_1, A_2, \ldots, A_n$ are mutually exclusive.

If a particular experiment has n outcomes $A_1, A_2, \ldots, A_n$ and no more, then these n events are said to be exhaustive events. We can easily show that if the n events are mutually exclusive and exhaustive, then

$$P(A_1) + P(A_2) + \cdots + P(A_n) = 1 \tag{2-5a}$$

This follows directly from the definition of 2-1. If the experiment is repeated N times and $m_1, m_2, \ldots, m_n$ are the number of outcomes which are favorable to events $A_1, A_2, \ldots, A_n$ respectively, then

$$P(A_i) = \lim_{N\to\infty} \frac{m_i}{N}$$

and

$$P(A_1) + P(A_2) + \cdots + P(A_n) = \lim_{N\to\infty} \frac{m_1 + m_2 + \cdots + m_n}{N}$$

But since the events are mutually exclusive, the sum $m_1 + m_2 + \cdots + m_n$ must be equal to N, the total number of trials. Hence the sum of all probabilities must be unity as in Eq. 2-5a.

Since $A_1, A_2, \ldots, A_n$ are mutually exclusive, from Eq. 2-3 it follows that

$$P(A_1 + A_2 + \cdots + A_n) = 1 \tag{2-5b}$$

The event $(A_1 + A_2 + \cdots + A_n)$ is the event which occurs when either one of the event $A_1, A_2, \ldots, A_n$ materializes. But since $A_1, A_2, \ldots, A_n$ are assumed to be exhaustive events, the event $(A_1 + A_2 + \cdots + A_n)$ is a certainty, denoted by event $\mathcal{S}$. Thus

$$P(\mathcal{S}) = 1 \tag{2-5c}$$

This result is intuitively obvious.

2-3. AXIOMATIC APPROACH TO PROBABILITY

The relative-frequency definition discussed so far has a great intuitive appeal for the beginner. From the mathematical point of view, however, it leaves much to be desired. In any physical experiment the number of trials N may be large but is always finite. How large a value of N must we choose in Eq. 2-1 to obtain the correct value of $P(A)$? This question cannot be answered easily. In fact it is possible that N may be very very large and yet the right-hand side of Eq. 2-1 may not approach the limit. In a coin-tossing experiment there is a possibility, albeit small, that all the tosses may result in heads (or tails). The foundation provided by the relative-frequency definition is too weak (imprecise) to allow one to build a strong (precise) mathematical structure upon it.

It is important here to inquire into the nature of the mathematical discipline. Broadly speaking, the field of mathematics may be summed up in one aphorism, "*This* implies *that*." In other words, if we are given a certain set of axioms (hypotheses) then, based upon this set of axioms alone, what else can be deduced?

Thus we begin with certain undefined concepts. Then we assume certain properties and the relationships between these concepts. These are the axioms. Then based upon these axioms alone, various propositions are obtained. These are called theorems. This is what the mathematics is all about. It is a pure deductive logic. The basic axioms are at the heart of any branch of mathematics. These axioms need satisfy only one requirement—they must be consistent with each other; that is, one or more axioms in the set taken individually or collectively should not contradict any other axioms in the set.

However, if a branch of mathematics is to be of some practical use, the axioms must closely conform to some observed phenomenon. This leads us directly into the notion of a mathematical model of a real phenomenon. The abstract objects about which the axioms are proposed must be identified with some real phenomenon, and the axioms must then closely conform to the observed relationships of the real phenomenon. The theorems of the mathematical theory will then depict the real phenomenon to the extent to which the axioms conform to the observed behavior of the phenomenon.

With this background, we are ready to discuss the axiomatic approach to probability. In this approach the undefined concept or the abstract object is probability. We must now postulate the axioms about probability so that they conform to the observed relationships in practice. The observed relationships are closely related to the relative-frequency definition and its consequences in Eqs. 2-2, 2-3, and 2-5. We may, therefore, consider these equations as the axioms. Note that we are not using the relative-frequency definition (Eq. 2-1) to define the probability. We start with a concept (as yet undefined) of probability and postulate three axioms (Eq. 2-2, 2-3, and 2-5) about this concept. Then based upon these axioms alone, we inquire what else is true. This is the essence of modern theory of probability. We can summarize this approach as follows. The probability of an event A is a number $P(A)$ assigned to this event. The only restriction upon the probability function is that it obeys the following three postulates (or axioms).

$$P(A) \geq 0 \tag{2-6}$$

$$P(\mathcal{S}) = 1 \tag{2-7}$$

where $\mathcal{S}$ is a certain event made up of all mutually exclusive and exhaustive events.

$$P(A + B) = P(A) + P(B) \tag{2-8}$$

if A and B are mutually exclusive.

Note that in the relative-frequency approach the probability $P(A)$ is defined as in Eq. 2-1 and Eq. 2-6, 2-7, and 2-8 follow as a consequence of this definition. In the axiomatic approach, on the other hand, we do not say anything about how to assign a probability $P(A)$ to an event A but we postulate that the probability function must obey the three postulates or axioms in Eq. 2-6, 2-7, and 2-8.

The skeptical reader may wonder whether in the axiomatic approach we are not ignoring the sticky problem of assigning probability to an event. This is quite true. The axiomatic approach washes its hands of assigning probability to an event. It does not interest itself with assigning probabilities. It assumes that somehow the probabilities are assigned to each event on an a priori basis. The only condition is that the probabilities must obey the three postulates (Eqs. 2-6, 2-7, and 2-8). This is the starting point of the axiomatic theory. From our earlier discussion it is obvious that if the mathematical model is to conform to a real phenomenon we must assign these probabilities consistent with the physical realities. Thus the probabilities may be assigned from the physical laws (e.g., mechanics of a coin, throw of a die, shuffling and dealing of cards, etc.) or by the behavior as determined by experiments. Usually the burden of assigning probabilities falls upon the relative frequency in Eq. 2-1. We have already observed in Eqs. 2-2, 2-3, and 2-4 that the probability as defined by the relative frequency (Eq. 2-1) satisfies the axioms[1] 2-6, 2-7, and 2-8. It is, therefore, evident that the axiomatic theory uses the relative-frequency definition, but cleverly avoids taking responsibility for such an imprecise definition. It should be noted that the deduction of axiomatic theory will depict the physical phenomenon to the extent our assigning of probabilities is correct.[2]

We shall illustrate the method of assigning probabilities to few simple cases.

Example 2-1. Tossing a Coin. If we consider the head and the tail as two possible outcomes, it is obvious that for a fair coin each of the outcomes will occur with the same likelihood. Also, according to axiom 2-7, $P(\text{head}) + P(\text{tail}) = 1$. Hence

$$P(\text{head}) = P(\text{tail}) = 0.5 \tag{2-9}$$

[1]It is possible to assign probabilities other than those obtained by relative-frequency Eq. 2-1 and still satisfy the axioms 2-6, 2-7, and 2-8. But such a model does not conform to reality and will not be of much use from the physical point of view.

[2]For a deeper study of theory of probability, some understanding of set theory is essential. We shall, however, avoid any discussion on set theory here partly for the reason that it can be studied more profitably from standard texts on probability theory and partly because the discussion throughout this book is substantiated by intuitive approach so as to avoid such a need. For more details on set theory, see Refs. 2-1, 2-2, and 2-3.

Example 2-2. From a deck of cards a card is drawn randomly. Assign the probability to the event of drawing an ace.

This event can materialize by way of drawing any one of the four aces. These four events are mutually exclusive. Hence from axiom 2-8,

$$P(\text{ace}) = P(\text{spade ace}) + P(\text{heart ace}) + P(\text{diamond ace}) + P(\text{club ace})$$

The probability of drawing the spade ace is 1/52 since it is one of the cards in a deck of 52 cards. Similarly the probability of drawing any other ace is 1/52. Hence

$$P(\text{ace}) = {}^4/_{52} = {}^1/_{13}$$

Example 2-3. Assign the probability to an event of obtaining one head and 2 tails in three tosses of a fair coin.

For three tosses there are 8 possible outcomes as follows

(1) HHH	(2) HHT	(3) HTH	(4) HTT
(5) THH	(6) THT	(7) TTH	(8) TTT

For a fair coin, any one of the above 8 outcomes is equally likely. Hence the relative frequency of each of the outcomes is 1/8. In other words, the probability of any one of the above outcomes is 1/8. Moreover all the 8 outcomes are mutually exclusive. Out of these, the outcomes 4, 6, and 7 are favorable to the desired event, one head and 2 tails. Hence from axiom 2-8 we have

$$\text{P(1H, 2T)} = \text{P(HTT)} + \text{P(THT)} + \text{P(TTH)} = {}^3/_8$$

Note that in all three problems, we are implicitly using the relative frequency in assigning probabilities.

2-4. JOINT PROBABILITY, CONDITIONAL PROBABILITY, AND INDEPENDENT EVENTS

Joint Probability. We shall now consider the probability of a joint event. If in an experiment there are two sets of outcomes, then the probability of observing a particular outcome A from one set and some outcome B from the other set is called *joint probability* of an event AB. As an example, we may consider a joint experiment of drawing 2 cards in succession (with or without replacement) from a deck of cards. Drawing an ace in the first draw may be considered as an event A and drawing an ace in the second draw as the event B. Then the event AB is the joint event of drawing two aces in succession (with or without replacement as the case may be).

Conditional Probability. One often comes across a situation where the probability of one event is influenced by another event. As an example, consider drawing two cards in succession from a deck. Let A denote the event that the first card drawn is an ace. We do not replace the card drawn in the first trial. Let B denote the event that the second card drawn is an ace. It is evident that the probability of drawing an ace in the second trial will be influenced by the outcome of the first draw. If the first draw did not result in an ace, then the probability of obtaining an ace in the second trial is 4/51. The probability of event B thus depends upon whether the event A took place or not. We therefore introduce the concept of *conditional probability* which signifies the probability of an event B given that the event A has occurred. This will be denoted by $P(B \mid A)$. Thus the quantity $P(B \mid A)$ represents reevaluation of the probability of the event B when it is known that the event A has already occurred.

How shall we obtain the conditional probability? A glance at Eqs. 2-8, 2-9, and 2-10 show that it is not possible to derive the expression for conditional probability from the axioms. Hence we must define conditional probability. The definition must, however, be reasonable and conform to physical observation. Since the relative-frequency model is in close conformity with the physical model, we shall attempt to derive the expression for conditional probability based on this model.

Let the experiment yielding the two sets of outcomes be performed N times out of which n_1 trials are favorable to outcome A. Out of these n_1 trials let n_2 trials be favorable to the event B. It is then obvious that n_2 is the number of trials which are favorable to joint event AB. Hence

$$
\begin{aligned}
P(AB) &= \lim_{N \to \infty} \frac{n_2}{N} \\
&= \lim_{N \to \infty} \left(\frac{n_1}{N}\right)\left(\frac{n_2}{n_1}\right) \qquad \text{(2-10)}
\end{aligned}
$$

Note that $\lim_{N \to \infty} n_1/N = P(A)$. Also $\lim_{N \to \infty} n_2/n_1 = P(B \mid A)$, since n_2 trials are favorable to B out of n_1 trials which are favorable to A. This obviously represents the conditional probability of B given that the event A has occurred. Therefore

$$P(AB) = P(A)\,P(B \mid A) \qquad \text{(2-11)}$$

We shall use Eq. 2-11 to define conditional probability $P(B \mid A)$. Thus we define

$$P(B \mid A) = \frac{P(AB)}{P(A)} \qquad \text{provided } P(A) \neq 0 \qquad \text{(2-12a)}$$

and

$$P(A \mid B) = \frac{P(AB)}{P(B)} \qquad \text{provided } P(B) \neq 0 \qquad \text{(2-12b)}$$

It follows from Eq. 2-12 that[3]

$$P(A \mid B) = \frac{P(A)\,P(B \mid A)}{P(B)} \tag{2-13a}$$

and

$$P(B \mid A) = \frac{P(B)\,P(A \mid B)}{P(A)} \tag{2-13b}$$

Equations 2-13 are referred to as Bayes' rule. In this rule, one conditional probability is expressed in terms of the reversed conditional probability.

Independent Events. In the previous section we presented an example where the outcome of one event was influenced by the outcome of another event. There are, of course, many examples where the two events are entirely independent of each other and the outcome of one event in no way affects the outcome of the other event. To illustrate this point we again consider the example in the previous section of drawing two cards in succession. In this case, however, the card obtained in the first draw is replaced before the second draw. It is evident that now the outcome of the second draw is in no way influenced by the outcome of the first draw. Thus $P(B)$, the probability of drawing an ace in the second draw, is independent of whether the event A (drawing an ace in the first trial) occurred or not. The events A and B are therefore independent. The conditional probability $P(B \mid A)$ is evidently given by $P(B)$. Thus the event B is said to be independent of the event A if

$$P(B \mid A) = P(B) \tag{2-14a}$$

It can be seen from Eqs. 2-13a and 2-14a that if the event B is independent of event A, then the event A is also independent of B, that is

$$P(A \mid B) = P(A) \tag{2-14b}$$

Note that if the events A and B are independent, then from Eqs. 2-12a and 2-14a it follows that

$$P(AB) = P(A)\,P(B) \tag{2-15}$$

Example 2-4. If an event A is drawing an ace from a deck of cards and the event B is obtaining a head in the toss of a coin, find the probability of a joint event AB, that is, the probability of drawing an ace and obtaining a head.

The cards can be drawn in 52 ways and a toss of a coin yields two outcomes. Hence the total number of outcomes in experiment AB is $52 \times 2 = 104$

[3]Throughout the book, whenever expressions of the form $P(A \mid B)$ is used, it will be assumed that $P(B) \neq 0$.

which are equally likely. Also four of these outcomes are favorable to the joint event AB. Hence

$$P(AB) = 4/104 = 1/26$$

Note that $P(A) = 1/13$ and $P(B) = 1/2$. Hence $P(AB) = P(A)P(B)$. It, therefore, follows that events A and B are independent.

Example 2-5. An experiment consists of drawing two cards from a deck in succession, the second card being drawn without replacing the first card drawn. Find (i) the probability of obtaining two red aces in two draws, (ii) the probability of obtaining a red ace in the second draw, (iii) the conditional probability $P(A \mid B)$ where A is the event of obtaining a red ace in the first draw and B is the event of obtaining a red ace in the second draw.

$$P(AB) = P(A)P(B \mid A)$$

It is obvious that

$$P(A) = 2/52 = 1/26$$

and $P(B \mid A)$ is the probability of obtaining a red ace in a draw under the condition that one red ace is already drawn and no replacement has been made. Hence

$$P(B \mid A) = 1/51$$

and

$$P(AB) = (1/26)(1/51) = 1/1326$$

To find $P(B)$, we use axiom 2-8. The event B can materialize in two ways: (i) first-draw red ace (A) and the second-draw red ace; (ii) first draw not a red ace (call this event $\bar{A}$) and the second-draw red ace. These are mutually exclusive, hence

$$\begin{aligned} P(B) &= P(AB) + P(\bar{A}B) \\ &= P(AB) + P(\bar{A})P(B \mid \bar{A}) \\ &= 1/1326 + (50/52)(2/51) \\ &= 1/26 \end{aligned}$$

The quantity $P(A \mid B)$ represents the probability that A has occurred when it is known that B has occurred, and may be obtained from Bayes' rule (Eq. 2-13).

$$\begin{aligned} P(A \mid B) &= \frac{P(AB)}{P(B)} \\ &= (1/1326)(26) = 1/51 \end{aligned}$$

Note that

$$P(AB) \neq P(A)P(B)$$

Hence the two events are not independent. If in this experiment, the second draw were with replacement of the first card drawn, then

$$P(B \mid A) = P(B) = 1/26$$

and the events A and B would be independent.

For three events A, B, and C, superficial extension of Eq. 2-15 yields the condition of independence as

$$P(ABC) = P(A)\,P(B)\,P(C) \tag{2-16a}$$

However, this condition alone is inadequate. If the three events are independent, then we expect that any two of the events should also be independent. Thus if A, B, and C are independent then

$$\begin{aligned} P(AB) &= P(A)\,P(B) \\ P(BC) &= P(B)\,P(C) \\ P(AC) &= P(A)\,P(C) \end{aligned} \tag{2-16b}$$

However, condition 2-16a does not imply condition 2-16b, and is therefore inadequate to define independence of three events. In order to define independence of three events we must have both conditions 2-16a and 2-16b.

In general events $A_1, A_2, \ldots, A_n$ are independent if and only if

$$P(A_{i_1}A_{i_2}, \ldots, A_{i_m}) = P(A_{i_1})\,P(A_{i_2}), \ldots, P(A_{i_m}) \tag{2-16c}$$

where

$$m = 2, 3, \ldots, n$$

and

$$1 \le i_1 < i_2 < \cdots \le i_m < n$$

2-5. RANDOM VARIABLES: DISCRETE CASE

A given experiment may have a certain number of outcomes. Each of these outcomes may be considered as an element of a set (set is a collection of objects). Thus for a coin-tossing experiment, the set consists of two elements; head and tail (H, T). Similarly for a rolling die, the set consists of 6 elements (six outcomes). The set of elements consisting of all possible distinct outcomes of an experiment is called the *sample space* for the experiment. Here the term space is used synonymously for the word set. The individual elements or points of the sample space are called *sample points.* Thus each sample point corresponds to one distinct outcome of the experiment.

It should be noted that in order to be able to specify the sample space, we must first specify as to what we consider the outcomes. For example in the case of three tosses discussed in Ex. 2-3, we may consider the outcome to be a particular order of heads and tails as they appear in three tosses. For this case there are 8 outcomes as shown in Ex. 2-3. The sample space here has eight sample points. On the other hand, we may consider the outcome of the experiment to be "obtaining a total of k number of heads." In this case there are only four possible outcomes: obtaining a total of (i) zero heads, (ii) 1 head, (iii) 2 heads, and (iv) 3 heads. In this case the sample space has only 4 sample points.[4] Thus, in general, for a given experiment, the choice of a sample space is not unique but will primarily depend upon what we consider as the outcomes of the experiment.

Each of the experiments may have a number of outcomes. These outcomes may be described by descriptive phrases such as head or a tail and so on. It is, however, inconvenient to deal with these phrases mathematically. For the ease of manipulation we may assign a real number to each of the outcomes by some fixed rule. For example, there are two outcomes in the coin-tossing experiment; the head and the tail. We can assign the number -1 to the outcome head and the number 1 to the outcome tail. In a game of chance, a particular outcome may be assigned a value of gain or loss associated with that outcome. This arrangement is obviously very attractive from a mathematical point of view and meaningful from a practical point of view. We thus assign a real number to each sample point in the sample space according to some rule. Thus in the coin-tossing experiment the outcome can be described by a random variable which can assume numerical value -1 or 1. We shall use roman type to designate a random variable. Thus the outcome of a coin-tossing experiment may be described by a random variable x which can assume two discrete values -1 and 1. Later on we shall study continuous random variables which can assume any value in a given range.

In general we assign a real number to each outcome (or each sample point) of the experiment. If there are n outcomes, then we assign numbers $x_1, x_2, \ldots, x_n$ to these outcomes. The outcome of a random experiment is therefore a random variable x, which can assume any of the n discrete values $x_1, x_2, \ldots, x_n$. A random variable is actually a function in a conventional sense. A function $f(y)$ assigns values to each y according to certain rule. Similarly a random variable x assigns numerical values (real numbers) to each sample point. So in this sense the random variable is a function of sample points and assigns a real number to each sample point according to some rule. As an example, consider again the experiment of three tosses of a coin (Ex. 2-3). The desired outcome is the event of obtaining a total of k heads. For this

[4]In this experiment the first sample space (with 8 points) can also serve as the sample space for the second case with 4 outcomes. Each outcome in the second case is associated with one or more than one sample point in the first sample space. Thus the outcome of obtaining a total of 2 heads is associated with elements HHT, HTH, and THH in the first sample space.

case, the sample space has 4 sample points corresponding to outcomes: (i) 0 heads, (ii) 1 head, (iii) 2 heads, (iv) 3 heads. We may assign number 0, 1, 2, and 3 respectively to these sample points. In this case the random variable x can assume four values $x_1 = 0$, $x_2 = 1$, $x_3 = 2$, and $x_4 = 3$. Here the numbers are assigned according to the rule $x = k$ (number of heads). We may as well have used the rule $x = k^2$ of some other function of k such as sin k.

Thus, a random variable is a real-valued point function x(u) defined over the sample space $\mathcal{S}$ of a random experiment, where u is an arbitrary element of the sample space $\mathcal{S}$.

There is a probability assigned to each value of the random variable x. Thus $P_x(x_i)$ is the probability of the outcome or the event to which the number x_i is assigned. In this notation $P_x(x_i)$ the subscript refers to the random variable x and the argument is the particular value of the random variable. Similarly $P_y(y_j)$ is the probability that a random variable y assumes a value y_j. The subscript is essential to indicate the probability function associated with a particular random variable. The probability functions of two random variables x and y are different in general. To point out this distinction, the subscript is used. If, however, there is only one random variable, say x, involved in the discussion, then the subscript is understood, and the notation $P(x_i)$ is without ambiguity. Thus $P(2)$ in such a case is the probability that the random variable x assumes a value 2. But where there are more variables involved, omission of the subscript will cause confusion. If, for example we are dealing with two random variables x and y, $P(2)$ will have ambiguous meaning. It may mean the probability that x = 2 or y = 2. The difficulty arises because we are using the same functional notation $P(\cdot)$ for two different functions. The probability functions of variables x and y are two distinct functions which can be conveniently represented by using appropriate subscript. The subscript therefore identifies a particular function.

If there are a total of n mutually exclusive outcomes of the experiment, then from axiom 2-7 it follows that

$$\sum_{i=1}^{n} P_x(x_i) = 1 \tag{2-17}$$

This is because the summation on the left-hand side of Eq. 2-17 represents the probability of a certain event (sum of probabilities over all sample points).

As an example, consider again the experiment of three coin tosses in Ex. 2-3. If we consider the desired event as obtaining a total of k heads, then there are, in all, four possible outcomes: (i) obtaining 0 heads, (ii) one head, (iii) 2 heads, and (iv) 3 heads. Thus the sample space has 4 sample points. We shall assign four numbers x_0, x_1, x_2, and x_3 to these outcomes in that order. Thus the outcome of the experiment is a random variable x which can assume values x_0, x_1, x_2, and x_3. We can obtain probability of these outcomes from Ex. 2-3.

$$P_x(x_0) = 1/8 \qquad P_x(x_1) = 3/8$$
$$P_x(x_2) = 3/8 \qquad P_x(x_3) = 1/8$$

Note that

$$\sum_{i=0}^{4} P_x(x_i) = 1$$

These probabilities can be plotted as shown in Fig. 2-1a. In this case we have assigned numbers to the outcomes as $x_0 = 0$, $x_1 = 1$, $x_2 = 2$, and $x_3 = 3$ in this figure.

If a random experiment has two outcomes or outcomes described by two sets of numbers denoted by random variables x and y which can assume values $x_1, x_2, \ldots, x_n$ and $y_1, y_2, \ldots, y_m$ respectively, then the sample space has a total of *mn* number of points. The probability function $P_{xy}(x_i, y_j)$ represents the joint probability that the random variable x assumes a value x_i and the random variable y assumes a value y_j. The joint probability can be represented graphically in three dimensions. As an example consider the experiment of tossing two coins. There are two outcomes to this experiment. Let us designate the outcome of the first toss by a random variable x and that of the second toss by a random variable y. There are two possible outcomes for each variable (head and tail). We shall assign number -1 for head and the number 1 for tail. Thus

$$x_1 = y_1 = 1 \quad \text{and} \quad x_2 = y_2 = -1$$

There are a total of four sample points and

$$P_{xy}(1,1) = P_{xy}(-1,1) = P_{xy}(1,-1) = P_{xy}(-1,-1) = 1/4$$

The probability function $P_{xy}(x_i, y_j)$ is plotted in Fig. 2-1b.

According to axiom 2-7, we must have

$$\sum_{i} \sum_{j} P_{xy}(x_i, y_j) = 1 \tag{2-18}$$

The above discussion can be easily extended to any number of random variables.

Conditional Probabilities. If x and y are two random variables, then the conditional probability of $x = x_i$ given $y = y_j$ is denoted by $P_x(x_i \mid y = y_j)$.

If the random variables x and y assume values $x_1, x_2, \ldots, x_n$ and $y_1, y_2, \ldots, y_m$ respectively, then for a given value of $y = y_j$, it is

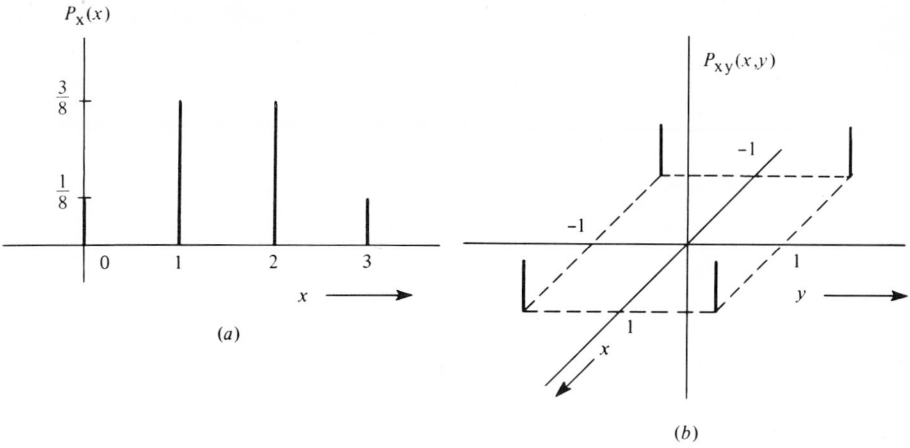

FIG. 2-1.

obvious that the random variable x must assume any one of the values $x_1, x_2, \ldots, x_n$. Thus the event that x assume any one of the n outcomes $x_1 \ldots x_n$ given that $y = y_j$ is a certainty. Hence

$$\sum_i P_x(x_i \mid y = y_j) = 1 \tag{2-19a}$$

Similarly,

$$\sum_j P_y(y_j \mid x = x_i) = 1 \tag{2-19b}$$

We also have from Eq. 2-12,

$$P_x(x_i \mid y = y_j) = \frac{P_{xy}(x_i, y_j)}{P_y(y_j)} \tag{2-19c}$$

$$P_y(y_j \mid x = x_i) = \frac{P_{xy}(x_i, y_j)}{P_x(x_i)} \tag{2-19d}$$

and (Bayes' rule),

$$P_x(x_i \mid y = y_j) = \frac{P_y(y_j \mid x = x_i)\, P_x(x_i)}{P_y(y_j)} \tag{2-19e}$$

2-6. RANDOM VARIABLES: CONTINUOUS CASE

In our discussion thus far we have restricted ourselves to the case where the sample space of an experiment consists of discrete elements (sample points). As a result the random variable describing the outcomes of this experiment is a discrete random variable which can assume only discrete values. There are many instances, where the sample space is not discrete but continuous. Consider for example the case of the temperature in a room. As we know the temperature varies randomly over some range (T_1, T_2) and can assume any value in this continuous range. There are nonenumerably infinite possible temperatures in this range that may prevail. The sample space therefore contains nonenumerably infinite (sample) points and cannot be represented by an enumerably infinite set (set of discrete numbers). The sample space is therefore continuous and the random variable defined over this sample space must necessarily be a continuous variable.

In the case of discrete sample space, all the values assumed by the corresponding random variable can be enumerated (finite or enumerably infinite sample points). To each of these can be assigned a number which is the probability that the random variable assumes that value. In the case of continuous sample space, however, the problem of assigning probability distribution to the corresponding random variable is somewhat complicated. A continuous sample space has nonenumerably infinite sample points and obviously the probability of observing any one point is zero. In the example of room temperature the temperature can assume nonenumerably infinite temperatures in a certain range (T_1, T_2) and the probability of observing any one particular temperature is zero. A more meaningful measure in this case is the probability of observing the temperature in some infinitesimal temperature interval ΔT.

A continuous random variable x may assume values in a certain continuous range (x_1, x_2). For the most general case obviously the range is ($-\infty, \infty$). As observed earlier the probability of a random variable x assuming a particular value x is generally zero and not very meaningful in the case of continuous random variable. Instead we inquire about the probability of observing a random variable x below some value x. Let us denote this probability by $F_{\mathrm{x}}(x)$

$$F_{\mathrm{x}}(x) = \text{probability}\ (\mathrm{x} \leq x) \tag{2-20}$$

As usual, we use the roman type to represent the random variable and the italic-face type to represent the value assumed by the random variable. Thus $F_{\mathrm{x}}(x)$ is the probability of observing the random variable x in the range ($-\infty, x$), or $\mathrm{x} \leq x$. We can at once see that $F_{\mathrm{x}}(-\infty) = 0$ since this is the probability that x assumes no value in the range ($-\infty, \infty$). Also $F_{\mathrm{x}}(\infty) = 1$, since this is the probability of observing x anywhere in the range $-\infty$ to ∞, which is a certainty. The function $F_{\mathrm{x}}(x)$ is called probability-distribution function (also *cumulative distribution* of x) and is a continuous function of x when x is a continuous random variable (Fig. 2-2a). In addition it can be seen that $F_{\mathrm{x}}(x)$ is a

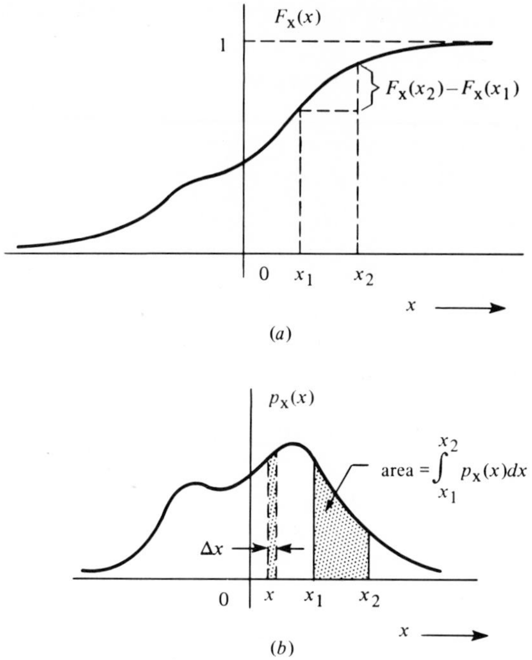

FIG. 2-2.

nondecreasing function of x. Thus if $x_2 > x_1$ then

$$F_x(x_2) \geq F_x(x_1)$$

This can be easily shown. By definition,

$$F_x(x_2) - F_x(x_1) = \text{probability}\,(\mathrm{x} \leq x_2) - \text{probability}\,(\mathrm{x} \leq x_1) \qquad (2\text{-}21)$$

Note that for $x_2 > x_1$ the events $\mathrm{x} \leq x_1$ and $x_1 < \mathrm{x} \leq x_2$ are mutually exclusive. Hence according to axiom 2-8

$$\text{Probability}\,(\mathrm{x} \leq x_2) = \text{probability}\,(\mathrm{x} \leq x_1) + \text{probability}\,(x_1 < \mathrm{x} \leq x_2)$$

Substitution of this equation in Eq. 2-21 yields

$$F_x(x_2) - F_x(x_1) = \text{probability}\,(x_1 < \mathrm{x} \leq x_2) \qquad (2\text{-}22)$$

The right-hand side being a probability of an event is a nonnegative quantity. Hence $F_x(x_2) \geq F_x(x_1)$.

Probability-Density Function. From Eq. 2-22, it follows that

$$F_x(x + \Delta x) = F_x(x) + \text{probability}\,(x < \mathrm{x} \leq x + \Delta x) \qquad (2\text{-}23a)$$

If $\Delta x \to 0$, then we can also express $F_x(x + \Delta x)$ by Taylor expansion as

$$F_x(x + \Delta x) = F_x(x) + \frac{dF_x(x)}{dx}\Delta x \qquad (2\text{-}23b)$$

From Eqs. 2-23a and 2-23b it follows that

$$\lim_{\Delta x \to 0} \frac{dF_x(x)}{dx} \Delta x = \text{probability}\,(x < \mathrm{x} \leq x + \Delta x) \tag{2-24}$$

We designate the derivative of $F_x(x)$ with respect to x by $p_x(x)$ as

$$\frac{dF_x(x)}{dx} = p_x(x) \tag{2-25}$$

The function $p_x(x)$ is called the *probability-density function* of the random variable x. It follows from Eq. 2-24, that the probability of observing the random variable x in the interval $(x, x + \Delta x)$ is $p_x(x)\,\Delta x(\Delta x \to 0)$. This is the area under the probability-density curve $p_x(x)$ over the interval Δx as shown in Fig. 2-2b.

We can express Eq. 2-25 alternately as

$$F_x(x) = \int_{-\infty}^{x} p_x(x)\,dx \tag{2-26}$$

Here we use the fact that $F_x(-\infty) = 0$. We also have from Eq. 2-22

$$\begin{aligned} \text{Probability}\,(x_1 < \mathrm{x} \leq x_2) &= F_x(x_2) - F_x(x_1) \\ &= \int_{-\infty}^{x_2} p_x(x)\,dx - \int_{-\infty}^{x_1} p_x(x)\,dx \\ &= \int_{x_1}^{x_2} p_x(x)\,dx \end{aligned} \tag{2-27}$$

Thus the probability of observing x in any interval is given by the area under probability-density function $p_x(x)$ over that particular interval.

Since $F_x(\infty) = 1$, it follows that

$$\int_{-\infty}^{\infty} p_x(x)\,dx = 1 \tag{2-28}$$

This is also obvious from the fact that the integral in Eq. 2-28 represents the probability of observing x in the interval $(-\infty, \infty)$ which is a certainty.

Every probability-density function must satisfy condition 2-28 to be a valid function.

It should be observed that the fact that the probability of observing x at a certain value x is zero does not necessarily mean that the random variable does not assume that particular value. It is also evident from axiom 2-6, that the probability-density function must be a nonnegative function

$$p_x(x) \geq 0$$

A discrete variable may be considered as a limiting case of continuous variable where the probability density is concentrated as impulses at some discrete points. If a discrete random variable assumes values $x_1, x_2, \ldots, x_n$ with probabilities $a_1, a_2, \ldots, a_n$, then the probability-density function can be expressed as a sequence of impulses

$$p_x(x) = a_1\delta(x - x_1) + a_2\delta(x - x_2) + \cdots + a_n\delta(x - x_n)$$

$$= \sum_{r=1}^{n} a_r\delta(x - x_r) \tag{2-29}$$

Observe that since

$$\sum_{r=1}^{n} a_r = 1$$

we have

$$\int_{-\infty}^{\infty} p_x(x)\,dx = 1$$

It is of course possible to have a situation where a probability-density function may be mixed, i.e., it may be a continuous curve and also may contain impulses. Consider the case of a random-noise signal whose amplitude probability-density function is shown in Fig. 2-3a. If such a signal is passed through a limiter which clips all the voltage levels greater than A, the new probability-density function will appear as shown in Fig. 2-3b. The impulse appearing at $x = A$ has a strength

$$k = \int_{A}^{\infty} p_x(x)\,dx$$

Similarly, if the limiter is used to clip both positive as well as negative amplitudes above A, the new probability-density function will appear as shown in Fig. 2-3c. The strength of the impulse at $x = -A$ is

$$k = \int_{-\infty}^{A} p_x(x)\,dx$$

It is obvious that the probability of observing any particular voltage amplitude x where $p_x(x)$ is continuous is zero. Such is the case for range $(-\infty < x < \infty)$ in Fig. 2-3a, $(-\infty < x < A)$ in Fig. 2-3b and

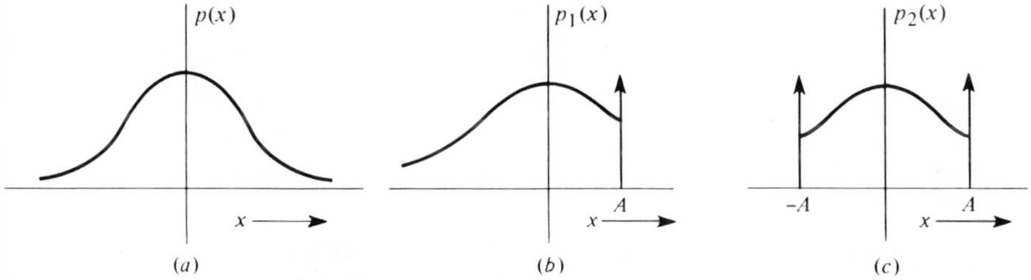

FIG. 2-3.

$(-A < \mathrm{x} < A)$ in Fig. 2-3c. However, in Fig. 2-3b, probability of observing the voltage amplitude A is k. Similarly in Fig. 2-3c, the probability of observing voltage amplitude A is k and that of observing amplitude $-A$ is also k.

We shall now illustrate an empirical method of determining probability-density function of a random variable. Consider a waveform $x(t)$ available over a very large time interval (Fig. 2-4). The amplitude of

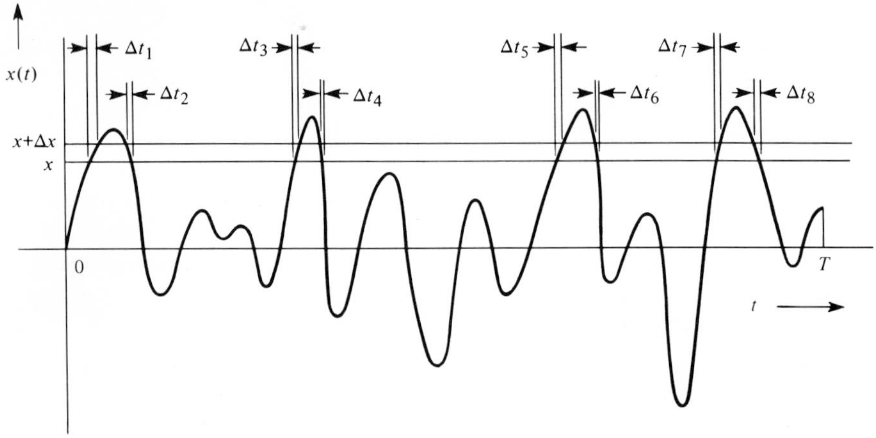

FIG. 2-4.

the waveform varies randomly with time and can assume any value in a certain range. The amplitude of the waveform can, therefore, be considered as a random variable x. We shall now determine the probability density function $p_{\mathrm{x}}(x)$ based upon relative-frequency interpretation. By definition,

$$\lim_{\Delta x \to 0} p_{\mathrm{x}}(x)\Delta x = \text{probability}\,(x < \mathrm{x} \leq x + \Delta x)$$

We now draw two horizontal lines at x and $x + \Delta x$ as shown in Fig. 2-4. According to relative-frequency interpretation, the probability that x assumes a value between x and $x + \Delta x$ is the ratio T_x/T, where T_x is the total amount of time that $x(t)$ falls inside the range $(x, x + \Delta x)$ during the observation time $T(T \to \infty)$. Thus

$$\lim_{\Delta x \to 0} p_{\mathrm{x}}(x)\Delta x = \lim_{T \to \infty} \frac{T_x}{T}$$

or

$$p_{\mathrm{x}}(x) = \lim_{\substack{\Delta x \to 0 \\ T \to \infty}} \frac{T_x}{T\Delta x} = \lim_{\substack{\Delta x \to 0 \\ T \to \infty}} \frac{\Sigma \Delta t_i}{T\Delta x}$$

The probability-distribution function (cumulative distribution) $F_{\mathrm{x}}(x)$ can also be obtained in a similar way. By definition,

$$F_{\mathrm{x}}(x) = \text{probability}\,(\mathrm{x} \leq x) = 1 - \text{probability}\,(\mathrm{x} > x)$$

We draw a horizontal line at x. The probability of observing x above x is the ratio T_a/T where T_a is the total amount of time that $x(t)$ lies above a value x during the observation time T. Hence

$$F_{\mathrm{x}}(x) = 1 - \lim_{T\to\infty} \frac{T_a}{T}$$

Example 2-6. As an example, consider a gaussian probability density. This density function is very frequently observed in nature.

$$p_{\mathrm{x}}(x) = \frac{1}{\sigma\sqrt{2\pi}}\, e^{-x^2/2\sigma^2}$$

This function is sketched in Fig. 2-5a. The reader can verify that this function satisfies condition 2-28. In other words, the area under $p_{\mathrm{x}}(x)$ is unity:

$$\frac{1}{\sigma\sqrt{2\pi}} \int_{-\infty}^{\infty} e^{-x^2/2\sigma^2}\, dx = 1$$

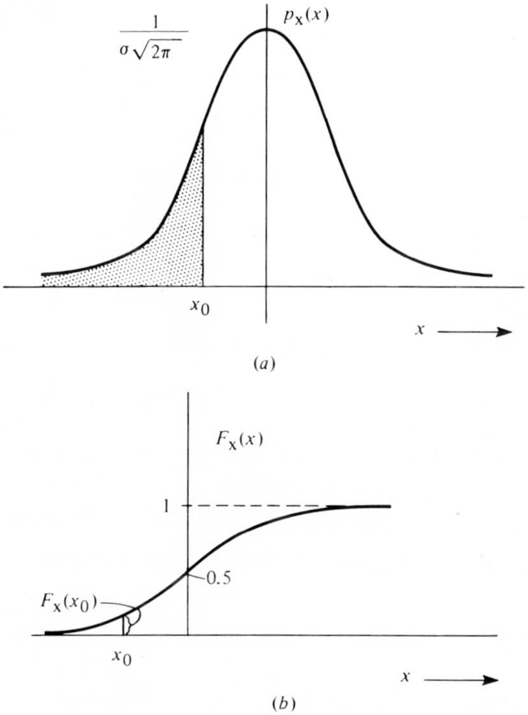

FIG. 2-5.

The cumulative-distribution function $F_x(x)$ is shown in Fig. 2-5b.

$$\begin{aligned} F_x(x) &= \int_{-\infty}^{x} p_x(x)\,dx \\ &= \int_{-\infty}^{x} \frac{1}{\sigma\sqrt{2\pi}} e^{-x^2/2\sigma^2}\,dx \\ &= \int_{-\infty}^{0} \frac{1}{\sigma\sqrt{2\pi}} e^{-x^2/2\sigma^2}\,dx + \int_{0}^{x} \frac{1}{\sigma\sqrt{2\pi}} e^{-x^2/2\sigma^2}\,dx \end{aligned}$$

The first integral represents the area of the probability density function over the interval $(-\infty, 0)$. This is obviously 0.5 since the total area over $(-\infty, \infty)$ is unity and the function is symmetrical about $x = 0$. Hence

$$F_x(x) = \frac{1}{2} + \int_{0}^{x} \frac{1}{\sigma\sqrt{2\pi}} e^{-x^2/2\sigma^2}\,dx \tag{2-30}$$

The integral on the right-hand side cannot be evaluated in a closed form. However, it is extensively tabulated in standard mathematical tables.[5]

Joint Distribution. A random experiment may have two outcomes. The sample points for such an experiment have two attributes. Consider the random experiment of throwing a dart on the target. The position of the hit is a random point and can be described by two numbers in a suitable coordinate system. Thus each sample point is described by a 2-tuple. We can associate two continuous random variables x and y over this sample space, and let the random variable x be the x coordinate and the variable y be the y coordinate of the hit point. Thus every point in the sample space can be described by a 2-tuple (x, y).

We now define a probability-distribution function (cumulative probability-distribution function) $F_{xy}(x, y)$ as follows:

$$\text{Probability}\,(\mathrm{x} \le x, \mathrm{y} \le y) = F_{xy}(x, y) \tag{2-32a}$$

and the joint probability-density function $p_{xy}(x, y)$ as

$$p_{xy}(x, y) = \frac{\partial^2}{\partial x\,\partial y} F_{xy}(x, y) \tag{2-32b}$$

From the limiting definition of the partial derivative, we can express Eq. 2-32b as

[5]The integral on the right-hand side of Eq. 2-30 can be expressed as an error function where we define

$$\operatorname{erf}(\alpha) = \left(\frac{1}{\sqrt{2\pi}}\right) \int_{-\infty}^{\alpha} e^{-x^2/2}\,dx \tag{2-31}$$

At present there exist in the literature, several definitions of the erf (x), which in essence are equivalent with minor differences.

$$p_{xy}(x, y) = \lim_{\substack{\Delta x \to 0 \\ \Delta y \to 0}} \frac{1}{\Delta x \Delta y} [F_{xy}(x + \Delta x, y + \Delta y)$$
$$- F_{xy}(x, y + \Delta y) - F_{xy}(x + \Delta x, y) + F_{xy}(x, y)]$$

The quantity inside the brackets is the probability of observing the outcome in the region described by

$$[(\mathrm{x} \leq x + \Delta x, \mathrm{y} \leq y + \Delta y) - (\mathrm{x} \leq x, \mathrm{y} \leq y + \Delta y)$$
$$- (\mathrm{x} \leq x + \Delta x, \mathrm{y} \leq y) - (\mathrm{x} \leq x, \mathrm{y} \leq y)]$$

This region is the shaded region shown in Fig. 2-6 and can be expressed as

$$(x \leq \mathrm{x} \leq x + \Delta x, y < \mathrm{y} \leq y + \Delta y)$$

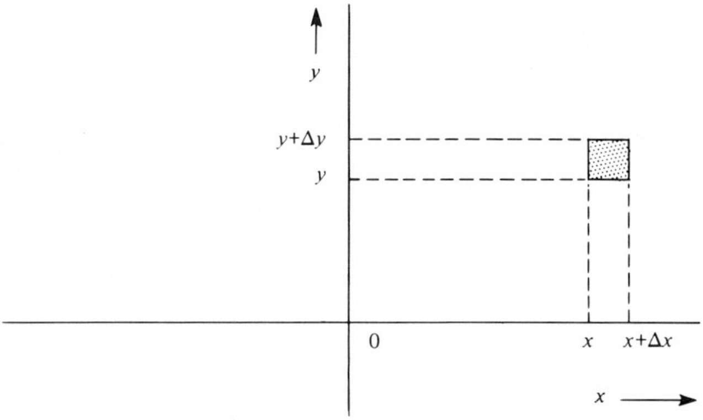

FIG. 2-6.

Thus the quantity in brackets is the probability of observing x and y jointly in the intervals $(x, x + \Delta x)$ and $y, y + \Delta y)$ respectively.

Therefore

$$\lim_{\substack{\Delta x \to 0 \\ \Delta y \to 0}} p_{xy}(x, y) \Delta x \Delta y = \text{probability} (x < \mathrm{x} \leq x + \Delta x, y < \mathrm{y} \leq y + \Delta y) \tag{2-33}$$

Hence the probability of observing variable x in the interval $(x, x + \Delta x)$ and y in the interval $(y, y + \Delta y)$ jointly is given by the volume under the joint probability-density function $p_{xy}(x, y)$ over the region bounded by $(x, x + \Delta x)$ and $(y, y + \Delta y)$ as shown in Fig. 2-7.

From Eq. 2-33 it follows that

$$\text{Probability} (x_1 < \mathrm{x} \leq x_2, y_1 < \mathrm{y} \leq y_2)$$
$$= \int_{x_1}^{x_2} \int_{y_1}^{y_2} p_{xy}(x, y)\, dx dy \tag{2-34}$$

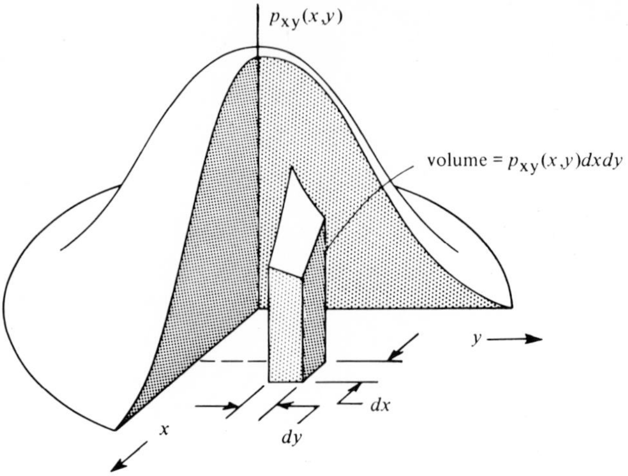

FIG. 2-7.

Thus the probability of jointly observing x in the interval (x_1, x_2) and y in the interval (y_1, y_2) is the volume under the probability-density function over the region bounded by (x_1, x_2) and (y_1, y_2).

The event of observing x in the interval $(-\infty, \infty)$ and observing y in the interval $(-\infty, \infty)$ is a certainty. Hence

$$\int_{-\infty}^{\infty} \int_{-\infty}^{\infty} p_{xy}(x, y)\, dxdy = 1 \tag{2-35}$$

Thus the total volume under the joint probability-density function must be unity.

When we are dealing with two random variables x and y, the individual probability densities $p_x(x)$ and $p_y(y)$ can be obtained from the joint density $p_{xy}(x, y)$. These densities are also called *marginal densities*. To obtain these densities, we note that $p_x(x)\Delta x$ is the probability of observing x in the interval $(x, x + \Delta x)$. The value of y is immaterial and may lie anywhere in the interval $(-\infty, \infty)$. Hence

$$\begin{aligned}\lim_{\Delta x \to 0} p_x(x)\Delta x &= \text{probability}\,(x < \mathrm{x} \leq x + \Delta x), (-\infty < \mathrm{y} \leq \infty) \\ &= \lim_{\Delta x \to 0} \int_x^{x+\Delta x} \int_{-\infty}^{\infty} p_{xy}(x, y)\, dxdy \\ &= \lim_{\Delta x \to 0} \int_x^{x+\Delta x} \left[\int_{-\infty}^{\infty} p_{xy}(x, y)\, dy\right] dx\end{aligned}$$

The integral inside the brackets is a function of x and can be assumed to be constant over the interval $(x, x + \Delta x)$ since $\Delta x \to 0$. Hence

$$\lim_{\Delta x \to 0} p_x(x)\Delta x = \left[\int_{-\infty}^{\infty} p_{xy}(x, y)\, dy\right]\left[\lim_{\Delta x \to 0} \int_x^{x+\Delta x} dx\right]$$

$$= \Delta x \int_{-\infty}^{\infty} p_{xy}(x, y)\, dy$$

Therefore

$$p_x(x) = \int_{-\infty}^{\infty} p_{xy}(x, y)\, dy \tag{2-36a}$$

Similarly,

$$p_y(y) = \int_{-\infty}^{\infty} p_{xy}(x, y)\, dx \tag{2-36b}$$

These results may be generalized for n random variables $x_1, x_2, \ldots, x_n$ as

$$\iint \cdots \int p_{x_1 x_2 \cdots x_n}(x_1, x_2, \ldots, x_n)\, dx_1\, dx_2 \cdots dx_n = 1 \tag{2-37a}$$

and

$$p_{x_1 x_2 \cdots x_{n-1}}(x_1, x_2, \ldots, x_{n-1}) = \int_{-\infty}^{\infty} p_{x_1 x_2 \cdots x_n}(x_1, x_2, \ldots, x_n)\, dx_n \tag{2-37b}$$

where $p_{x_1 x_2 \cdots x_n}(x_1, x_2, \ldots, x_n)$ is the joint probability density function of random variables $x_1, x_2, \ldots, x_n$.

The probability of observing variables $x_1, x_2, \ldots, x_n$ in the intervals $(\alpha_1, \beta_1), (\alpha_2, \beta_2), \ldots, (\alpha_n, \beta_n)$ respectively is given by

$$\int_{\alpha_1}^{\beta_1} \int_{\alpha_2}^{\beta_2} \cdots \int_{\alpha n}^{\beta n} p_{x_1 x_2 \cdots x_n}(x_1, x_2, \ldots x_n)\, dx_1\, dx_2 \cdots dx_n \tag{2-37c}$$

Conditional Densities. The concept of conditional probabilities can be easily extended for the case of continuous random variables. We define the conditional probability density $p_x(x \mid y = y_j)$ as the probability density of x given that y has a value y_j. This is equivalent to saying that $p_x(x \mid y = y_j)\, \Delta x$ is the probability of observing x in the range $(x, x + \Delta x)$, given that $y = y_j$. The probability density $p_x(x \mid y = y_j)$ is the intersection of a plane $y = y_j$ with the joint probability-density function $p_{xy}(x, y)$. This is shown in Fig. 2-8. Similarly, $p_y(y \mid x = x_i)$ is the intersection of a plane $x = x_i$ with the joint probability density function $p_{xy}(x, y)$.

Extension of the results derived for discrete case yields

$$p_x(x \mid y = y)\, p_y(y) = p_{xy}(x, y) \tag{2-38a}$$

and

$$p_y(y \mid x = x)\, p_x(x) = p_{xy}(x, y) \tag{2-38b}$$

These results can be obtained as follows. From the definition of conditional probability (Eq. 2-12) we have

$$\text{Probability } (x \le x \mid y < y \le y + \Delta y)$$

$$= \frac{\text{probability } (x \le x, y < y \le y + \Delta y)}{\text{probability } (y < y \le y + \Delta y)}$$

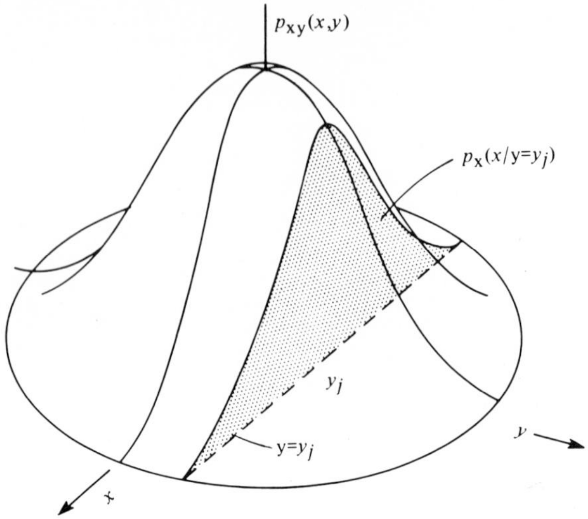

FIG. 2-8.

$$= \frac{\int_{-\infty}^{x} \int_{y}^{y+\Delta y} p_{xy}(x, y)\, dx dy}{\int_{y}^{y+\Delta y} p_y(y)\, dy}$$

In the limit as $\Delta y \to 0$, the numerator on the right-hand side becomes

$$\Delta y \left[\int_{-\infty}^{x} p_{xy}(x, y)\, dx \right]$$

and the denominator becomes $p_y(y)\,\Delta y$. Therefore

$$\lim_{\Delta y \to 0} \text{probability}\,(\mathrm{x} \leq x \mid y < \mathrm{y} \leq y + \Delta y) = \frac{\int_{-\infty}^{x} p_{xy}(x, y)\, dx}{p_y(y)}$$

As $\Delta y \to 0$, the left-hand term becomes probability $(\mathrm{x} \leq x \mid \mathrm{y} = y)$. This is a conditional cumulative distribution defined as

$$F_x(x \mid \mathrm{y} = y) = \text{probability}\,(\mathrm{x} \leq x \mid \mathrm{y} = y)$$

Hence

$$F_x(x \mid \mathrm{y} = y) = \frac{\int_{-\infty}^{x} p_{xy}(x, y)\, dx}{p_y(y)} \tag{2-39}$$

The derivative of $F_x(x \mid \mathrm{y} = y)$ with respect to x is the conditional density $p_x(x \mid \mathrm{y} = y)$

$$\frac{dF_x(x \mid \mathrm{y} = y)}{dx} = p_x(x \mid \mathrm{y} = y) \tag{2-40}$$

Differentiating both sides of Eq. 2-39 with respect to x and using the result in Eq. 2-40, we get

$$p_{\mathrm{x}}(x \mid \mathrm{y} = y) = \frac{p_{\mathrm{xy}}(x, y)}{p_{\mathrm{y}}(y)} \tag{2-41}$$

This is the desired result as shown in Eq. 2-38a.

Use of Eq. 2-36a in Eq. 2-41 yields

$$p_{\mathrm{x}}(x \mid \mathrm{y} = y) = \frac{p_{\mathrm{xy}}(x, y)}{\int_{-\infty}^{\infty} p_{\mathrm{xy}}(x, y)\, dx} \tag{2-42a}$$

Similarly,

$$p_{\mathrm{y}}(y \mid \mathrm{x} = x) = \frac{p_{\mathrm{xy}}(x, y)}{\int_{-\infty}^{\infty} p_{\mathrm{xy}}(x, y)\, dy} \tag{2-42b}$$

It follows from Eq. 2-38 that

$$p_{\mathrm{x}}(x \mid \mathrm{y} = y) = \frac{p_{\mathrm{y}}(y \mid \mathrm{x} = x)\, p_{\mathrm{x}}(x)}{p_{\mathrm{y}}(y)} \tag{2-43a}$$

and

$$p_{\mathrm{y}}(y \mid \mathrm{x} = x) = \frac{p_{\mathrm{x}}(x \mid \mathrm{y} = y)\, p_{\mathrm{y}}(y)}{p_{\mathrm{x}}(x)} \tag{2-43b}$$

This is Bayes' rule for continuous random variables.

In certain instances, one may have to deal with discrete and continuous random variables jointly. If x is a discrete random variable and y is a continuous random variable, then it is evident that the joint probability of $\mathrm{x} = x$ and $y < \mathrm{y} \le y + \Delta y$ is given by

$$\begin{aligned} &\text{Probability } (\mathrm{x} = x, y < \mathrm{y} \le y + \Delta y) \\ &\qquad = P_{\mathrm{x}}(x \mid y < \mathrm{y} \le y + \Delta y)\, p_{\mathrm{y}}(y)\, \Delta y \\ &\qquad = P_{\mathrm{x}}(x)\, p_{\mathrm{y}}(y \mid \mathrm{x} = x)\, \Delta y \end{aligned}$$

Note that

$$\lim_{\Delta y \to 0} P_{\mathrm{x}}(x \mid y < \mathrm{y} \le y + \Delta y) = P_{\mathrm{x}}(x \mid \mathrm{y} = y)$$

Hence

$$P_{\mathrm{x}}(x \mid \mathrm{y} = y)\, p_{\mathrm{y}}(y) = P_{\mathrm{x}}(x)\, p_{\mathrm{y}}(y \mid \mathrm{x} = x) \tag{2-43c}$$

This is the mixed form of Bayes' rule.

Note that the event of observing x in the interval $(-\infty, \infty)$ given that $\mathrm{y} = y$ is a certainty. Hence

$$\int_{-\infty}^{\infty} p_{\mathrm{x}}(x \mid \mathrm{y} = y)\, dx = 1 \tag{2-44a}$$

and similarly

$$\int_{-\infty}^{\infty} p_y(y \mid x = x)\, dy = 1 \tag{2-44b}$$

The continuous random variables x and y are said to be statistically independent if

$$p_x(x \mid y = y) = p_x(x) \tag{2-45a}$$

From Eq. 2-45a and Eq. 2-43b, it follows that

$$p_y(y \mid x = x) = p_y(y) \tag{2-45b}$$

This implies that

$$p_{xy}(x, y) = p_x(x)\, p_y(y) \tag{2-46a}$$

In general, n random variables $x_1, x_2, \ldots, x_n$ are independent if and only if

$$p_{x_1 x_2 \cdots x_n}(x_1, x_2, \ldots, x_n) = p_{x_1}(x_1)\, p_{x_2}(x_2) \cdots p_{x_n}(x_n) \tag{2-46b}$$

Example 2-7. Jointly gaussian density: given

$$p_{xy}(x, y) = \frac{1}{2\pi\sqrt{M}}\, e^{-(K_{22}x^2 + K_{11}y^2 - 2K_{12}xy)/2M}$$

where $M = K_{11}K_{22} - K_{12}^2$. Find

$$p_x(x), \quad p_y(y), \quad p_x(x \mid y = y) \quad \text{and} \quad p_y(y \mid x = x)$$

We have

$$\begin{aligned} p_x(x) &= \int_{-\infty}^{\infty} p_{xy}(x, y)\, dy \\ &= \frac{1}{2\pi\sqrt{M}} \int_{-\infty}^{\infty} e^{-(K_{22}x^2 + K_{11}y^2 - 2K_{12}xy)/2M}\, dy \\ &= \frac{1}{2\pi\sqrt{M}}\, e^{-(K_{22}x^2)/2M} \int_{-\infty}^{\infty} e^{-(K_{11}y^2 - 2K_{12}xy)/2M}\, dy \end{aligned}$$

From standard tables we obtain[6]

$$\int_{-\infty}^{\infty} e^{-py^2 + qy}\, dy = \sqrt{\frac{\pi}{p}}\, e^{q^2/4p}$$

Using this result in our integral, we obtain

$$p_x(x) = \frac{1}{\sqrt{2\pi K_{11}}}\, e^{-x^2/2K_{11}}$$

[6]This integral can be evaluated by rewritting the integral as

$$\int_{-\infty}^{\infty} e^{-py^2 + qy}\, dy = e^{\alpha^2} \int_{-\infty}^{\infty} e^{-(\sqrt{p}y + \alpha)^2}\, dy \quad \text{where} \quad \alpha = \frac{q}{2\sqrt{p}}$$

Now if we let $x/\sqrt{2} = \sqrt{p}y + \alpha$, the integrand reduces to standard Gaussian form.

Similarly we can obtain

$$p_y(y) = \frac{1}{\sqrt{2\pi K_{22}}} e^{-y^2/2K_{22}}$$

Also

$$p_x(x \mid y = y) = \frac{p_{xy}(x, y)}{p_y(y)} = \sqrt{\frac{K_{22}}{2\pi M}}\, e^{-K_{22}[x - (K_{12}/K_{22})\, y]^2/2M}$$

and

$$p_y(y \mid x = x) = \frac{p_{xy}(x, y)}{p_x(y)} = \sqrt{\frac{K_{11}}{2\pi M}}\, e^{-K_{11}[\, y - (K_{12}/K_{11}) x]^2/2M}$$

2-7. FUNCTIONS OF RANDOM VARIABLES

We are often interested in obtaining a probability density function after a random variable has gone through some transformation. If, for example, the probability density function of a random variable x is given to be $p_x(x)$ we are interested in finding the probability density function of a new variable y related to x as

$$y = f(x)$$

Let the desired probability density function of y by $p_y(y)$. The two probability densities are sketched in Fig. 2-9. If x changes by $\Delta x (\Delta x \to 0)$, and the corresponding change in y is Δy, then it is obvious that the probability of observing x between x and $x + \Delta x$ is the same as the probability of observing the corresponding y between y and $y + \Delta y$. But these probabilities are given by $p_x(x)\,\Delta x$ and $p_y(y)\,\Delta y$ respectively. Hence

$$\lim_{\Delta x \to 0} p_x(x)\,\Delta x = p_y(y)\,\Delta y \qquad \text{(2-47a)}$$

properly speaking the above equation should be expressed as

$$\lim_{\Delta x \to 0} p_x(x)\,|\,\Delta x\,| = p_y(y)\,|\,\Delta y\,| \qquad \text{(2-47b)}$$

This is because the probabilities are equal to the magnitudes of the areas under Δx and Δy respectively. Hence

$$p_y(y) = \frac{p_x(x)}{|\,dy/dx\,|} = \frac{p_x(x)}{|f'(x)|} \qquad \text{(2-48a)}$$

Observe that $p_y(y)$ is a function of y. Hence on the right-hand side in Eq. 2-48a, the variable x must be expressed in terms of y. Assuming that $y = f(x)$ has an inverse $x = f^{-1}(y)$, we have

$$p_y(y) = \left.\frac{p_x(x)}{|f'(x)|}\right|_{x = f^{-1}(y)} \qquad \text{(2-48b)}$$

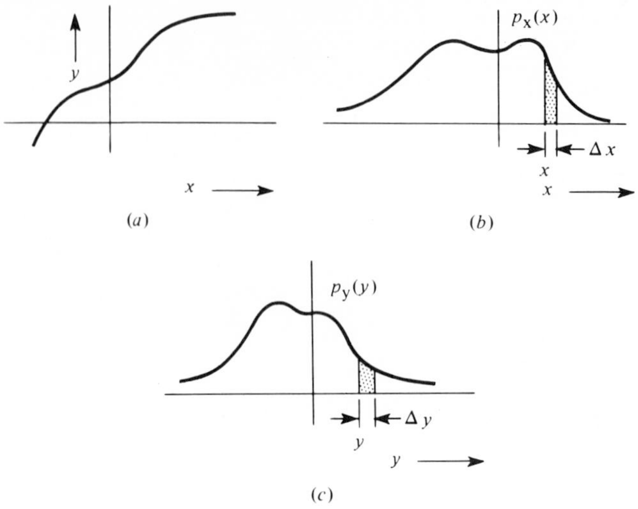

FIG. 2-9.

Example 2-8. Given

$$p_x(x) = \frac{1}{\sqrt{2\pi}} e^{-x^2/2}$$

Find $p_y(y)$ if $y = -3x + 2$. We have

$$\left|\frac{dy}{dx}\right| = 3$$

and

$$x = \frac{2 - y}{3}$$

Hence

$$p_y(y) = \frac{p_x(x)}{3}\bigg|_{x=(2-y)/3}$$

$$= \frac{1}{3\sqrt{2\pi}} e^{-(2-y)^2/18}$$

It has been implicitly assumed in Eqs. 2-48a and 2-48b that there is a one-to-one correspondence between x and y, that is there is only one value of x for a given y, and vice versa. This amounts to the assumption that both $f(x)$ and $f^{-1}(y)$ are single-valued functions (Fig. 2-9a). If this is not true, say for example, a given value of y corresponds to more than one value of x, then Eq. 2-48 must be modified. As an example, consider the relationship $y = f(x)$ as shown in Fig. 2-10. In this figure, for a given value of y there are two corresponding values of x. Thus the equation $y = f(x)$ has two roots, x_1 and x_2. Let us break up the function into two single valued functions as shown in Fig. 2-10.

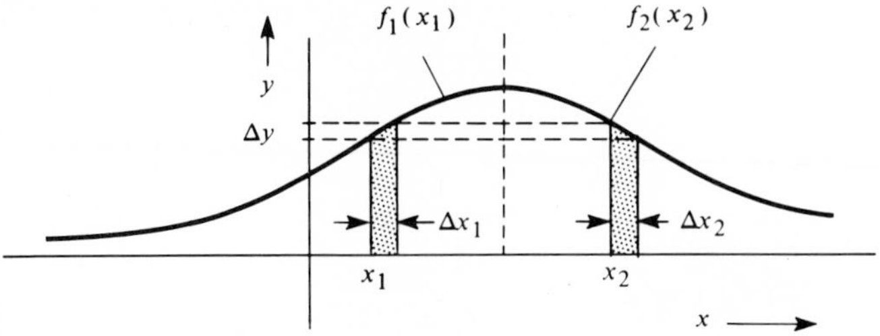

FIG. 2-10.

$$y = f_1(x_1) \qquad \text{and} \qquad y = f_2(x_2)$$

Note that there is one to one correspondence between x and y in each of these functions. Thus x_1 and x_2 are single valued functions of y. Let the inverse relationship be given by

$$x_1 = f_1^{-1}(y) \qquad \text{and} \qquad x_2 = f_2^{-1}(y)$$

From Fig. 2-10 it follows that y lies in the interval $(y, y + \Delta y)$ when x_1 lies in the interval $(x_1, x_1 + \Delta x_1)$ or when x_2 lies in the interval $(x_2, x_2 + \Delta x_2)$. The latter two events are mutually exclusive since x can either be x_1 or x_2 but not both. Hence we have

$$\lim_{\substack{\Delta y \to 0 \\ \Delta x_1 \to 0 \\ \Delta x_2 \to 0}} p_y(y)\,|\Delta y| = p_x(x_1)\,|\Delta x_1| + p_x(x_2)\,|\Delta x_2| \tag{2-49}$$

and

$$p_y(y) = \frac{p_x(x_1)}{|f_1'(x_1)|} + \frac{p_x(x_2)}{|f_2'(x_2)|} \tag{2-50a}$$

If there are n corresponding values of x for a given y, then extension of Eq. 2-50a yields

$$p_y(y) = \frac{p_x(x_1)}{|f_1'(x_1)|} + \frac{p_x(x_2)}{|f_2'(x_2)|} + \cdots + \frac{p_x(x_n)}{|f_n'(x_n)|} \tag{2-50b}$$

where $x_1, x_2, \ldots, x_n$ are the values of x when y $= y$.

Example 2-9. Find $p_y(y)$ if $y = f(x)$ is as shown in Fig. 2-11a and

$$p_x(x) = \frac{1}{\sigma\sqrt{2\pi}}\, e^{-x^2/2\sigma^2}$$

It is obvious from Fig. 2-11a that for each value of y, there corresponds two values of x. Hence there are two components of $p_y(y)$ corresponding to each value of x. We shall evaluate these two components separately:

$$p_y(y)_1 = \frac{p_x(x_1)}{\left|\dfrac{dy}{dx_1}\right|} \quad \text{for positive } x$$

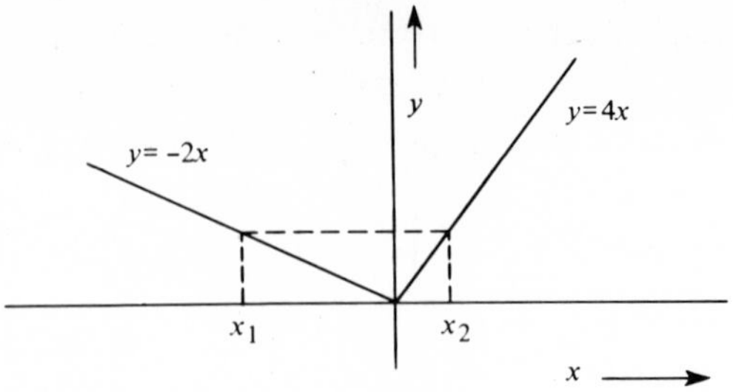

(a)

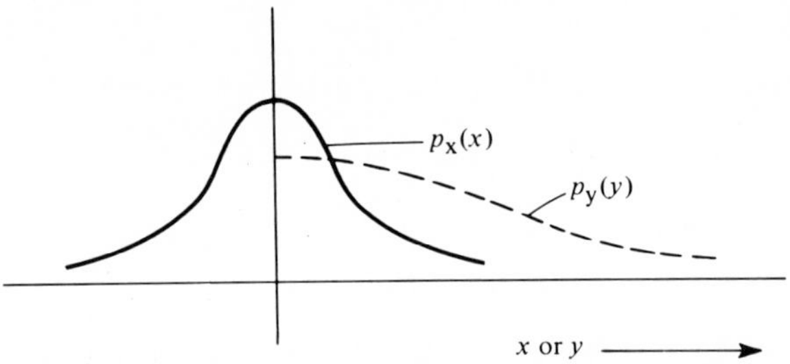

(b)

FIG. 2-11.

Note that $dy/dx = 4$ for this component. Hence

$$p_y(y)_1 = \frac{p_x(x)}{4}$$

$$= \frac{1}{4\sigma\sqrt{2\pi}} e^{-x^2/2\sigma^2}$$

On this branch $x = y/4$. Hence

$$p_y(y)_1 = \frac{1}{4\sigma\sqrt{2\pi}} e^{-y^2/32\sigma^2} u(y)$$

Similarly,

$$p_y(y)_2 = \frac{1}{2\sigma\sqrt{2\pi}} e^{-y^2/8\sigma^2} u(y)$$

Hence

$$p_y(y) = p_y(y)_1 + p_y(y)_2 = \frac{1}{4\sigma\sqrt{2\pi}} [e^{-y^2/32\sigma^2} + 2e^{-y^2/8\sigma^2}] u(y)$$

Similar relations may be derived for probability densities of more than one variable. Consider a joint probability-density function $p_{xy}(x, y)$. Let u and v be some other random variables related to x and y by

$$u = u(x, y) \qquad \text{and} \qquad v = v(x, y)$$

Assume that u and v are both single-valued functions of x and y, and vice versa. To obtain $p_{uv}(u, v)$ from $p_{xy}(x, y)$, we observe that

$$p_{uv}(u, v) \mid dudv \mid = p_{xy}(x, y) \mid dxdy \mid$$

Hence

$$p_{uv}(u, v) = \frac{p_{xy}(x, y)}{\left| \dfrac{dudv}{dxdy} \right|}$$

The relation between the area elements in two coordinate systems can be expressed in terms of the Jacobian as (Ref. 2-4)

$$dudv = J\left(\frac{u, v}{x, y}\right) dxdy \tag{2-51}$$

where J is the Jacobian of the transformation given by the determinant

$$J\left(\frac{u, v}{x, y}\right) = \begin{vmatrix} \dfrac{\partial u}{\partial x} & \dfrac{\partial u}{\partial y} \\ \dfrac{\partial v}{\partial x} & \dfrac{\partial v}{\partial y} \end{vmatrix} \tag{2-52}$$

Hence

$$p_{xy}(x, y) = \frac{p_{xy}(x, y)}{\left| J\left(\dfrac{u, v}{x, y}\right) \right|} \tag{2-53}$$

Note that for the Jacobian to exist the partial derivatives of u and v with respect to x and y must exist.

If x and y are multiple-valued functions, that is, if

$$(x_1, y_1), (x_2, y_2) \cdots (x_n, y_n)$$

are the solutions of equations

$$u = u(x, y) \qquad \text{and} \qquad v = v(x, y)$$

then

$$p_{uv}(u, v) = \frac{p_{xy}(x_1, y_1)}{\left| J\left(\dfrac{u, v}{x_1, y_1}\right) \right|} + \frac{p_{xy}(x_2, y_2)}{\left| J\left(\dfrac{u, v}{x_2, y_2}\right) \right|} + \cdots + \frac{p_{xy}(x_n, y_n)}{\left| J\left(\dfrac{u, v}{x_n, y_n}\right) \right|} \tag{2-54}$$

The above result can be extended to any number of variables. Suppose we have n random variables $x_1, x_2, \ldots, x_n$ with a joint probability-density function $p_{x_1 x_2 \cdots x_n}(x_1, x_2, \ldots, x_n)$. We now desire to find the joint probability-density function $p_{y_1 y_2 \cdots y_n}(y_1, y_2, \ldots, y_n)$ of n random variables $y_1, y_2, \ldots, y_n$ related to $x_1, x_2, \ldots, x_n$ by

$$y_i = y_i(x_1, x_2, \ldots, x_n) \qquad i = 1, 2, \ldots, n \tag{2-55a}$$

and

$$\mathrm{x}_j = \mathrm{x}_j(\mathrm{y}_1, \mathrm{y}_2, \ldots, \mathrm{y}_n) \qquad j = 1, 2, \ldots, n \tag{2-55b}$$

All these functions are assumed to be single-valued functions and with continuous partial derivatives everywhere. We have

$$p_{\mathrm{y}_1\mathrm{y}_2\cdots\mathrm{y}_n}(y_1, y_2, \ldots, y_n)\,|\,dy_1\,dy_2\cdots dy_n\,| = p_{\mathrm{x}_1\mathrm{x}_2\cdots\mathrm{x}_n}(x_1, x_2, \ldots, x_n)\,|\,dx_1\,dx_2\cdots dx_n\,| \tag{2-56}$$

Hence

$$p_{\mathrm{y}_1\mathrm{y}_2\cdots\mathrm{y}_n}(y_1, y_2, \ldots, y_n) = \frac{p_{\mathrm{x}_1\mathrm{x}_2\cdots\mathrm{x}_n}(x_1, x_2, \ldots, x_n)}{\left|\dfrac{dy_1dy_2\cdots dy_n}{dx_1dx_2\cdots dx_n}\right|} \tag{2-57}$$

The ratio of area elements is given by the Jacobian of transformation $J(y_1, y_2, \ldots, y_n/x_1, x_2, \ldots, x_n)$

$$dy_1dy_2\cdots dy_n = J\left(\frac{y_1, y_2, \ldots, y_n}{x_1, x_2, \ldots, x_n}\right) dx_1dx_2\cdots dx_n \tag{2-58}$$

where

$$J\left(\frac{y_1, y_2, \ldots, y_n}{x_1, x_2, \ldots, x_n}\right) = \begin{vmatrix} \dfrac{\partial y_1}{\partial x_1} & \dfrac{\partial y_1}{\partial x_2} & \cdots & \dfrac{\partial y_1}{\partial x_n} \\ \dfrac{\partial y_2}{\partial x_1} & \dfrac{\partial y_2}{\partial x_2} & \cdots & \dfrac{\partial y_2}{\partial x_n} \\ \cdots & \cdots & \cdots & \cdots \\ \dfrac{\partial y_n}{\partial x_1} & \dfrac{\partial y_n}{\partial x_2} & \cdots & \dfrac{\partial y_n}{\partial x_n} \end{vmatrix} \tag{2-59}$$

Hence

$$p_{\mathrm{y}_1\mathrm{y}_2\cdots\mathrm{y}_n}(y_1, y_2, \ldots, y_n) = \frac{p_{\mathrm{x}_1\mathrm{x}_2\cdots\mathrm{x}_n}(x_1, x_2, \ldots, x_n)}{\left|J\left(\dfrac{y_1, y_2, \ldots, y_n}{x_1, x_2, \ldots, x_n}\right)\right|} \tag{2-60}$$

It can be shown that

$$J\left(\frac{y_1, y_2, \ldots, y_n}{x_1, x_2, \ldots, x_n}\right) = \frac{1}{J\left(\dfrac{x_1, x_2, \ldots, x_n}{y_1, y_2, \ldots, y_n}\right)} \tag{2-61}$$

If $y_1, y_2, \ldots, y_n$ are multiple-valued functions of $x_1, x_2, \ldots, x_n$, then an equation similar to 2-54 should be used.

Example 2-10. To illustrate the example of transformation of a second-order probability-density function, consider the case of throwing a dart. Assume that both the variables x and y that describe coordinates of a point where the dart hits are independent and have normal (Gaussian) probability-density functions.

$$p_x(x) = \frac{1}{\sigma\sqrt{2\pi}} e^{-x^2/2\sigma^2}$$

$$p_y(y) = \frac{1}{\sigma\sqrt{2\pi}} e^{-y^2/2\sigma^2}$$

We shall now find the probability-density function $p_{r\theta}(r, \theta)$ where r is the distance of the hit from the origin and θ is the angle of the hit with respect to the x axis.

$$r = \sqrt{x^2 + y^2} \quad \text{and} \quad \theta = \tan^{-1}\left(\frac{y}{x}\right)$$

Hence

$$J\left(\frac{r, \theta}{x, y}\right) = \begin{vmatrix} \dfrac{\partial r}{\partial x} & \dfrac{\partial r}{\partial y} \\[2ex] \dfrac{\partial \theta}{\partial x} & \dfrac{\partial \theta}{\partial y} \end{vmatrix}$$

$$= \begin{vmatrix} \dfrac{x}{\sqrt{x^2 + y^2}} & \dfrac{y}{\sqrt{x^2 + y^2}} \\[2ex] \dfrac{-y}{x^2 + y^2} & \dfrac{x}{x^2 + y^2} \end{vmatrix}$$

$$= \frac{1}{\sqrt{x^2 + y^2}}$$

$$= \frac{1}{r}$$

Hence

$$p_{r\theta}(r, \theta) = \frac{p_{xy}(x, y)}{\left|J\left(\frac{r, \theta}{x, y}\right)\right|}$$

$$= \frac{r}{2\pi\sigma^2} e^{-(x^2 + y^2)/2\sigma^2}$$

$$= \frac{r}{2\pi\sigma^2} e^{-r^2/2\sigma^2} \qquad \text{(2-62)}$$

The variable Θ does not appear in Eq. 2-62. Hence the random variables r and Θ must be independent and $p_\Theta(\theta)$ must be a constant. Since Θ varies over 0 to 2π, it is evident that $p_\Theta(\theta)$ is a constant $1/2\pi$ in order to have

$$\int_0^{2\pi} p_\Theta(\theta)\, d\theta = 1$$

Thus

$$p_{r\Theta}(r, \theta) = \left(\frac{1}{2\pi}\right)\left(\frac{r}{\sigma^2} e^{-r^2/2\sigma^2}\right)$$

$$= p_\Theta(\theta)\, p_r(r)$$

where

$$p_\Theta(\theta) = \begin{cases} \dfrac{1}{2\pi} & 0 < \Theta < 2\pi \\ 0 & \text{otherwise} \end{cases}$$

and

$$p_r(r) = \frac{r}{\sigma^2} e^{-r^2/2\sigma^2} \tag{2-63}$$

This is known as the Rayleigh density function (Fig. 2-12).

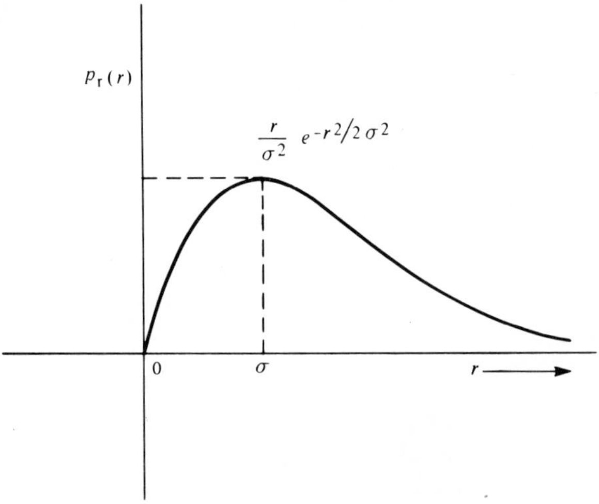

FIG. 2-12.

2-8. STATISTICAL AVERAGES (MEANS)

The concept of averages assumes an extremely important position in random processes. As mentioned earlier, the random processes are characterized by statistical regularity. By use of the term *statistical regularity* we indicate that the process cannot be predicted in specifics, but can be predicted on an average basis. For example, in a coin-tossing experiment it is not possible to predict the outcome of any particular toss. But on the average we can be confident that half the tosses will be heads and the other half tails, provided we average over a sufficiently large number of tosses. Indeed, the empirical definition of probability given in Eq. 2-1 itself represents a form of average.

We shall now define the mean (or average) of a random variable. The definition must be in conformity with our empirical notion of the mean. Consider a random variable x which can assume n values

$x_1, x_2, \ldots, x_n$. Let the experiment (represented by x) be repeated N times ($N \to \infty$) and let $m_1, m_2, \ldots, m_n$ be the number of trials favorable to outcomes $x_1, x_2, \ldots, x_n$ respectively. Then the mean value of x, according to traditional concept is $\overline{\mathrm{x}}$ given by

$$\overline{\mathrm{x}} = \frac{1}{N}(m_1x_1 + m_2x_2 + \cdots + m_nx_n)$$

$$= \frac{m_1}{N}x_1 + \frac{m_2}{N}x_2 + \cdots + \frac{m_n}{N}x_n$$

In the limit as $N \to \infty$, the ratio m_i/N tends to $P_{\mathrm{x}}(x_i)$ according to the relative-frequency definition of the probability. Hence

$$\overline{\mathrm{x}} = \sum_{i=1}^{n} x_i P_{\mathrm{x}}(x_i) \tag{2-64}$$

Thus Eq. 2-64 is a meaningful definition of the mean of a random variable. Note that we define the mean of a random variable by Eq. 2-64. The equation can be derived if we define the probability by Eq. 2-1 (relative-frequency definition). But since we are using an axiomatic approach to probability, we cannot derive the expression for the mean but must define it. As seen above, Eq. 2-64 represents a meaningful and practical definition of the mean and hence we accept it as a definition.

The mean value is also called the *expected value* of the random variable x and is denoted by $E[\mathrm{x}]$. Thus

$$\overline{\mathrm{x}} = E[\mathrm{x}] = \sum_{i} x_i P_{\mathrm{x}}(x_i) \tag{2-65}$$

We shall use both these notations depending upon the circumstances. It should henceforth be remembered that a straight horizontal bar on top of any quantity implies the statistical average (the expected value).

If the random variable is continuous, we must use considerations similar to those used in arriving at Eq. 2-65. This leads us to the definition

$$\overline{\mathrm{x}} = E[\mathrm{x}] = \int_{-\infty}^{\infty} x p_{\mathrm{x}}(x)\, dx \tag{2-66}$$

The reader can easily show that this definition conforms to our traditional notion of the mean when relative-frequency interpretation of the probability is used. This can be done by approximately the continuous variable x by a discrete one by quantizing it in steps of Δx and in the limit letting $\Delta x \to 0$.

Definition 2-66 is more general and includes definition 2-65. This is because the discrete random variable can be considered as a continuous random variable with impulsive distribution. In such case, Eq. 2-66 reduces to Eq. 2-65.

Example 2-11. A general Gaussian density is given by

$$p_{\mathrm{x}}(x) = \frac{1}{\sigma\sqrt{2\pi}} e^{-(x-m)^2/2\sigma^2}$$

This distribution is the same as shown in Fig. 2-5a except that it is shifted along the positive x axis by amount m as shown in Fig. 2-13. For this distribution

$$\bar{\mathrm{x}} = \frac{1}{\sigma\sqrt{2\pi}} \int_{-\infty}^{\infty} x e^{-(x-m)^2/2\sigma^2}\, dx$$

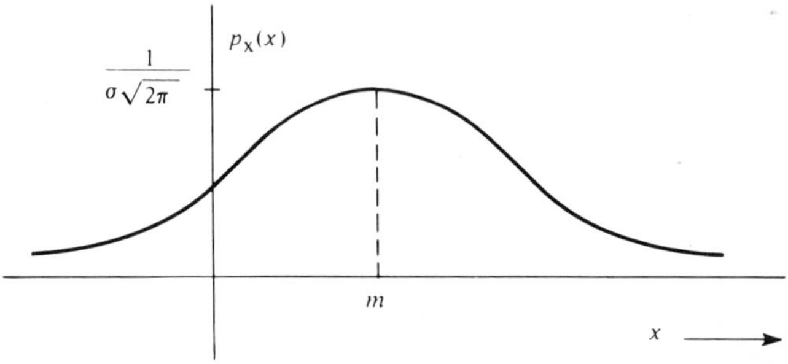

FIG. 2-13.

Change of variable $x = y + m$ yields

$$\bar{\mathrm{x}} = \frac{1}{\sigma\sqrt{2\pi}} \int_{-\infty}^{\infty} (y+m) e^{-y^2/2\sigma^2}\, dy$$

$$= \frac{1}{\sigma\sqrt{2\pi}} \left[\int_{-\infty}^{\infty} y e^{-y^2/2\sigma^2}\, dy + m \int_{-\infty}^{\infty} e^{-y^2/2\sigma^2}\, dy\right]$$

The first integral inside the bracket is zero because the integrand is an odd function of y. The second integral is found from standard tables to be $\sigma\sqrt{2\pi}$. Hence

$$\bar{\mathrm{x}} = m \tag{2-67}$$

Mean of a Function of a Random Variable. Often it is desired to find the mean value of a certain function of a random variable rather than that of the random variable itself. As a simple example of this, we give here a case of a noise signal whose amplitude is represented by a random variable x. In practice we are more interested in the mean-square value of the signal rather than the mean value of the signal. The mean-square value in this case is obviously the mean of the square of x$(=\overline{\mathrm{x}^2})$. In general, we wish to obtain the expression of the mean value of a random variable y which is a function of the variable x.

$$\mathrm{y} = g(\mathrm{x}) \tag{2-68}$$

By definition,

$$\overline{\mathrm{y}} = E(\mathrm{y}) = \int_{-\infty}^{\infty} y p_{\mathrm{y}}(y)\, dy \tag{2-69}$$

We shall now show that

$$\overline{\mathrm{y}} = \int_{-\infty}^{\infty} y p_{\mathrm{y}}(y)\, dy = \int_{-\infty}^{\infty} g(x) p_{\mathrm{x}}(x)\, dx \tag{2-70}$$

Figure 2-14 shows y as a function of x. In general y is a multiple-value function of x. Figure 2-14 shows three roots x_1, x_2, and x_3 for a given value of y. Reasoning as in section 2-7, we have

$$p_{\mathrm{y}}(y)\, dy = p_{\mathrm{x}}(x_1)\, dx_1 + p_{\mathrm{x}}(x_2)\, dx_2 + p_{\mathrm{x}}(x_3)\, dx_3$$

and

$$y p_{\mathrm{y}}(y)\, dy = g(x_1) p_{\mathrm{x}}(x_1)\, dx_1 + g(x_2) p_{\mathrm{x}}(x_2)\, dx_2 + g(x_3) p_{\mathrm{x}}(x_3)\, dx_3 \tag{2-71}$$

where all differentials are assumed positive as shown in Fig. 2-14. We now integrate both sides of Eq. 2-71. To each differential element there are three nonoverlapping differential elements dx_1, dx_2, and dx_3. As dy covers the y axis, the interval $(-\infty, \infty)$, the corresponding dx's cover disjoint regions along x axis. They scan the entire x axis $(-\infty, \infty)$ among them. Hence

$$\int_{-\infty}^{\infty} y p_{\mathrm{y}}(y)\, dy = \int_{-\infty}^{\infty} g(x) p_{\mathrm{x}}(x)\, dx$$

For a discrete random variable x, Eq. 2-70 reduces to

$$\overline{\mathrm{y}} = \overline{g(\mathrm{x})} = \sum_i g(x_i)\, P_{\mathrm{x}}(x_i) \tag{2-72}$$

Example 2-12. We shall find the mean-square value of the gaussian distribution in Example 2-11:

$$\mathrm{y} = \mathrm{x}^2$$

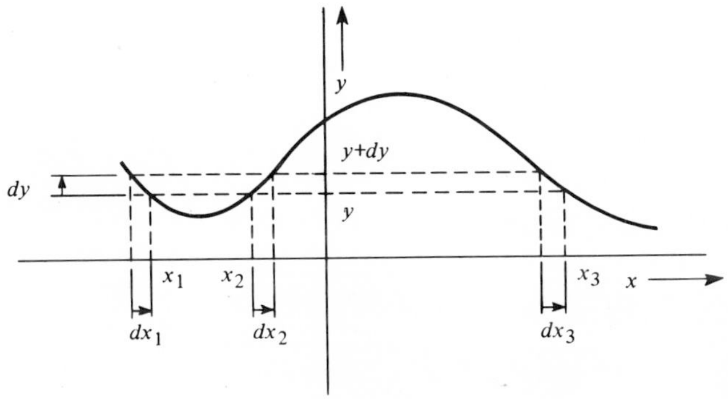

FIG. 2-14.

$$\overline{y} = \frac{1}{\sigma\sqrt{2\pi}} \int_{-\infty}^{\infty} x^2 e^{-(x-m)^2/2\sigma^2}\,dx$$

$$= \frac{1}{\sigma\sqrt{2\pi}} \int_{-\infty}^{\infty} (u+m)^2 e^{-u^2/2\sigma^2}\,du$$

The integral can be separated into three integrals and each component can be obtained from standard tables. The final result is

$$\overline{y} = \overline{x^2} = \sigma^2 + m^2 \tag{2-73}$$

Multiple Variables. If a random variable z is a function of two random variables x and y,

$$z = g(x, y) \tag{2-74}$$

Then

$$\overline{z} = E[z] = \int_{-\infty}^{\infty} z p_z(z)\,dz$$

One can determine $p_z(z)$ from Eq. 2-74 and the knowledge of the joint density $p_{xy}(x, y)$. However one can determine $E[z]$ directly from the joint density $p_{xy}(x, y)$ by using a fundamental theorem called the theorem of expectation which states that

$$\overline{z} = E[z] = \int_{-\infty}^{\infty}\int_{-\infty}^{\infty} g(x, y)\, p_{xy}(x, y)\,dxdy \tag{2-75a}$$

The proof of this relationship is similar to that of Eq. 2-70. If the variable z lies in the interval $(z, z + \Delta z)$, then variable x and y lies in the region bounded by $(x, x + \Delta x)$ and $(y, y + \Delta y)$. The area of this region is obviously $\Delta x\,\Delta y$. It also follows that

$$p_z(z)\,\Delta z = p_{xy}(x, y)\,\Delta x\,\Delta y$$

Equation 2-75a now follows directly from this equation.

For discrete variables Eq. 2-75a becomes

$$\overline{z} = \sum_i \sum_j g(x_i, y_j)\, P_{xy}(x_i, y_j) \tag{2-75b}$$

We can easily extend Eq. 2-75a to the case of a function of n random variables:

$$z = g(x_1, x_2, \ldots, x_n)$$

and

$$\overline{z} = E[z]$$

$$= \int_{-\infty}^{\infty}\int_{-\infty}^{\infty} \cdots \int_{-\infty}^{\infty} g(x_1, x_2, \ldots, x_n)\, p_{x_1x_2\cdots x_n}(x_1, x_2, \ldots, x_n)\,dx_1dx_2\cdots dx_n \tag{2-76}$$

If some of the random variables are discrete, Eq. 2-76 is still valid since the discrete distribution is considered as a limiting case of continuous distribution by use of impulse functions.

Example 2-13. Mean of the Sum. We can show that if $f_1(x,y)$, $f_2(x,y)$, $\ldots, f_n(x,y)$ are functions of x and y, then

$$\overline{f_1(x,y) + f_2(x,y) + \cdots + f_n(x,y)} = \overline{f_1(x,y)} + \overline{f_2(x,y)} + \cdots + \overline{f_n(x,y)} \tag{2-77a}$$

The proof is trivial and follows directly from the definition of the mean. Thus the mean (expectation) of the sum is equal to the sum of the means. Some simple examples of this are

$$\overline{x+y} = \overline{x} + \overline{y} \tag{2-77b}$$

$$\overline{x^2 + y^2} = \overline{x^2} + \overline{y^2} \tag{2-77c}$$

The results in Eq. 2-77a can be extended to functions of any number of random variables.

Example 2-14. Mean of the Product of Two Independent Variables. If

$$z = xy$$

Then

$$\overline{z} = \int_{-\infty}^{\infty} \int_{-\infty}^{\infty} xy\,p_{xy}(x,y)\,dxdy$$

If x and y are independent, then (by Eq. 2-46)

$$p_{xy}(x,y) = p_x(x)\,p_y(y)$$

and

$$\overline{z} = \int_{-\infty}^{\infty} x p_x(x)\,dx \int_{-\infty}^{\infty} y p_y(y)\,dy$$

$$= \overline{x}\,\overline{y}$$

Thus if x and y are independent random variables,

$$\overline{xy} = \overline{x}\,\overline{y} \tag{2-78a}$$

For an independent random variable the mean of the product is equal to the product of the individual means. This result can be extended to any number of variables.

Actually the result 2-78a is a special case of a more general result which states that for if x and y are independent, then for

$$z = g_1(x)g_2(y)$$

$$\overline{z} = \overline{g_1(x)}\,\overline{g_2(y)}$$

In other words,

$$\overline{g_1(x)g_2(y)} = \overline{g_1(x)}\,\overline{g_2(y)} \tag{2-78b}$$

This result can be easily proved along lines similar to that used for Eq. 2-78a. It can now be seen that Eq. 2-78a is a special case of Eq. 2-78b.

The Mean Square of the Sum. The mean square value of $z = x + y$ is given by

$$\overline{z^2} = \overline{(x + y)^2} = \int_{-\infty}^{\infty}\int_{-\infty}^{\infty} (x + y)^2 p_{xy}(x, y)\, dxdy$$

$$= \int_{-\infty}^{\infty}\int_{-\infty}^{\infty} x^2 p_{xy}(x, y)\, dxdy + \int_{-\infty}^{\infty}\int_{-\infty}^{\infty} y^2 p_{xy}(x, y)\, dxdy + 2\int_{-\infty}^{\infty}\int_{-\infty}^{\infty} xy p_{xy}(x, y)\, dxdy$$

If the random variables x and y are independent, then

$$\int_{-\infty}^{\infty}\int_{-\infty}^{\infty} x^2 p_{xy}(x, y)\, dxdy = \int_{-\infty}^{\infty} x^2 p_x(x)\, dx \int_{-\infty}^{\infty} p_y(y)\, dy = \int_{-\infty}^{\infty} x^2 p_x(x)\, dx = \overline{x^2}$$

Similarly

$$\int_{-\infty}^{\infty}\int_{-\infty}^{\infty} y^2 p_{xy}(x, y)\, dxdy = \overline{y^2}$$

and from 2-78a

$$\int_{-\infty}^{\infty}\int_{-\infty}^{\infty} xy p_{xy}(x, y)\, dxdy = \overline{xy} = \bar{x}\bar{y}$$

Hence for independent random variables x and y

$$\overline{(x + y)^2} = \overline{x^2} + \overline{y^2} + 2\bar{x}\bar{y} \tag{2-79a}$$

If either $\bar{x}$ or $\bar{y}$ or both are zero, then

$$\overline{(x + y)^2} = \overline{x^2} + \overline{y^2} \tag{2-79b}$$

Conditional Mean. The conditional mean (or conditional expected value) of a random variable x given that another random variable $y = y$ is denoted by $E[x \mid y = y]$ and is defined as

$$E[x \mid y = y] = \int_{-\infty}^{\infty} x p_x(x \mid y = y)\, dx \tag{2-80}$$

This follows from the basic definition of the mean.

2-9. MOMENTS

The nth moment of a random variable x is defined as the expected value of the nth power of x. Thus the nth moment is defined as

$$E[\mathrm{x}^n] = \int_{-\infty}^{\infty} x^n p_{\mathrm{x}}(x)\,dx \tag{2-81a}$$

The nth central moment of the variable x is its moment about its mean value m and is given by

$$E[(\mathrm{x} - m)^n] = \int_{-\infty}^{\infty} (x - m)^n p_{\mathrm{x}}(x)\,dx \tag{2-81b}$$

The second central moment about the mean is given the special name *variance* and denoted σ_{x}^2. Thus

$$\begin{aligned}\sigma_{\mathrm{x}}^2 &= E[(\mathrm{x} - m)^2] \\ &= E[\mathrm{x}^2 - 2m\mathrm{x} + m^2]\end{aligned}$$

From Eq. 2-77 and its generalization, it follows that

$$\sigma_{\mathrm{x}}^2 = E[\mathrm{x}^2] - E[2m\mathrm{x}] + E[m^2]$$

Note that m being the mean of x ($m = \overline{\mathrm{x}}$) is a constant. Hence

$$\begin{aligned}\sigma_{\mathrm{x}}^2 &= E[\mathrm{x}^2] - 2mE[\mathrm{x}] + m^2 \\ &= E[\mathrm{x}^2] - 2m^2 + m^2 \\ &= E[\mathrm{x}^2] - m^2 \\ &= \overline{\mathrm{x}^2} - m^2\end{aligned} \tag{2-82a}$$

Thus the variance of a random variable is equal to its mean of its square minus the square of its mean.

Example 2-15. Find the variance σ_{x}^2 for the gaussian density in Example 2-11.

$$p_{\mathrm{x}}(x) = \frac{1}{\sigma\sqrt{2\pi}} e^{-(x-m)^2/2\sigma^2} \tag{2-82b}$$

In Ex. 2-11, we found that

$$\overline{\mathrm{x}} = m$$

and in Ex. 2-12, it was found that

$$\overline{\mathrm{x}^2} = E[\mathrm{x}^2] = \sigma^2 + m^2$$

Hence from Eq. 2-82, we have

$$\sigma_{\mathrm{x}}^2 = [\sigma^2 + m^2] - m^2 = \sigma^2$$

Thus the variance of a gaussian distribution in Eq. 2-83 is σ^2.

Variance of Sum of Independent Random Variables. We shall now show that the variance of the sum of independent random variables is equal to the sum of the variances of the variables. Thus if x and y are independent random variables and

$$z = x + y$$

then

$$\sigma_z^2 = \sigma_x^2 + \sigma_y^2 \tag{2-83a}$$

This can be shown as follows

$$\begin{aligned}\sigma_z^2 &= \overline{[x + y - (\bar{x} + \bar{y})]^2} \\ &= \overline{[(x - \bar{x}) + (y - \bar{y})]^2} \\ &= \overline{(x - \bar{x})^2} + \overline{(y - \bar{y})^2} + 2\overline{(x - \bar{x})(y - \bar{y})} \\ &= \sigma_x^2 + \sigma_y^2 + 2\overline{(x - \bar{x})(y - \bar{y})}\end{aligned} \tag{2-83b}$$

Since x and y are independent random variables, $(x - \bar{x})$ and $(y - \bar{y})$ are also independent random variables and hence from Eq. 2-78a, we have

$$\overline{(x - \bar{x})(y - \bar{y})} = \overline{(x - \bar{x})}\,\overline{(y - \bar{y})}$$

But the means of $(x - \bar{x})$ and $(y - \bar{y})$ are zero. Hence

$$\sigma_z^2 = \sigma_x^2 + \sigma_y^2$$

This result can be extended to any number of variables.

2-10. THE TRANSFORM OF PROBABILITY-DENSITY FUNCTIONS

From Eq. 2-70, the mean of $g(x)$, a function of random variable x is given by

$$\overline{g(x)} = \int_{-\infty}^{\infty} g(x)\,p_x(x)\,dx$$

An interesting situation occurs if we let $g(x) = e^{-j\omega x}$. It is obvious that the mean value of $e^{-j\omega x}$ is given by

$$\overline{\exp(-j\omega x)} = \int_{-\infty}^{\infty} p_x(x)\,e^{-j\omega x}\,dx$$

Note that the mean value of $e^{-j\omega x}$ defines the direct Fourier transform of the probability-density function $p_x(x)$. By our convention $\overline{\exp(-j\omega x)}$ will be denoted by $P_x(\omega)$, the Fourier transform $p_x(x)$. Thus

$$\overline{\exp(-j\omega x)} = P_x(\omega) = \int_{-\infty}^{\infty} p_x(x)\,e^{-j\omega x}\,dx \tag{2-84}$$

In the literature similar transforms are defined; the more common ones being the *characteristic function* and the *moment-generating function*. The characteristic function is defined as the mean value of $e^{j\omega x}$ and is equivalent to 2π times the inverse Fourier transform of the probability-density function $p_x(x)$. The moment-generating function is defined as the mean value of e^{sx} and is equal to $2\pi j$ times the inverse (bilateral) Laplace transform of the probability-density function $p_x(x)$. We shall, however, for convenience, use the direct Fourier transform of $p_x(x)$ as given by $\overline{\exp(-j\omega x)}$. This is because we are already familiar with the Fourier transform and have transform tables available. These transforms prove very helpful in obtaining the probability-density function of a sum of a number of independent random variables. It will be shown in the next section that the probability-density function of a random variable z obtained by adding two independent random variables x and y ($z = x + y$) is given by the convolution of the probability-density functions of x and y. The significance of the Fourier transform (or the characteristic function or moment-generating function) now at once becomes obvious. Note that $p_x(x)$ is the inverse Fourier transform of $P_x(\omega)$. Hence

$$p_x(x) = \frac{1}{2\pi}\int_{-\infty}^{\infty} P_x(\omega)\, e^{j\omega x}\, d\omega \tag{2-85}$$

Probability-Density Function of a Sum of Two Independent Random Variables. Let x and y be two statistically independent random variables. Let z be a new random variable obtained by adding x and y:

$$z = x + y$$

Let $p_x(x)$, $p_y(y)$, and $p_z(z)$ be the probability-density functions of x, y, and z respectively and let $P_x(\omega)$, $P_y(\omega)$, and $P_z(\omega)$ be the Fourier transforms of $p_x(x)$, $p_y(y)$, and $p_z(z)$ respectively. By definition,

$$P_z(\omega) = \overline{\exp(-j\omega z)} \tag{2-86}$$

Since z is a function of x and y, we use Eq. 2-75 to express $\overline{\exp(-j\omega z)}$

$$P_z(\omega) = \overline{\exp(-j\omega z)} = \int_{-\infty}^{\infty}\int_{-\infty}^{\infty} e^{-j\omega(x+y)} p_{xy}(x, y)\, dx dy \tag{2-87}$$

where $p_{xy}(x, y)$ is the joint probability-density function of x and y. Since x and y are assumed to be independent,

$$p_{xy}(x, y) = p_x(x)\, p_y(y)$$

Substitution of this result in Eq. 2-87 yields

$$P_z(\omega) = \int_{-\infty}^{\infty}\int_{-\infty}^{\infty} e^{-j\omega(x+y)} p_x(x)\, p_y(y)\, dx dy$$

$$= \int_{-\infty}^{\infty} p_x(x)\, e^{-j\omega x}\, dx \int_{-\infty}^{\infty} p_y(y)\, e^{-j\omega y}\, dy$$

$$= P_x(\omega)\, P_y(\omega) \tag{2-88}$$

We therefore reach a very important conclusion.

The Fourier transform of a probability-density function of a sum of two independent random variables is equal to the product of the Fourier transforms of the individual probability-density functions of the two variables. Use of the frequency-convolution theorem now yields

$$\begin{aligned} p_z(z) &= p_x(x) * p_y(y) \\ &= p_y(y) * p_x(x) \end{aligned} \tag{2-89}$$

This is a very significant result. It states that the probability-density function of a sum of two independent random variables is equal to the convolution of the probability-density functions of the two random variables. This result can be extended to any number of independent random variables. Thus if $x_1, x_2, \ldots, x_n$ are n independent random variables with probability-density functions $p_{x_1}(x_1), p_{x_2}(x_2), \ldots, p_{x_n}(x_n)$ respectively, and if

$$z = x_1 + x_2 + \cdots + x_n$$

then

$$p_z(z) = p_{x_1}(x_1) * p_{x_2}(x_2) * \cdots * p_{x_n}(x_n) \tag{2-90}$$

The results derived in this section also apply to discrete random variables which can be considered as a limiting case of continuous random variables with impulse distributions. We can write Eq. 2-89 explicitly as

$$p_z(z) = \int_{-\infty}^{\infty} p_x(\lambda) p_y(z - \lambda)\, d\lambda \tag{2-91}$$

Example 2-16. The two independent random variables x and y have probability densities given by

$$p_x(x) = e^{-x}\, u(x)$$
$$p_y(y) = 2e^{-2y}\, u(y)$$

We shall determine the probability-density function of a new random variable z given by

$$z = x + y$$

Since x and y are independent,

$$p_z(z) = p_x(x) * p_y(y)$$

If $P_x(\omega)$, $P_y(\omega)$, and $P_z(\omega)$ are the Fourier transforms of $p_x(x)$, $p_y(y)$, and $p_z(z)$ respectively, then

$$P_z(\omega) = P_x(\omega)\, P_y(\omega)$$

From the table of transforms (Table 1-1), we have

$$P_x(\omega) = \frac{1}{1 + j\omega}$$

$$P_y(\omega) = \frac{1}{2 + j\omega}$$

Hence

$$P_z(\omega) = \frac{2}{(1 + j\omega)(2 + j\omega)}$$
$$= \frac{2}{1 + j\omega} - \frac{2}{2 + j\omega}$$

Hence

$$p_z(z) = \mathfrak{F}^{-1}[P_z(\omega)]$$
$$= 2[e^{-z} - e^{-2z}]\,u(z)$$

The three distributions are shown in Fig. 2-15. This result can also be obtained directly by convolving $p_x(x)$ and $p_y(y)$:

$$p_z(z) = p_x(x) * p_y(y)$$
$$= 2\int_0^z e^{-\lambda}\, e^{-2(z-\lambda)}\, d\lambda$$
$$= 2e^{-2z}\int_0^z e^{\lambda}\, d\lambda$$
$$= 2e^{-2z}(e^z - 1)\,u(z)$$
$$= 2[e^{-z} - e^{-2z}]\,u(z)$$

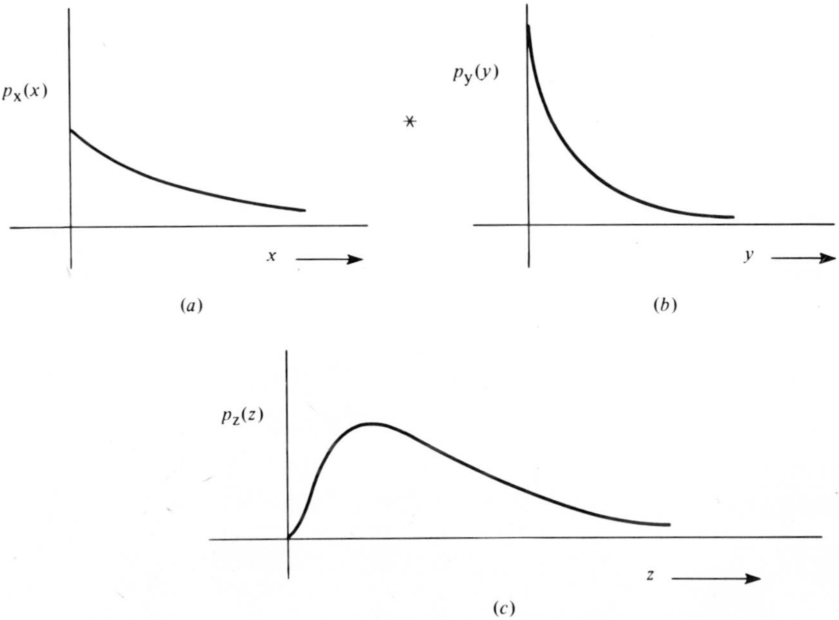

FIG. 2-15.

Example 2-17. The two independent random variables x and y both have gaussian probability-density functions given by

$$p_x(x) = \frac{1}{\sigma_x\sqrt{2\pi}}\, e^{-(x-m_x)^2/2\sigma_x^2}$$

$$p_y(y) = \frac{1}{\sigma_y\sqrt{2\pi}}\, e^{-(y-m_y)^2/2\sigma_y^2}$$

We shall determine the probability density function of the sum variable $z = x + y$.

As in Ex. 2-16, we have

$$P_z(\omega) = P_x(\omega)\, P_y(\omega)$$

From Table 1-1, and the time-shifting property, we get

$$P_x(\omega) = [e^{-(\sigma_x\omega)^2/2}]\, e^{-j\omega m_x}$$

$$P_y(\omega) = [e^{-(\sigma_y\omega)^2/2}]\, e^{-j\omega m_y}$$

Therefore

$$P_z(\omega) = [e^{-(\sigma_x^2+\sigma_y^2)\omega^2/2}][e^{-j\omega(m_x+m_y)}]$$

Hence

$$p_z(z) = \mathfrak{F}^{-1}[P_z(\omega)] = \frac{1}{\sqrt{2\pi(\sigma_x^2+\sigma_y^2)}}\, e^{-[z-(m_x+m_y)]^2/2(\sigma_x^2+\sigma_y^2)}$$

$$= \frac{1}{\sigma_z\sqrt{2\pi}}\, e^{-(z-m_z)^2/2\sigma_z^2} \tag{2-92}$$

where

$$\sigma_z^2 = \sigma_x^2 + \sigma_y^2 \tag{2-93a}$$

and

$$m_z = m_x + m_y \tag{2-93b}$$

It is obvious from Eq. 2-92 that the probability-density function of a sum of two independent gaussian random variables is also gaussian. It is easy to see from this example that this conclusion can be extended to any number of normally distributed variables. The mean and the variance of the sum variable is equal to the sum of means and the sum of variances respectively of the individual random variables.

2-11. THE CENTRAL LIMIT THEOREM

In the Ex. 2-17, we observed that the sum of two independent gaussian random variables is also a gaussian random variable. It can be easily seen that the sum of any number of independent Gaussian random variables will be a gaussian random variable. This result is not very surprising. What is surprising is that under certain conditions, the sum of large number of independent random variables tends to be a gaussian random variable, independent of the probability densities of the variables added.[7] This fact can be intuitively appreciated if we realize that addition of independent variables is equivalent to convolving their density functions. The operation of convolution tends to smooth out the functions. If a large number of functions are convolved in a sequence, the resultant faction tends to be a gaussian function. Alternatively if $p_{x_1}(x_1), p_{x_2}(x_2) \cdots p_{x_n}(x_n)$ are the probability-density functions of n random variables and $P_{x_1}(\omega)$, $P_{x_2}(\omega) \cdots P_{x_n}(\omega)$ are their Fourier transforms respectively, then the product $P_z(\omega)$ given by

$$P_z(\omega) = P_{x_1}(\omega)\,P_{x_2}(\omega) \cdots P_{x_n}(\omega)$$

tends to a gaussian function as $n \to \infty$. The rigorous statement of this tendency is what is known as the central limit theorem. The general proof of this theorem is rather involved and will not be given here.

The tendency toward a gaussian distribution when a large number of functions are convolved is shown in Fig. 2-16. For simplicity we

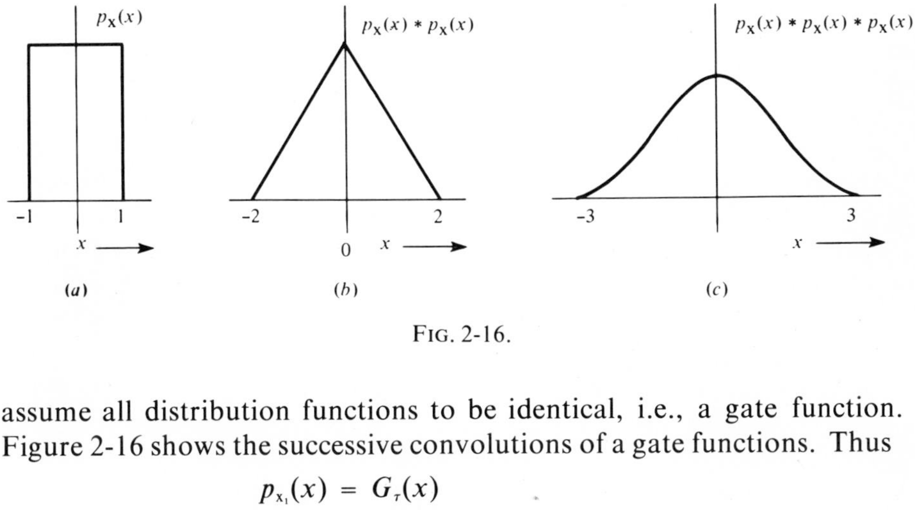

FIG. 2-16.

assume all distribution functions to be identical, i.e., a gate function. Figure 2-16 shows the successive convolutions of a gate functions. Thus

$$p_{x_1}(x) = G_\tau(x)$$
$$p_{x_2}(x) = G_\tau(x) * G_\tau(x)$$
$$p_{x_3}(x) = G_\tau(x) * G_\tau(x) * G_\tau(x)$$

The tendency toward a gaussian distribution is evident.

2-12. CORRELATION

Consider a random experiment whose outcome is specified by two random variables x and y. Assume that we have no other data about

[7]If the variables are gaussian, this applies even if the variables are not independent.

the experiment. Under this condition, is it possible to obtain any information regarding the dependence (or independence) of the variables x and y by examining the outcomes of large number of trials of the experiment? It appears reasonable that one should be able to draw some inference regarding the dependence of x and y from observing the outcomes of the large number of trials. We shall now show that a simple test of correlation does indeed yield some information about the dependence of variables.

Let us consider the case where variables x and y are dependent (or related) such that they tend to vary in harmony. In other words, if x increases, y increases and if x decreases, y also decreases.

As an example, consider the relationship between the number of publications by a person and his grade point average (GPA) in the school. It is reasonable to expect that there is a relation between the two quantities. We may study a large group of scientists and engineers and tabulate the GPA of and the number of publications by each person. This can be considered as a random experiment with outcomes x (the GPA) and y (the number of publications). The examination of each individual corresponds to one trial in the experiment. We shall plot the points (x, y) for each individual. This plot, known as the *scatter diagram*, will have an appearance as in Fig. 2-17a. This plot shows that for a large value of x, y is likely to be large. Note the use of the word likely. It is not always true that y will be large if x is large, but it is true most of the time. In other words, there will be few cases where a student with a lower GPA produces a large number of publications and one with a higher GPA produces fewer publications. This is quite obvious from the scatter diagram.

The variable $x - \bar{x}$ represents the difference between the actual and the average GPA and $y - \bar{y}$ represents the difference between the actual and the average number of publications. In general, a person with an above average GPA is likely to produce larger than the average number of publications and a person with below average GPA is likely to produce fewer than the average number of publications. Obviously if $x - \bar{x}$ is positive, $y - \bar{y}$ is more likely to be positive and if $x - \bar{x}$ is negative (below average GPA), $y - \bar{y}$ is more likely to be negative (below average number of publications). Thus the quantity $(x - \bar{x}) \cdot (y - \bar{y})$ will be positive for most of the trials. We compute this product for every person, add them and then divide it by the number of persons. This gives us the mean value of $(x - \bar{x})(y - \bar{y})$, that is, $\overline{(x - \bar{x})(y - \bar{y})}$. It is evident that this quantity (the mean) will be positive for the example under consideration. In such a case, we say that a *positive correlation* exists between variables x and y. The positive correlation,

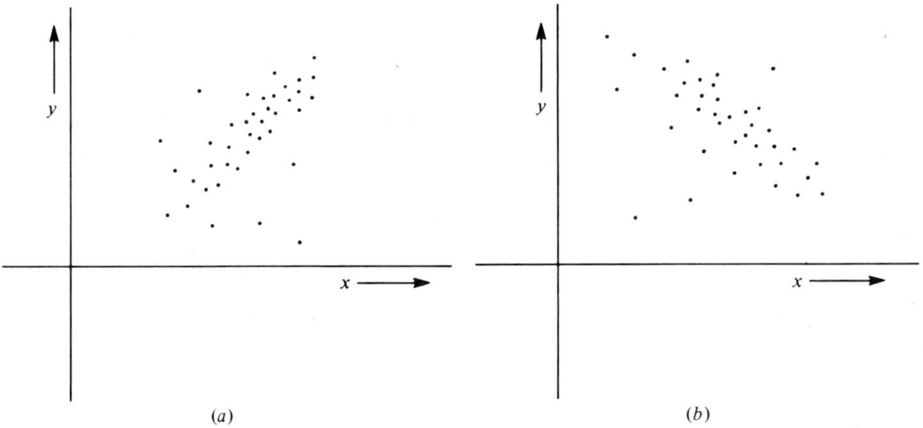

FIG. 2-17.

therefore implies variation of two variables in harmony (in the same direction—up or down).

Next, we consider the case where the two variables are related, but vary in opposite direction. As an example of this type of correlation consider the relationship between the GPA (variable x) and the class absenteeism (variable y) of the students. In this case, the variables will show dependence in a negative sense, that is, the higher the GPA, the lower the absenteeism. The scatter diagram for this experiment is shown in Fig. 2-17b. Thus if $x - \bar{x}$ is positive (above average GPA), $y - \bar{y}$ is likely to be negative (below average absenteeism) and if $x - \bar{x}$ is negative (below average GPA), $y - \bar{y}$ is likely to be positive (above average absenteeism). Thus the product $(x - \bar{x})(y - \bar{y})$ will be negative for most of the trials and the mean $\overline{(x - \bar{x})(y - \bar{y})}$ will be negative. In such a case, we say that *negative correlation* exists between x and y. It should be stressed here that negative correlation does *not* mean that x and y are unrelated. It means that they are dependent, but when one increases, the other decreases and vice versa.

Next, consider the case where the variables x and y are such that the value of x has no bearing on the value of y. As an example, consider the relationship between the GPA (variable x) and the number of children (variable y) a person has. It is reasonable to expect that the variations in x and y will not show any pattern of dependence. The scatter diagram will appear as shown in Fig. 2-18. If the quantity $x - \bar{x}$ is positive, $y - \bar{y}$ is equally likely to be positive as negative. The product $(x - \bar{x}) \cdot (y - \bar{y})$, therefore is equally likely to be positive as negative, and the mean $\overline{(x - \bar{x})(y - \bar{y})}$ will evidently be zero. In such a case, we say that the variables x and y are *uncorrelated*. The quantity $\overline{(x - \bar{x})(y - \bar{y})}$

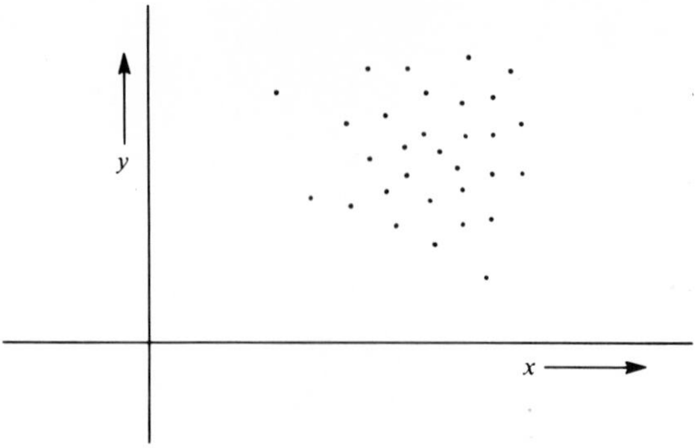

FIG. 2-18.

is the mean or the expectation of $(x - \bar{x})(y - \bar{y})$ and is called the *covariance* of x and y and will be denoted by K_{xy}.

$$K_{xy} = E[(x - \bar{x})(y - \bar{y})] = \overline{(x - \bar{x})(y - \bar{y})} \tag{2-94a}$$

Note that the variance σ_x^2 of x is defined as

$$\sigma_x^2 = \overline{(x - \bar{x})^2} = \overline{(x - \bar{x})(x - \bar{x})}$$

Covariance is an extension of the concept of variance as applied to two variables.

To reiterate, if K_{xy} is positive (negative) then x and y have a positive (negative) correlation and if $K_{xy} = 0$, the variables x and y are uncorrelated.

From the above discussion, it is evident that covariance is a measure of the relatedness between two variables. It does not, however, reveal the exact nature of dependence and cannot, in general, give complete information about the relatedness of the two variables. It does provide some information about the interdependence of the two variables and proves useful in a number of applications.

The covariance K_{xy} may be expressed in an alternate form. By definition

$$\begin{aligned} K_{xy} &= \overline{(x - \bar{x})(y - \bar{y})} \\ &= \overline{xy} - \overline{\bar{x}y} - \overline{x\bar{y}} + \overline{\bar{x}\bar{y}} \\ &= \overline{xy} - \bar{x}\bar{y} - \bar{x}\bar{y} + \bar{x}\bar{y} \\ &= \overline{xy} - \bar{x}\bar{y} \end{aligned} \tag{2-94b}$$

From Eq. 2-94b, it follows that the variables x and y are uncorrelated ($K_{xy} = 0$) if

$$\overline{xy} = \bar{x}\bar{y} \tag{2-95}$$

Note that for independent random variables (Eq. 2-78a)

$$\overline{xy} = \bar{x}\bar{y}$$

and

$$K_{xy} = 0$$

Hence independent random variables are uncorrelated. This supports the heuristic argument presented earlier. It should be noted that while independent variables are necessarily uncorrelated the converse is not true—uncorrelated variables are not necessarily independent. As an example, consider the random variable x and y given by

$$\mathrm{x} = \sin \Theta \qquad \text{and} \qquad \mathrm{y} = \cos \Theta \tag{2-96}$$

where Θ is a random variables uniformly distributed in the range $(0, 2\pi)$

$$p_\Theta(\theta) = \begin{cases} 1/2\pi & 0 < \Theta < 2\pi \\ 0 & \text{Otherwise} \end{cases}$$

Note that

$$\overline{\mathrm{x}} = \overline{\sin \Theta} = \int_{-\infty}^{\infty} \sin \theta \, p_\Theta(\theta) \, d\theta$$

$$= \int_0^{2\pi} \frac{1}{2\pi} \sin \theta \, d\theta = 0$$

Similarly $\overline{\mathrm{y}} = 0$ and (Eq. 2-94b)

$$K_{xy} = \overline{\mathrm{xy}} = \overline{\sin \Theta \cos \Theta}$$

$$= \frac{1}{2} \overline{\sin 2\Theta}$$

$$= \frac{1}{2} \int_0^{2\pi} \frac{1}{2\pi} \sin 2\theta \, d\theta$$

$$= 0$$

Hence x and y are uncorrelated. But as seen from Eq. 2-96

$$\mathrm{x}^2 + \mathrm{y}^2 = 1$$

Hence the variables x and y are not independent.

To reiterate, random variables which are uncorrelated may not be independent. Statistical independence is a much stronger and more restrictive condition than uncorrelatedness. For independent variables, we have shown (Eq. 2-78b) that

$$\overline{g_1(\mathrm{x}) g_2(\mathrm{y})} = \overline{g_1(\mathrm{x})}\, \overline{g_2(\mathrm{y})}$$

for any functions g_1 and g_2 whereas for uncorrelatedness the only requirement is that

$$\overline{\mathrm{xy}} = \overline{\mathrm{x}}\overline{\mathrm{y}}.$$

The amount of correlation between the variables x and y can conveniently be defined by a coefficient of correlation ρ_{xy} which is the normalized value of the covariance K_{xy}.

$$\rho_{xy} = \frac{K_{xy}}{\sigma_x \sigma_y} \tag{2-97}$$

where σ_x^2 and σ_y^2 are the variances of x and y respectively.

It can be easily shown that the magnitude of ρ_{xy} is always less than or equal to unity ($-1 \leq \rho_{xy} \leq 1$). This can be shown as follows:

For any real number a,

$$E\{[a(x - \overline{x}) - (y - \overline{y})]^2\}$$

must be nonnegative. Hence

$$a^2 \underbrace{E[(x - \overline{x})^2]}_{\sigma_x^2} - 2a \underbrace{E[(x - \overline{x})(y - \overline{y})]}_{K_{xy}} + \underbrace{E[(y - \overline{y})^2]}_{\sigma_y^2} \geq 0$$

or

$$a^2\sigma_x^2 - 2aK_{xy} + \sigma_y^2 \geq 0$$

This is a quadratic in a and is nonnegative. Hence its discriminant must be nonpositive. Therefore

$$K_{xy}^2 \leq \sigma_x^2 \sigma_y^2$$

From this equation and Eq. 2-97, it follows that

$$|\rho_{xy}| \leq 1$$

or

$$-1 \leq \rho_{xy} \leq 1$$

Note that

$$\rho_{xy} = 1$$

if

$$K_{xy} = \sigma_x \sigma_y$$

This is possible only if $x = k_1 y + k_2$ where k_1, k_2 are arbitrary constants. In this case

$$\sigma_x^2 = k_1^2 \sigma_y^2$$

and

$$K_{xy} = E[(x - \overline{x})(k_1 x - k_1 \overline{x})] = k_1 \sigma_x^2 = \sigma_x \sigma_y$$

Similarly, $\rho_{xy} = -1$ when $x = -k_1 y + k_2$. Thus the maximum amount of correlation occurs (negative or positive) if the variables are linearly related.

The covariance K_{xy} can be expressed in terms of joint probability density function $p_{xy}(x, y)$. By definition

$$K_{xy} = \overline{(x - \overline{x})(y - \overline{y})} = \int_{-\infty}^{\infty} \int_{-\infty}^{\infty} (x - \overline{x})(y - \overline{y}) p_{xy}(x, y)\, dx\, dy$$

If the random variables x, y are discrete, then

$$K_{xy} = \sum_i \sum_j (x_i - \overline{x})(y_j - \overline{y}) P_{xy}(x_i, y_j) \tag{2-98b}$$

The quantity $\overline{xy}$ (Eq. 2-94) will be denoted by R_{xy} for convenience. We shall call R_{xy} the *correlation* of variables x and y (note that this is not the same as the correlation coefficient in Eq. 2-97). Thus we define

$$R_{xy} = \overline{xy} \tag{2-98c}$$

It is evident from Eq. 2-94b that

$$K_{xy} = R_{xy} - \overline{x}\overline{y}$$

It follows that the variables x, y are uncorrelated ($K_{xy} = 0$) if

$$R_{xy} = \overline{x}\overline{y}$$

If

$$R_{xy} = 0$$

we say that variables x and y are *orthogonal.* Note that if either one or both of $\overline{x}$ and $\overline{y}$ is zero, then orthogonality implies uncorrelatedness.

It is evident from this discussion that covariance K_{xy} (or correlation R_{xy}) is a measure of the relatedness of the two variables x and y.

If

$$z = x + y$$

then from Eq. 2-83b, we have

$$\sigma_z^2 = \sigma_x^2 + \sigma_y^2 + 2K_{xy}$$

If the variables x and y are uncorrelated, $K_{xy} = 0$, and

$$\sigma_z^2 = \sigma_x^2 + \sigma_y^2 \tag{2-98d}$$

This result can be extended to any number of random variables. We therefore conclude that for uncorrelated variables, the variance of the sum is equal to the sum of variances.

2-13. SOME USEFUL PROBABILITY DISTRIBUTIONS

We have already come across certain probability distributions such as the normal or gaussian density function, the Rayleigh density function, etc. Another important distribution is the Poisson distribution, which will also be examined in this section.

Rayleigh Density. The Rayleigh density function is given by

$$p_x(x) = \frac{x}{\sigma^2} e^{-x^2/2\sigma^2} u(x) \tag{2-99}$$

The mean value x is given by

$$\overline{x} = \frac{1}{\sigma^2} \int_0^\infty x^2 e^{-x^2/2\sigma^2} dx$$

$$= \sqrt{\frac{\pi}{2}}\, \sigma$$

The mean-square value is given by

$$\overline{\mathrm{x}^2} = \frac{1}{\sigma^2}\int_0^\infty x^3 e^{-x^2/2\sigma^2}\,dx$$

$$= 2\sigma^2 \tag{2-100a}$$

The variance σ_{x}^2 is given by

$$\sigma_{\mathrm{x}}^2 = \overline{\mathrm{x}^2} - \overline{\mathrm{x}}^2$$

$$= \left(2 - \frac{\pi}{2}\right)\sigma^2 \tag{2-100b}$$

Poisson Distribution. This is one of the very important discrete distribution functions in probability theory. This distribution is associated with the following situation. A certain event occurs randomly on average α times per second (averaged over a large time interval) and the occurrence of n such event in any interval $(t_0, t_0 + \tau)$ is independent of the number of occurrences of the event before t_0 and depends only upon the length of interval τ. For an infinitesimal interval of length $\Delta\tau$, the probability that one event occurs within $\Delta\tau$ is proportional to $\Delta\tau$ and the probability that more than one event occurs within $\Delta\tau$ is an infinitesimal of order $(\Delta\tau)^2$ or higher. In this case, the probability of k events occurring in τ seconds is given by the Poisson's distribution.

$$P_k(\tau) = \frac{(\alpha\tau)^k e^{-\alpha\tau}}{k!} \tag{2-101}$$

This distribution is extremely common in nature. A familiar example is the random emission of electrons from the filament of a vacuum tube.[8] The probability of emission of exactly k electrons in τ seconds by a cathode is given by Eq. 2-101. Similarly the spontaneous decomposition of radioactive atomic nucleii follow Poisson distribution.

To obtain the Poisson distribution in Eq. 2-101, we consider the process of random emission of electrons from the filament of a vacuum tube. Let us examine the time interval $(0, \tau + \Delta\tau)$. It is reasonable to expect that this process satisfies the condition of Poisson's distribution mentioned above. Hence the probability of emission of one electron in this interval $\Delta\tau$ will be $\alpha\Delta\tau$.

$$P_1(\Delta\tau) = \alpha\Delta\tau \tag{2-102}$$

where α is a constant (it will be shown later that α is the average number of electrons emitted per second). We shall make $\Delta\tau$ so small $(\Delta\tau \to 0)$,

[8] This is true for temperature-limited operation, where all the electrons emitted previously are attracted to the plate and no space charge exists to influence the future emission.

that the probability of emission of more than one electron is negligible. Hence in $\Delta\tau$ seconds, there may be no emission or only one emission. Thus

$$P_0(\Delta\tau) + P_1(\Delta\tau) = 1 \tag{2-103}$$

Since the emission in the interval $\Delta\tau$ is independent of that in the previous interval of τ seconds, the probability of no emission in 0 to $\tau + \Delta\tau$ seconds is the product of the probability of no emission 0 to τ seconds and the probability of no emission in $\Delta\tau$ seconds.

$$P_0(\tau + \Delta\tau) = P_0(\tau)\,P_0(\Delta\tau)$$

Substituting Eqs. 2-102 and 2-103 in the above equation, we get

$$\frac{P_0(\tau + \Delta\tau) - P_0(\tau)}{\Delta\tau} = -\alpha P_0(\tau) \tag{2-104}$$

As $\Delta\tau \to 0$, Eq. 2-104 becomes

$$\frac{dP_0(\tau)}{d\tau} = -\alpha P_0(\tau) \tag{2-105}$$

Hence

$$P_0(\tau) = Ae^{-\alpha\tau} \tag{2-106}$$

However, $P_0(0)$ is the probability of no emission at $t = 0$; this being a certainty

$$P_0(0) = 1$$

Substitution of this initial condition in Eq. 2-106 yields

$$P_0(\tau) = e^{-\alpha\tau} \tag{2-107}$$

The probability of no emission in τ seconds is given by Eq. 2-107. (Fig. 2-19). Next we consider the probability of exactly k electrons being emitted in 0 to τ seconds. As before we consider the probability of emission of exactly k electrons in 0 to $(\tau + \Delta\tau)$ seconds. This event can occur in two ways. There may be $(k - 1)$ electrons emitted in τ seconds and 1 electron emitted in next $\Delta\tau$ seconds, or there may be k electrons emitted in first τ seconds and no emission in the next $\Delta\tau$ second. From the theorem of total probability (Eq. 2-3) and the statistical independence of emission in τ seconds and $\Delta\tau$ seconds, we obtain

$$P_k(\tau + \Delta\tau) = P_{k-1}(\tau)\,P_1(\Delta\tau) + P_k(\tau)\,P_0(\Delta\tau) \tag{2-108}$$

Substituting Eqs. 2-102 and 2-103 in Eq. 2-108, we obtain

$$\frac{P_k(\tau + \tau) - P_k(\tau)}{\Delta\tau} + \alpha P_k(\tau) = \alpha P_{k-1}(\tau) \tag{2-109}$$

As $\Delta\tau \to 0$, Eq. 2-109 becomes

$$\frac{dP_k(\tau)}{d\tau} + \alpha P_k(\tau) = \alpha P_{k-1}(\tau) \tag{2-110}$$

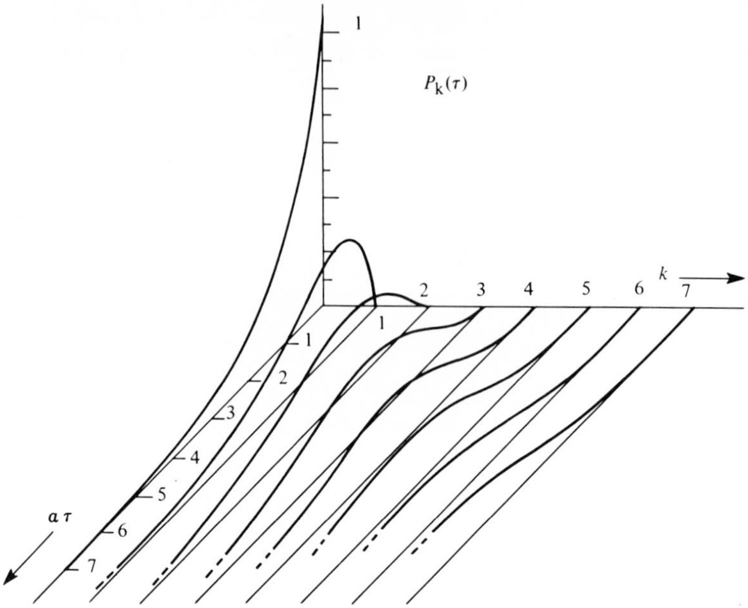

FIG. 2-19.

This is the first-order, linear-differential equation of the type

$$\frac{dy}{dx} + f(x)y = g(x)$$

whose solution is given by

$$ye^{\int f(x)dx} = \int g(x)e^{\int f(x)dx}\,dx \tag{2-111}$$

Comparison of Eqs. 2-110 and 2-111 gives

$$y \sim P_k(\tau), \quad f(x) \sim \alpha \quad \text{and} \quad g(x) \sim \alpha P_{k-1}(\tau)$$

Thus

$$\int f(x)\,dx \sim \int \alpha d\tau = \alpha\tau$$

and the solution of Eq. 2-110 is given by

$$P_k(\tau)e^{\alpha\tau} = \int \alpha P_{k-1}(\tau)e^{\alpha\tau}\,d\tau \tag{2-112}$$

For $k = 1$, this becomes

$$P_1(\tau)e^{\alpha\tau} = \int \alpha P_0(\tau)e^{\alpha\tau}\,d\tau \tag{2-113}$$

Substituting $P_0(\tau)$ from Eq. 2-107 in Eq. 2-113, we get

$$P_1(\tau)e^{\alpha\tau} = \alpha\tau + C \tag{2-114}$$

But since $P_1(0)$, the probability of one electron being emitted in zero second is zero, the constant of integration C in Eq. 2-114 must be zero. Hence

$$P_1(\tau) = \alpha\tau e^{-\alpha\tau} \tag{2-115}$$

Using Eq. 2-115 in Eq. 2-112, we proceed in a similar way to obtain $P_2(\tau)$ and so on. In general, we obtain

$$P_k(\tau) = \frac{(\alpha\tau)^k e^{-\alpha\tau}}{k!} \tag{2-116}$$

This distribution function is sketched in Fig. 2-19 for various values of k. The mean value of k (in τ seconds) is given by

$$\begin{aligned}
\bar{k} &= \sum_{k=0}^{\infty} kP_k(\tau) \\
&= \sum_{k=0}^{\infty} k\,\frac{(\alpha\tau)^k e^{-\alpha\tau}}{k!} \\
&= \alpha\tau e^{-\alpha\tau}\left[1 + \alpha\tau + \frac{(\alpha\tau)^2}{2!} + \frac{(\alpha\tau)^3}{3!} + \cdots\right] \\
&= \alpha\tau e^{-\alpha\tau} e^{\alpha\tau} \\
&= \alpha\tau
\end{aligned} \tag{2-117}$$

This result states that the average number of electrons emitted in τ seconds is $\alpha\tau$. In other words the average number of electrons emitted per second is α.

Example 2-18. Computing the Conductance of a Metal. Find the conductivity of a material which has n free electrons per unit volume and each free electron experiences an average of α collisions per second with the lattice structure. Let us assume that a uniform electric field E is applied in the $-y$ direction to a unit cube of the metal as shown in Fig. 2-20. Each electron with

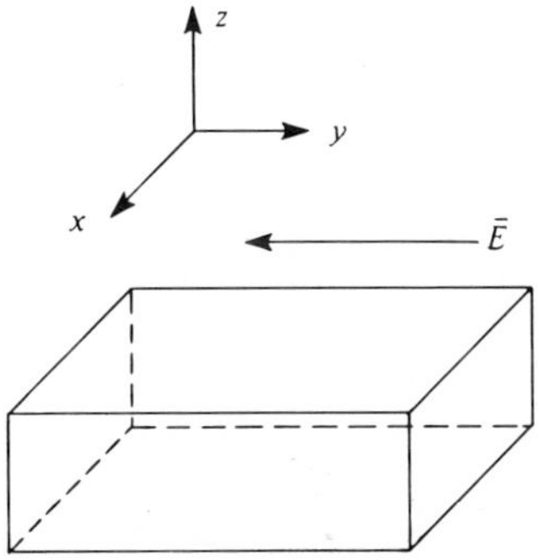

FIG. 2-20.

charge q, experiences a force qE in the y direction. Hence $\ddot{y}$, the acceleration of an electron, is given by

$$\ddot{y} = \frac{qE}{m}$$

where m is the mass of an electron. The velocity $\dot{y}$ is given by (neglecting the thermal velocity)

$$\dot{y} = \frac{qE}{m} t$$

and y, the distance traversed, is given by

$$y = \frac{qE}{2m} t^2$$

The motion of the electron is impeded by the lattice structure. The electron experiences collisions with the lattice structure at which time it is brought to rest. The electron starts moving in the y direction again until it is brought to rest by next collision, and so on. The time elapsed between successive collisions is a random variable T. The average distance $\overline{\mathrm{y}}$ traversed by the electron between successive collisions is given by

$$\overline{\mathrm{y}} = \frac{qE}{2m} \overline{\mathrm{T}^2} \tag{2-118}$$

where $\overline{\mathrm{T}^2}$ represents the mean-square value of the collision interval and is given by

$$\overline{\mathrm{T}^2} = \int_0^\infty T^2 p_{\mathrm{T}}(T)\, dT \tag{2-119}$$

The limits in the above integral are $(0, \infty)$ because the collision interval can never be negative. In Eq. 2-119, $p_{\mathrm{T}}(T)$ is the probability density of the collision interval T, that is $p_{\mathrm{T}}(T)\, dT$ represents the probability that the collision interval lies between T and $T + dT$. It is reasonable to assume that the occurrence of a collision in any interval is independent of previous collisions. This is precisely the condition for Poisson distribution. Hence the collision times are Poisson distributed. Thus $P_k(\tau)$, the probability of k collision in τ seconds is given by

$$P_k(\tau) = \frac{(\alpha\tau)^k}{k!} e^{-\alpha\tau} \tag{2-120}$$

where α is the average collisions per second. As mentioned earlier $p_{\mathrm{T}}(T)\, dT$ represents the probability of the event that a collision interval lies between T and $T + dT$. This event is possible only if no collisions occur for T seconds and one collision occurs during the next dT seconds. Hence

$$\begin{aligned} p_{\mathrm{T}}(T)\, dT &= P_0(T)\, P_1(dT) \\ &= e^{-\alpha T}(\alpha\, dT) \qquad \text{(assuming } dT \rightarrow 0\text{)} \end{aligned}$$

Hence

$$p_{\mathrm{T}}(T) = \alpha e^{-\alpha T}$$

and

$$\overline{T^2} = \int_0^{\infty} \alpha T^2 e^{-\alpha T}\, dT = \frac{2}{\alpha^2}$$

and

$$\overline{y} = \frac{qE}{m\alpha^2}$$

The average drift velocity $\overline{v}_y$ of the electron is equal to the average distance traversed per collision multiplied by the average number of collisions per second. Hence

$$\overline{v}_y = \alpha\overline{y} = \frac{qE}{m\alpha}$$

The conduction current I through unit area equals the charge passing through this cross section per unit time. Hence

$$I = nq\overline{v}_y$$
$$= \frac{nq^2E}{m\alpha}$$

The conductivity σ by definition is

$$\sigma = \frac{I}{E} = \frac{nq^2}{m\alpha} \tag{2-121}$$

The resistivity ρ is the reciprocal of σ:

$$\rho = \frac{m\alpha}{nq^2} \tag{2-122}$$

Gaussian (Normal) Distribution.[9] The random variable x is said to have a Gaussian (or normal) distribution if

$$p_x(x) = \frac{1}{\sigma\sqrt{2\pi}} e^{-(x-m)^2/2\sigma^2} \tag{2-123}$$

It has already been shown (Examples 2-11, 2-12, and 2-15) that the mean value, the variance, and the mean-square value are given by

$$\begin{aligned} \overline{x} &= m \\ \overline{x^2} &= \sigma^2 + m^2 \\ \sigma_x^2 &= \sigma^2 \end{aligned} \tag{2-124}$$

Multidimensional Gaussian Distribution: Jointly Gaussian Variables. The distribution in Eq. 2-123 represents the distribution of a single

[9]It is common practice to refer to probability density function by a general term distribution. This, however, should not be confused with the probability-distribution (cumulative-distribution) function $F_x(x)$.

variable. We shall now consider the case of several variables which are jointly gaussian. For convenience, the case of two variables will be considered first.

The random variables x_1 and x_2 are said to have a joint gaussian distribution (or are jointly gaussian) if

$$p_{x_1 x_2}(x_1, x_2) = \frac{1}{2\pi\sqrt{M}} e^{-(K_{22}x_1^2 + K_{11}x_2^2 - 2K_{12}x_1x_2)/2M} \tag{2-125}$$

where

$$K_{11} = \overline{x_1^2}, \quad K_{22} = \overline{x_2^2}$$
$$K_{12} = \overline{x_1 x_2} \quad \text{and} \quad M = K_{11}K_{22} - K_{12}^2$$

The reader can easily verify that for this distribution the means of x_1 and x_2 are zero.

$$\overline{x_1} = \overline{x_2} = 0$$

The quantities K_{11}, K_{22} are the mean-square values of x_1 and x_2 respectively. Similarly, K_{12} is the correlation of x_1 and x_2. Since the means of x_1 and x_2 are zero, K_{11}, K_{22} (the mean-square values) are identical to variances $\sigma_{x_1}^2$ and $\sigma_{x_2}^2$ respectively. Similarly, K_{12} is covariance of x_1 and x_2. It is evident that the joint probability density function for jointly gaussian variables is completely determined from the knowledge of variances of x_1 and x_2 and the covariance of x_1 and x_2. This is not true in general for random variables that are not jointly gaussian.

The general expression for two jointly gaussian variables with nonzero means is given by

$$p_{x_1 x_2}(x_1, x_2) = \frac{1}{2\pi\sqrt{M}} e^{-[K_{22}(x_1 - m_1)^2 + K_{11}(x_2 - m_2)^2 - 2K_{12}(x_1 - m_1)(x_2 - m_2)]/2M} \tag{2-126}$$

where m_1 and m_2 are the means of x_1 and x_2 respectively, K_{11}, K_{22} are the variances of x_1 and x_2 respectively and K_{12} is the covariance of x_1 and x_2

$$m_1 = \overline{x_1}, \qquad K_{11} = \sigma_{x_1}^2 = \overline{(x_1 - m_1)^2}$$
$$m_2 = \overline{x_2}, \qquad K_{22} = \sigma_{x_2}^2 = \overline{(x_2 - m_2)^2}$$

and

$$K_{12} = \overline{(x_1 - m_1)(x_2 - m_2)} \quad \text{and} \quad M = K_{11}K_{22} - K_{12}^2$$

In our future discussion we can assume the variables to have zero means. This does not entail any loss of generality since we can always form a random variable with a zero mean from the one with nonzero mean merely by subtracting the mean value. Thus if a random variable x has a mean m, then we can form a new random variable $x' = x - m$ which has a zero mean. It can be easily seen that Eq. 2-126 reduces to the form of Eq. 2-125 if we substitute $x_1' = x_1 - m_1$ and $x_2' = x_2 - m_2$

in Eq. 2-126. The new variables x_1' and x_2' are jointly Gaussian of the form in Eq. 2-125.

We can find the individual (marginal) densities $p_{x_1}(x_1)$, $p_{x_2}(x_2)$ and conditional densities $p_{x_1}(x_1 \mid x_2 = x_2)$, $p_{x_2}(x_2 \mid x_1 = x_1)$ from Eqs. 2-125, 2-36, and 2-42. This has already been done in example 2-7. We obtained

$$
\begin{aligned}
p_{x_1}(x_1) &= \frac{1}{\sqrt{2\pi K_{11}}} e^{-x_1^2/2K_{11}} \\
p_{x_2}(x_2) &= \frac{1}{\sqrt{2\pi K_{22}}} e^{-x_2^2/2K_{22}} \\
p_{x_1}(x_1 \mid x_2 = x_2) &= \frac{K_{22}}{\sqrt{2\pi M}} e^{-K_{22}[x_1-(K_{12}/K_{22})x_2]/2M} \\
p_{x_2}(x_2 \mid x_1 = x_1) &= \frac{K_{11}}{\sqrt{2\pi M}} e^{-K_{11}[x_2-(K_{12}/K_{11})x_1]^2/2M}
\end{aligned}
\qquad (2\text{-}127)
$$

It can be seen from these equations that all the marginal and conditional densities are gaussian. This is an important property of jointly gaussian distribution.

Another very important property of jointly gaussian variables is that if such variables are uncorrelated, they are also independent. This can be easily seen from the fact that if the variables x_1 and x_2 are uncorrelated, then the covariance $K_{12} = 0$. In this case the joint density in Eq. 2-125 reduces to the product of two individual densities.

$$
\begin{aligned}
p_{x_1x_2(x_1,x_2)} &= \frac{1}{2\pi\sqrt{M}} e^{-(K_{22}x_1^2+K_{11}x_2^2)/2K_{11}K_{22}} \\
&= \left(\frac{1}{\sqrt{2\pi K_{11}}} e^{-x_1^2/2K_{11}}\right)\left(\frac{1}{\sqrt{2\pi K_{22}}} e^{-x_2^2/2K_{22}}\right)
\end{aligned}
$$

The right-hand side of the above equation is $p_{x_1}(x_1)p_{x_2}(x_2)$ as seen from Eq. 2-127. Obviously the two variables x_1 and x_2 are independent. Thus the jointly gaussian variables are independent if they are uncorrelated. This is not true for a general distribution. We have seen in Sec. 2-12 that if the two variables are independent, they are necessarily uncorrelated but the converse is not true—that is, if the variables are uncorrelated they are not necessarily independent. Independence is much stronger condition than uncorrelatedness. However for gaussian variables, the uncorrelatedness implies independence.

Linear Combination of Jointly Gaussian Variables. Jointly gaussian variables have another very important property. New variables formed by the linear combination (linear transformation) of jointly gaussian variables are also jointly gaussian.[10] Let us consider variables x_3, x_4 formed by linear combination of jointly gaussian variables x_1 and x_2.

$$\begin{aligned} \mathrm{x}_3 &= a_{11}\mathrm{x}_1 + a_{12}\mathrm{x}_2 \\ \mathrm{x}_4 &= a_{21}\mathrm{x}_1 + a_{22}\mathrm{x}_2 \end{aligned} \tag{2-128}$$

In matrix form, we have

$$\begin{bmatrix} \mathrm{x}_3 \\ \mathrm{x}_4 \end{bmatrix} = \begin{bmatrix} a_{11} & a_{12} \\ a_{21} & a_{22} \end{bmatrix} \begin{bmatrix} \mathrm{x}_1 \\ \mathrm{x}_2 \end{bmatrix}$$

This yields

$$\begin{aligned} \begin{bmatrix} \mathrm{x}_1 \\ \mathrm{x}_2 \end{bmatrix} &= \begin{bmatrix} a_{11} & a_{12} \\ a_{21} & a_{22} \end{bmatrix}^{-1} \begin{bmatrix} \mathrm{x}_3 \\ \mathrm{x}_4 \end{bmatrix} \\ &= \begin{bmatrix} b_{11} & b_{12} \\ b_{21} & b_{22} \end{bmatrix} \begin{bmatrix} \mathrm{x}_3 \\ \mathrm{x}_4 \end{bmatrix} \end{aligned} \tag{2-129}$$

where

$$\begin{aligned} b_{11} &= \frac{a_{22}}{c}, \qquad b_{12} = \frac{-a_{12}}{c} \\ b_{21} &= \frac{-a_{21}}{c}, \qquad b_{22} = \frac{a_{11}}{c} \end{aligned} \qquad c = a_{11}a_{22} - a_{12}a_{21} \tag{2-130}$$

We now use Eq. 2-53 to obtain $p_{\mathrm{x}_3\mathrm{x}_4}(x_3, x_4)$

$$p_{\mathrm{x}_3\mathrm{x}_4}(x_3, x_4) = \frac{p_{\mathrm{x}_1\mathrm{x}_2}(x_1, x_2)}{J\left(\dfrac{x_3, x_4}{x_1, x_2}\right)}$$

But

$$J\left(\frac{x_3, x_4}{x_1, x_2}\right) = \begin{vmatrix} \dfrac{\partial x_3}{\partial x_1} & \dfrac{\partial x_3}{\partial x_2} \\ \dfrac{\partial x_4}{\partial x_1} & \dfrac{\partial x_4}{\partial x_2} \end{vmatrix} = \begin{vmatrix} a_{11} & a_{12} \\ a_{21} & a_{22} \end{vmatrix} = c \tag{2-131}$$

Hence

$$p_{\mathrm{x}_3\mathrm{x}_4}(x_3, x_4) = \frac{1}{c}\, p_{\mathrm{x}_1\mathrm{x}_2}(x_1, x_2)\Bigg|_{\substack{x_1 = b_{11}x_3 + b_{12}x_4 \\ x_2 = b_{21}x_3 + b_{22}x_4}}$$

For convenience, we shall assume $\mathrm{x}_1, \mathrm{x}_2$ to have zero means (this can be done without loss of generality). The joint density in this case is given by Eq. 2-125. Thus

[10] This is true only if the transformation is reversible; that is, the transformation matrix is nonsingular. This assumption is implied throughout the discussion.

$$p_{x_3x_4}(x_3, x_4) = \frac{1}{2\pi c\sqrt{M}} \exp\left[\frac{-(K_{22}x_1^2 + K_{11}x_2^2 - 2K_{12}x_1x_2)}{2M}\right]\Bigg|_{\substack{x_1 = b_{11}x_3 + b_{12}x_4 \\ x_2 = b_{21}x_3 + b_{22}x_4}}$$

$$= \frac{1}{2\pi\sqrt{M'}} \exp\left[\frac{-(K_{44}x_3^2 + K_{33}x_4^2 - 2K_{34}x_3x_4)}{2M'}\right] \qquad (2\text{-}132)$$

where

$$K_{33} = a_{11}^2K_{11} + a_{12}^2K_{22} + 2a_{12}a_{11}K_{12}$$
$$K_{44} = a_{22}^2K_{22} + a_{21}^2K_{22} + 2a_{21}a_{22}K_{12}$$
$$K_{34} = K_{12}(a_{11}a_{22} + a_{12}a_{21}) + a_{11}a_{21}K_{11} + a_{12}a_{22}K_{22}$$
$$M' = K_{33}K_{44} - K_{34}^2$$

It is evident from Eq. 2-132, that x_3, x_4 are jointly gaussian. The results can be extended to variables which are linear combinations of any number of gaussian variables.

In general, n random variables $x_1, x_2, \ldots, x_n$ are said to be jointly gaussian if

$$p_{x_1x_2\cdots x_n}(x_1, x_2, \ldots, x_n)$$
$$= \frac{1}{(2\pi)^{n/2}\sqrt{|\mathbf{K}|}} \exp\left(-\frac{1}{2|\mathbf{K}|}\sum_{i=1}^{n}\sum_{j=1}^{n} \Delta_{ij}(x_i - \overline{x}_i)(x_j - \overline{x}_j)\right) \qquad (2\text{-}133)$$

where $\mathbf{K}$ is the covariance matrix:

$$\mathbf{K} = \begin{bmatrix} K_{11} & K_{12} & \cdots & K_{1n} \\ K_{21} & K_{22} & \cdots & K_{2n} \\ \cdots & \cdots & \cdots & \cdots \\ K_{n1} & K_{n2} & \cdots & K_{nn} \end{bmatrix} \qquad (2\text{-}134a)$$

$$K_{ij} = \overline{(x_i - \overline{x}_i)(x_j - \overline{x}_j)} \quad \text{covariance of } x_i \text{ and } x_j \qquad (2\text{-}134b)$$

$|\mathbf{K}|$ is the determinant of matrix $\mathbf{K}$ and Δ_{ij} is the cofactor of the element K_{ij}.

If the random variables $x_1, x_2, \ldots, x_n$ have zero means, then Eq. 2-133 reduces to

$$p_{x_1x_2\cdots x_n}(x_1, x_2, \ldots, x_n)$$
$$= \frac{1}{(2\pi)^{n/2}\sqrt{|\mathbf{K}|}} \exp\left(-\frac{1}{2|\mathbf{K}|}\sum_{i=1}^{n}\sum_{j=1}^{n} \Delta_{ij}x_ix_j\right) \qquad (2\text{-}135)$$

Note that in this case

$$K_{ij} = \overline{x_1x_j}$$
$$= R_{ij} \quad (\text{correlation of } x_i \text{ and } x_j)$$

For two jointly gaussian variables (bivariate distribution) we have proved certain properties earlier in this section. These properties hold in general for n jointly gaussian variables (multivariate distribution). These properties are not true in general for random variables that are not jointly gaussian. These important properties are extremely useful and are responsible for making the jointly gaussian distribution easy to handle compared to other kinds of distributions. These properties (four in all) will now be discussed.

Property 1. The jointly gaussian distribution is completely specified by various second moments (variances and covariances) and the mean values of the variables. This property is evident from Eq. 2-133.

Property 2. If the variables $\mathrm{x}_1, \mathrm{x}_2, \ldots, \mathrm{x}_n$ are uncorrelated, they are also independent.

If all variables are uncorrelated, then

$$K_{ij} = 0 \qquad i \neq j$$

The joint distribution in Eq. 2-133 reduces to the product of n one-dimensional normal density functions.

$$p_{\mathrm{x}_1\mathrm{x}_2\cdots\mathrm{x}_n}(x_1, x_2, \ldots, x_n) = \prod_{i=1}^{n} \left(\frac{1}{\sqrt{2\pi K_{ii}}} e^{-(x_i - \bar{\mathrm{x}}_i)^2/2K_{ii}} \right) \tag{2-136}$$

It is evident from Eq. 2-136, that the variables are rendered independent.

Property 3. All individual (or marginal) and conditional densities are also gaussian. This result was proved in Example 2-7 for $n = 2$. It is true for any value of n in general.

Property 4. Linear transformations on jointly gaussian variables yield variables that are also jointly gaussian. Thus if we form m variables $\mathrm{y}_1, \mathrm{y}_2, \ldots, \mathrm{y}_m$ $(m < n)$ such that

$$\mathrm{y}_i = \sum_{k=1}^{n} a_{ik}\mathrm{x}_k$$

then $\mathrm{y}_1, \mathrm{y}_2, \ldots, \mathrm{y}_m$ are jointly gaussian. We have proved this result for $n = 2$. For a general case the reader is referred to Ref. 2-5.

2-14. VECTOR OF RANDOM VARIABLES

When an outcome of a random experiment can be specified by n random variables $\mathrm{x}_1, \mathrm{x}_2, \ldots, \mathrm{x}_n$, it is very convenient to use the notion of a n-dimensional vector associated with these outcomes. We may consider the outcome of such an experiment to be a n-dimensional vector whose components are $\mathrm{x}_1, \mathrm{x}_2, \ldots, \mathrm{x}_n$. In short, the vector is such that its components are random variables. We shall denote this random vector by $\mathbf{x}(\mathrm{x}_1, \mathrm{x}_2, \ldots, \mathrm{x}_n)$.

The joint probability density function $p_{\mathrm{x}_1\mathrm{x}_2\cdots\mathrm{x}_n}(x_1, x_2, \ldots, x_n)$ will be expressed as $p_{\mathbf{x}}(\boldsymbol{x})$.

$$p_{\mathbf{x}}(\boldsymbol{x}) = p_{\mathrm{x}_1\mathrm{x}_2\cdots\mathrm{x}_n}(x_1, x_2, \ldots, x_n)$$

The vector concept offers a convenient geometrical interpretation for joint probability density function. For simplicity, consider a two-dimensional case of a random vector $\mathbf{x}(\mathrm{x}_1, \mathrm{x}_2)$, with the joint probability density $p_{\mathrm{x}_1\mathrm{x}_2}(x_1, x_2)$. The probability of observing $x_1 < \mathrm{x}_1 \leq x_1 + \Delta x_1$ and $x_2 < \mathrm{x}_2 \leq x_2 + \Delta x_2$ is $p_{\mathrm{x}_1\mathrm{x}_2}(x_1, x_2)\,\Delta x_1 \Delta x_2$. This is given by the volume under $p_{\mathrm{x}_1\mathrm{x}_2}(x_1, x_2)$ over the region bounded by $x_1 < \mathrm{x}_1 \leq x_1 + \Delta x_1$ and $x_2 < \mathrm{x}_2 \leq x_2 + \Delta x_2$ as shown in Fig. 2-21a. Using the vector concept, it is obvious that $p_{\mathrm{x}_1\mathrm{x}_2}(x_1, x_2)\,\Delta x_1 \Delta x_2$ represents the probability of observing the (tip of the) vector $\mathbf{x}(\mathrm{x}_1, \mathrm{x}_2)$ in the shaded region. This result now can be generalized to arbitrary region I as shown in Fig. 2-21b. The volume under the joint probability density function $p_{\mathrm{x}_1\mathrm{x}_2}(x_1, x_2)$ over the region I is the probability of observing the (tip of the) vector $\mathbf{x}(\mathrm{x}_1, \mathrm{x}_2)$ in the region I. This probability is given by

$$\text{Probability } (\mathbf{x} \text{ in } I) = \iint_I p_{\mathrm{x}_1\mathrm{x}_2}(x_1, x_2)\, dx_1 dx_2$$

The integral on the right-hand side can be conveniently expressed as

$$\int_I p_{\mathbf{x}}(\boldsymbol{x})\, \boldsymbol{dx}$$

Thus

$$\text{Probability } (\mathbf{x} \text{ in } I) = \iint_I p_{\mathrm{x}_1\mathrm{x}_2}(x_1, x_2)\, dx_1 dx_2 \qquad \text{(2-137a)}$$

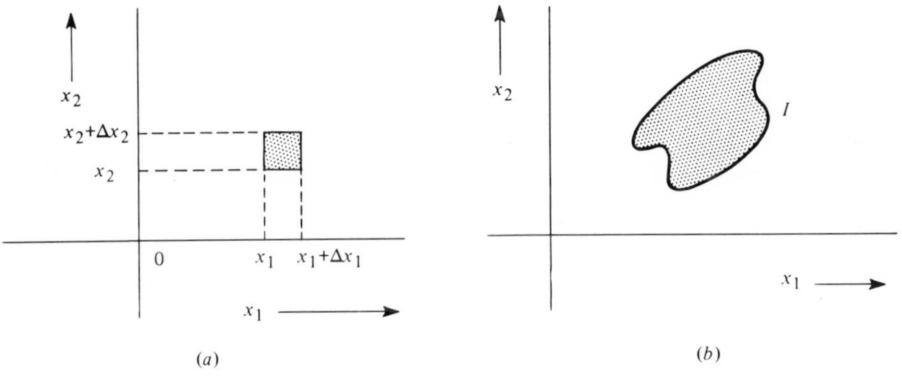

FIG. 2-21.

$$= \int_I p_{\mathbf{x}}(\boldsymbol{x})\, d\boldsymbol{x} \tag{2-137b}$$

The vector notation for the integral is merely a matter of convenience and such an integral should be interpreted properly as given in Eq. 2-137. For a general n-dimensional case, the region I is n-dimensional and

$$\int_I p_{\mathbf{x}}(\boldsymbol{x})\, d\boldsymbol{x} = \iint_I \cdots \int p_{\mathrm{x}_1\mathrm{x}_2\cdots\mathrm{x}_n}(x_1, x_2, \ldots, x_n)\, dx_1 dx_2 \cdots dx_n \tag{2-138}$$

The integral in Eq. 2-138 represents the probability of observing vector $\mathbf{x}(\mathrm{x}_1, \mathrm{x}_2, \ldots, \mathrm{x}_n)$ in the region I.

A trial of the random experiment (two-dimensional) yields the outcomes (x_1, x_2). This is represented by a point (x_1, x_2) or a vector $\boldsymbol{x}(x_1, x_2)$ in the x_1x_2 plane. Since the experiment is random, each trial yields different point $\boldsymbol{x}(x_1, x_2)$. The random point $\mathbf{x}(\mathrm{x}_1, \mathrm{x}_2)$ or the random vector $\mathbf{x}$ with its probability distribution in x_1x_2 plane describes the random experiment completely. Geometrically, we may visualize a region in x_1x_2 plane where the outcome may lie. This region may be shaded to indicate the relative frequency of observing the outcome $\mathbf{x}$ at a particular point. Thus darker shading will indicate that the outcome is more likely to be observed in that region and the lighter shading will imply that the outcome will be observed with relatively less frequency in that region. Note that this diagram is essentially a scatter diagram (Figs. 2-17 and 2-18). It is evident that the relative intensity of the shading is proportional to the joint probability density of x_1 and x_2 at that point. Such a representation is shown in Fig. 2-22 for a two-dimensional case. One may extend such representation in n-dimensional hyperplane. This type of visual representation proves very helpful in the problems of signal detection and will be considered again in detail in Chapter 6.

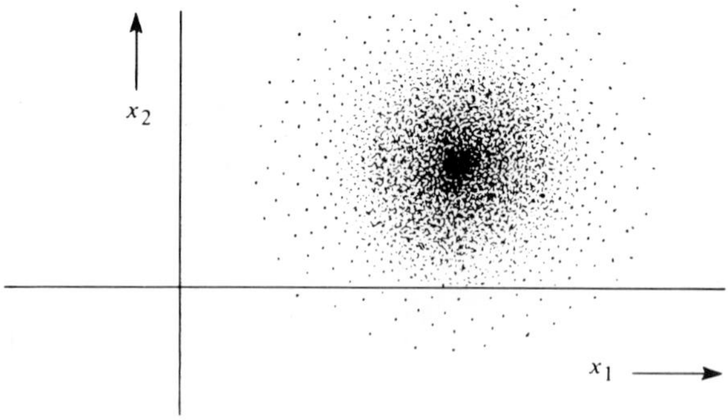

FIG. 2-22.

Jointly Gaussian Variables as a Random Vector. The joint probability-density function of n jointly gaussian random variables (Eq. 2-133) can be expressed very concisely by using vector notation. If we represent vector $\mathbf{x}(\mathrm{x}_1, \mathrm{x}_2, \ldots, \mathrm{x}_n)$ as a row vector, then $\mathbf{x}'$, the transpose of $\mathbf{x}$, is a column vector.

$$\mathbf{x}' = \begin{bmatrix} \mathrm{x}_1 \\ \mathrm{x}_2 \\ \vdots \\ \mathrm{x}_n \end{bmatrix}$$

Note that we represent a random vector by roman boldface type, and a particular value assumed by the random vector in some trial is represented by boldface italic type. Thus $\boldsymbol{x}$ represents the value assumed by the random vector $\mathbf{x}$ in some trial, where

$$\boldsymbol{x} = (x_1, x_2, \ldots, x_n)$$

The covariance matrix $\mathbf{K}$ is given by Eq. 2-134. The inverse of $\mathbf{K}$ is given by

$$\mathbf{K}^{-1} = \frac{1}{|\mathbf{K}|} \begin{bmatrix} \Delta_{11} & \Delta_{21} & \cdots & \Delta_{n1} \\ \Delta_{12} & \Delta_{22} & \cdots & \Delta_{n2} \\ \cdots & \cdots & \cdots & \cdots \\ \Delta_{1n} & \Delta_{2n} & \cdots & \Delta_{nn} \end{bmatrix} \tag{2-139}$$

where Δ_{ij} is the cofactor of K_{ij}. Using this notation, we observe that

$$\boldsymbol{x}\mathrm{K}^{-1}\boldsymbol{x}' = \frac{1}{|\mathbf{K}|} \sum_{i=1}^{n} \sum_{j=1}^{n} \Delta_{ij} x_i x_j$$

and

$$(\boldsymbol{x} - \overline{\mathbf{x}})\mathrm{K}^{-1}(\boldsymbol{x}' - \overline{\mathbf{x}}') = \frac{1}{|\mathbf{K}|} \sum_{i=1}^{n} \sum_{j=1}^{n} \Delta_{ij}(x_i - \overline{\mathrm{x}}_i)(x_j - \overline{\mathrm{x}}_j)$$

Hence the joint probability-density function of n jointly gaussian variables (Eq. 2-133) can be expressed as

$$p_{\mathbf{x}}(\boldsymbol{x}) = \frac{1}{(2\pi)^{n/2}\sqrt{|\mathbf{K}|}} \exp\left[-\frac{1}{2}(\boldsymbol{x} - \overline{\mathbf{x}})\mathbf{K}^{-1}(\boldsymbol{x}' - \overline{\mathbf{x}}')\right] \tag{2-140}$$

where

$$\overline{\mathbf{x}}' = \begin{bmatrix} \overline{\mathrm{x}}_1 \\ \overline{\mathrm{x}}_2 \\ \vdots \\ \overline{\mathrm{x}}_n \end{bmatrix}$$

and

$$\bar{\mathbf{x}} = [\bar{\mathrm{x}}_1, \bar{\mathrm{x}}_2, \ldots, \bar{\mathrm{x}}_n]$$

If the random variables $\mathrm{x}_1, \mathrm{x}_2, \ldots, \mathrm{x}_n$ have zero means, then

$$p_{\mathrm{x}}(x) = \frac{1}{(2\pi)^{n/2}\sqrt{|\mathbf{K}|}} \exp\left(-\frac{1}{2}\, x\mathbf{K}^{-1}x'\right) \tag{2-141}$$

If the random variables $\mathrm{x}_1, \mathrm{x}_2, \ldots, \mathrm{x}_n$ are uncorrelated, then

$$K_{ij} = 0 \qquad i \neq j$$

and the matrix **K** becomes a diagonal matrix.

REFERENCES

1. P. Pfeiffer, *Concepts of Probability Theory*, McGraw-Hill, New York, 1965.

2. E. Parzen, *Modern Probability Theory and Its Applications*, Wiley, New York, 1960.

3. A. Papoulis, *Probability, Random Variables, and Stochastic Processes*, McGraw-Hill, New York, 1965.

4. W. Kaplan, *Advanced Calculus*, Addison-Wesley, Reading, Mass., 1953.

5. J. M. Wozencraft and I. M. Jacobs, *Principles of Communication Engineering*, Wiley, New York, 1965.

PROBLEMS

2-1. Four coins are tossed simultaneously. (a) How many outcomes are possible? (b) Assign suitable probabilities to the events of obtaining all tails, only one head, only two heads, only three heads and all heads in this experiment.

2-2. Three dice are thrown. Assign probabilities to the events of obtaining a sum of 8, 9, and 10 points.

2-3. A certain town has 8 traffic lights randomly located, four of which stay on for one-half minute for the E-W direction and one-half minute for the N-S direction, three stay on for 1/4 minute for the E-W direction and 3/4 minute for the N-S direction, and the one stays on 3/4 minute for the E-W and 1/4 minute for the N-S. Assume that all the traffic lights are independent—that is, there is no synchronization between any of them. An automobile is wandering randomly through the town. Find the probability that it will find an on signal on E-W street. Find the same for N-S street. What is the probability that the automobile wandering randomly will observe an on signal?

2-4. An urn contains 3 red balls and 2 white balls. Two balls are drawn in succession; the first ball being replaced before the second ball is drawn. (a) How many outcomes are possible? (b) Assign probability to each of these outcomes.

2-5. Repeat Prob. 2-4 if the first ball is not replaced before the second draw.

2-6. In Prob. 2-5, if it is known that the first draw is a white ball, what is the probability that the second draw is also a white ball?

2-7. In Prob. 2-5, if it is known that the second draw is a red ball, what is the probability that the first ball drawn is also red? What is the probability that the first ball drawn is white?

2-8. An urn contains 3 red balls, 5 white balls and 8 black balls. Another urn has 6 red balls, 7 white balls and 4 black balls. One ball is drawn from each urn. Find the probability of obtaining both balls of the same color.

2-9. It is given that the probability that a woman will get married and will also subsequently give birth to a child is 0.90, the probability that a married woman will give birth to a child is 0.97, and the probability that a woman will give birth to a child is 0.95. (a) Find the probability of a woman getting married. (b) Find the probability that a woman who gives birth to a child is not married.

2-10. It is given that the probability that a woman will give birth to a child before marriage and will also get married (later) is 0.1, the probability that a woman will give birth to a child is 0.95 and the probability that a woman will get married is 0.927. (a) Find the probability that a married woman will have given birth to a child before marriage. (b) Find the probability that a woman who has a child will have married after this child was born.

2-11. The voltage-amplitude probability-density function of a certain signal is given by

$$p_x(x) = xe^{-x}u(x)$$

What is the probability that the amplitude of the signal will be greater than 1 volt? What is the probability of observing the signal amplitude in the range 1 to 2 volts?

2-12. A voltage waveform $x(t)$ is applied to the input terminals of a rectifier circuit as shown in Fig. P. 2-12. The probability-density function of the input voltage amplitude is given by

$$p_x(x) = \frac{1}{\sigma\sqrt{2\pi}} e^{-x^2/2\sigma^2}$$

(a) Determine and sketch $p_y(y)$ the output-voltage probability-density function. (b) Determine and sketch the distribution function $F_y(y)$ which represents the probability of observing the output-voltage amplitude below y volts.

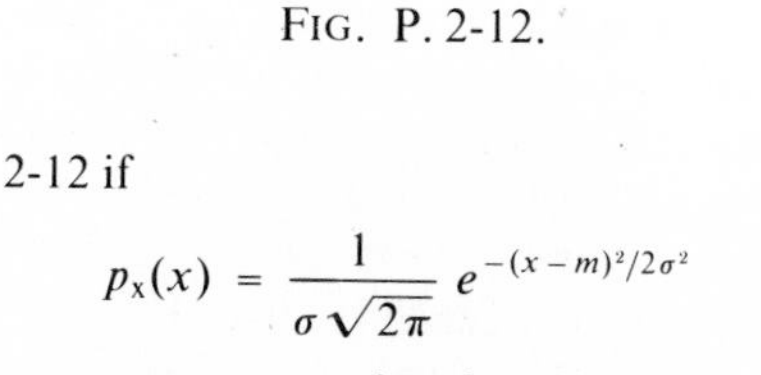

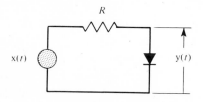

FIG. P. 2-12.

2-13. Repeat Prob. 2-12 if

$$p_x(x) = \frac{1}{\sigma\sqrt{2\pi}} e^{-(x-m)^2/2\sigma^2}$$

2-14. A certain source generates a noise signal which has a gaussian distribution with zero mean value and 2 volts rms value. Find the probability that the signal amplitude will exceed 5 volts.

2-15. Repeat Prob. 2-14 if the mean value of the signal is 1 volt.

2-16. The joint probability-density function $P_{xy}(x, y)$ of two continuous random variables x and y is given by

$$p_{xy}(x, y) = xye^{-(x^2+y^2)/2}u(x)u(y)$$

(a) Find the probability densities $p_x(x)$, $p_y(y)$, $p_x(x \mid y = y)$ and $p_y(y \mid x = x)$. (b) Are the random variables x and y independent?

2-17. The joint probability-density function of two random variables x and y is given by

$$p_{xy}(x, y) = ke^{-(x^2 + 2xy + 2y^2)}$$

(a) Determine the constant k. (b) Determine the probability-density functions $p_x(x)$ and $p_y(y)$. (c) Determine the conditional densities $p_x(x \mid y = y)$ and $p_y(y \mid x = x)$. (d) Are the two variables independent?

2-18. Determine the mean value, the mean-square value and the variance of the signal in Prob. 2-11.

2-19. Determine the mean value and the mean-square value of the output voltage in Fig. P. 2-12.

2-20. The probability-density function of the signal amplitude of a certain source is shown in Fig. P. 2-20. (a) Find the mean value, mean-square value and the variance of this signal. (b) This signal is passed through an ideal rectifier. Determine the mean-square value and the variance of the rectifier output.

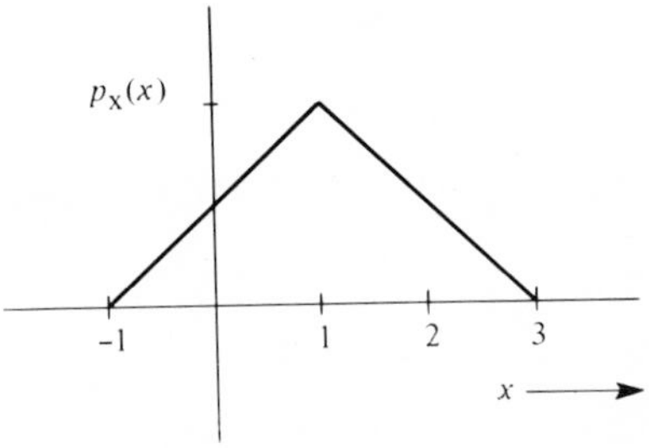

FIG. P. 2-20.

2-21. The input signal x and the output signal y of a half-wave rectifier are related by

$$y = \begin{cases} x^2 & x > 0 \\ 0 & x \leq 0 \end{cases}$$

The input-signal amplitude probability density is given by

$$p_x(x) = \frac{1}{\sigma\sqrt{2\pi}} e^{-x^2/2\sigma^2}$$

Find $p_y(y)$.

2-22. The input signal x and the output signal y of a full-wave rectifier are related by

$$y = x^2$$

The input-signal amplitude probability density is given by

$$p_x(x) = \frac{1}{\sigma\sqrt{2\pi}} e^{-x^2/2\sigma^2}$$

Determine $p_y(y)$.

2-23. In Prob. 2-22,

$$p_x(x) = \frac{1}{\sigma\sqrt{2\pi}} e^{-(x-1)^2/2\sigma^2}$$

Determine $p_y(y)$.

2-24. The probability-density function of a random variable x is shown in Fig. P. 2-24a. A random variable y is related to x as shown in Fig. P. 2-24b. Determine $p_y(y)$.

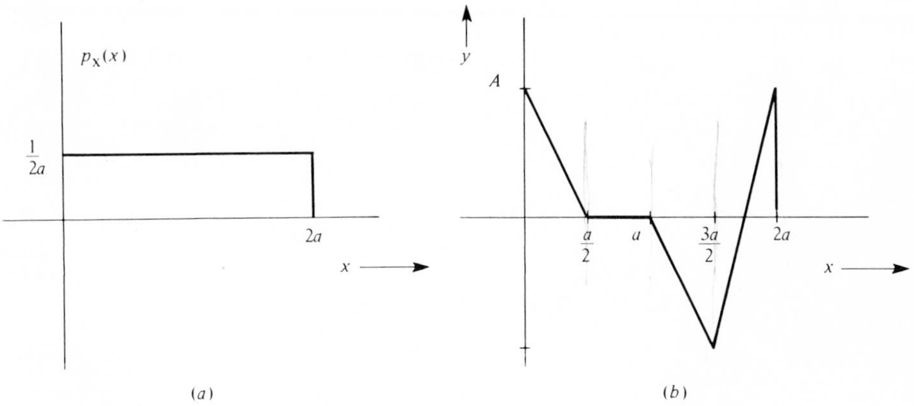

FIG. P. 2-24.

2-25. A source generates a pure sinusoidal signal voltage 2 cos $2\pi t$. The voltage is observed at a random instant t. (a) Show that the amplitude probability-density function of the voltage x is given by

$$p_x(x) = \begin{cases} \dfrac{1}{\pi\sqrt{1 - x^2}} & -1 < x < 1 \\ 0 & \text{Otherwise} \end{cases}$$

(b) Determine the probability that the voltage will be greater than 0.5. (c) Determine the mean and the mean-square value of the voltage observed from the knowledge of its probability-density function obtained in part (a). (*Hint*: The signal being periodic, one may consider the random variable only in the range of one period, viz., 1 sec. Thus the random variable t has a uniform distribution over the range 0, 1 ($p_t(t) = 1$). The amplitude x is a function of random variable t, x = cos 2πt. Now use Eq. 2-50 to obtain $p_x(x)$. (Note that x is a double-valued function of t.)

2-26. The joint probability-density function of random variables x and y is given as

$$p_{xy}(x, y) = \frac{1}{2\pi\sigma^2} e^{-(x^2+y^2)/2\sigma^2}$$

Find the probability-density function of the random variables z and w given by

$$\text{(a) } z = ax + by, \quad w = cx + dy$$

$$\text{(b) } z = yx^2u(x), \quad w = xy^2u(y)$$

2-27. The joint probability-density function of two random variables x and y is given by

$$p_{xy}(x, y) = ke^{-(x^2+2xy+2y^2)}$$

(a) Determine the mean value and the mean-square value of the random variable z = xy. (b) Determine the mean value and the mean-square value of the random variable z = $2x^2 + 3y$.

2-28. If $P_x(\omega)$ is the Fourier transform of a probability-density function $p_x(x)$ and m_n represents the nth moment of the variable x,

$$m_n = \int_{-\infty}^{\infty} x^n p_x(x)\, dx$$

(a) Then show that

$$m_n = (j)^n \left. \frac{d^n P_x(\omega)}{d\omega^n} \right|_{\omega=0}$$

(b) Show that if $P_x(\omega)$ is expanded by Taylor's series, then

$$P_x(\omega) = m_0 - jm_1\omega - \frac{m_2\omega^2}{2!} + \frac{jm_3\omega^3}{3!} + \cdots + = \sum_{n=0}^{\infty} (-j)^n m_n \left(\frac{\omega^n}{n!}\right)$$

2-29. Use the results in Prob. 2-28 to determine the mean value and the mean-square value of (a) A gaussian signal. (b) The signal in Prob. 2-11. (c) The input signal in Prob. 2-20. (*Hint*: Find $P_x(\omega)$ and expand in power series as in 2-28b. The second and the third coefficients represent the mean and the mean-square values).

2-30. Find the moments of random variable x uniformly distributed as shown in Fig. P. 2-30. Use results in Prob. 2-28b to obtain $P_x(\omega)$ for $p_x(x)$ in Fig. P. 2-30.

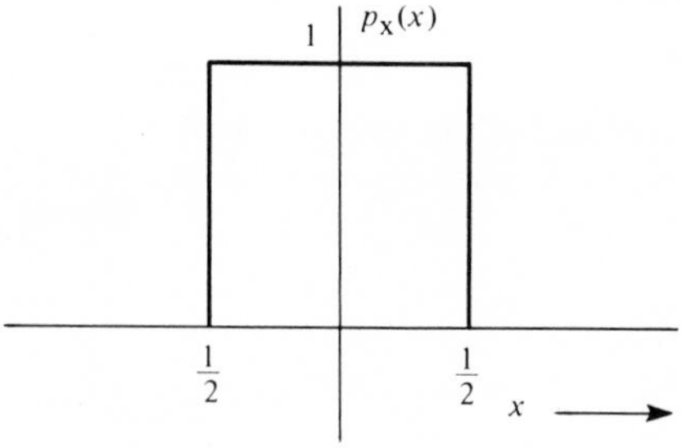

FIG. P. 2-30.

2-31. Signal $x(t)$ is a gaussian random signal with a mean value of 1 and mean-square value of 4. The signal $y(t)$ represents a random telegraph signal which can assume only two discrete values +1 and −1 with equal probability (Fig. P. 2-31). A new signal $z(t)$ is formed by adding together $x(t)$ and $y(t)$:

$$z(t) = x(t) + y(t)$$

Determine the probability density function, the mean value and the mean-square value of the signal $z(t)$. (b) If signal $x(t)$ is a random telegraph signal which can assume only two values +1 and −1 with probability p and $1 - p$ respectively, and $y(t)$ is similar to $x(t)$ but with probabilities q and $1 - q$, determine the probability density of $z(t)$.

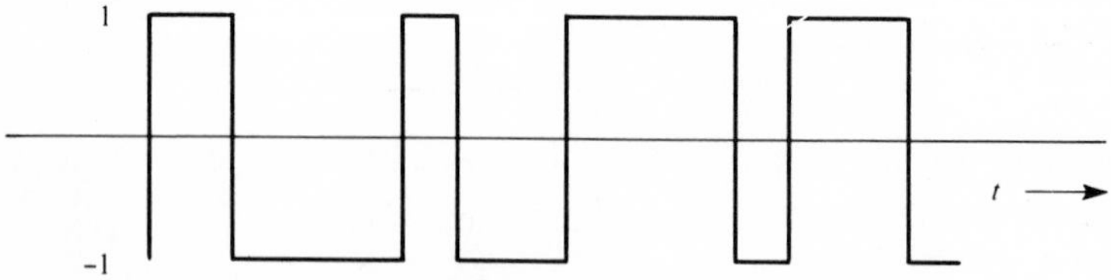

FIG. P. 2-31.

2-32. The amplitude probability-density function of signal $x(t)$ is shown in Fig. P. 2-32. Determine the probability-density function of the signal $z(t)$ given by

$$z(t) = x(t) + x^2(t)$$

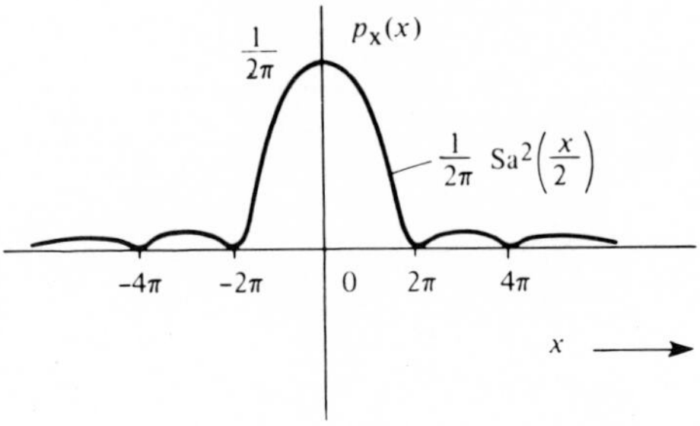

FIG. P. 2-32.

2-33. The amplitude probability-density functions of the signals $x(t)$ and $y(t)$ are given by

$$\begin{aligned}
&(a)\ p(x) = ae^{-x}u(x); && p(y) = be^{y}u(-y)\\
&(b)\ p(x) = ae^{-|x|}; && p(y) = be^{y}u(-y)\\
&(c)\ p(x) = ae^{-|x|}; && p(y) = bye^{-y}u(y)
\end{aligned}$$

A new signal $z(t)$ is formed:

$$z(t) = x(t) + y(t)$$

Determine the probability-density function of $z(t)$ in each case.

3

Characterization of Random Signals

3-1. INTRODUCTION

In this chapter we shall study various methods of describing random signals. As mentioned in the preceding chapter, it is not possible to completely specify a random process. Hence there is an inherent limit to the amount of knowledge we can obtain from such a study, due to a lack of complete information. The description of random processes is statistical in nature—that is, the characterization is in terms of certain averages. When we apply random signals thus specified at the input terminals of a system, we hope to obtain a similar statistical characterization of the output signals. There are, of course, a large number of statistical averages or parameters of the input signal which can be studied. We must, however, restrict ourselves to those parameters that are relatively simple to derive from the theoretical model or the empirical data available of the random process. These parameters should also be such that a knowledge of their values at the input of a system should enable us to calculate their values at the output. In addition, the knowledge of these parameters corresponding to the input and the output signals of a system should yield significant information regarding the quality of a system.

3-2. THE CONCEPT OF A RANDOM PROCESS

In Chapter 2 we associated one sample point ζ with each random outcome of the experiment. The collection of all sample points was called the sample space $\mathcal{S}$ for the experiment. To each sample point ζ over this space we assigned a real number x according to some rule.

$$x = x(\zeta)$$

This function defined over the sample space was called the random variable x.

A random process is an extension of the concept of a random variable. In the case of random variables, we assigned a real number to a sample point over the sample space of a given experiment. In the case of random processes, to each sample point ζ, we assign a waveform $x(t)$ (which is a function of time)[1], according to some rule

$$\mathrm{x}(t, \zeta)$$

Thus each sample space will have a certain collection of waveforms, each assigned to one sample point. The collection of waveforms is called an *ensemble*. The individual waveforms in the ensemble are known as *sample functions*. Thus each sample function corresponds to one sample point. The probability distribution of the sample points over the sample space will determine the probability distribution of sample functions over the ensemble. The probability system, comprising the sample space, the ensemble (set of waveforms) and the probability-distribution function is called a *random process*.

As an example, consider the case of two successive tosses of a coin. If we consider a particular order of appearance of heads and tails as our outcome, then there are four outcomes HH, HT, TH, and TT. There are four sample points ζ_1, ζ_2, ζ_3, and ζ_4, associated with these outcomes respectively. The probability assigned to each sample point is 1/4. We now assign a waveform (or a function of time) to each of the sample points. We can assign four waveforms $x(t, \zeta_1)$, $x(t, \zeta_2)$, $x(t, \zeta_3)$, and $x(t, \zeta_4)$ according to some rule. For example, we may have

$$x(t, \zeta_k) = \sin kt$$

The ensemble in this case has 4 sample functions $\sin t$, $\sin 2t$, $\sin 3t$, and $\sin 4t$. The probability assigned to each waveform is 1/4.

The number of sample points over a given sample space may be finite, enumerably infinite, or nonenumerably infinite. Correspondingly, the ensemble may consist of finite, enumerably infinite, or nonenumerably infinite sample functions.

A random process is also called a *stochastic process*. The notation $\mathrm{x}(t, \zeta)$ represents a random process. It is, however, common practice to omit ζ and express the random process simply by $\mathrm{x}(t)$, meaning a collection of waveforms occurring with a certain measure of probability. We shall use roman $\mathrm{x}(t)$ to represent a random process. A particular sample function of this random process will be denoted by italics $x(t)$. For convenience the sample space variable ζ is omitted in this notation.

The reader should note carefully the three concepts; (i) the outcome of the experiment, (ii) the sample point associated with the outcome, and (iii) the sample function (waveform) assigned to that particular sample point. In practice we often come across a situation where the direct out-

[1] In general the function of time $\mathrm{x}(t)$ may be complex.

come of the random experiment is a waveform. In such case, the concept of the outcome and the sample function assigned to that outcome tend to fuse together. The outcome itself may be considered to be the sample function assigned to that outcome. As an example, consider the noise signal at the output of any electrical system. Each time the system is turned on it will yield a different noise waveform. The collection (which will be nonenumerably infinite) of all such noise waveforms can be taken as the ensemble for this random process. Such an ensemble is shown in Fig. 3-1. We may obtain this ensemble by repeating the observations over the same system or observing the outcomes of several identical systems simultaneously.

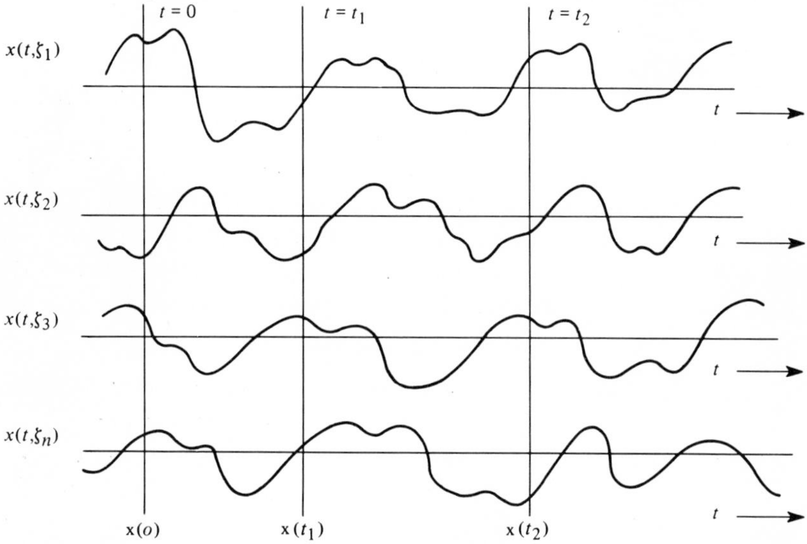

FIG. 3-1.

It is important to realize that various waveforms (sample function) are assigned to sample points (or the outcomes). Hence the waveforms themselves are assumed to be deterministic. There is no randomness associated with the waveform of any sample function. The randomness is associated with the occurrence of a particular sample function (assigned to a particular outcome). The probability distribution over the sample space is the probability distribution associated with observing a particular sample function.

Some random processes may be too complicated to be described by an analytical expression. The noise signal of an electrical system is one such case. In other cases one may be able to describe a random process by an analytical expression and the appropriate probability measure. Consider for example, a large collection of receivers which are receiving a carrier signal transmitted by some station. The path traveled by the carrier from the transmitter to the receiver will vary with each receiver (since the receivers are located at different positions). This will cause the phase of the received signal to be different at each receiver. Also the gain of antennae will be different for each receiver. As a result, the collection of received waveforms can be described as

$$x(t) = A\cos(\omega_0 t + \Theta) \tag{3-1}$$

where the amplitude A and the phase Θ are both random variables. We can determine the probability distributions of A and Θ empirically. We may have, for example, a gaussian probability density for the amplitude A, and Θ may be uniformly distributed over the range $(0, 2\pi)$; that is,

$$p_A(A) = \frac{1}{\sigma\sqrt{2\pi}}\, e^{-(A-m)^2/2\sigma^2} \tag{3-2}$$

$$p_\Theta(\theta) = \begin{cases} \dfrac{1}{2\pi} & 0 < \theta < 2\pi \\ 0 & \text{Otherwise} \end{cases}$$

The amplitude A and phase Θ are constant for a given sample function but vary from one sample function to the other. The probability distri-

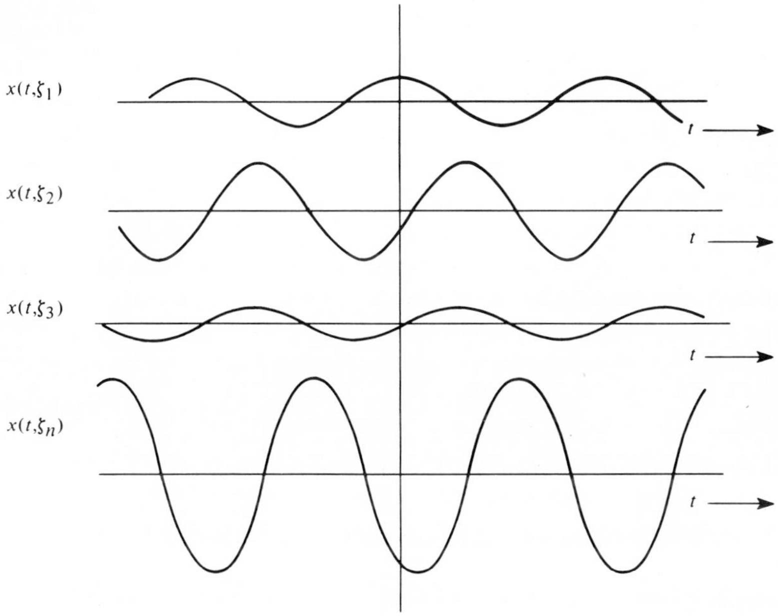

FIG. 3-2.

bution of these variables over the ensemble is given by Eq. 3-2. The ensemble x(t) in Eq. 3-1 along with the probability measure in Eq. 3-2 is an example of a random process. The ensemble for this process is shown in Fig. 3-2.

Another example of a random process which can be described analytically is

$$\mathrm{x}(t) = \mathrm{a}t + \mathrm{b} \tag{3-3}$$

where a and b are random variables with some given probability distribution over the ensemble. For a given sample function, a and b are constants but vary from one sample function to the other.

3-3. DESCRIPTION OF AN ENSEMBLE

A random process is described by its ensemble and the probability measure over the ensemble. From this information we derive the secondary statistics of the process, which prove very useful in describing and specifying the process.

Consider again a process x(t) whose ensemble is shown in Fig. 3-1. For convenience we shall here use the notation $\mathrm{x}(t, \zeta)$ instead of x(t) to represent the process. At any instant $t = t_1$ the amplitudes of various sample functions over the ensemble can be represented by a collection of numbers $\mathrm{x}(t_1, \zeta)$, where ζ is some sample point (Fig. 3-1). This collection has a certain probability distribution, that is, the amplitudes of various sample functions at $t = t_1$ has a certain probability distribution. Obviously $\mathrm{x}(t_1, \zeta)$ is a random variable[2] according to the concepts in Chapter 2. The random variable $\mathrm{x}(t_1, \zeta)$ is usually denoted by roman $\mathrm{x}(t_1)$. Thus roman x(t) represents the random variable $\mathrm{x}(t, \zeta)$, the amplitudes of sample functions at instant t. Note that roman x(t) is used to represent the random process as well as the random variable representing sample function amplitudes at instant t. The specific interpretation of x(t) should be clear from the context.

The notation will now be further simplified in accordance with common practice, by using the symbol x_1 instead of $\mathrm{x}(t_1)$ to represent the random variable. Henceforth in our discussion, it will be understood that x_1 represents the random variable defined by the value of all sample functions over the ensemble at a particular instant $t = t_1$. Similarly, x_2 is the random variable defined by sample function values at $t = t_2$ and so on. In this way we can define an infinite number of random variables

[2]This condition is imposed upon a random process. In other words, for a random process, we require that for any $t = t_0$, $\mathrm{x}(t_0, \zeta)$, must form a random variable over the ensemble.

over the ensemble. The statistics of each of these variables individually are called first-order statistics of the random process. The joint statistics of two such variables, (x_1 and x_2 for instance), are called *second-order statistics*. Similarly the joint statistics of n such variables are the nth order statistics. We shall start by discussing the first order statistics.

First-Order Statistics. The first-order statistics are completely specified by the probability distribution of the random variable—that is by specifying the probability density function of the amplitudes of sample functions at instant t (for all values of t). This random variable $\mathrm{x}(t)$ will be denoted by x for convenience. The probability density function of the variable x is $p_{\mathrm{x}}(x)$. In this notation, we must however indicate somewhere the fact that the random variable x is defined by amplitude of sample functions at instant t. For this reason the notation $p_{\mathrm{x}}(x;t)$ is more appropriate. In practice the subscript x is generally dropped for convenience. Thus we shall use the notation $p(x;t)$ to represent the probability-density function of the random variable $\mathrm{x}(t)$ or x defined by amplitudes of the sample functions of the process at instant t. Note that $p(x;t)$ is not a joint density function of two variables x and t. The parameter t indicates the instant at which the sample-function amplitudes define the variable $\mathrm{x}(t)$. Once the probability-density function $p(x;t)$ is determined, the mean value and the mean-square value of the random variable $\mathrm{x}(t)$ can be determined according to the equations

$$\overline{\mathrm{x}(t)} = \int_{-\infty}^{\infty} x p(x;t)\,dx \tag{3-4a}$$

$$\overline{\mathrm{x}^2(t)} = \int_{-\infty}^{\infty} x^2 p(x;t)\,dx \tag{3-4b}$$

Example 3-1. Find the first-order probability-density function $p(x;t)$ for a random process $\mathrm{x}(t)$ described by

$$\mathrm{x}(t) = \mathrm{a}t + b \tag{3-5}$$

where b is constant and a is a random variable over the ensemble having a probability density

$$p_{\mathrm{a}}(a) = \frac{1}{\sqrt{2\pi}}\, e^{-a^2/2} \tag{3-6}$$

We wish to find the probability-density function of x(i.e. $\mathrm{x}(t)$) which is a random variable formed by sample-function amplitudes at instant t. The variable x which is a function of the random variable a is given by

$$\mathrm{x}(t) = \mathrm{a}t + b$$

so

$$\mathrm{a} = \frac{1}{t}(\mathrm{x} - b)$$

Also,

$$\frac{dx}{da} = t$$

Hence from the results of Sec. 2-7 we have

$$p_x(x) = \frac{p_a(a)}{dx/da} = \frac{1}{t\sqrt{2\pi}} e^{-a^2/2}$$

$$= \frac{1}{t\sqrt{2\pi}} e^{-(x-b)^2/2t^2}$$

or

$$p(x;t) = \frac{1}{t\sqrt{2\pi}} e^{-(x-b)^2/2t^2} \tag{3-7}$$

This is the desired first-order probability-density function for the process x(t) in Eq. 3-5. The amplitudes of sample functions at instant t will have the probability-density function as given by Eq. 3-7. Thus the amplitudes of sample functions in the ensemble at $t = 2$ will have the probability density

$$p(x;2) = \frac{1}{2\sqrt{2\pi}} e^{-(x-b)^2/8}$$

The functions $p_a(a)$ and $p(x;2)$ are shown in Fig. 3-3. The variable x(t) has a gaussian distribution. The mean and the mean-square values of x can be easily seen to be (see Ex. 2-11 and 2-12).

$$\overline{x(t)} = b$$
$$\overline{x^2(t)} = b^2 + t^2$$

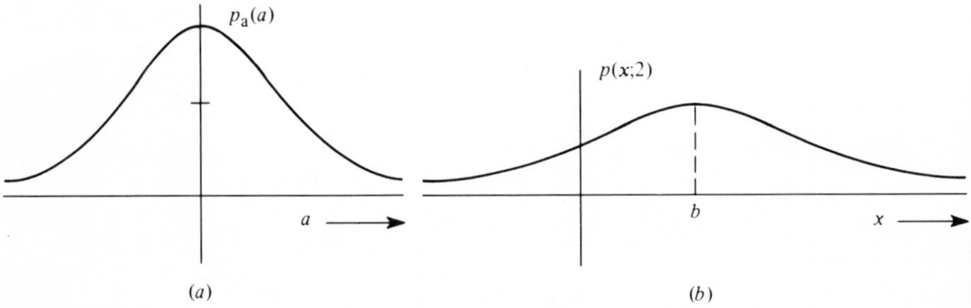

FIG. 3-3.

Example 3-2. Find the first-order probability-density function $p(x;t)$ for a random process

$$x(t) = k \cos(\omega_0 t + \Theta)$$

where Θ is a random variable over the ensemble and is uniformly distributed over the interval $(0, 2\pi)$.

$$p_\Theta(\theta) = \begin{cases} \dfrac{1}{2\pi} & 0 < \theta < 2\pi \\ 0 & \text{otherwise} \end{cases}$$

The random variable x (x(t)) is given by

$$\mathrm{x} = k\cos(\omega_0 t + \Theta) \tag{3-8}$$

Thus x is a function of another random variable Θ. As before we have

$$p(x;t) = \frac{p_\Theta(t)}{dx/d\theta}$$

Once again it should be remembered that the parameter t is a constant here. Note, however, that x is a double-valued function of θ—that is, for each value x, there are two values of θ as seen from Fig. 3-4a. Hence from Eq. 2-50a,

$$p(x;t) = \frac{p_\Theta(\theta_1)}{|dx/d\theta_1|} + \frac{p_\Theta(\theta_2)}{|dx/d\theta_2|}$$

for both roots θ_1 and θ_2, $p_\Theta(\theta) = \dfrac{1}{2\pi}$ and $x = k\cos(\omega_0 t + \theta)$,

$$\frac{d\theta}{dx} = \frac{1}{dx/d\theta} = k\sin(\omega_0 t + \theta)$$

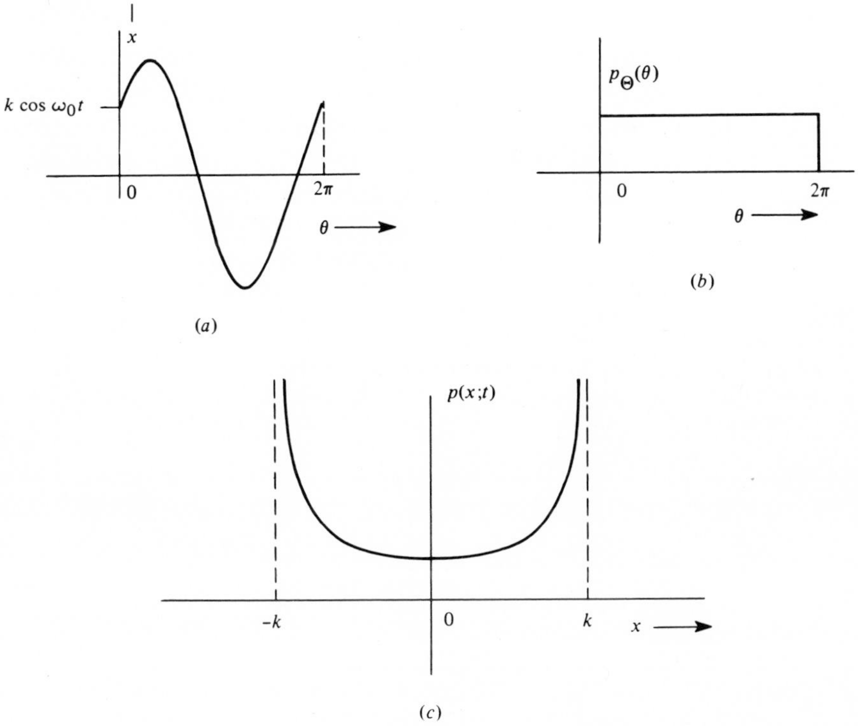

FIG. 3-4.

$$= k\left[\sqrt{1 - \cos^2(\omega_0 t + \theta)}\right]$$

$$= k\sqrt{1 - \frac{x^2}{k^2}}$$

$$= \sqrt{k^2 - x^2}$$

and hence

$$p(x;t) = \frac{1}{2\pi\sqrt{k^2 - x^2}} + \frac{1}{2\pi\sqrt{k^2 - x^2}}$$

$$= \frac{1}{\pi\sqrt{k^2 - x^2}}$$

The random variable x as given in Eq. 3-8, exists in the range $(-k, k)$ and is zero outside this range.

$$p(x;t) = \begin{cases} \dfrac{1}{\pi\sqrt{k^2 - x^2}} & -k \le x \le k \\ 0 & \text{Otherwise} \end{cases} \tag{3-9}$$

This density is shown in Fig. 3-4c.

Observe that $p(x;t)$ in Eq. 3-9 is independent of t; that is, the amplitude distribution of the ensemble is the same for all values of t. If the random variable Θ in this problem were distributed uniformly over the range $(0, \pi)$ instead of $(0, 2\pi)$, then the probability density function $p(x;t)$ will depend upon t. This is left as an exercise for the reader. The reader can easily show using Eq. 3-4 and 3-9 that in this example

$$\overline{\mathrm{x}} = 0$$

and

$$\overline{\mathrm{x}^2} = \frac{k^2}{2}$$

Relative-Frequency Interpretation of $p(x;t)$. The first-order density $p(x;t)$ is the probability density of the sample-function amplitude at instant t. Thus $p(x;t)\,dx$ represents the probability that a sample-function amplitude will lie in the interval $(x, x + dx)$ at instant t. This is obviously a probability that a sample function will pass through a slit in the xt plane as shown in Fig. 3-5. If there are a total of N sample functions, of which m sample functions pass through the slit, then according to the relative-frequency interpretation,

$$p(x;t)\,dx = \frac{m}{N}$$

$$p(x;t) = \frac{m}{N\,dx}$$

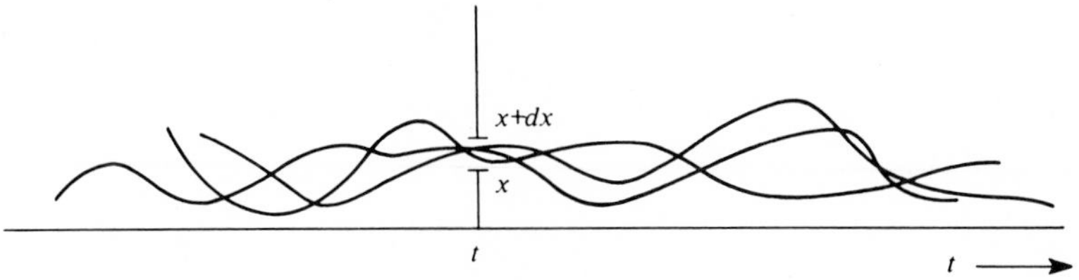

FIG. 3-5.

Second- and Higher-Order Statistics. The first-order statistics give us the amplitude distributions of the sample functions for all values of t. We now ask ourselves whether the probability-density function $p(x;t)$ gives all the statistical information about the random process, or whether we need to obtain other probability-density functions to have a complete statistical knowledge of the process. It is not difficult to see that the knowledge of $p(x;t)$ is very inadequate in describing the process. Suppose that the process represents electrical signals and it is desired to draw an inference regarding the frequency content of these signals. If a signal contains predominantly low-frequency components, then the signal changes very slowly and $x(t_1)$, the signal value of a generator at $t = t_1$ and $x(t_1 + \tau)$, the value at $t = t_1 + \tau$ are not much different. It is evident that the random variables $\mathrm{x}(t_1)$ and $\mathrm{x}(t_1 + \tau)$ are not statistically independent, provided that τ is small enough. The knowledge of one gives a certain amount of information regarding the other. (See Fig. 3-6a.) On the other hand, if the random signals contain predominantly high frequencies, the signal changes very rapidly and the values of the signal separated by the same interval τ may show very little resemblance or dependence. (Fig. 3-6b.) Hence the correlation between signal values at various time intervals yields useful information regarding its frequency contents. This will be shown more rigorously later. For the moment, we observe that the first-order probability-density function cannot yield information regarding the extent of statistical dependence of signal values at two instants and hence is inadequate in describing the process. We know that a measure of the statistical relatedness of two random variables is obtained from their correlation, which, in turn can be obtained from their joint probability-density function. Thus in addition to the first-order statistics we must also obtain the joint statistics of two variables $\mathrm{x}(t_1)$ and $\mathrm{x}(t_2)$ which are defined by the amplitude of sample functions at instants t_1 and t_2 respectively. We shall denote these variables by x_1 and x_2 respectively. The joint probability-density function of these variables is $p_{\mathrm{x}_1\mathrm{x}_2}(x_1,x_2)$. Following our previous practice of dropping the subscript $\mathrm{x}_1\mathrm{x}_2$ and indicating the parameter t_1 and t_2 we shall denote this joint probability

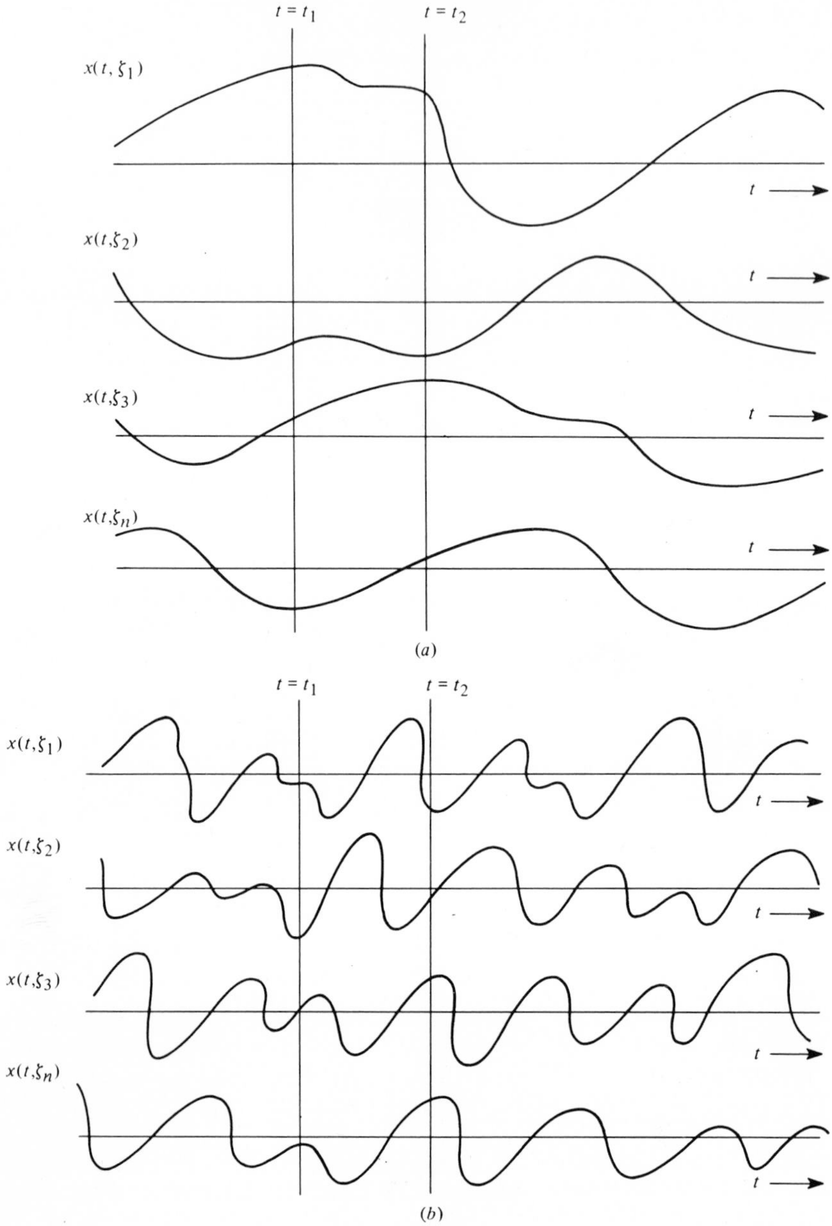

FIG. 3-6.

density by $p(x_1,x_2;t_1,t_2)$. This is also known as the second-order probability-density function of the process. The probability of the joint event that a sample function amplitude will be in the range $(x_1, x_1 + dx_1)$ at $t = t_1$ and in the range $(x_2, x_2 + dx_2)$ at $t = t_2$ is given by $p(x_1,x_2;t_1,t_2)\,dx_1\,dx_2$ (in the limit $dx_1, dx_2 \to 0$). This probability-density function can be interpreted according to relative-frequency definition in a manner similar to that given for the first-order probability-density function. We plot all the N sample functions on the same system of coordinates, and if the number of sample functions passing simultaneously through two slits of widths dx_1 and dx_2, located at $t = t_1$ and t_2 respectively (Fig. 3-7) is m_{12}, then

$$p(x_1,x_2;t_1,t_2)\,dx_1\,dx_2 = \frac{m_{12}}{N}$$

and

$$p(x_1,x_2;t_1,t_2) = \frac{m_{12}}{N\,dx_1\,dx_2}$$

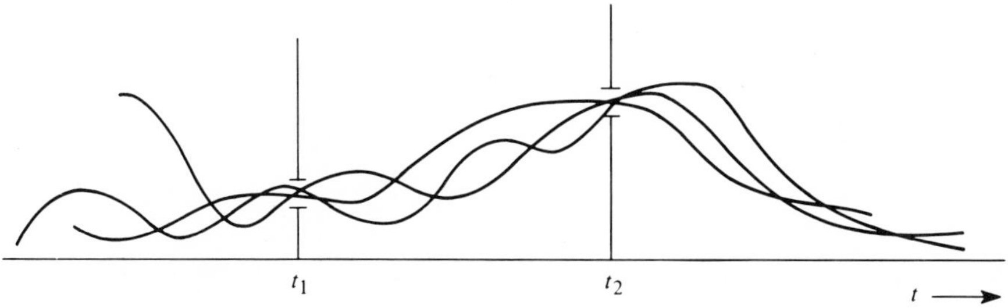

FIG. 3-7.

Note that the second-order probability density is a function of two independent variables x_1, x_2 and two parameters t_1, and t_2. The nth-order probability-density function is $p(x_1,x_2, \ldots, x_n;t_1,t_2, \ldots, t_n)$ and represents a joint probability-density function of sample function values at $t = t_1, t_2, \ldots, t_n$. Here x_k represents the value of the random variable defined by sample function amplitudes at $t = t_k$.

It is interesting to note that an nth-order probability-density function contains all the information in the first $(n - 1)$ probability-density functions. We can easily show that a lower-order probability-density function can be obtained by successive integration of a higher-order probability-density function. This point will now be illustrated by a second-order probability-density function. We have

$$p(x_1,x_2) = p(x_1)\,p(x_2 \mid x_1)$$

For the sake of clarity we shall use the notation for probability densities with time parameters. Thus the equation above can be written as

$$p(x_1,x_2;t_1,t_2) = p(x_1;t_1)p(x_2;t_2 \mid x_1;t_1) \tag{3-10}$$

Here $p(x_2;t_2 \mid x_1;t_1)$ is the conditional probability density associated with the event that a sample function has an amplitude at x_2 at t_2, given that the amplitude at t_1 is x_1. Integrating both sides of Eq. 3-10 with respect to x_2, we have

$$\int_{-\infty}^{\infty} p(x_1,x_2;t_1,t_2)\,dx_2 = \int_{-\infty}^{\infty} p(x_1;t_1)\,p(x_2;t_2 \mid x_1;t_1)\,dx_2$$

$$= p(x_1;t_1) \int_{-\infty}^{\infty} p(x_2;t_2 \mid x_1;t_1)\,dx_2$$

The integral on the right-hand side of the above equation represents the probability of observing any value of x_2 in the range $-\infty, \infty$ at $t = t_2$ given that the signal value is x_1 at t_1. This is obviously unity since x_2 must assume some value in the range $(-\infty, \infty)$. Hence we obtain

$$p(x_1;t_1) = \int_{-\infty}^{\infty} p(x_1,x_2;t_1,t_2)\,dx_2 \tag{3-11}$$

Similarly,

$$p(x_2;t_2) = \int_{-\infty}^{\infty} p(x_1,x_2;t_1,t_2)\,dx_1 \tag{3-12}$$

It is evident from Eqs. 3-11 and 3-12 that integration of a probability-density function yields a lower-order probability-density function. Hence if we have an nth order probability-density function it is possible to obtain all $(n - 1)$ lower probability-density functions by successive integration.

A random process is completely specified by the joint probability density function $p(x_1, x_2, \ldots, x_n; t_1, t_2, \ldots, t_n)$ for any finite set observation instants $t_1, t_2, \ldots, t_n$ and for any value of n. To determine such a density is a lot of work and may sound very discouraging. Fortunately for our purpose we do not need all this information. We shall be concerned mainly with the transmission of random signals through linear systems. It will be shown in a later section that the first- and second-order probability-density functions suffice to determine the power density spectrum, the mean and the mean-square value of a process. In particular, the correlation between variables x_1 and x_2 is sufficient to determine the power-density spectrum.[3] The correlation between x_1 and x_2, that is, $\overline{\mathrm{x}_1\mathrm{x}_2}$ (which is a significant parameter) is obtained from the second-order statistics. The correlation of random variables x_1 and x_2 is denoted by $R_{\mathrm{x}_1\mathrm{x}_2}$ according to our convention. However, to bring out the fact clearly that x_1 and x_2 are the random variables defined by sample-function amplitudes at t_1, and t_2 respectively, we denote this correlation by $R_{\mathrm{x}}(t_1,t_2)$. This is called *autocorrelation function* of the

[3]For the case of transmission of random signals through nonlinear systems, in general, higher orders of probability density functions are required.

process $\mathrm{x}(t)$, so named because it is the correlation between two variables, both defined on the same process. It is a measure of the dependence of the amplitudes of the sample functions at $t = t_1$ on the amplitudes of the same (auto) functions at $t = t_2$. Later on we shall define crosscorrelation as a correlation between sample function amplitudes of two difference processes.

The autocorrelation function is an ensemble (statistical) average of the product of two random variables. Hence we can express it in terms of the joint probability-density function $p(x_1,x_2;t_1,t_2)$

$$R_{\mathrm{x}}(t_1,t_2) = \overline{\mathrm{x}_1\mathrm{x}_2} = \int_{-\infty}^{\infty}\int_{-\infty}^{\infty} x_1 x_2\, p(x_1,x_2;t_1,t_2)\,dx_1\,dx_2 \tag{3-13}$$

Example 3-3. Determine the correlation of variables x_1 and x_2 for the random process $\mathrm{x}(t)$ in Example 3-1. We have

$$\mathrm{x} = \mathrm{a}t + b$$
$$\mathrm{x}_1 = \mathrm{x}(t_1) = \mathrm{a}t_1 + b$$
$$\mathrm{x}_2 = \mathrm{x}(t_2) = \mathrm{a}t_2 + b$$

The autocorrelation function $R_{\mathrm{x}}(t_1,t_2)$ is given by

$$\begin{aligned} R_{\mathrm{x}}(t_1,t_2) &= E[\mathrm{x}(t_1)\mathrm{x}(t_2)] \\ &= E[(\mathrm{a}t_1 + b)(\mathrm{a}t_2 + b)] \\ &= E[\mathrm{a}^2 t_1 t_2 + \mathrm{a}b(t_1 + t_2) + b^2] \\ &= t_1 t_2 E[\mathrm{a}^2] + (t_1 + t_2) b\, E[\mathrm{a}] + b^2 \end{aligned}$$

Since a is a gaussian variable with zero mean and unity variance (see Eq. 3-6)

$$E[\mathrm{a}] = \bar{\mathrm{a}} = 0 \quad \text{and} \quad E[\mathrm{a}^2] = \overline{\mathrm{a}^2} = 1$$

Hence

$$R_{\mathrm{x}}(t_1,t_2) = t_1 t_2 + b^2 \tag{3-14}$$

Example 3-4. Determine the autocorrelation function $R_{\mathrm{x}}(t_1,t_2)$ for the random process in Example 3-2.

$$\mathrm{x} = k \cos(\omega_0 t + \Theta)$$
$$\mathrm{x}_1 = \mathrm{x}(t_1) = k\cos(\omega_0 t_1 + \Theta)$$
$$\mathrm{x}_2 = \mathrm{x}(t_2) = k\cos(\omega_0 t_2 + \Theta)$$

Hence

$$\begin{aligned} R_{\mathrm{x}}(t_1,t_2) &= E[\mathrm{x}(t_1)\mathrm{x}(t_2)] \\ &= k^2 E[\cos(\omega_0 t_1 + \Theta)\cos(\omega_0 t_2 + \Theta)] \\ &= \frac{k^2}{2} E[\cos(\omega_0 t_1 - \omega_0 t_2) + \cos(\omega_0 t_1 + \omega_0 t_2 + 2\Theta)] \\ &= \frac{k^2}{2}\{E[\cos(\omega_0 t_1 - \omega_0 t_2)] + E[\cos(\omega_0 t_1 + \omega_0 t_2 + 2\Theta)]\} \end{aligned}$$

Note that $\cos(\omega_0 t_1 - \omega_0 t_2)]$ is not a random quantity; hence

$$E[\cos(\omega_0 t_1 - \omega_0 t_2)] = \cos(\omega_0 t_1 - \omega_0 t_2)$$

Also

$$\begin{aligned} E[\cos(\omega_0 t_1 + \omega_0 t_2 + 2\Theta)] &= \int_{-\infty}^{\infty} \cos(\omega_0 t_1 + \omega_0 t_2 + 2\theta) p_\Theta(t)\, d\theta \\ &= \frac{1}{2\pi} \int_{-0}^{2\pi} \cos(\omega_0 t_1 + \omega_0 t_2 + 2\theta)\, d\theta \\ &= 0 \end{aligned}$$

and

$$R_x(t_1,t_2) = \frac{k^2}{2} \cos[\omega_0(t_1 - t_2)] \tag{3-15}$$

3-4. STATIONARY AND NONSTATIONARY RANDOM PROCESSES

In Ex. 3-1, the probability-density function $p(x;t)$ was a function of parameter t (Eq. 3-7) whereas in Ex. 3-2, $p(x;t)$ was found to be independent of t (Eq. 3-9). For the random process in Ex. 3-1, the first-order statistics are different at different instants whereas for the random process in Ex. 3-2, the first-order statistics are identical at every instant. A shift of time origin will be indistinguishable for the second random process (example 3-2) so far as the first-order statistics are concerned. This is not true for the first random process (example 3-1). The second process therefore appears to have a stationary nature in the sense that its (first-order) statistics remain unchanged with time. In general we call a process a *stationary* process if all its statistics (all orders of probability functions) exhibit this invariance with time shift. Thus we define a process to be a stationary (or a *strictly stationary*) if none of its statistics are affected by a shift in the time origin.

It is obvious that for a stationary process the probability density $p(x;t)$ is independent of t and, therefore, can be expressed as

$$p(x;t) = p(x)$$

However, since the probability densities are unaffected by change in time origin, the second probability density $p(x_1,x_2;t_1,t_2)$ must depend upon t_1 and t_2 through the difference $t_1 - t_2$ alone. This is the only possible way in which the second-order statistics can be independent of time origin, as can easily be seen in the following. The second-only probability-density function $p(x_1,x_2;t_1,t_2)$ become $p(x_1,x_2;t_1 + T,t_2 + T)$ when the origin is shifted by T seconds (along the negative direction). Stationarity demands that for all values of T

$$p(x_1,x_2;t_1,t_2) = p(x_1,x_2;t_1 + T,t_2 + T)$$

This is possible only if

$$p(x_1,x_2;t_1,t_2) = p(x_1,x_2;t_2 - t_1)$$

If we let $t_2 - t_1 = \tau$, then

$$p(x_1,x_2;t_1,t_2) = p(x_1,x_2;\tau)$$

In a similar way we can show that the nth-order probability-density function is independent of the time origin if and only if

$$p(x_1,x_2,\ldots,x_n;t_1,t_2,\ldots,t_n) = p(x_1,x_2,\ldots,x_n;\tau_1,\tau_2,\ldots,\tau_{n-1}) \quad (3\text{-}16)$$

where $\tau_k = t_{k+1} - t_1$.

We note that the autocorrelation $R_x(t_1,t_2)$ is one of the second-order statistics parameter. Hence for a stationary process

$$R_x(t_1,t_2) = R_x(\tau); \quad \tau = t_2 - t_1$$

We observe in Eq. 3-14, that $R_x(t_1,t_2) = 2t_1t_2 + b^2$. Obviously the random process in Ex. 3-1 is not stationary. On the other hand, we have in Eq. 3-15

$$R_x(t_1,t_2) = \frac{k^2}{2}\cos\omega_0\tau = R_x(\tau)$$

We cannot, however, conclude that this process (Ex. 3-2) is stationary just from this information. For stationarity we must ascertain that all higher-order statistics are also independent of time origin—that is, Eq. 3-16 must be satisfied for all values of n.

It is not easy to determine whether or not a process is stationary. If the probability densities depend upon the choice of a time origin, then the process is nonstationary. As an example, consider a vacuum-tube-amplifier noise signal. Just as the power supply is turned on, there will be little noise because the filaments of the tubes are not heated and as a result, shot noise is absent. The amplifier will eventually warm up and reach a steady state. In a strict sense the noise generation here represents a case of a nonstationary process. However, when we consider the amplifier in the steady state, the process can reasonably be assumed stationary.[4] It is often possible to make an assessment of whether a process is stationary or not from the mechanism generating the process. If there is no change in the source or in its environment, it is likely that the process is stationary. This criterion cannot, however, be relied upon without additional empirical analysis.

Some processes are stationary in a limited sense. If a process has probability density functions, up to the kth order, independent of the shift in the time origin, then the process is said to be a process stationary of the order k. Note that the kth-order probability density determines

[4]Strictly speaking no physical process is stationary because every physical process must begin and end at some finite instants. Obviously the statistics cannot be independent of the time origin in such a case. A stationary process is an idealized model.

all the probability densities of order lower than k. Hence it is necessary only for the kth-order probability-density function to be independent of time shift in order for a process to qualify as a process stationary of order k.

Wide-Sense Stationary Process. We shall see in a later section that the power-density spectrum is completely determined by the autocorrelation function. Hence for the purpose of analysis of random signals in linear systems the parameters that are important are the correlation and the mean of the random process. For this reason it is important that at least these parameters be independent of time origin for ease of analysis. A process satisfying these conditions is known as a *wide-sense stationary* process. Thus we define a process to be stationary in the wide sense (also known as *weakly stationary*) if

$$\overline{\mathrm{x}(t)} = \text{constant for all } t \tag{3-17a}$$

$$R_{\mathrm{x}}(t_1,t_2) = R_{\mathrm{x}}(\tau); \quad \tau = t_2 - t_1 \tag{3-17b}$$

Note that wide-sense stationarity is a weaker condition than stationarity of order 2. A second-order stationary process must have all possible second-order statistics independent of time origin. On the other hand, for a wide-sense stationary process only one second-order statistics [autocorrelation function $R_{\mathrm{x}}(t_1,t_2)$] need be independent of time origin. Thus a process that is strictly stationary or stationary of order 2 is also wide-sense stationary. But the converse is not necessarily true.

Turning back to the random process in Example 3-2,

$$\mathrm{x}(t) = k\cos(\omega_0 t + \Theta)$$

It was shown that $\overline{\mathrm{x}(t)} = 0$, and from Eq. 3-15, we have

$$R_{\mathrm{x}}(t_1,t_2) = \frac{k^2}{2}\cos\omega_0\tau; \quad \tau = t_2 - t_1 \tag{3-18}$$

Thus the autocorrelation function is a function of τ only. Obviously the process is stationary at least in the wide sense.

It is evident from the above discussion that the number of statistics required to specify a stationary process are much fewer than that required to specify a nonstationary process. In general, the analysis of nonstationary processes is much more difficult because their statistics depend upon the time origin. In this book we shall be concerned mainly with stationary and wide-sense stationary processes in particular. In all our future discussions, random processes will be assumed to be stationary (at least in the wide-sense) unless otherwise stated.

3-5. ERGODIC AND NONERGODIC PROCESSES

We have observed that in order to obtain the complete statistics of a process, we need an ensemble of sample functions. In certain processes, called *ergodic* processes, the complete statistics can be determined from any one sample function. In other words, every sample function carries an identical statistical information and hence any one sample function describes (statistically) the entire random process.

Since it is possible to determine only time averages from a single record (single sample function), it is obvious that in an ergodic process the time averages must be identical with ensemble averages. This is the most significant property of an ergodic process. The amplitude distribution of the ensemble at any instant t is identical to the amplitude distribution of a single sample function. The first probability-density function, for example, can be determined empirically as shown in Fig. 3-8.

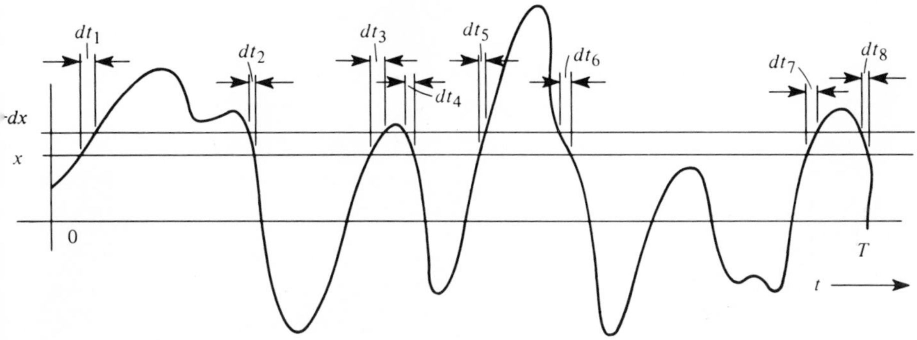

FIG. 3-8.

In this figure a sample function $x(t)$ is shown as a function of time. Two horizontal lines representing magnitudes x and $x + dx$ are drawn. The total time interval over which the function $x(t)$ lies in the range of x and $x + dx$ is given by $\Sigma\, dt_i$, as shown in the figure. If the calculations are performed over T seconds ($T \rightarrow \infty$), then

$$p(x)\,dx = \frac{\Sigma\, dt_i}{T}$$

and

$$p(x) = \frac{\Sigma\, dt_i}{T\,dx}$$

The ergodic assumption implies that any given member of the ensemble (the sample function) takes on all possible values in time with the same relative frequency that an ensemble will take at any given instant.

It should now become evident that an ergodic process is necessarily a stationary process. There is only one set of probability-density functions that can be obtained from a single record. If this record contains all the statistical information of the process, it is obvious that the process must be stationary. A nonstationary process has infinite sets of probability-density functions depending upon the choice of the time origin. Obviously for such a process it is impossible for one sample function to contain all the statistical information of the process. *Hence we conclude that an ergodic process is necessarily a stationary process.* A stationary process, however, may or may not be an ergodic process. Thus an ergodic process is a subclass of stationary processes which in turn form a subclass of random processes in general. This fact is represented graphically in Fig. 3-9.

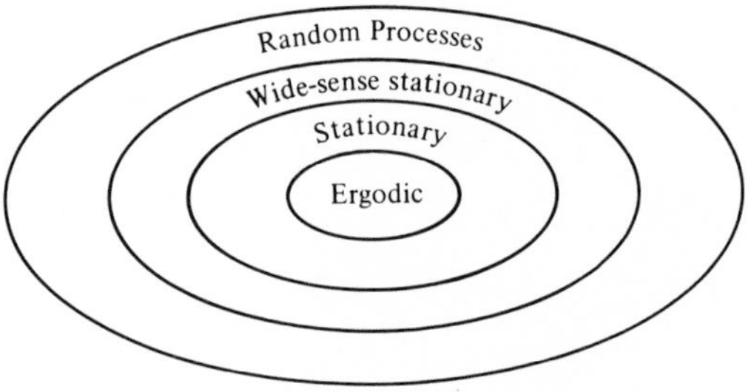

FIG. 3-9.

An example of ergodicity can be illustrated by an example of traffic lights in a city. Let us assume that there is a large number of traffic lights in a certain city, and each light stays green for $^3/_4$ sec in the E–W direction and stays green for $^1/_4$ sec in the N–S direction. Let us further assume that the switching instants of all the traffic lights are random and independent. If we consider a certain person driving a car arriving at any traffic light randomly in the E–W direction, the probability that he will have a green signal is $^3/_4$—that is, on the average 75 percent of the time he will observe a green light. On the other hand if we consider a large number of drivers each of them arriving at any random traffic light in E–W direction simultaneously, then 75 percent of the drivers will observe green lights and the remaining 25 percent will observe red lights. Thus the experience of a single driver arriving randomly many times at the same traffic light will contain the same statistical information (sample-function statistics) as a large number of drivers arriving simultaneously at various traffic lights (ensemble statistics at one instant).

Moreover, if a large number of drivers arrive randomly many times at their chosen traffic light, the experience of each driver will have the same statistical information as any other driver.

As mentioned earlier for an ergodic process, the ensemble averages (statistical averages) are equal to time averages. Thus the ensemble average of the random variable $\mathrm{x}(t)$ will be the same as the time average of any one sample function $x(t)$:

$$\overline{\mathrm{x}} = \widetilde{x(t)} \tag{3-19}$$

where the straight overline indicates ensemble (statistical) average as usual and the wavy overline indicates the time average.

$$\overline{\mathrm{x}} = \int_{-\infty}^{\infty} x\,p(x)\,dx \tag{3-20a}$$

$$\widetilde{x(t)} = \lim_{T\to\infty} \frac{1}{T} \int_{-T/2}^{T/2} x(t)\,dt \tag{3-20b}$$

where $p(x)$ is the first-order probability-density function of x. Hence for an ergodic process,

$$\lim_{T\to\infty} \frac{1}{T} \int_{-T/2}^{T/2} x(t)\,dt = \int_{-\infty}^{\infty} x\,p(x)\,dx \tag{3-21}$$

Indeed, since all time averages are equivalent to statistical averages for an ergodic process, the time average of any function $\phi[x(t)]$ of a sample function $x(t)$ must be equal to the statistical (ensemble) average of $\phi(\mathrm{x})$. The time average of $\phi[x(t)]$ is given by

$$\widetilde{\phi[x(t)]} = \lim_{T\to\infty} \frac{1}{T} \int_{-T/2}^{T/2} \phi[x(t)]\,dt \tag{3-22a}$$

and the statistical average of $\phi(\mathrm{x})$ is given by Eq. 2.70.

$$\overline{\phi(\mathrm{x})} = \int_{-\infty}^{\infty} \phi(x)p(x)\,dx \tag{3-22b}$$

Hence, for an ergodic process in general, we have

$$\lim_{T\to\infty} \frac{1}{T} \int_{-T/2}^{T/2} \phi[x(t)]\,dt = \int_{-\infty}^{\infty} \phi(x)p(x)\,dx \tag{3-23}$$

This equation expresses the equality of the first-order time averages and ensemble (statistical) averages. This is obvious from the fact that Eq. 3-23 involves only the first-order probability-density function. For a strictly ergodic process the equality of time and ensemble averages holds for any order of averages. Consider for example the second-order averages. Let $\mathrm{x}(t)$ and $\mathrm{x}(t + \tau)$ be the two random variables defined over the ensemble at instants t and $t + \tau$ respectively. For convenience, we shall denote these variables by x_1 and x_2 respectively. If $\phi(\mathrm{x}_1,\mathrm{x}_2)$ is a general function of these two variables, then the statistical mean of $\phi(\mathrm{x}_1,\mathrm{x}_2)$ is given by

$$\overline{\phi(\mathrm{x}_1,\mathrm{x}_2)} = \int_{-\infty}^{\infty}\int_{-\infty}^{\infty} \phi(x_1,x_2)p(x_1,x_2;\tau)\,dx_1\,dx_2$$

where $p(x_1,x_2;\tau)$ is the joint probability-density function of x_1 and x_2. This is obviously the second-order probability-density function of the process. Also

$$\widetilde{\phi[x(t), x(t+\tau)]} = \lim_{T\to\infty} \frac{1}{T}\int_{-T/2}^{T/2} \phi[x(t),x(t+\tau)]\,dt$$

For an ergodic process the two averages are identical. Hence

$$\int_{-\infty}^{\infty}\int_{-\infty}^{\infty} \phi(x_1,x_2)p(x_1,x_2;\tau)\,dx_1\,dx_2 = \lim_{T\to\infty} \frac{1}{T}\int_{-T/2}^{T/2} \phi[x(t),x(t+\tau)]\,dt \tag{3-24}$$

These results may be generalized for averages of any order.[5] It is possible to have a process for which only the first k order time averages and ensemble averages are equal. Such processes are said to have ergodicity of the kth order.

The autocorrelation function $R_{\mathrm{x}}(\tau)$ is a second-order average defined by

$$R_{\mathrm{x}}(\tau) = \overline{\mathrm{x}(t)\mathrm{x}(t+\tau)}$$

The corresponding time average is the time autocorrelation function $\mathcal{R}_{\mathrm{x}}(\tau)$ defined in Chapter 1:

$$\mathcal{R}_{\mathrm{x}}(\tau) = \widetilde{x(t)x(t+\tau)}$$

where $x(t)$ is any sample function of the process. Obviously for an ergodic process[6] the autocorrelation function is equivalent to the time autocorrelation function of any sample function of the process

[5]Thus for a strictly ergodic process,

$$\int_{-\infty}^{\infty}\int_{-\infty}^{\infty}\cdots\int_{-\infty}^{\infty} \phi(x_1,x_2,\ldots,x_n)p(x_1,x_2,\ldots,x_n;\tau_1,\tau_2,\ldots,\tau_{n-1})\,dx_1\,dx_2\ldots dx_n$$

$$= \lim_{T\to\infty}\frac{1}{T}\int_{-T/2}^{T/2} \phi[x(t),x(t+\tau_1),\ldots x(t+\tau_{n-1})]\,dt$$

for all integral values of n.

[6]Note that this property holds for strictly ergodic process. However, it also holds for a process of ergodicity of order k where $k \geq 2$.

$$R_x(\tau) = \mathcal{R}_x(\tau)$$
$$= \lim_{T\to\infty} \frac{1}{T} \int_{-T/2}^{T/2} x(t)x(t+\tau)\,dt \tag{3-25}$$

This equation expresses the autocorrelation function in terms of a single sample function and is applicable to ergodic processes only. For stationary processes in general, the autocorrelation function is given by Eq. 3-18.

An important question immediately arises. Is there a test for determining whether a given process is ergodic or not? Unfortunately it is not very easy to determine whether a process is ergodic or not. First we must ascertain that the process is stationary. This needs extensive ensemble calculations at various instants. In practice, one may determine only the mean value and the mean-square value of the ensemble at various instants. If these averages are time-invariant, then one may proceed to determine whether the process is ergodic or not. This is done by calculating certain time averages (such as mean value and mean-square value) for various sample functions. If these values agree with each other within an accepted statistical error and with those obtained previously by ensemble averaging, then the process may be justifiably classed as being ergodic to the order of averages calculated.

For two random processes $\mathrm{x}(t)$ and $\mathrm{y}(t)$, we define cross-correlation function $R_{xy}(t_1, t_2)$ as

$$R_{xy}(t_1, t_2) = \overline{\mathrm{x}(t_1)\mathrm{y}(t_2)} \tag{3-26}$$

The processes $\mathrm{x}(t)$ and $\mathrm{y}(t)$ are said to be jointly stationary if the joint probability density function of variables $\mathrm{x}(t_1)$, $\mathrm{x}(t_2), \ldots \mathrm{x}(t_j)$, $\mathrm{y}(t_1')$, $\mathrm{y}(t_2'), \ldots \mathrm{y}(t_k')$ is invariant under the shift of time origin for all values of $t_1, t_2, \ldots t_j, t_1', t_2', \ldots t_k'$. The processes $\mathrm{x}(t)$ and $\mathrm{y}(t)$ are said to be jointly stationary in wide sense if only their means and cross-correlation function is invariant under the shift of time origin. Thus, for wide sense jointly stationary processes

$$R_{xy}(t_1,t_2) = R_{xy}(\tau); \qquad \tau = t_1 - t_2$$
$$= \overline{\mathrm{x}(t)\mathrm{y}(t+\tau)} \tag{3-27}$$

If both $\mathrm{x}(t)$ and $\mathrm{y}(t)$ are jointly ergodic (at least of the order 2), then the cross-correlation function $R_{xy}(\tau)$ is equal to $\mathcal{R}_{xy}(\tau)$, the time cross-correlation function.

$$R_{xy}(\tau) = \mathcal{R}_{xy}(\tau) = \overline{\mathrm{x}(t)\mathrm{y}(t+\tau)}$$
$$= \lim_{T\to\infty} \frac{1}{T} \int_{-T/2}^{T/2} x(t)y(t+\tau)\,dt$$

3-6. FREQUENCY CONTENT OF A RANDOM PROCESS: THE POWER-DENSITY SPECTRUM

An electrical engineer instinctively thinks of signals in terms of their frequency content. Linear systems are characterized by their frequency response (the transfer function) and signals are expressed in terms of the relative amplitudes of their frequency components (Fourier or Laplace transforms). From a knowledge of these two quantities, the response to a given signal can be easily obtained in terms of its frequency content. This is the procedure followed for nonrandom signals. We wonder if similar concepts may not be applied to random signals. Ideally, all the sample functions of a random process are assumed to exist over the entire time interval $(-\infty, \infty)$ and thus are power signals. We may therefore inquire about the possible existence of a power-density spectrum. Can there be a power-density spectrum for a random process? Superficially, the concept of a power-density spectrum of a random process may appear ridiculous for the following reasons. In the first place, a random signal cannot be described analytically. There is a great deal of ignorance about the signal, which can be expressed only in terms of certain averages. Secondly, for a given process every sample function is different from another. Hence even if a power-density spectrum does exist for a sample function, it may be different for each sample functions. How can we overcome these difficulties? Fortunately, however, both these problems can be neatly resolved and it is possible to define a power-density spectrum for a random process which is meaningful and significant.

For the sake of clarity, we shall first start with a simpler case, that is the ergodic random process and then go on to general stationary processes. It was shown in Chapter 1 (Eq. 1-127) that for a signal $x(t)$, the power-density spectrum $\mathcal{S}_x(\omega)$ is given by the Fourier transform of the time autocorrelation function $\mathcal{R}_x(\tau)$, which, in turn is the time average $\widetilde{x(t)x(t+\tau)}$. For an ergodic process, the time average is equal to the ensemble average. Hence the time autocorrelation function $\mathcal{R}_x(\tau)$ is equal to the autocorrelation function $R_x(\tau)$ which is an ensemble average of $x(t)x(t+\tau)$. Thus for an ergodic process, the power-density spectrum of any sample function is given by the Fourier transform of the autocorrelation function $R_x(\tau)$. Hence the first of the two questions is clearly answered—that is, even if the random signals are not known completely but are known in terms of certain statistical averages, the power-density spectrum can be obtained from a knowledge of one of the average parameters [viz., $R_x(\tau)$]. It should be realized that the power-density spectrum is not a complete measure of a signal but is just an average parameter of the signal. This is because the power-density spectrum by definition is a power density averaged over a large time interval T (see Eq. 1-120). Indeed, the power-density spectrum of a signal is the Fourier transforms of its time autocorrelation function $\mathcal{R}_x(\tau)$ which is a time average of $x(t)x(t + \tau)$. Thus the power-density spectrum of a signal is one of its average parameters and it should come as no surprise that a meaningful power-density spectrum may exist for a random signal which is known only in terms of its average parameters.

For ergodic signals all the sample functions have the same averages and hence the power-density spectra of all sample functions are identical. From the earlier discussion it follows that the power-density spectrum of an ergodic random process is given by the Fourier transform of the autocorrelation function:

$$\mathcal{S}_x(\omega) = \mathcal{F}[R_x(\tau)] \tag{3-28a}$$

$$= \int_{-\infty}^{\infty} R_x(\tau) e^{-j\omega\tau} d\tau \tag{3-28b}$$

Next we come to a slightly more general case of a random process, namely the stationary random process. As we know, a stationary process may or may not be an ergodic. If the process is not ergodic, then the sample functions are not statistically equivalent and each sample function has a different time autocorrelation and hence a different power-density spectrum. Can we define a power-density spectrum that will be meaningful in this situation? At the outset we must admit that we will have to accept something less than what was possible in the ergodic case. Can we find the power-density spectra of various sample functions and obtain the average of these spectra? This would be the ensemble or statistical average of the power-density spectrum. One must not be too greatly disappointed with the prospect of an average power-density spectrum. We have done this all along for random signals in general. For a random signal or a variable, we do not have complete information and hence we are content to describe it in terms of its averages. When tossing a coin, the best we can say about the outcome is that on the average the outcome will be heads half the time and tails half the time. It is not possible to transcend this limit of knowledge because of our ignorance about the process. With this thought, it is hoped that the concept of an ensemble average of a power-density spectrum will not be so unpalatable. The ergodic random process, of course lends itself to a great deal of simplification, since each of its sample functions has the same power-density spectrum. Hence the ensemble average of the power density is the same as that of any of the sample functions. It should be realized that for a general process the ensemble average of the power-density spectrum is the only reasonable method of approach, because we never know which of the sample functions we are dealing with. In designing a system to perform certain functions one must optimize it with respect to not one sample function but the whole ensemble.

We now define the power-density spectrum of a random process as the ensemble average of the power-density spectra of all the sample functions. The power-density spectrum of a sample function $x(t)$ is given (see Eq. 1-120) by

$$\lim_{T \to \infty} \frac{|X_T(\omega)|^2}{T}$$

where $X_T(\omega)$ is the Fourier transform of the truncated signal $x_T(t)$:

$$X_T(\omega) = \int_{-\infty}^{\infty} x_T(t) e^{-j\omega t}\, dt = \int_{-T/2}^{T/2} x(t) e^{-j\omega t}\, dt \tag{3-29}$$

Thus for a random process $\mathrm{x}(t)$ we define power-density spectrum $\mathcal{S}_{\mathrm{x}}(\omega)$ as

$$\mathcal{S}_{\mathrm{x}}(\omega) = \lim_{T \to \infty} \frac{\overline{|X_T(\omega)|^2}}{T} \tag{3-30}$$

From Eq. 3-29 it follows that when $x(t)$ is real,

$$|X_T(\omega)|^2 = X_T(\omega) X_T(-\omega) = \int_{-T/2}^{T/2} x(t_1) e^{j\omega t_1}\, dt_1 \int_{-T/2}^{T/2} x(t_2) e^{-j\omega t_2}\, dt_2$$

$$= \int_{-T/2}^{T/2} \int_{-T/2}^{T/2} x(t_1) x(t_2) e^{-j\omega(t_2 - t_1)}\, dt_1\, dt_2$$

and

$$\mathcal{S}_{\mathrm{x}}(\omega) = \lim_{T \to \infty} \frac{1}{T} \overline{|X_T(\omega)|^2}$$

$$= \lim_{T \to \infty} \frac{1}{T} \overline{\int_{-T/2}^{T/2} \int_{-T/2}^{T/2} x(t_1) x(t_2) e^{-j\omega(t_2 - t_1)}\, dt_1\, dt_2} \tag{3-31}$$

In Eq. 3-31 the operation of ensemble averaging is performed after the operation of integration. We shall now interchange these operations[7]—

[7]In future we shall often come across such an interchange of integration and ensemble averaging. It should be realized that the operation of ensemble averaging is also an operation of integration. Thus

$$\bar{\mathrm{x}} = \int_{-\infty}^{\infty} x p(x)\, dx$$

Hence the interchange of integration and ensemble averaging is really equivalent to interchanging the order of integration. It can be shown rigorously that if $x(t)$ is a sample function (or some function of a sample function) and $f(t)$ is a nonrandom time function, that is, the integration will be performed after the operation of ensemble averaging. This yields

$$\mathcal{S}_x(\omega) = \lim_{T\to\infty} \frac{1}{T} \int_{-T/2}^{T/2} \int_{-T/2}^{T/2} \overline{x(t_1)\,x(t_2)}\, e^{-j\omega(t_2 - t_1)}\, dt_1\, dt_2$$

$$= \lim_{T\to\infty} \frac{1}{T} \int_{-T/2}^{T/2} \int_{-T/2}^{T/2} R_x(t_2 - t_1)\, e^{-j\omega(t_2 - t_1)}\, dt_1\, dt_2$$

Let

$$R_x(t_2 - t_1)\, e^{-j\omega(t_2 - t_1)} = \phi(t_2 - t_1)$$

Then

$$\mathcal{S}_x(\omega) = \lim_{T\to\infty} \frac{1}{T} \int_{-T/2}^{T/2} \int_{-T/2}^{T/2} \phi(t_2 - t_1)\, dt_1\, dt_2 \tag{3-33}$$

The integral on the right-hand side is a double integral over the range $\left(-\frac{T}{2}, \frac{T}{2}\right)$ on the variables t_1 and t_2. The square region of integration in the $t_1 t_2$ plane is shown in Fig. 3-10. The integral in Eq. 3-33 represents

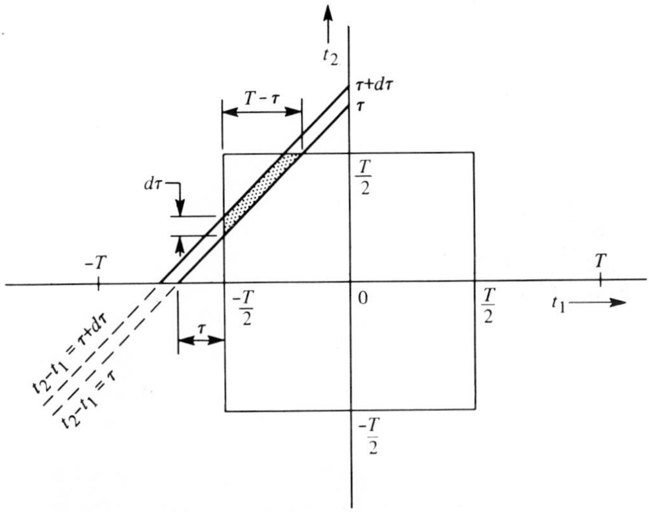

FIG. 3-10.

---

then

$$\overline{\int_{t_1}^{t_2} x(t)\, f(t)\, dt} = \int_{t_1}^{t_2} \overline{x(t)}\, f(t)\, dt \tag{3-32}$$

provided $x(t)$ is bounded on the interval (t_1, t_2) where t_1 and t_2 may be infinite and

$$\int_{t_1}^{t_2} \overline{|x(t)|}\, |f(t)|\, dt < \infty$$

the volume under the surface $\phi(t_2 - t_1)$ above the square region shown in Fig. 3-10. This integral can be converted into a single integral by observing that the function $\phi(t_2 - t_1)$ is constant over any line in the t_1-t_2 plane for which $t_2 - t_1 =$ constant. It is obvious that these lines have a slope of 1 in the $t_1 t_2$ plane. Let $t_2 - t_1 = \tau$. Then Eq. 3-33 reduces to

$$\mathcal{S}_x(\omega) = \lim_{T\to\infty} \int_{-T/2}^{T/2} \int_{-T/2}^{T/2} \phi(\tau)\, dt_1\, dt_2$$

Two lines $t_2 - t_1 = \tau$ and $t_2 - t_1 = \tau + d\tau$ are shown in Fig. 3-10. The function $\phi(\tau)$ is constant over the line $t_2 - t_1 = \tau$. The function is also constant over the line $t_2 - t_1 = \tau + d\tau$ and has a value $\phi(\tau) + d\phi$. If $d\tau \to 0$, then the function can be assumed to have a constant value $\phi(\tau)$ over the entire shaded area. The volume under the shaded area is therefore $\phi(\tau)$ times the area of the shaded region. It can easily be seen that the area of the shaded region is $(T - \tau)\,d\tau$. For negative values of τ, this becomes $(T + \tau)\,d\tau$. Hence the area is $(T - |\tau|)\,d\tau$. The volume under the shaded region is $\phi(\tau)(T - |\tau|)\,d\tau$. The desired volume is obtained by integrating over the range of τ. From Fig. 3-10 it is evident that when t_1 and t_2 have a range $\frac{-T}{2}$ to $\frac{T}{2}$, the range of τ is $-T$ to T. Hence

$$\begin{aligned}\mathcal{S}_x(\omega) &= \lim_{T\to\infty} \frac{1}{T} \int_{-T}^{T} \phi(\tau)(T - |\tau|)\,d\tau \\ &= \lim_{T\to\infty} \int_{-T}^{T} \phi(\tau)\left(1 - \frac{|\tau|}{T}\right) d\tau\end{aligned}$$

The function $1 - \frac{|\tau|}{T}$ is a triangular function shown in Fig. 3-11. It is evident that in the limit as $T \to \infty$, the value of this function tends to unity. Hence we have

$$\mathcal{S}_x(\omega) = \int_{-\infty}^{\infty} \phi(\tau)\, d\tau$$

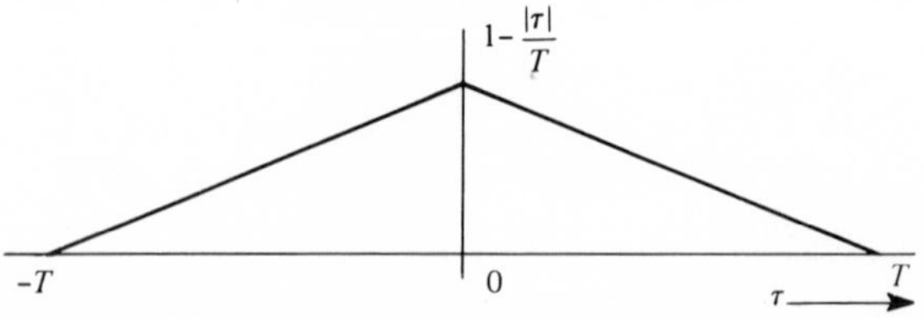

FIG. 3-11.

$$= \int_{-\infty}^{\infty} R_x(\tau) e^{-j\omega\tau}\, d\tau$$

This is the desired result. We have shown that the power density spectrum of a random process is given by the Fourier transform of its autocorrelation function $R_x(\tau)$:

$$R_x(\tau) \leftrightarrow \mathcal{S}_x(\omega)$$

This fact can be expressed as a Fourier transform pair:

$$\mathcal{S}_x(\omega) = \int_{-\infty}^{\infty} R_x(\tau) e^{-j\omega\tau}\, d\tau \tag{3-34a}$$

$$R_x(\tau) = \frac{1}{2\pi} \int_{-\infty}^{\infty} \mathcal{S}_x(\omega) e^{j\omega\tau}\, d\omega \tag{3-34b}$$

This is precisely the result we obtained earlier (Eq. 3-28) for the special case of ergodic processes. The results in Eq. 3-33 are general, however, and applicable to all processes where $R_x(t_1, t_2) = R_x(\tau)$; $\tau = t_2 - t_1$.

In other words, these results apply to all stationary as well as wide-sense stationary processes.

The pair of Eqs. 3-34a and 3-34b are known as the *Wiener-Khinchin relations.* The autocorrelation function $R_x(\tau)$ is an even function of τ. This follows from the fact that

$$R_x(\tau) = \overline{x(t)\,x(t + \tau)}$$

and for a stationary process, the ensemble averages are independent of the time origin. If we let $t + \tau = \sigma$, then

$$\begin{aligned} R_x(\tau) &= \overline{x(\sigma)\,x(\sigma - \tau)} \\ &= R_x(-\tau) \end{aligned} \tag{3-35}$$

Obviously $R_x(\tau)$ is an even function of τ. The power density spectrum $\mathcal{S}_x(\omega)$ is given by

$$\mathcal{S}_x(\omega) = \lim_{T \to \infty} \frac{\overline{X_T(\omega) X_T(-\omega)}}{T}$$

It is evident from this equation that $\mathcal{S}_x(\omega)$ is a positive, real and even function of ω. The Wiener-Khinchin relations (Eq. 3-34) in this case reduce to (see problem 1-7)

$$\mathcal{S}_x(\omega) = \int_{-\infty}^{\infty} R_x(\tau) \cos \omega\tau\, d\tau$$

$$= 2 \int_{0}^{\infty} R_x(\tau) \cos \omega\tau\, d\tau \tag{3-36a}$$

and

$$R_x(\tau) = \frac{1}{2\pi}\int_{-\infty}^{\infty} \mathcal{S}_x(\omega)\cos\omega\tau\, d\omega$$

$$= \frac{1}{\pi}\int_{0}^{\infty} \mathcal{S}_x(\omega)\cos\omega\tau\, d\omega \tag{3-36b}$$

Cross-Power-Density Spectrum. We define the cross-power-density spectrum $\mathcal{S}_{xy}(\omega)$ for two real random processes $x(t)$ and $y(t)$ as

$$\mathcal{S}_{xy}(\omega) = \lim_{T\to\infty} \frac{\overline{X_T(-\omega)\,Y_T(\omega)}}{T} \tag{3-37}$$

where $X_T(\omega)$ and $Y_T(\omega)$ are the Fourier transforms of the truncated sample functions $x_T(t)$ and $y_T(t)$ of the processes $x(t)$ and $y(t)$ respectively. Proceeding along the lines of the derivation of Eq. 3-34a, we obtain

$$R_{xy}(\tau) \leftrightarrow \mathcal{S}_{xy}(\omega) \tag{3-38}$$

where $R_{xy}(\tau)$ is the cross-correlation function of $x(t)$ and $y(t)$ as defined in Eq. 3-27.

It can be seen from Eq. 3-37 defining the cross-power-density spectrum that

$$\mathcal{S}_{xy}(\omega) = \mathcal{S}_{yx}(-\omega) \tag{3-39}$$

Complex Random Processes. For complex random processes, one can derive the Wiener-Khinchin relations along lines similar to those used for real processes. It was indicated in Chapter 1 that for a complex signal $x(t)$, the power-density spectrum is given by

$$\lim_{T\to\infty} \frac{X_T^*(\omega)\,X_T(\omega)}{T} = \lim_{T\to\infty} \frac{|X_T(\omega)|^2}{T}$$

For a complex random process, the power-density spectrum is defined as an ensemble average of the power-density spectra of all the sample functions. Hence

$$\mathcal{S}_x(\omega) = \lim_{T\to\infty} \frac{\overline{X_T^*(\omega)\,X_T(\omega)}}{T} = \lim_{T\to\infty} \frac{\overline{|X_T(\omega)|^2}}{T} \tag{3-40}$$

The reader can easily show along lines similar to those used in deriving Eq. 3-34a, that the Wiener-Khinchin relations are valid for complex processes provided we define the autocorrelation function as

$$R_x(\tau) = \overline{x^*(t)\,x(t+\tau)} \tag{3-41a}$$

$$= \overline{x(t)\,x^*(t-\tau)} \tag{3-41b}$$

The cross-power-density spectrum $\mathcal{S}_{xy}(\omega)$ for complex random processes $x(t)$ and $y(t)$ is defined by

$$\mathcal{S}_{xy}(\omega) = \lim_{T \to \infty} \frac{1}{T} \overline{X_T^*(\omega) Y_T(\omega)} \tag{3-42}$$

It can be shown that

$$R_{xy}(\tau) \leftrightarrow \mathcal{S}_{xy}(\omega)$$

where

$$R_{xy}(\tau) = \overline{x^*(t) y(t + \tau)} = \overline{x(t) y^*(t - \tau)} \tag{3-43}$$

A Comment on Autocorrelation Function and Power Density Spectrum. At this point it is important to realize that for a given process there is a unique autocorrelation function but the converse is not true. A given autocorrelation function may correspond to a large number of different processes. This is because an autocorrelation function is not a complete measure of a random process but is one of the average parameters. As is well known, a given value of a certain average can correspond to infinitely large number of situations. For example, if we are given a certain mean value (or a mean square value) of a signal, it does not uniquely specify the signal. There are many (infinitely many) signals which may have the same mean value (or a mean square value). Similar conclusions apply to power density spectrum. We therefore conclude that a given autocorrelation function or a power density spectrum does not uniquely specify the process. There are, in general, large numbers of different processes which may have the same autocorrelation function (and the power density spectrum). Similar conclusion applies to cross-correlation function and cross- power-density spectrum.

3-7. PHYSICAL SIGNIFICANCE OF THE AUTOCORRELATION FUNCTION

It is evident from the discussion in the previous section that the autocorrelation function is the most significant quantity in the spectral analysis of random signals. It is worthwhile to understand qualitatively the reason behind this. We have already indicated in Sec. 3-3 the possibility of the dependence of the frequency content of a random process upon the autocorrelation function. The autocorrelation function is a measure of the rapidity of variation of a given signal. Consider two random variables x_1 and x_2, where x_1 represents the value of the sample functions at some instant t_1 and x_2 represents the values of the same sample functions at the instant $t_2 = t_1 + \tau$, as shown in Fig. 3-12. The

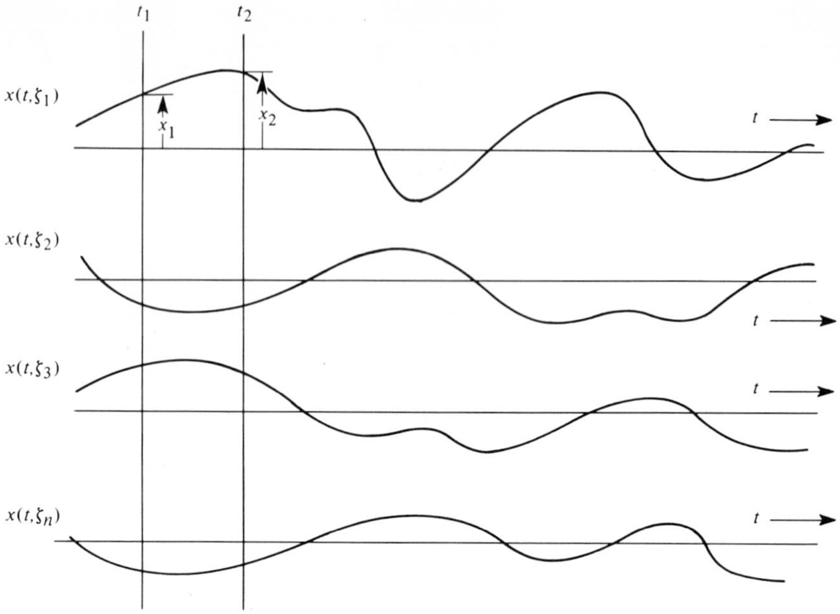

FIG. 3-12.

autocorrelation function is the correlation between x_1 and x_2, i.e., the ensemble average of the product of x_1 and x_2.

$$R_x(\tau) = \overline{x_1 x_2} = \overline{x(t)x(t + \tau)}$$

Consider a sample function $x(t)$, and let us choose an instant t_1. The value of the signal at the instant t_1 is x_1 and that at τ seconds later—that is, at the instant $t_1 + \tau$ is x_2. If the signal contains predominantly low frequencies it varies slowly and hence we expect that the value of the signal will not change much in τ seconds. Thus x_1 and x_2 have a measure of similarity, and it is possible to obtain some information about one from a knowledge of the other. This is, of course, true for reasonable values of τ. If we make τ large, the interdependence of x_1 and x_2 will be reduced. However, the more slowly does the signal vary the larger will be the value of τ for which x_1 and x_2 retain their relatedness. If a signal contains predominantly high frequencies, it varies rapidly and hence x_1 and x_2 lose their relatedness for relatively small values of τ.

We have already investigated the physical significance of the correlation of two random variables x_1 and x_2. It is a measure of the dependence of two variables—that is, the larger the dependence the larger the value of the correlation. It is therefore obvious that for slowly varying signals the autocorrelation function $R_x(\tau)$, which is the correlation of x_1 and x_2, will be sizable even for large values of τ. For rapidly varying signals, however, the dependence becomes weaker for relatively small values of

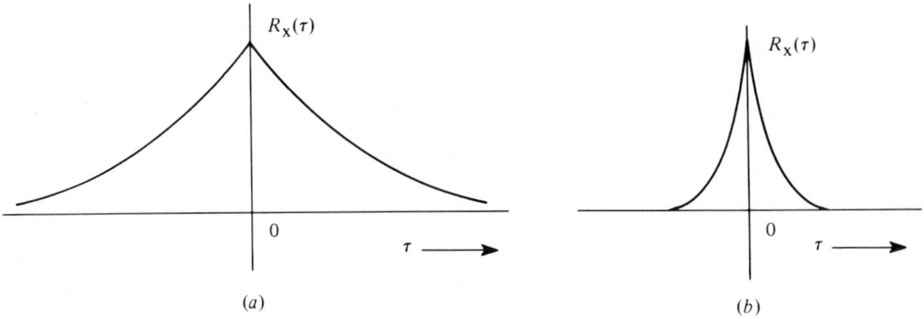

FIG. 3-13. Autocorrelation function. (a) for slowly varying signals, (b) for rapidly varying signals.

τ and hence the correlation between x_1 and x_2 will become small even for relatively small values of τ. Hence $R_x(\tau)$ in this case will die out much earlier. This is shown in Fig. 3-13. Note that in either case the autocorrelation function decays with τ and eventually goes to zero. This is true for all signals (except periodic signals and signals with a d-c component) because no matter how slowly the signal varies, for very large values of τ the relatedness must come to an end for a random signal. For periodic signals it is obvious that the relatedness is also of periodic nature and hence $R_x(\tau)$ will be periodic. If the signal has a d-c component, then both x_1 and x_2 have nonzero mean values and as $\tau \rightarrow \infty$, x_1 and x_2 tend to become uncorrelated. Hence the autocorrelation function approaches the value $\bar{x}_1\bar{x}_2$. This follows from the result in Sec. 2-12 which states that if x_1 and x_2 are uncorrelated

$$\overline{x_1x_2} = \bar{x}_1\,\bar{x}_2 \tag{3-44}$$

Thus we conclude that the autocorrelation function is compressed in the τ domain for random signals containing high-frequency components, whereas it is spread out in the τ domain for random signals with low-frequency components. We know from Sec. 1-8 that the compression of a signal in the time domain is equivalent to expansion in the frequency domain and vice versa. Hence the Fourier transform of $R_x(\tau)$ for slowly varying signals is compressed in the frequency domain (containing lower frequencies), whereas that for a rapidly varying signal is spread out in the frequency domain (contains higher frequencies). It therefore appears plausible that the Fourier transform of the autocorrelation function be proportional to some measure in the frequency domain. This measure happens to be the power-density spectrum.

3-8. DETERMINATION OF THE AUTOCORRELATION FUNCTION

The autocorrelation function $R_x(\tau)$ for a process $x(t)$ is defined as

$$R_x(\tau) = \overline{x(t)\,x(t + \tau)} \tag{3-45}$$

The most important quantity required in determining the autocorrelation function is the second-order probability-density function $p(x_1,x_2;\tau)$, which is the joint probability-density function of random variables x_1 and x_2, that is, $\mathrm{x}(t)$ and $\mathrm{x}(t+\tau)$ respectively. Once this density function is determined, the autocorrelation function can be obtained as

$$R_{\mathrm{x}}(\tau) = \int_{-\infty}^{\infty}\int_{-\infty}^{\infty} x_1 x_2\, p(x_1,x_2;\tau)\, dx_1\, dx_2 \tag{3-46}$$

If the process is a discrete random process, then the probability density $p(x_1,x_2;\tau)$ consists of impulses and the integral in Eq. 3-46 reduces to summation

$$R_{\mathrm{x}}(\tau) = \sum\sum x_1 x_2 P(x_1,x_2;\tau) \tag{3-47}$$

where $P(x_1,x_2;\tau)$ is the probability of observing $\mathrm{x}_1 = x_1$ and $\mathrm{x}_2 = x_2$ jointly.

If the random process is ergodic, then the autocorrelation function is also equal to the time autocorrelation of a sample function. Hence

$$\begin{aligned} R_{\mathrm{x}}(\tau) &= \mathcal{R}_{\mathrm{x}}(\tau) = \widetilde{x(t)x(t+\tau)} \\ &= \lim_{T\to\infty}\frac{1}{T}\int_{-T/2}^{T/2} x(t)x(t+\tau)\,dt \end{aligned}$$

The power-density spectrum is of course the Fourier transform of $R_{\mathrm{x}}(\tau)$.

$$\mathcal{S}_{\mathrm{x}}(\omega) = \mathcal{F}\,[R_{\mathrm{x}}(\tau)]$$

We shall now derive the autocorrelation function and the power-density spectrum for a number of random processes.

Example 3-5. Here we shall determine the autocorrelation function and the power-density spectrum of the random processes of Example 3-2:

$$\mathrm{x}(t) = k\cos(\omega_0 t + \Theta)$$

where Θ is a random variable over the ensemble and is uniformly distributed over the range $(0, 2\pi)$.

$$p_{\Theta}(\theta) = \begin{cases} \dfrac{1}{2\pi} & 0 < \theta < 2\pi \\ 0 & \text{otherwise} \end{cases}$$

We have already derived [Example 3-4 (Eq. 3-15)] the autocorrelation function for this process.

$$R_{\mathrm{x}}(\tau) = \frac{k^2}{2}\cos\omega_0\tau \tag{3-48a}$$

Hence

$$\mathcal{S}_{\mathrm{x}}(\omega) = \frac{\pi k^2}{2}\,[\delta(\omega-\omega_0) + \delta(\omega+\omega_0)] \tag{3-48b}$$

Example 3-6. We shall find the autocorrelation function of the Poisson-distributed unit impulses shown in Fig. 3-14a. The unit impulses are randomly distributed according to the Poisson distribution function.

The characteristic of the Poisson distribution is that the probability of observing an event in an interval $d\tau$ is given by $\alpha d\tau$ when α is the average number of occurrences per second (See Eq. 2-102).

In order to solve this problem, we shall consider the limiting form of the impulse. Each unit impulse will be assumed to be a rectangular pulse of height h and infinitesimal width ϵ such that the area $h\epsilon = 1$ (Fig. 3-14b).

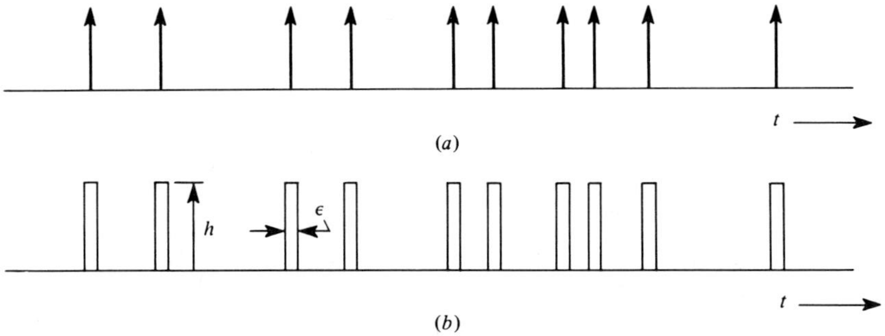

FIG. 3-14.

To calculate $R_x(\tau)$, we observe that the random variable $x(t)$ for this process (in the limiting form) can assume only two values: 0 and h. Hence both x_1 $(= x(t_1))$ and x_2 $(= x(t_2))$ are discrete random variables. In this case, the autocorrelation function is given by Eq. 3-47.

$$R_x(\tau) = \sum_{x_1} \sum_{x_2} x_1 x_2 P_{x_1x_2}(x_1,x_2) \tag{3-49}$$

Let us first consider the region $\tau > \epsilon$. It can easily be seen that in this case x_1 can be either 0 or h and x_2 can also assume the values 0 or h. Hence the summation in Eq. 3-49 will consist of four terms:

$$\begin{aligned} R_x(\tau) &= (h \times 0)P_{x_1x_2}(h,0) + (h \times h)P_{x_1x_2}(h,h) + (0 \times h)P_{x_1x_2}(0,h) \\ &\qquad + (0 \times 0)P_{x_1x_2}(0,0) \\ &= h^2 p_{x_1x_2}(h,h) \end{aligned} \tag{3-50}$$

where $P_{x_1x_2}(h,h)$ is the probability of observing $x_1 = h$ and $x_2 = h$ jointly. We thus need to calculate the probability of observing the amplitude h at both t_1 and at $t_2 = t_1 + \tau$. We have

$$P_{x_1x_2}(h,h) = P_{x_1}(h)P_{x_2}(h \mid x_1 = h) \tag{3-51}$$

If $\tau > \epsilon$, then the same impulse cannot exist at both t_1 and $t_1 + \tau$ (because the rectangular pulse is of width ϵ). Hence if $x_1 = h$ and $x_2 = h$, then obviously at t_1 and $t_1 + \tau$ two different impulses will be observed. But the Poisson distribution is characterized by the fact that the occurrence of one event (that is, the impulse observed at t_1) is independent of the occurrence of the other event (the impulse observed at $t_1 + \tau$). Hence for $\tau > \epsilon$

$$P_{x_2}(h \mid x_1 = h) = P_{x_2}(h)$$

and

$$R_x(\tau) = h^2 P_{x_1 x_2}(h,h) = h^2 P_{x_1}(h) P_{x_2}(h)$$

To calculate $P_{x_1}(h)$, we use the fact that α pulses (each of width ϵ), per second, occupy a total time interval of $\alpha\epsilon$ per second. In other words, the ratio of the time interval over which the impulses exist to the total interval is $\alpha\epsilon$. Hence the probability of observing an impulse at any random instant is $\alpha\epsilon$. Thus

$$P_{x_1}(h) = P_{x_2}(h) = \alpha\epsilon$$

and

$$\begin{aligned} R_x(\tau) &= h^2 \alpha^2 \epsilon^2 \\ &= \alpha^2 \quad (\text{since } h\epsilon = 1) \end{aligned} \tag{3-52}$$

Next, consider the case $\tau < \epsilon$. Since $\epsilon \longrightarrow 0$, the values of τ in this range are extremely small, and the event $x_1 = h$ and $x_2 = h$ can occur only if the same impulse is observed at t_1 and $t_1 + \tau$. In this case, the two events $x_1 = h$ and $x_2 = h$ cannot be independent. The width of the pulse is ϵ. Hence if it is given that a pulse is observed at t_1, the probability of observing the same pulse at $t_1 + \tau$ is $1 - \frac{\tau}{\epsilon}$ (Fig. 3-15). Hence

$$P_{x_2}(h \mid x_1 = h) = 1 - \frac{\tau}{\epsilon}$$

and

$$P_{x_1 x_2}(h,h) = P_{x_1}(h) P_{x_2}(h \mid x_1 = h) = \alpha\epsilon \left(1 - \frac{\tau}{\epsilon}\right)$$

and, by Eq. 3-50,

$$R_x(\tau) = h^2 \alpha\epsilon \left(1 - \frac{\tau}{\epsilon}\right) \qquad \text{for } \tau < \epsilon$$

Since $R_x(\tau)$ is an even function of τ,

$$R_x(\tau) = h^2 \alpha\epsilon \left(1 - \frac{|\tau|}{\epsilon}\right) \qquad \text{for } |\tau| < \epsilon$$

Thus for small values of τ (around the origin), $R_x(\tau)$ is a triangular pulse of area α (Fig. 3-16). In the limit as $\epsilon \longrightarrow 0$, the triangular pulse becomes an impulse of strength α. Hence

$$R_x(\tau) = \alpha\delta(\tau), \qquad \tau = 0$$

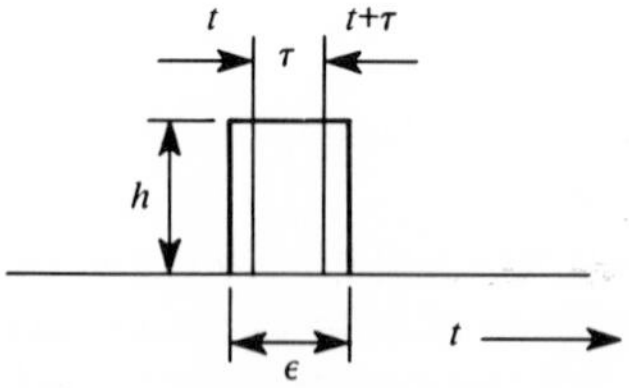

FIG. 3-15.

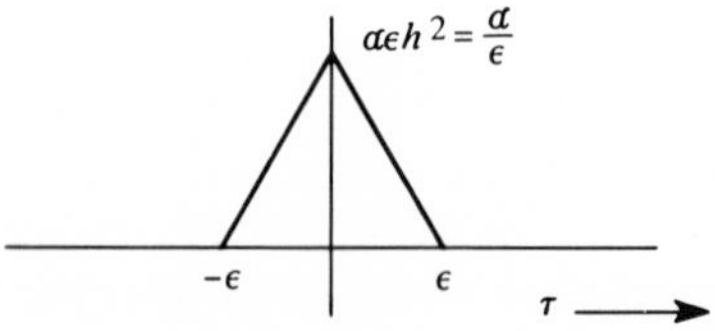

FIG. 3-16.

Thus $R_x(\tau)$ consists of an impulse of strength α at the origin and a constant term α^2 (Eq. 3-52). Hence

$$R_x(\tau) = \alpha\delta(\tau) + \alpha^2 \tag{3-53a}$$

The power-density spectrum is the Fourier transform of $R_x(\tau)$. Hence

$$\mathcal{S}_x(\omega) = \alpha + 2\pi\alpha^2\delta(\omega) \tag{3-53b}$$

The autocorrelation function and the corresponding power density spectrum are shown in Fig. 3-17.

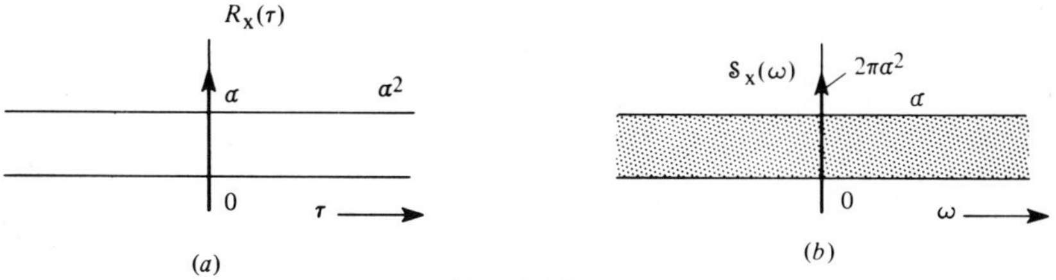

FIG. 3-17.

Example 3-7. White Noise. If the impulses had the same distribution as Example 3-6 (Poisson), but could assume positive and negative values with equal probability (Fig. 3-18), the resultant signal is one form of white noise.

The autocorrelation function of this process can be derived easily by a method similar to that used in Example 3-6. In this problem, x_1 and x_2 can each assume three values: 0, h, and $-h$. Also,

$$P_{x_1}(h) = P_{x_1}(-h) = P_{x_2}(h) = P_{x_2}(-h) = \frac{\alpha\epsilon}{2}$$

and (3-54)

$$P_{x_1}(0) = P_{x_2}(0) = 1 - \alpha\epsilon$$

and for $\tau > \epsilon$,

$$R_x(\tau) = \sum_{x_1}\sum_{x_2} x_1 x_2 P_{x_1x_2}(x_1,x_2) = \sum_{x_1}\sum_{x_2} x_1 x_2 P_{x_1}(x_1)P_{x_2}(x_2) \tag{3-55}$$

where x_1 and x_2 can assume values 0, h, and $-h$ with the probabilities given in Eq. 3-54. Substitution of these values in Eq. 3-55 yields

$$R_x(\tau) = 0 \qquad \tau > \epsilon$$

For $\tau < \epsilon$, the value of $R_x(\tau)$ is the same as that in Example 3-6. This is because an impulse is either positive or negative and since t_1 and $t_1 + \tau$ fall upon the same impulse x_1 and x_2 are both either h or $-h$, and

$$R_x(\tau) = h^2\alpha\epsilon\left(1 - \frac{|\tau|}{\epsilon}\right)$$

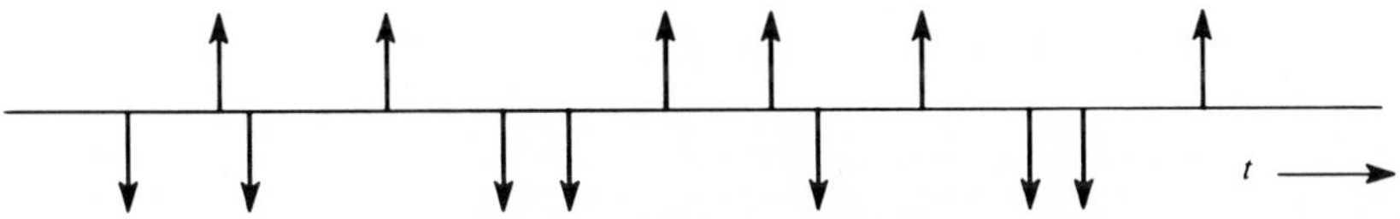

FIG. 3-18. A form of white noise.

In the limit as $\epsilon \to 0$,

$$R_x(\tau) = \alpha\delta(\tau) \tag{3-56a}$$

and

$$\mathcal{S}_x(\omega) = \alpha \tag{3-56b}$$

The power-density spectrum in this case is uniform over the entire frequency range. This is the reason why this signal is called *white noise*, "white" implying that it contains all "colors" or frequencies in equal strength. The autocorrelation function and the power-density spectrum for this signal is shown in Fig. 3-19.

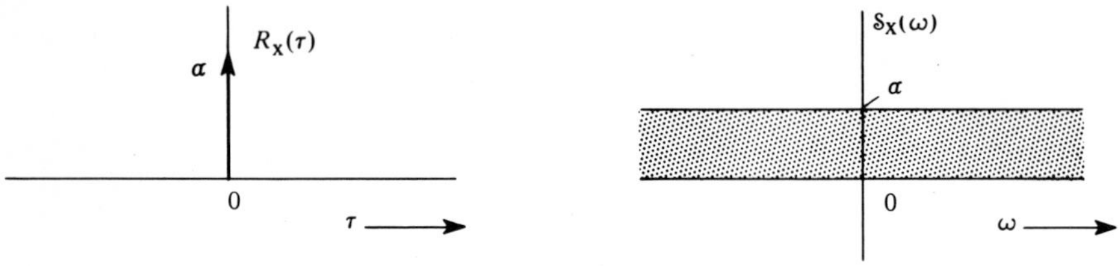

FIG. 3-19.

Example 3-8. Random Binary Transmission. In this example we shall consider a binary random process whose typical sample function is shown in Fig. 3-20. The signal can assume only two states (values) 1 or -1. The transition from one state to another can take place every b second. Points where such transitions may occur will be called *nodes*. The probability of transition from one state to the other[8] is 0.5.

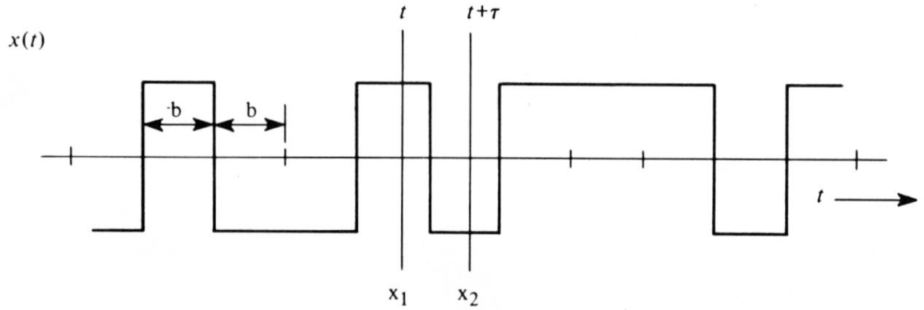

FIG. 3-20.

The amplitude at t represents x_1 and that at $t + \tau$ represents x_2. Note that the random variables x_1 and x_2 are discrete and each can assume only two values; -1 and 1. Hence

$$\begin{aligned} R_x(\tau) &= \sum_{x_1}\sum_{x_1} x_1 x_2 P_{x_1x_2}(x_1,x_2) \\ &= P_{x_1x_2}(1,1) + P_{x_1x_2}(-1,-1) - P_{x_1x_2}(-1,1) \\ &\qquad - P_{x_1x_2}(1,-1) \end{aligned} \tag{3-57}$$

[8]In addition, it is implied that the distance of the first node from the origin is a random variable with a uniform probability distribution over the range $(0,b)$.

We have

$$P_{x_1x_2}(1,1) = P_{x_1}(1)\,P_{x_2}(1 \mid x_1 = 1) \tag{3-58}$$

Since both states are equiprobable,

$$P_{x_1}(1) = P_{x_1}(-1) = 0.5 \tag{3-59}$$

Also,

$$P_{x_2}(1 \mid x_1 = 1) = 1 - P_{x_2}(-1 \mid x_2 = 1)$$

where $P_{x_2}(-1 \mid x_1 = 1)$ is the probability of observing $x_2 = -1$, given that $x_1 = 1$.

We shall first consider the region $\tau < b$. In this case x_2 can be -1 when $x_1 = 1$ only if a node lies in the interval $(t, t + \tau)$ and if the state changes at this node. Thus

$$P_{x_2}(-1 \mid x_1 = 1) = \text{probability (that a node lies in } t \text{ to } t + \tau) \quad \text{probability (state change)}$$
$$= \tfrac{1}{2}\,\text{probability (that a node lies in } t \text{ to } t + \tau)$$

Since a node can be anywhere with equal probability, and there must be one node in every interval of b seconds, we must have

$$\text{Probability (that a node lies in } t \text{ to } t + \tau) = \frac{\tau}{b}$$

Hence

$$P_{x_2}(-1 \mid x_1 = 1) = \frac{\tau}{2b}$$

-and

$$P_{x_2}(1 \mid x_1 = 1) = 1 - \frac{\tau}{2b} \tag{3-60}$$

Substituting Eqs. 3-59 and 3-60 in Eq. 3-58, we get

$$P_{x_1x_2}(1,1) = \frac{1}{2}\left(1 - \frac{\tau}{2b}\right) \qquad \tau < b \tag{3-61a}$$

Similarly we can show that

$$P_{x_1x_2}(-1,-1) = \frac{1}{2}\left(1 - \frac{\tau}{2b}\right) \qquad \tau < b \tag{3-61b}$$

Also,

$$P_{x_1x_2}(1,-1) = P_{x_1}(1)\,P_{x_2}(-1 \mid x_1 = 1)$$
$$= \frac{1}{2}\left(\frac{\tau}{2b}\right) = \frac{\tau}{4b} \tag{3-61c}$$

Similarly,

$$P_{x_1x_2}(-1,1) = \frac{\tau}{4b} \tag{3-61d}$$

Substituting Eqs. 3-61a, 3-61b, 3-61c, and 3-61d in Eq. 3-57, we have

$$R_x(\tau) = \left(1 - \frac{\tau}{b}\right) \quad \text{for } \tau < b$$

Since $R_x(\tau)$ is an even function of τ, we have

$$R_x(\tau) = 1 - \frac{|\tau|}{b} \qquad |\tau| < b \tag{3.62}$$

Next consider the range $b < \tau < 2b$. In this case there is at least one node and at most 2 nodes in the interval $(t, t + \tau)$. It can easily be seen that once a node change occurs it is equally likely that x_2 is 1 or -1 given that $x_1 = 1$ (or -1). Hence all the four probabilities derived above are equal ($= \frac{1}{4}$) and

$$R_x(\tau) = 0 \qquad b < |\tau| < 2b$$

Similar reasoning applies for $|\tau| > 2b$. Hence we have

$$R_x(\tau) = \begin{cases} 1 - \dfrac{|\tau|}{b} & |\tau| < b \\ 0 & |\tau| > b \end{cases} \tag{3-63a}$$

The power-density spectrum $\mathcal{S}_x(\omega)$ is given by

$$\begin{aligned} \mathcal{S}_x(\omega) &= \mathcal{F}\left[\left(1 - \frac{|\tau|}{b}\right)\right] \\ &= b\left[\mathrm{Sa}\left(\frac{\omega b}{2}\right)\right]^2 \end{aligned} \tag{3-63b}$$

The autocorrelation function and the power-density spectrum are shown in Fig. 3-21.

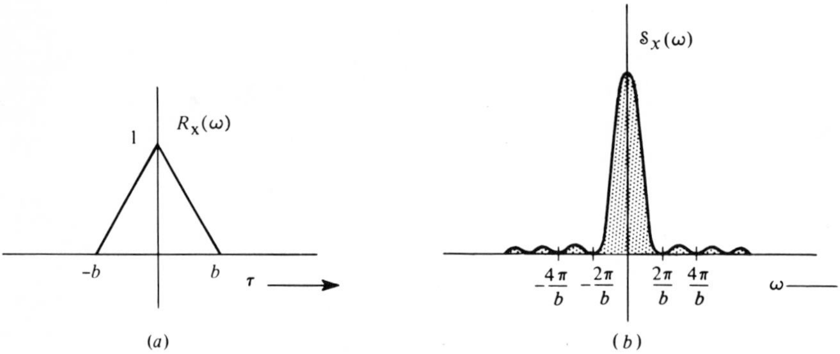

FIG. 3-21.

Example 3-9. Thermal Noise. We shall now consider a random process which represents the waveform of thermal noise in conductors. A typical sample function of this process is shown in Fig. 3-22 (see appendix at the end of the chapter). The signal $x(t)$ changes abruptly in amplitude at random instants. The average number of changes or shifts in amplitudes are α per second and the number of changes are Poisson-distributed. The amplitude after a shift is independent of the amplitude prior to the shift. The first-order probability density of the process is $p(x;t)$. It can be shown that this process is stationary of order 2. Hence $p(x;t)$ can be expressed as $p(x)$. We have

$$R_x(\tau) = \int_{-\infty}^{\infty}\int_{-\infty}^{\infty} x_1 x_2\, p_{x_1 x_2}(x_1, x_2)\, dx_1\, dx_2$$

$$= \int_{-\infty}^{\infty} \int_{-\infty}^{\infty} x_1 x_2 p_{x_1}(x_1) p_{x_2}(x_2 \mid x_1 = x_1) \, dx_1 \, dx_2 \tag{3-64}$$

To calculate $p_{x_2}(x_2 \mid x_1 = x_1)$, we observe that in τ seconds (interval between x_1 and x_2), there are two mutually exclusive possibilities; either there may be no

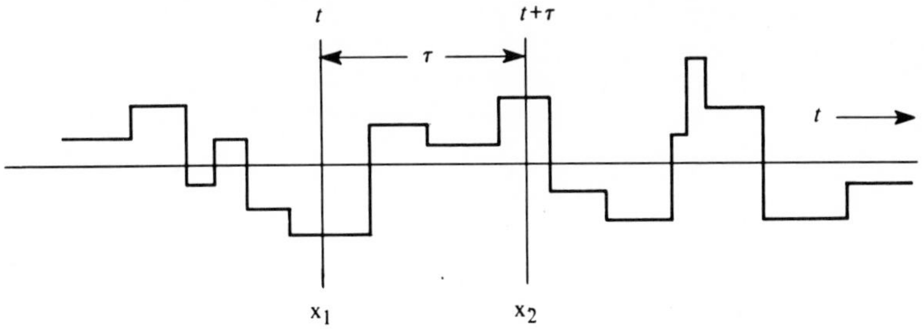

FIG. 3-22.

amplitude shift ($x_2 = x_1$), or there may be an amplitude shift ($x_2 \neq x_1$). We can therefore express $p_{x_2}(x_2 \mid x_1 = x_1)$ as

$$\begin{aligned} p_{x_2}(x_2 \mid x_1 = x_1) = {} & p_{x_2}(x_2 \mid x_1 = x_1, \text{no amplitude shift}) \\ & P(\text{no amplitude shift}) + \\ & p_{x_2}(x_2 \mid x_1 = x_1, \text{amplitude shift}) \, P(\text{amplitude shift}) \end{aligned}$$

The number of amplitude shifts are given to have Poisson distribution. The probability of k shifts in τ seconds is given by Eq. 2-116:

$$P_k(\tau) = \frac{(\alpha \tau)^k}{k!} e^{-\alpha \tau} \tag{3-65}$$

where there are on the average α shifts per second. The probability of no shift is obviously $P_0(\tau)$, where

$$P_0(\tau) = e^{-\alpha \tau}$$

The probability of amplitude shift $= 1 - P_0(\tau) = 1 - e^{-\alpha\tau}$. Hence

$$\begin{aligned} p_{x_2}(x_2 \mid x_1 = x_1) = {} & e^{-\alpha\tau} p_{x_2}(x_2 \mid x_1 = x_1, \text{no amplitude shift}) + \\ & (1 - e^{-\alpha\tau}) p_{x_2}(x_2 \mid x_1 = x_1, \text{amplitude shift}) \end{aligned} \tag{3-66}$$

when there is no shift, $x_2 = x_1$ and the probability density of x_2 is concentrated at the single value x_1. This is obviously an impulse located at $x_2 = x_1$. Thus

$$p_{x_2}(x_2 \mid x_1 = x_1, \text{no amplitude shift}) = \delta(x_2 - x_1) \tag{3-67}$$

whenever there are one or more shifts involved, in general, $x_2 \neq x_1$. Moreover, we are given that the amplitudes before and after a shift are independent. Hence

$$p_{x_2}(x_2 \mid x_1 = x_1, \text{amplitude shift}) = p_{x_2}(x_2) = p(x) \tag{3-68}$$

where $p_{x_2}(x_2)$ is the first-order probability density of the process. This is obviously $p(x)$. Substituting Eqs. 3-67 and 3-68 in Eq. 3-66, we get

$$\begin{aligned} p_{x_2}(x_2 \mid x_1 = x_1) &= e^{-\alpha\tau} \delta(x_2 - x_1) + (1 - e^{-\alpha\tau}) p_{x_2}(x_2) \\ &= e^{-\alpha\tau} [\delta(x_2 - x_1) + (e^{\alpha\tau} - 1) p_{x_2}(x_2)] \end{aligned}$$

Substituting this equation in Eq. 3-64, we get

$$R_x(\tau) = e^{-\alpha\tau} \int_{-\infty}^{\infty}\int_{-\infty}^{\infty} x_1 x_2 p_{x_1}(x_1)[\delta(x_2 - x_1) + (e^{\alpha\tau} - 1)p_{x_2}(x_2)]dx_1 dx_2$$

$$= e^{-\alpha\tau}\left[\int_{-\infty}^{\infty}\int_{-\infty}^{\infty} x_1 x_2 p_{x_1}(x_1)\delta(x_2 - x_1)\,dx_1 dx_2 + \int_{-\infty}^{\infty}\int_{-\infty}^{\infty} x_1 x_2 (e^{\alpha\tau} - 1)p_{x_1}(x_1)p_{x_2}(x_2)\,dx_1 dx_2\right]$$

$$= e^{-\alpha\tau}\left[\int_{-\infty}^{\infty} x_1^2 p_{x_1}(x_1)\,dx_1 + (e^{\alpha\tau} - 1)\int_{-\infty}^{\infty} x_1 p_{x_1}(x_1)\,dx_1 \int_{-\infty}^{\infty} x_2 p_{x_2}(x_2)\,dx_2\right] \quad (3\text{-}69)$$

$$= e^{-\alpha\tau}[\overline{x^2} + (e^{\alpha\tau} - 1)\overline{x}^2]$$

where $\overline{x}$ and $\overline{x^2}$ are the mean and the mean-square value of the process. For a thermal noise x = 0 (see Appendix 3) and Eq. 3-69 becomes

$$R_x(\tau) = \overline{x^2}e^{-\alpha\tau} \qquad \tau > 0$$

Since autocorrelation is an even function of τ, we have

$$R_x(\tau) = \overline{x^2}e^{-\alpha|\tau|} \tag{3-70a}$$

and

$$\mathcal{S}_x(\omega) = \frac{2\alpha\overline{x^2}}{\alpha^2 + \omega^2} \tag{3-70b}$$

It is shown in the Appendix that $\overline{x^2}$ for thermal-noise current is $kTG\alpha$ where k is the Boltzmann constant (1.38×10^{-23}), T is the ambient temperature in degrees Kelvin, G (= $1/R$) is the conductance of the resistor, and α is the average number of collisions per second of an electron with the lattice structure ($\alpha \simeq 10^{14}$). Hence

$$R_x(\tau) = kTG\alpha e^{-\alpha|\tau|} \tag{3-71a}$$

and

$$\mathcal{S}_x(\omega) = \frac{2kTG\alpha^2}{\alpha^2 + \omega^2} \tag{3-71b}$$

The autocorrelation function and the corresponding power-density spectrum are shown in Fig. 3-23. Note that α is of the order of 10^{14} and hence $\mathcal{S}_x(\omega)$ is

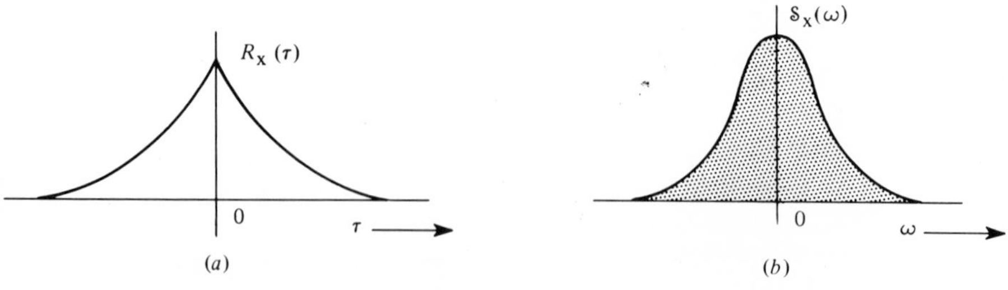

FIG. 3-23.

flat up to radian frequencies $\omega \cong 10^{13}$ (10^{12} Hz). Hence for all practical purposes thermal noise can be treated as white noise:

$$\mathcal{S}_x(\omega) \cong 2kTG \tag{3-72}$$

3-9. AUTOCOVARIANCE AND CROSS-COVARIANCE FUNCTIONS

We have already defined the covariance of two random variables x and y (Eq. 2-96) by the equation

$$\text{Covariance}\,(x,y) = \overline{(x - \overline{x})(y - \overline{y})} \tag{3-73}$$

The extension of this definition to random processes gives the autocovariance function and the cross-covariance function. For a random process $x(t)$, we define the autocovariance function $K_x(\tau)$ as

$$\begin{aligned} K_x(\tau) &= \overline{(x_1 - \overline{x}_1)(x_2 - \overline{x}_2)} \\ &= \overline{[x(t) - \overline{x(t)}][x(t + \tau) - \overline{x(t + \tau)}]} \end{aligned}$$

For a stationary process $\overline{x(t)} = \overline{x(t + \tau)} = \overline{x}$, and

$$\begin{aligned} K_x(\tau) &= \overline{[x(t) - \overline{x}][x(t + \tau) - \overline{x}]} \\ &= R_x(\tau) - \overline{x}[\overline{x(t)} + \overline{x(t + \tau)}] + \overline{x}^2 \\ &= R_x(\tau) - 2\overline{x}^2 + \overline{x}^2 \\ &= R_x(\tau) - \overline{x}^2 \end{aligned}$$

In a similar way, we define the cross-covariance function $K_{xy}(\tau)$ of two random processes $x(t)$ and $y(t)$ as

$$K_{xy}(\tau) = \overline{(x(t) - \overline{x})(y(t + \tau) - \overline{y})} \tag{3-75a}$$

$$= R_{xy}(\tau) - \overline{x}\overline{y} \tag{3-75b}$$

3-10. UNCORRELATED, ORTHOGONAL, AND INDEPENDENT PROCESSES

Two processes $x(t)$ and $y(t)$ are said to be uncorrelated if their cross-correlation function is equal to the product of their means; that is,

$$R_{xy}(\tau) = \overline{x}\overline{y} \tag{3-76}$$

Therefore for uncorrelated processes

$$R_{xy}(\tau) = \overline{x(t)\,y(t + \tau)} = \overline{x}\,\overline{y}$$

This implies that a random variable defined on $x(t)$ at any instant t and a random variable defined on $y(t)$ at another instant $t + \tau$ (for all values of τ) are uncorrelated if Eq. 3-76 is satisfied.

From Eq. 3-75 it follows that the cross-covariance function of uncorrelated processes vanishes; that is,

$$K_{xy}(\tau) = 0 \tag{3-77}$$

Hence the processes are said to be uncorrelated if their cross-covariance function vanishes.

Processes are said to be orthogonal if their cross-correlation function vanishes; that is, if

$$R_{xy}(\tau) = 0 \tag{3-78}$$

Note that if either one of the processes has a zero mean, uncorrelatedness implies orthogonality and vice versa.

It is evident from Eq. 3-78 that for orthogonal processes, the cross-power-density spectrum $\mathcal{S}_{xy}(\omega)$ vanishes.

We define the processes $x(t)$ and $y(t)$ to be independent, if the group of variables

$$x(t_1), x(t_2), \ldots, x(t_n)$$

is independent of the group of variables

$$y(t'_1), y(t'_2), \ldots, y(t'_n)$$

for any $t_1, t_2, \ldots, t_n$ and $t'_1, t'_2, \ldots, t'_n$ and for any n where independence implies that the joint probability-density function of the variables is the product of the probability-density functions of the individual variables.

It should be noted that independent processes are necessarily uncorrelated, but the converse is not true. Uncorrelated (or orthogonal) processes are not necessarily independent. It also follows from the definition of independent processes that if the two random processes are generated by independent sources that are not synchronized or related in any way, the processes must be independent.

3-11. PROPERTIES OF CORRELATION FUNCTIONS

In previous discussions, we have come across several significant properties of the autocorrelation function. Here is the summary of some of the significant properties of the autocorrelation function.

Property 1.

$$R_x(\tau) = R_x(-\tau) \tag{3-79}$$

This relation indicates that the autocorrelation function is an even function of τ. It has already been shown that this property follows from the fact that the second-order statistics are invariant under a change of the time origin for processes which are stationary or wide-sense stationary.

Property 2.

$$R_x(0) = \overline{x^2} \tag{3-80}$$

The value of the autocorrelation function at the origin is equal to the mean-square value of the random process $x(t)$.

This property follows from the definition of the autocorrelation function

$$R_x(\tau) = \overline{x(t)x(t+\tau)}$$

Hence

$$R_x(0) = \overline{x^2(t)} = \overline{x^2}$$

Property 3. If

$$z(t) = x(t) + y(t)$$

Then

$$R_z(\tau) = R_x(\tau) + R_y(\tau) + R_{xy}(\tau) + R_{yx}(\tau) \tag{3-81}$$

Proof:

$$\begin{aligned} R_z(\tau) &= \overline{z(t)z(t+\tau)} \\ &= \overline{[x(t)+y(t)][x(t+\tau)+y(t+\tau)]} \\ &= \overline{x(t)x(t+\tau)} + \overline{y(t)y(t+\tau)} + \overline{x(t)y(t+\tau)} + \overline{y(t)x(t+\tau)} \\ &= R_x(\tau) + R_y(\tau) + R_{xy}(\tau) + R_{yx}(\tau) \end{aligned}$$

Property 4. If a random process has a periodic component, of period T, then the autocorrelation function also has a periodic component of period T.

This property can easily be proved by considering only the periodic component of the process. For this component, the amplitudes at instants separated in time by integral multiples of the period T are equal. Hence

$$x(t) = x(t+T) = x(t+nT)$$

But

$$\begin{aligned} R_x(\tau) &= \overline{x(t)x(t+\tau)} \\ &= \overline{x(t)x(t+\tau+nT)} \\ &= R_x(\tau+nT) \end{aligned} \tag{3-82}$$

It is evident from this equation that $R_x(\tau)$ has a period T. From Property 3 it now follows that the autocorrelation will contain this periodic component.

Property 5.

$$\lim_{\tau\to\infty} R_x(\tau) = \overline{x}^2 \tag{3-83}$$

provided $x(t)$ does not contain any periodic components. This follows from the fact that for large values of τ, $x(t)$ and $x(t+\tau)$ are so dissimilar, that they lose their relatedness and tend to become uncorrelated.

Hence for $|\tau| \longrightarrow \infty$

$$\lim_{|\tau|\to\infty} R_x(\tau) = \overline{x(t)x(t+\tau)}$$
$$= \overline{x(t)}\,\overline{x(t+\tau)}$$
$$= \overline{x}^2$$

If the process has a zero mean value, then

$$\lim_{|\tau|\to\infty} R_x(\tau) = 0$$

Property 6.

$$R_x(0) \geq |R_x(\tau)| \qquad \tau \neq 0 \tag{3-84}$$

This property states that the maximum value of the autocorrelation function occurs at $\tau = 0$. For values of $\tau \neq 0$, the autocorrelation function may be as great as but not larger than $R_x(0)$.

This statement may be proved as follows: The mean-square value of any variable is nonnegative. Hence

$$\overline{(x_1 \pm x_2)^2} = \overline{x_1^2} + \overline{x_2^2} \pm \overline{2x_1x_2} \geq 0$$

or

$$\overline{x_1^2} + \overline{x_2^2} \geq \pm\, \overline{2x_1x_2} \tag{3-85}$$

But

$$x_1 = x(t_1) \text{ and } x_2 = x(t_1 + \tau)$$

Hence

$$\overline{x_1^2} = \overline{x_2^2} = \overline{x^2}$$

and

$$\overline{x_1x_2} = \overline{x(t_1)x(t_1+\tau)} = R_x(\tau)$$

Using this result and Eq. 3-80 in Eq. 3-85, we have

$$2R_x(0) \geq \pm\, 2R_x(\tau)$$

or

$$R_x(0) \geq |R_x(\tau)|$$

If a random process is periodic, the autocorrelation function is also periodic. Hence $R_x(\tau)$ will be equal to $R_x(0)$ for $\tau = T, 2T, \ldots, nT$, etc., where T is the period of the process.

Property 7. If the mean value of the process $x(t)$ is nonzero, then $R_x(\tau)$ will have a d-c component.

If the mean value of the process is $\overline{x}$, then the random variable $x(t)$ can be expressed as

$$x(t) = \overline{x} + y(t)$$

where $y(t)$ is a random variable with zero mean ($\overline{y} = 0$). Also

$$x(t+\tau) = \overline{x} + y(t+\tau)$$

and

$$\begin{aligned} R_x(\tau) &= \overline{[\bar{x} + y(t)][\bar{x} + y(t+\tau)]} \\ &= \bar{x}^2 + \bar{x}[\overline{y(t)} + \overline{y(t+\tau)}] + \overline{y(t)y(t+\tau)} \\ &= \bar{x}^2 + R_y(\tau) \quad (\text{since } \bar{y} = 0) \end{aligned} \tag{3-86}$$

Note that $R_y(\tau) \to 0$ as $\tau \to \infty$ because the variable $y(t)$ has a zero mean. (See Property 5.) Hence $R_y(\tau)$ does not have a d-c component, but $R_x(\tau)$ has a d-c component $\bar{x}^2$.

Properties of Cross-correlation Functions. The reader may easily prove the following properties of the cross-correlation function, viz.,

(1)
$$R_{xy}(\tau) = R_{yx}(-\tau) \tag{3-87}$$

(2)
$$|R_{xy}(\tau)| \le [R_x(0)R_y(0)]^{1/2} \tag{3-88}$$

(3) If the two processes $x(t)$ or $y(t)$ are independent, then

$$R_{xy}(\tau) = R_{yx}(\tau) = \bar{x}\bar{y} \tag{3-89}$$

3-12. DETECTION OF A PERIODIC SIGNAL IN THE PRESENCE OF NOISE BY CORRELATION

Message-bearing signals are corrupted by noise in the process of transmission. The noise may be so strong as to mask the desired message. In such circumstances it is not possible to recognize the message signal from a direct examination of the received waveform. Correlation techniques are extensively employed in detecting signals embedded in noise. This topic will be discussed in more detail in Chapters 4 and 6. Here we shall study a very simple application of correlation techniques in detecting a special class of signals (periodic signals) masked by noise. These techniques are used in such fields as detection of radar signals, detection of periodic components in brain waves, and detection of a cyclical component in ocean-wave analysis. In all these cases the important problem is to decide whether the received signal (which is heavily masked by noise) contains a periodic signal.

Suppose a periodic random process $s(t)$ is mixed with a random-noise process $n(t)$. We shall assume processes $s(t)$ and $n(t)$ to be statistically independent and hence uncorrelated. The autocorrelation function of the process $f(t)$ formed by the addition of the signal and noise $[f(t) = s(t) + n(t)]$, is given by

$$\begin{aligned} R_f(\tau) &= \overline{[s(t) + n(t)][s(t+\tau) + n(t+\tau)]} \\ &= R_s(\tau) + R_n(\tau) + R_{ns}(\tau) + R_{sn}(\tau) \end{aligned}$$

Since $s(t)$ and $n(t)$ are uncorrelated,

$$R_{ns}(\tau) = R_{sn}(\tau) = \bar{s}\bar{n}$$

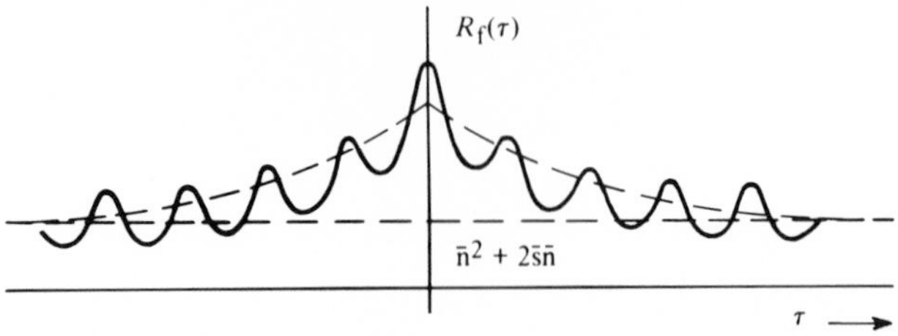

FIG. 3-24.

Remember that $R_s(\tau)$ is periodic, as it is the autocorrelation function of a periodic signal. (See Property 4 of autocorrelation functions.) Also, from Property 5 (Eqs. 3-83):

$$\lim_{\tau \to \infty} R_n(\tau) = \overline{n}^2$$

Hence

$$\lim_{\tau \to \infty} R_f(\tau) = R_s(\tau) + 2\overline{s}\overline{n} + \overline{n}^2$$

As $\overline{n}^2$ and $\overline{s}\overline{n}$ are constants, it follows that for large values of τ, $R_f(\tau)$ exhibits a periodic behavior of the same period as that of $s(t)$. This is shown in Fig. 3-24. Hence if the autocorrelation function of the received signal $f(t)$ shows a periodic behavior, $f(t)$ must contain a periodic component.

Alternatively we may cross-correlate $f(t)$ with a locally generated periodic process $c(t)$ which has the same period as that of the periodic component of $f(t)$. We have

$$\begin{aligned} R_f(\tau) &= \overline{f(t)c(t+\tau)} \\ &= \overline{[s(t) + n(t)]c(t+\tau)} \\ &= R_{sc}(\tau) + R_{nc}(\tau) \end{aligned}$$

Since $n(t)$ and $c(t)$ are uncorrelated $R_{nc}(\tau) = \overline{n}\overline{c}$. Hence

$$R_{fc}(\tau) = R_{sc}(\tau) + \overline{n}\overline{c}$$

As the process $s(t)$ and $c(t)$ have periodic components and have equal periods, $R_{sc}(\tau)$ will also have a periodic component of the same period. Hence $R_{fc}(\tau)$ exhibits a periodic behavior if $f(t)$ contains a periodic component of the same period as that of $c(t)$. If $c(t)$ is chosen as a periodic-impulse train signal, then it can be shown that $R_{fc}(\tau)$ yields the signal $s(\tau)$. Hence this method not only detects the presence of a periodic waveform, but also reveals its exact waveform (Refs. 3-1, 3-2):

3-13. CALCULATION OF THE MEAN-SQUARE VALUE OF A RANDOM PROCESS

The mean-square value of a random process is a very important parameter. It can be calculated in a number of ways. There are of course two different mean squares; the time mean square of a given sample function and the ensemble mean square of the process. For an ergodic process, the two means are identical. The time mean square of a sample function $x(t)$ is given by

$$\widetilde{x^2} = \lim_{T \to \infty} \frac{1}{T} \int_{-T/2}^{T/2} x^2(t)\,dt$$

$$= \mathcal{R}_x(0) \tag{3-90a}$$

Hence the time mean square is given by the value of the time autocorrelation function at the origin. The ensemble mean-square value of the process $x(t)$ is given by

$$\overline{x^2} = \int_{-\infty}^{\infty} x^2 p(x)\,dx \tag{3-90b}$$

Alternatively (Eq. 3-80),

$$\overline{x^2} = R_x(0) \tag{3-90c}$$

We can also evaluate $\overline{x^2}$ from the power-density spectrum using the Wiener-Khinchin relation given in Eq. 3-34b:

$$\overline{x^2} = R_x(0) = \frac{1}{2\pi} \int_{-\infty}^{\infty} \mathcal{S}_x(\omega)\,d\omega \tag{3-90d}$$

In the special case when $\mathcal{S}_x(\omega)$ is rational, the integral in Eq. 3-90d can be conveniently evaluated by using the residue theorem. The details are given in appendix 2 at the end of this chapter.

3-14. THE GAUSSIAN RANDOM PROCESS

A process $x(t)$ is said to be a gaussian random process if the random variables $x(t_1), x(t_2), \ldots, x(t_n)$ are jointly gaussian (see Eq. 2-133) for every n and for every set $(t_1, t_2, \ldots, t_n)$.

Gaussian random processes are very important in practice. Many noise (and other random) processes approximate reasonably well to gaussian random processes. As mentioned in Chapter 2, a random variable which depends upon the sum of a large number of independent random variables tends to be gaussian. This is the result of the one-dimensional central limit theorem. The multidimensional central limit theorem justifies the result that a set of n samples observed at any set of instants $(t_1, t_2, \ldots, t_n)$ tend to become jointly gaussian if the values of these samples depend upon a sum of large number of relatively independent perturbations. This condition is satisfied in many cases such as thermal noise in resistors, shot noise in vacuum tubes, diffusion noise in semiconductors, intergalactic noise in radio astronomy, etc. For this reason the gaussian random processes are of great practical importance. Fortunately, they are also relatively easier to handle compared to other classes of random processes. The reasons for this will become clear as our discussion progresses.

For convenience, the random variables $x(t_1), x(t_2), \ldots, x(t_n)$ will as usual be denoted by $x_1, x_2, \ldots, x_n$. The joint probability-density function of these variables is given in Eq. 2-133. It can be seen that the process is completely specified by the mean values of the variables $x_1, x_2, \ldots, x_n$ and the various covariance functions K_{ij}. This discussion applies to stationary as well as nonstationary gaussian processes. It is evident from the definition of a gaussian process, that its nth-order probability-density function is given by (Eq. 2-133).

$$p_{x_1 x_2 \ldots x_n}(x_1, x_2, \ldots, x_n) = \frac{1}{(2\pi)^{n/2}\sqrt{|K|}} \exp\left(\frac{-1}{2|K|}\sum_{i=1}^{n}\sum_{j=1}^{n} \Delta_{ij}(x_i - \overline{x_i})(x_j - \overline{x_j})\right) \tag{3-91a}$$

where $|K|$ is the determinant of the covariance matrix K given by

$$K = \begin{bmatrix} K_{11} & K_{12} & \ldots & K_{1n} \\ K_{21} & K_{22} & \ldots & K_{2n} \\ \ldots & \ldots & \ldots & \ldots \\ K_{n1} & K_{n2} & \ldots & K_{nn} \end{bmatrix} \tag{3-91b}$$

and Δ_{ij} is the cofactor of K_{ij}.

Note that K_{ij} is the covariance of x_i and x_j:

$$K_{ij} = \overline{(x_i - \overline{x}_i)(x_j - \overline{x}_j)}$$

Since $x_i = x(t_i)$ and $x_j = x(t_j)$,

$$K_{ij} = K_x(t_i, t_j)$$

where $K_x(t_i, t_j)$ is the autocovariance function of the process.

Note that

$$\begin{aligned} K_{ij} = K_x(t_i, t_j) &= \overline{(x_i - \overline{x}_i)(x_j - \overline{x}_j)} \\ &= \overline{x_i x_j} - \overline{x}_i \overline{x}_j - \overline{x}_j \overline{x}_i + \overline{x}_i \overline{x}_j \\ &= \overline{x_i x_j} - \overline{x}_i \overline{x}_j \\ &= R_x(t_i, t_j) - \overline{x}_i \overline{x}_j \end{aligned} \tag{3-92}$$

At this point we shall introduce the use of the n-dimensional random vector $\mathbf{x}$. We can represent n random variables $x_1, x_2, \ldots, x_n$ by an n-dimensional row vector $\mathbf{x}$ $(x_1, x_2, \ldots, x_n)$. This can be represented as

$$\mathbf{x} = (x_1, x_2, \ldots, x_n)$$

when the variables $x_1, x_2, \ldots, x_n$ assume values $x_1, x_2, \ldots, x_n$, the random variable $\mathbf{x}$ assumes the value $\boldsymbol{x}$ given by

$$x = (x_1, x_2, \ldots, x_n)$$

The joint probability density $p_{x_1x_2\ldots x_n}(x_1, x_2, \ldots, x_n)$ will henceforth be represented by $p_{\mathbf{x}}(x)$. Using this notation, Eq. 3-91 becomes

$$p_{\mathbf{x}}(x) = \frac{1}{(2\pi)^{n/2}\sqrt{|\mathrm{K}|}} \exp\left[-\frac{1}{2}(x - \overline{\mathbf{x}})\mathrm{K}^{-1}(x' - \overline{\mathbf{x}'})\right] \quad (3\text{-}93a)$$

where $\mathbf{x}'$ is the transpose of $\mathbf{x}$ and is a column vector

$$\mathbf{x}' = \begin{bmatrix} \mathrm{x}_1 \\ \mathrm{x}_2 \\ \cdot \\ \cdot \\ \cdot \\ \mathrm{x}_n \end{bmatrix}$$

For a random process with zero mean, $\overline{\mathbf{x}} = \overline{\mathbf{x}'} = 0$, and

$$p_{\mathbf{x}}(x) = \frac{1}{(2\pi)^{n/2}\sqrt{|\mathrm{K}|}} \exp\left(-\frac{1}{2}x\,\mathrm{K}^{-1}x'\right) \quad (3\text{-}93b)$$

The Stationary Gaussian Random Process. So far the discussion on gaussian processes has been general and applies to stationary as well as nonstationary processes. We have shown that, in general, the gaussian process is completely specified by its autocorrelation function $R_{\mathrm{x}}(t_i, t_j)$ and its mean value function.[9] An important corollary of this statement is that if the autocorrelation function $R_{\mathrm{x}}(t_i, t_j)$ and the means are unaffected by a shift of the time origin, then the entire statistics of the process are unaffected by a shift of the time origin. In other words, the process is stationary. Thus if

$$\begin{aligned} R_{\mathrm{x}}(t_i, t_j) &= R_{\mathrm{x}}(t_i - t_j) \\ \overline{\mathrm{x}(t)} &= \text{const. for all } t \end{aligned} \quad (3\text{-}94)$$

the process is stationary if it is gaussian. But condition 3-94 defines wide-sense stationarity. Hence, for a gaussian process, wide-sense stationarity implies stationarity in the strict sense. Note that this conclusion is not true in general of processes that are not gaussian. In general, if the condition of Eq. 3-94 is satisfied, we can say for certain only that the process is wide-sense stationary. For a gaussian process, however, wide-sense stationarity implies stationarity in the strict sense.

For a stationary gaussian process, the covariance function K_{ij} becomes (Eq. 3-92)

$$K_{ij} = R_{\mathrm{x}}(t_j - t_i) - \overline{\mathrm{x}}^2 \quad (3\text{-}95)$$

[9]For nonstationary processes (or processes which are not wide-sense stationary), the mean value is a function of t.

We shall once again stress the point which distinguishes the gaussian from the nongaussian process. The complete statistics of a gaussian process are determined from its autocorrelation function (which is a second-order statistics parameter) and its mean value. This is the property that simplifies the study of gaussian processes and systems involving them. In general, for nongaussian processes higher-order statistics cannot be determined from the lower-order statistics. For the gaussian process, however, the second-order statistics parameter, viz., K_{ij} determines all the higher-order statistics. This very property also enables us to state that a gaussian process is strictly stationary if it is wide-sense stationary.

Transmission of a Gaussian Process Through a Linear System. Another significant property of a gaussian process is that the response of a linear system to it is also a gaussian process. This can be shown as follows. Let $x(t)$ be a gaussian process applied at the input of a linear system whose unit impules response is $h(t)$. If $y(t)$ is the output (response) process, then

$$y(t) = \int_{-\infty}^{\infty} x(t - \tau)h(\tau)\,d\tau \tag{3-96}$$

It is evident from Eq. 3-96 that the random variable $y(t)$ can be expressed as a (continuous) sum of random variables $x(t)$ over the interval $-\infty < t < \infty$. This statement can be made explicit by approximating the integral in Eq. 3-96 with a discrete sum

$$y(t) = \lim_{\Delta\tau \to 0} \sum_{k=-\infty}^{\infty} x(t - \tau_k)h(\tau_k)\Delta\tau \qquad \tau_k = k\,\Delta\tau$$

Since $x(t)$ is a gaussian process, all the variables $x(t - \tau_k)$ are jointly gaussian (by definition). Hence the variables $y(t_1), y(t_2), \ldots, y(t_n)$ for all n and every set $(t_1, t_2, \ldots, t_n)$ are linear combinations of variables which are jointly gaussian. Therefore the variables $y(t_1), y(t_2), \ldots, y(t_n)$ must be jointly gaussian according to the discussion in Sec. 2-13. It therefore follows that the process $y(t)$ is a gaussian process.

To summarize, the gaussian random process has the following properties:

1. The complete statistics of the gaussian random process are specified by its autocorrelation function and its mean values.
2. If a gaussian random process is wide-sense stationary, then it is stationary in the strict sense.
3. The response of a linear system to a gaussian random process is also a gaussian random process. Since it is a gaussian process, it

can be completely specified by its autocorrelation function and the mean value. It will be shown in Chapter 4, that the autocorrelation function and the mean of the response process can be determined from a knowledge of the autocorrelation function and the mean of the input process. Hence the complete statistics of the output process are determined from a knowledge of the input process, if the input process is gaussian. This is not true in general of processes that are not gaussian.

3-15. GAUSSIAN WHITE NOISE

As explained earlier "white noise" is a term used to describe processes whose power-density spectra is uniform over the entire frequency interval $(-\infty, \infty)$ (Fig. 3-19). White-noise processes which are also gaussian are called *gaussian white-noise* processes. Thus a gaussian white noise process is defined as a gaussian process with a uniform power-density spectrum over the entire frequency range.

For a general white-noise process,

$$\mathcal{S}_x(\omega) = \frac{\mathfrak{N}}{2} \tag{3-97a}$$

The power contribution by frequency components in a certain frequency range is given by the area of $\mathcal{S}_x(\omega)$ over the frequency range (integrated with respect to f over the negative and positive frequencies). Hence the power per unit bandwidth (in Hertz) is given by $2\left(\frac{\mathfrak{N}}{2}\right) = \mathfrak{N}$ watts. The autocorrelation is the inverse Fourier transform of $\mathcal{S}_x(\omega)$ and is given by

$$R_x(\tau) = \frac{\mathfrak{N}}{2}\,\delta(\tau)$$

It can be easily seen that a white-noise process is an idealization and cannot exist in practice because the mean-square value of such a process is infinity.

$$\overline{x^2} = \frac{1}{2\pi}\int_{-\infty}^{\infty} \mathcal{S}_x(\omega)\,d\omega = \infty$$

Although true white noise is a mathematical fiction, it proves very useful in practical applications. We often encounter situations where a signal is contaminated by wideband noise which has a uniform power-density spectrum over a band much larger than the spectrum of the desired signal. The effect of noise is generally reduced by passing the corrupted signal through a linear filter whose bandwidth is of the same order of magnitude as that of the signal. In such case the effect of the (wideband) noise will be the same as that of idealized white noise. Wideband noise can, therefore, be treated like white noise for all practical purposes. In the τ-domain the autocorrelation function of white noise is an impulse, whereas that of wideband noise will be a narrow pulse of some finite width. The reader can easily see that replacing wideband white noise by true white noise is much the same type of approximation as is made when replacing a narrow-pulse signal by an idealized impulse in cases where the system time constant is much larger than the width of the pulse. Since the system's time constant is much larger than the pulse width, its response to the pulse is almost identical with its response to an equivalent impulse (whose strength is equal to the area of the pulse).

Since the autocorrelation function $R_x(\tau) = 0$ for all τ (except $\tau = 0$), it is evident that for white noise any two random variables $x(t_i)$ and $x(t_j)$ are uncorrelated, no matter how close t_i and t_j may be. If the process is gaussian white noise, then zero correlation implies independence (Sec. 2-13). Hence for white gaussian noise any two samples of the process are independent no matter how closely together in time they are. This is obviously the ultimate in randomness when the amplitudes of sample functions have no relatedness whatsoever, however short the time interval between observation of these amplitudes.

Band-Limited Gaussian White Noise. Band-limited gaussian white noise is a gaussian noise process which has a uniform power-density spectrum of constant magnitude $\frac{\mathfrak{N}}{2}$ over a band $(-W, W)$ rps and zero outside this band (Fig. 3-25a).

$$\mathcal{S}_x(\omega) = \begin{cases} \dfrac{\mathfrak{N}}{2} & |\omega| < W \\ 0 & \text{Otherwise} \end{cases} \tag{3-98}$$

The autocorrelation function $R_x(\tau)$ is the inverse Fourier transform of $\mathcal{S}_x(\omega)$.

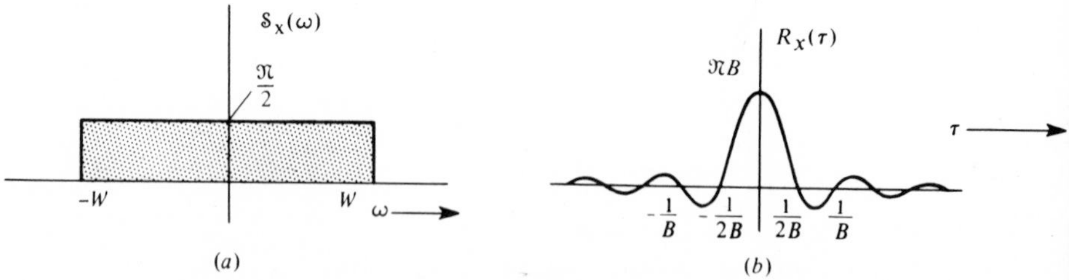

FIG. 3-25.

$$R_x(\tau) = \mathfrak{N} B \, \mathrm{Sa}(W\tau) \tag{3-99}$$

where B is the bandwidth in Hertz ($B = W/2\pi$).

This autocorrelation function is shown in Fig. 3-25b. This process can also be obtained by passing white gaussian noise through an ideal low-pass filter whose cutoff frequency is B Hz. The mean-square value of this process is given by

$$\overline{\mathrm{x}^2} = R_x(0) = \mathfrak{N} B \tag{3-100}$$

Note that this process has a zero mean value, since

$$R_x(\tau) \longrightarrow 0 \text{ as } \tau \longrightarrow \infty \quad \text{(see Property 5)}$$

It was shown in Chapter 1 that a signal $x(t)$ band-limited to B Hz can be completely specified by its samples taken $1/2B$ seconds apart[10] (Eq. 1-130)

$$x(t) = \sum_{k=-\infty}^{\infty} x_k \, \mathrm{Sa}(Wt - k\pi) \tag{3-101a}$$

where $x_k = x(kT_s)$, $T_s = 1/2B$ (the sampling interval). The coefficient x_K is different for each sample function and hence $\mathrm{x}_k (k = 0, \pm 1, \pm 2 \ldots)$ is a random variable. We can express Eq. 3-101a for the process as

$$\mathrm{x}(t) = \sum_{k=-\infty}^{\infty} \mathrm{x}_k \, \mathrm{Sa}(Wt - k\pi) \tag{3-101b}$$

The variables $\mathrm{x}_1, \mathrm{x}_2, \ldots, \mathrm{x}_k \ldots$ are random variables. It is interesting to observe that the complete process can be specified by the set of random variables x_k, $k = 0, \pm 1, \pm 2, \ldots$ etc. This is a very significant result. It allows us to replace continuous signals with a discrete sequence of random variables without any loss of information. A random waveform or a random process can be represented by a point $(\mathrm{x}_1, \mathrm{x}_2, \ldots)$ or a vector in the corresponding "geometrical" space. Such a representation proves to be very convenient in the area of signal detection and is extensively utilized in Chapter 6.

At this point we make an interesting observation about $R_x(\tau)$ (see Fig. 3-25b). The autocorrelation function is zero for $\tau = 1/2B$ and any integral multiple of $1/2B$. Hence samples of this process taken $1/2B$ seconds apart are uncorrelated. The random variables $\mathrm{x}_1, \mathrm{x}_2, \ldots, \mathrm{x}_k \ldots$, etc. are the Nyquist samples taken at intervals $1/2B$ seconds apart. Obviously all these random variables are uncorrelated. Since the process is gaussian, these samples (the variables $\mathrm{x}_1, \mathrm{x}_2, \ldots$ etc.), are also gaussian. But if the gaussian variables are uncorrelated, they are necessarily inde-

[10]This result was derived for energy signals (signals with finite energy). It can be extended to power signals. See Refs. 3-4 and 3-5.

pendent. Hence all the variables $x_1, x_2, \ldots$ etc., are independent. Thus the Nyquist samples of a gaussian process are independent. The joint probability density function of any n of these variables can be expressed as a product of n individual probability densities of these variables. For all these variables, since the mean values are zero, $K_{ij} = R_{ij}$ and

$$K_{ij} = R_{ij} = R_x(t_i - t_j) = \begin{cases} 0 & i \neq j \\ R_x(0) = \mathfrak{N}B & i = j \end{cases}$$

Hence we can express Eq. 3-91a as (see Eq. 2-136)

$$p_{x_1 x_2 \ldots x_n}(x_1, x_2, \ldots, x_n) = \left(\frac{1}{\sqrt{2\pi\mathfrak{N}B}} e^{-x_1^2/2\mathfrak{N}B}\right) \cdots \left(\frac{1}{\sqrt{2\pi\mathfrak{N}B}} e^{-x_n^2/2\mathfrak{N}B}\right)$$

$$= \frac{1}{(2\pi\mathfrak{N}B)^{n/2}} e^{-(x_1^2 + x_2^2 + \cdots + x_n^2)/2\mathfrak{N}B} \tag{3-102}$$

3-16. GEOMETRICAL REPRESENTATION OF A RANDOM PROCESS

The geometrical representation of a signal has been extensively discussed in Chapter 1 (Sec. 1-15 and 1-16). A signal can be expressed in terms of a complete set of orthogonal functions[11] $\{\phi_k(t)\}$ as

$$x(t) = x_1\phi_1(t) + x_2\phi_2(t) + \cdots + x_n\phi_n(t) \tag{3-103}$$

$$= \sum_{k=1}^{n} x_k \phi_k(t) \tag{3-104}$$

where n may be finite or infinite. The coefficient x_k is given by (Eq. 1-129b)

$$x_k = \int_{-\infty}^{\infty} x(t)\phi_k(t)\,dt$$

when the set $\{\phi_k(t)\}$ is orthonormal. In general, for an orthogonal set $\{\phi_k(t)\}$

$$x_k = \frac{\int_{-\infty}^{\infty} x(t)\phi_k(t)\,dt}{\int_{-\infty}^{\infty} \phi_k^2(t)\,dt}$$

A random process $x(t)$ may be expressed as

$$x(t) = x_1\phi_1(t) + x_2\phi_2(t) + \cdots + x_n\phi_n(t)$$

[11] It is not necessary for the set to be orthogonal. In our discussion, however, we shall always consider an orthogonal basis set. If a set is not orthogonal, one can always orthogonalize it by using the Gram-Schmidt procedure.

$$= \sum_{k=1}^{n} x_k \phi_k(t)$$

Here $x_1, x_2 \ldots, x_n$ are random variables. For each sample function of the process, variables $x_1, x_2 \ldots x_n$ assume a certain value. The random process $x(t)$ therefore can be represented by a random point $\mathbf{x}(x_1, x_2, \ldots x_n)$. For each sample function, $\mathbf{x}$ assume a certain value.

To get a better understanding of this picture, we observe that each sample function of the process can be represented by a point in space. A random process (which consists of an ensemble of sample functions) therefore can be represented by an ensemble of points in the hyperspace (Fig. 3-26). This ensemble may be considered as a random point in the

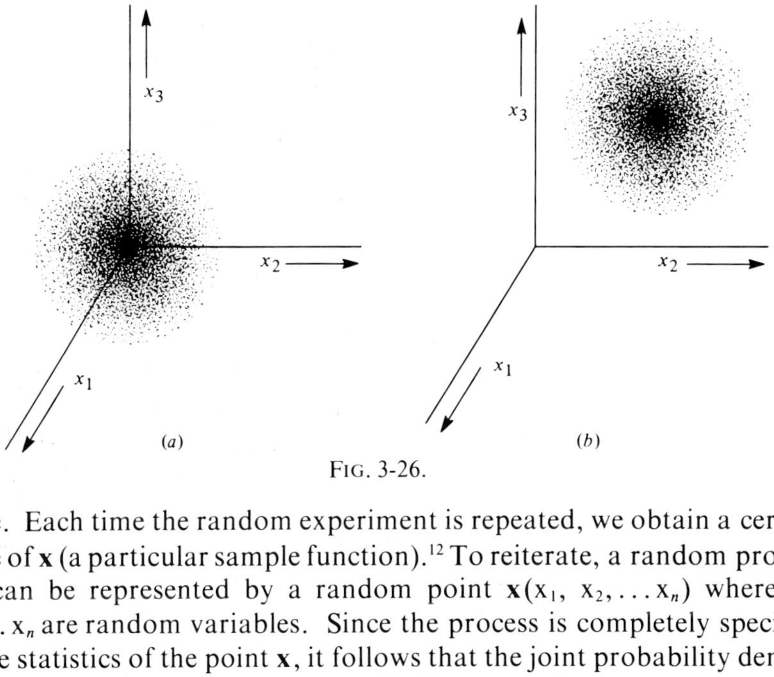

FIG. 3-26.

space. Each time the random experiment is repeated, we obtain a certain value of $\mathbf{x}$ (a particular sample function).[12] To reiterate, a random process $x(t)$ can be represented by a random point $\mathbf{x}(x_1, x_2, \ldots x_n)$ where $x_1, x_2, \ldots x_n$ are random variables. Since the process is completely specified by the statistics of the point $\mathbf{x}$, it follows that the joint probability density function of variables $x_1, x_2, \ldots, x_n$ describes the process completely. This probability density will be denoted by $p_{\mathbf{x}}(x)$ as usual

$$p_{\mathbf{x}}(x) = p_{x_1 x_2 \ldots x_n}(x_1, x_2, \ldots x_n)$$

The ensemble of points representing a random process has a useful interpretation. The density of points (number of points per unit volume) will

[12] It should be once again stressed here that the waveforms of various sample functions are assigned to the sample points (outcomes) of the experiment and are, therefore, assumed to be deterministic. There is no randomness associated with the waveforms of sample functions. The randomness, however, is associated with the occurrence of a particular sample function (outcome).

in general vary over the entire region. Since each point represents a particular sample function, the likelihood of observing the outcome of the random process trial in the high density region will be higher (Fig. 3-26). Indeed the probability of observing the random point $\mathbf{x}$ in any region will be proportional to the density of the points in that region. But the probability of observing $\mathbf{x}$ in a certain region is proportional to $p_{\mathbf{x}}(x)$. Hence the density of points is directly proportional to $p_{\mathbf{x}}(x)$.

Geometrical Representation of a Band-Limited Gaussian Process. Consider a band-limited gaussian process $\mathrm{x}(t)$. Such a process can be represented in the form shown in Eq. 3-101b. The variable x_k is $\mathrm{x}(kT_s)$ where T_s is the sampling interval ($T_s \leq 1/2B$). But by definition, all the random variables x_k of a gaussian process are jointly gaussian. Hence the variables $\mathrm{x}_1, \mathrm{x}_2, \ldots, \mathrm{x}_n$ are jointly gaussian and we can (assuming the process to have zero mean[13]) express $p_{\mathbf{x}}(x)$ as

$$p_{\mathbf{x}}(x) = \frac{1}{(2\pi)^{n/2}\sqrt{|\mathrm{K}|}}\, e^{-(x\mathrm{K}^{-1}x')/2} \tag{3-105}$$

Note the difference between Eqs. 3-93b and 3-105. Eq. 3-93b gives the probability-density function of a random vector $\mathbf{x}(\mathrm{x}_1, \mathrm{x}_2, \ldots, \mathrm{x}_n)$ formed by n random variables $\mathrm{x}(t_1), \mathrm{x}(t_2), \ldots, \mathrm{x}(t_n)$ taken over the process. The values $t_1, t_2, \ldots, t_n$ are arbitrary. Hence Eq. 3-93b represents a probability density function of arbitrary random vector formed by taking any n random variables over the process. In contrast to this, Eq. 3-105 represents the probability density of a particular random vector $\mathbf{x}(\mathrm{x}_1, \mathrm{x}_2, \ldots, \mathrm{x}_n)$ where $\mathrm{x}_1, \mathrm{x}_2, \ldots, \mathrm{x}_n$ are the random variables formed by the samples (taken at a uniform rate greater than or equal to Nyquist sampling rate) over the random process. Since these samples describe the process completely, the random vector formed by all sample variables in Eq. 3-105 describes the process completely and hence the random vector $\mathbf{x}$ in Eq. 3-105 represents the process $\mathrm{x}(t)$ itself.

Band-Limited White Gaussian Process. If a band-limited gaussian process has a uniform (white) power-density spectrum, then Eq. 3-105 can be considerably simplified. We have shown (see Fig. 3-25b or Eq. 3-102) that the variables x_i and x_j are uncorrelated for a band-limited gaussian process if the samples are taken at the Nyquist rate.

$$K_{ij} = \begin{cases} 0 & i \neq j \\ \mathfrak{N}B & i = j \end{cases} \tag{3-106}$$

[13]This assumption can be made without loss of generality. If the mean of the process is nonzero, then one can define a new variable $\mathbf{y} = \mathbf{x} - \overline{\mathbf{x}}$. It is evident that $\overline{\mathbf{y}} = 0$. Alternately one may replace x and x' by $(x - \overline{\mathbf{x}})$ and $x' - \overline{\mathbf{x}}'$ in Eq. 3-105.

Hence K becomes a diagonal matrix:

$$\mathrm{K} = \begin{bmatrix} \mathfrak{N}B & 0 & . & . & 0 \\ 0 & \mathfrak{N}B & . & . & 0 \\ . & . & . & . & . \\ . & & & & \\ 0 & 0 & . & . & \mathfrak{N}B \end{bmatrix}$$

$$= (\mathfrak{N}B) \begin{bmatrix} 1 & 0 & 0 & . & . & 0 \\ 0 & 1 & 0 & . & . & 0 \\ 0 & 0 & 0 & . & . & 0 \\ . & . & . & . & . & . \\ 0 & 0 & 0 & . & . & 1 \end{bmatrix}$$

$$= (\mathfrak{N}B)\mathrm{I}$$

Where I is the identity matrix. Obviously

$$|\mathrm{K}| = (\mathfrak{N}B)^n$$

$$\mathrm{K}^{-1} = \frac{1}{(\mathfrak{N}B)} \mathrm{I}$$

and

$$x\mathrm{K}^{-1}x' = \frac{1}{\mathfrak{N}B} [x_1 \ x_2 \ \cdots x_n] \begin{bmatrix} 1 & 0 & 0 & . & . & 0 \\ 0 & 1 & 0 & . & . & 0 \\ . & . & . & . & . & . \\ . & . & . & . & . & . \\ . & . & . & . & . & . \\ 0 & 0 & 0 & . & . & 1 \end{bmatrix} \begin{bmatrix} x_1 \\ x_2 \\ . \\ . \\ . \\ x_n \end{bmatrix}$$

$$= \frac{1}{(\mathfrak{N}B)} (x_1^2 + x_2^2 + \cdots + x_n^2)$$

$$= \frac{|\mathbf{x}|^2}{\mathfrak{N}B}$$

Hence

$$p_{\mathbf{x}}(x) = \frac{1}{(2\pi\mathfrak{N}B)^{n/2}} e^{-|\mathbf{x}|^2/2\mathfrak{N}B} \tag{3-107}$$

This result shows that the probability-density function $p_{\mathbf{x}}(x)$ depends only upon the magnitude of the vector **x** in hyperspace. It is evident that the probability-density function is spherically symmetrical if plotted in the hyperspace with $x_1, x_2, \ldots, x_n$ as coordinate axes. The shaded diagram for this probability density (for three dimensions) is shown in Fig. 3-26a. Observe that the intensity of shading (which is proportional to the probability density) is maximum at the center and diminishes exponentially as the square of the distance from the center. We also have the mean-square value of white gaussian process band-limited to B Hz (Eq. 3-100):

$$\overline{\mathrm{x}^2} = R_{\mathrm{x}}(0) = \mathfrak{N}B$$

If σ^2 denotes the mean-square value (or variance) of this signal, then

$$\sigma^2 = \mathfrak{N}B$$

Equation 3-107 can then be expressed as

$$p_{\mathrm{x}}(x) = \frac{1}{(2\pi\sigma^2)^{n/2}} e^{-|x|^2/2\sigma^2} \qquad (3\text{-}108)$$

Equation 3-108 represents the probability-density function of a white gaussian process with mean-square value (variance) σ^2. It is assumed that the mean value of the process is zero. If the process has a nonzero mean, then Eq. 3-108 becomes[14]

$$p_{\mathrm{x}}(x) = \frac{1}{(2\pi\sigma^2)^{n/2}} e^{-|x-\overline{\mathbf{x}}|^2/2\sigma^2} \qquad (3\text{-}109)$$

where σ^2 is the variance and $\overline{\mathbf{x}}$ is the mean value of the process. The probability density for a process with nonzero mean is shown in Fig. 3-26b for a three-dimensional case.

3-17. MEASUREMENT OF AUTOCORRELATION FUNCTIONS:

In the earlier section the method of calculating autocorrelation functions based upon second-order probability density has been illustrated by various examples. In practice, however, with few exceptions, the complexity of the many factors that cause random variations to occur does not easily permit one to determine the second-order probability-density function. In such cases the autocorrelation function is estimated from experimental measurements (instead of deriving it from the second-order probability density). If the process is ergodic, it is necessary to

[14]Equation 3-105 is simplified to Eq. 3-108 (or Eq. 3-109) because the Nyquist samples of a white band-limited process are uncorrelated. If the process is not a band-limited white process, then the Nyquist samples are not uncorrelated. In such cases, it is possible to expand a random process by using different orthogonal basis set which will yield uncorrelated coefficients of expansion. The Karhunen-Loéve theorem states that a general random process $\mathrm{x}(t)$ can be expanded over a time interval (a,b) by orthogonal series with $\{\Theta_k(t)\}$ as the basis (Refs. 3-5, 3-6):

$$\mathrm{x}(t) = \sum_n \mathrm{c}_n \Theta_n(t) \qquad a \le t \le b$$

The coefficients c_n's are uncorrelated provided the set $\Theta_n(t)$ is a solution of equation

$$\int_a^b \Theta_n(\alpha) R_{\mathrm{x}}(t-\alpha)\,d\alpha = \lambda_n \Theta_n(t) \qquad a \le t \le b$$

have only one sample function for such a calculation. The autocorrelation function is identical with the time autocorrelation function. Time averaging is performed over a finite but large value of T.

$$R_x(\tau) = \mathcal{R}_x(\tau) = \frac{1}{T}\int_{-T/2}^{T/2} x(t)x(t-\tau)\,dt \qquad T \gg \tau \qquad (3\text{-}110)$$

Ideally $T \to \infty$. But in practice T is finite and this introduces a random error in the measurement. One can estimate the value of T required for a given mean-square error. For example, it can be shown that for white gaussian noise band-limited to B Hz the mean-square error $\overline{\epsilon^2}$ in estimating $R_x(\tau)$ is given by (Ref. 3-7).

$$\overline{\epsilon^2} \cong \frac{1}{2TB}[R_x^2(\tau) + R_x^2(0)] \qquad (3\text{-}111)$$

Example 3-10. An autocorrelation function of an ergodic white gaussian process band-limited to B Hz is determined from its sample function of duration T seconds. Estimate the minimum value of T required to achieve the normalized rms square error of 0.1 ($\sqrt{\overline{\epsilon^2}/R_x^2(\tau)} = 0.1$) at a value of τ where $R_x(\tau)$ is 0.2 $R_x(0)$. In this problem

$$\frac{\overline{\epsilon^2}}{R_x^2(\tau)} = 10^{-2} \text{ and } \frac{R_x(\tau)}{R_x(0)} = 0.2$$

From Eq. 3-111, it follows that

$$T \cong \frac{1}{2B}\left(\frac{R_x^2(\tau)}{\overline{\epsilon^2}}\right)\left(1 + \frac{R_x^2(0)}{R_x^2(\tau)}\right)$$
$$= 0.13 \text{ sec}$$

It can be seen from Eq. 3-111 that to reduce the rms error below 10 percent at $\tau = 0$, we have

$$\frac{\overline{\epsilon^2}}{R_x^2(0)} = (0.1)^2$$

Hence, by Eq. 3-111,

$$T = \frac{1}{2B\overline{\epsilon^2}}[R_x^2(\tau) + R_x^2(0)]$$
$$= \frac{2R_x^2(0)}{2B\overline{\epsilon^2}} \qquad \text{for } \tau = 0$$
$$= \frac{100}{B}$$

Note that the width of the autocorrelation function $R_x(\tau)$ is about $1/2B$ (see Fig. 3-25). Hence T, the length of the record, in this case should be about 200 times the width of the autocorrelation function.

The averaging operation described in Eq. 3-110 can be accomplished by using of a finite time-averaging filter whose unit impulse response is

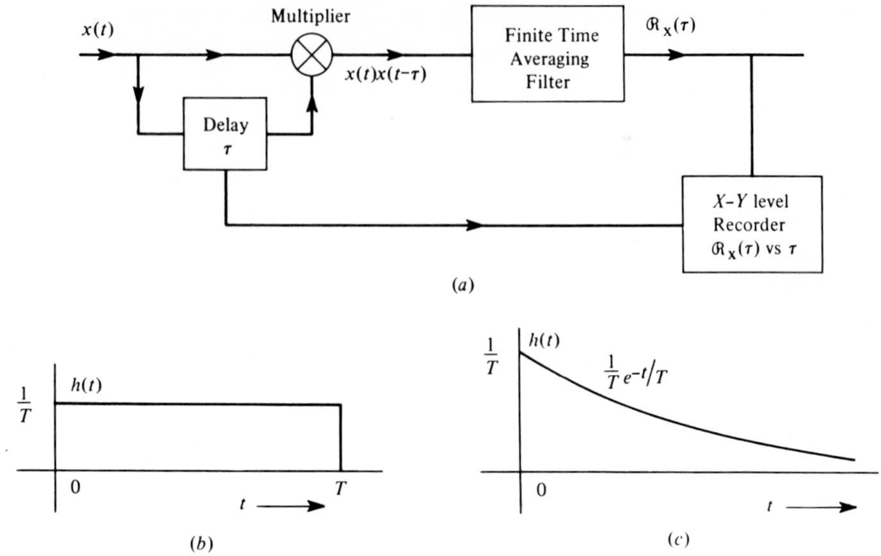

FIG. 3-27.

shown in Fig. 3-27b. If a signal $f(t)$ is applied at the input terminals of the finite time-averaging filter, its output $r(t)$ is given by

$$r(t) = \int_{-\infty}^{\infty} f(\alpha)h(t - \alpha)\,d\alpha$$

$$= \frac{1}{T}\int_{t-T}^{t} f(\alpha)\,d\alpha$$

It is evident that the filter averages the signal $f(t)$ over the interval $(t - T, t)$. If the input to this filter is $x(t)x(t - \tau)$ as shown in Fig. 3-27a, then it follows that the output is given by

$$\frac{1}{T}\int_{t-T}^{t} x(\alpha)x(\alpha - \tau)\,d\alpha$$

which we recognize as the time autocorrelation function of $x(t)$. To obtain $R_x(\tau)$, the product $x(t)x(t - \tau)$ is passed through a finite time-averaging filter as shown in Fig. 3-27a.

In practice, it is difficult to build an ideal finite time-averaging filter. One can, however, approximate such a filter with one whose impulse response is a slowly decaying exponential (Fig. 3-27c). The slower the rate of decay the better the approximation. This type of averaging is called *exponentially weighted past* (EWP) averaging (Ref. 3-8). It is obvious that if $x(t)x(t - \tau)$ is the input signal to an EWP averaging filter, the response will be given by

$$\frac{1}{T}\int_{-\infty}^{t} x(\alpha)x(\alpha - \tau)e^{-(t-\alpha)/T}\,d\alpha$$

Here t is the instant of observation at the output (Fig. 3-27a) and should be about 3 to 4 times the value of T. This is the weighted time average in which the recent past has been weighted more heavily than the remote past. It is obvious that it approaches a true-time average if T is made sufficiently large. Note that the method of determining $R_x(\tau)$ as described here applies to ergodic processes where the time averages are equal to ensemble averages. If the process is not ergodic, then several sample functions must be examined. The interested reader is referred to Bendat and Piersol (Ref. 3-7).

The above is an analog method to estimate the autocorrelation function. It is possible to estimate autocorrelation using digital techniques by sampling $x(t)$ at intervals of α seconds (α being less than the Nyquist interval). If these samples are labeled as $x_0, x_1, x_2, \ldots, x_N$., then

$$R_x(n\alpha) = \frac{1}{N-n}\sum_{k=0}^{N-n} x_k x_{k+n} \qquad n = 0, 1, 2, \ldots, M \qquad M \ll N$$

Note that $N\alpha \simeq T$ (the length of the record). The required length T (and hence the sample size N) can be estimated assuming a certain allowable error (Ref. 3-7). One may use Eq. 3-111 to estimate T.

Similar techniques can be used to estimate crosscorrelation functions.

3-18. MEASUREMENT OF THE POWER-DENSITY SPECTRUM

The power-density spectrum of a single record $x(t)$ can be determined by a wave analyzer or a tunable narrow-band filter as shown in Fig. 3-28a. The narrow-band filter centered at frequency ω and bandwidth $\Delta\omega$ allows only the frequency components in this band to pass. Hence the mean-square value of the output of this filter is approximately $2\mathcal{S}_x(\omega)\Delta f$, as Δf is small.

The squarer and the finite time-averaging filter perform the operation of mean-squaring. Obviously the final output is $2\mathcal{S}_x(\omega)\Delta f$. Hence the final output when divided by $2\Delta f$ yields the value of the power density spectrum at frequency ω. The center frequency of the tunable filter is varied until all the frequencies in the frequency range of the spectrum are scanned.

This method, however, has a serious drawback. The bandwidth of the narrow-band filter changes as the center frequency is changed. For this reason, in practice a slight modification to the above procedure is commonly made. The input signal is heterodyned with a variable-frequency oscillator to translate by required amount the entire power-density spectrum. The heterodyned output is then passed through a

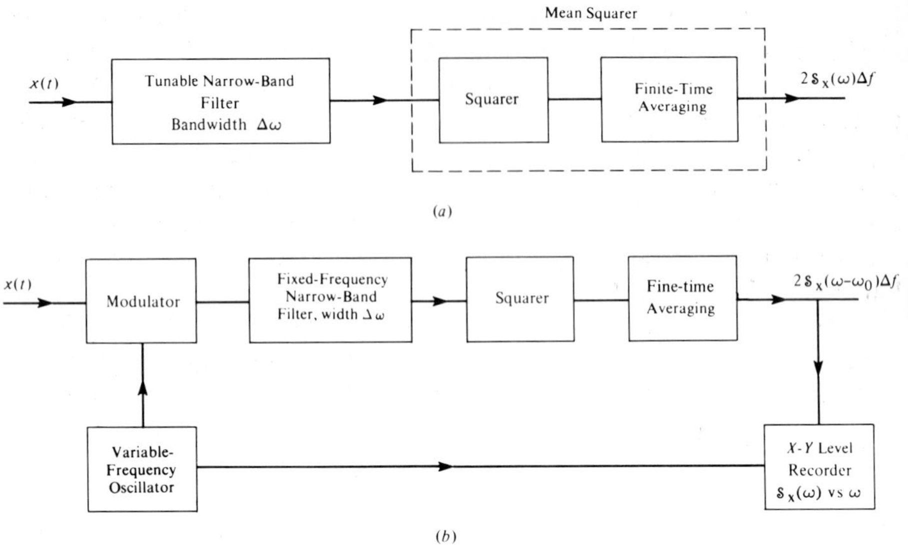

FIG. 3-28.

fixed-frequency, narrow-band filter and a mean squarer (Fig. 3-28b). The oscillator frequency is varied in steps so that the different frequency bands of the original signal $x(t)$ are translated to the passband frequencies and are consequently transmitted through the fixed-frequency, narrow-band filter. Thus, instead of scanning the whole frequency range by varying the center frequency of the narrow-band filter, we shift the entire power-density spectrum by heterodyning the signal with a variable-frequency signal and passing it through a fixed-frequency narrow-band filter. This arrangement is shown in Fig. 3-28b.

For an alternative method of determining the power-density spectrum using EWP averages, the reader is referred to Otterman (Ref. 3-8). For a general discussion on measurement of power density spectra, the reader should consult Refs. 3-7 and 3-10.

APPENDIX 3-1. THERMAL NOISE

This type of noise occurs due to random motion of free electrons in a conducting medium such as a resistor. Each free electron inside the resistor is in motion due to its thermal energy. The path of electron motion is random and zigzag due to collisions with the lattice centers. The motion of electrons induces a random (noise) current in the external terminals.

In order to study the thermal-noise current consider a slab of conductor of length d, cross section area A and whose ends are shorted as shown in Fig. A3-1. The free electrons in this slab move with random

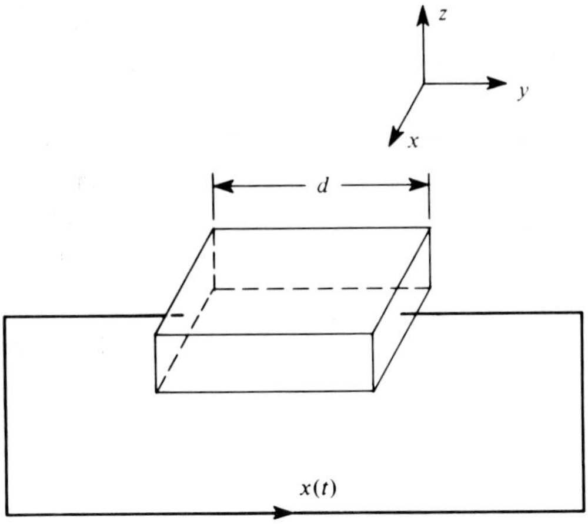

FIG. A3-1.

thermal velocities. It is reasonable to assume that (a) the direction of velocities of electrons before and after collision are independent and randomly oriented, (b) each electron moves with constant velocity between collisions, since no electric field is impressed upon the slab. If v_x, v_y, and v_z are the three components of the velocity of an electron, then from assumption (a) it follows that the mean kinetic energies associated with each velocity component must be equal; that is,

$$\tfrac{1}{2} m\overline{v_x^2} = \tfrac{1}{2} m\overline{v_y^2} = \tfrac{1}{2} m\overline{v_z^2}$$

where m is the mass of an electron.

Here we use an important result from statistical mechanics which states that the average kinetic energy of each electron associated with motion in a given direction is equal to $\tfrac{1}{2} kT$, where k is the Boltzmann constant (1.38×10^{-23}) and T is the ambient temperature in degrees Kelvin. Hence

$$\tfrac{1}{2} m\overline{v_x^2} = \tfrac{1}{2} m\overline{v_y^2} = \tfrac{1}{2} m\overline{v_z^2} = \tfrac{1}{2} k\,T$$

or

$$\overline{v_x^2} = \overline{v_y^2} = \overline{v_z^2} = \frac{kT}{m} \tag{A3-1}$$

The current $x(t)$ in the lead exists solely due to the existence of v_y, the velocity component in the y direction. It can be shown that the induced current is given by (Ref. 3-9)

$$x(t) = \frac{q}{d} v_y(t) \tag{A3-2}$$

where q is the charge on the electron, v_y is its velocity, and d is the distance between the plates.

The velocity v_y abruptly changes at each collision, and has a constant value [assumption (b)] between collisions. Hence the induced current has the waveform shown in Fig. A3-2. The length of the time interval between two successive collisions is random, and the occurrence of any collision during any interval is independent of the previous collision. Occurrence of an event independent of previous events is precisely the requirement that the occurrence be Poisson-distributed. Hence the abrupt amplitude changes of $x(t)$ are Poisson-distributed (see Example 2-18). The autocorrelation function and the power density spectrum of this random process have been derived in Example 3-9. From the symmetry of the problem it is evident that the mean value of $x(t)$ must be zero.

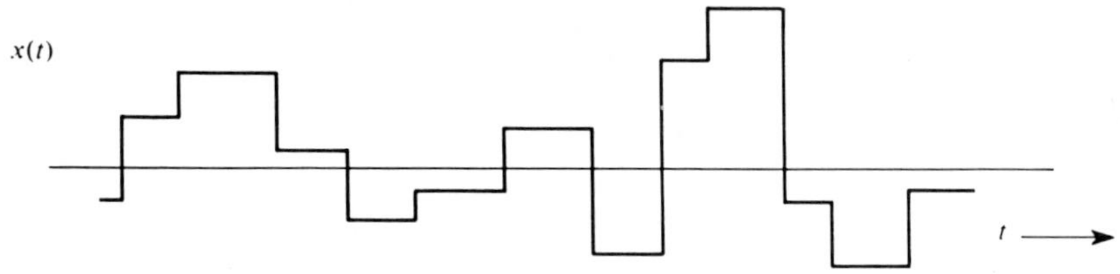

FIG. A3-2.

Also from Eq. A3-1 and A3-2,

$$\overline{x^2} = \frac{q^2}{d^2}\frac{kT}{m}$$

Substituting these results in Eq. 3-70 we obtain

$$R_x(\tau) = \frac{q^2}{d^2}\frac{kT}{m} e^{-\alpha|\tau|} \tag{A3-3}$$

This represents the autocorrelation function of the current induced by a single electron. If there are n free electrons per unit volume of the conductor, then there is total of nAd number of electrons in the rectangular section of the conductor shown in Fig. A3-1. Each electron induces a current $x(t)$ as shown in Fig. A3-2. The motion of each electron is independent of that of the other electrons. Under these conditions we shall now show that the autocorrelation function of the total current $i(t)$ is equal to the autocorrelation function $R_x(\tau)$ due to a single electron multiplied by the total number of electrons (nAd).

$$\begin{aligned} R_i(\tau) &= nAd\,R_x(\tau) \\ &= \frac{nAq^2kT}{md} e^{-\alpha|\tau|} \end{aligned}$$

To prove this result, let us first consider two free electrons producing two current waveforms $x(t)$ and $y(t)$. The total current $z(t)$ is given by

$$z(t) = x(t) + y(t)$$

$$\begin{aligned} R_z(\tau) &= \overline{z(t)\,z(t+\tau)} \\ &= \overline{[x(t)+y(t)][x(t+\tau)+y(t+\tau)]} \\ &= R_x(\tau)+R_y(\tau)+R_{xy}(\tau)+R_{xy}(\tau) \end{aligned}$$

But since the paths of two electrons are assumed independent, the processes x(t) and y(t) are independent, and

$$R_{xy}(\tau) = R_{yx}(\tau) = \overline{x}\,\overline{y} = 0$$

Also,

$$R_z(\tau) = R_x(\tau) + R_y(\tau)$$

It is evident that the autocorrelation function of the total current is equal to the sum of the autocorrelation functions of the currents induced by individual electrons. The autocorrelation function of the current induced by a single electron is given by Eq. A3-3. Hence the autocorrelation function of the net current i(t) (due to nAd number of free electrons) is given by

$$R_i(\tau) = \frac{nAq^2kT}{md}\,e^{-\alpha|\tau|} \tag{A3-4}$$

We have shown in Ex. 2-18 (Eq. 2-121) that σ, the conductivity of a metal is

$$\sigma = \frac{nq^2}{m\alpha}$$

Hence the conductance G of the slab in Fig. A3-1 is

$$G = \frac{\sigma A}{d} = \frac{nAq^2}{m\alpha d}$$

Substituting this result in Eq. A3-4, we obtain

$$R_i(\tau) = kTG\alpha\,e^{-\alpha|\tau|} \tag{A3-5}$$

and

$$\mathcal{S}_i(\omega) = \frac{2kTG\alpha^2}{\alpha^2+\omega^2} \tag{A3-6}$$

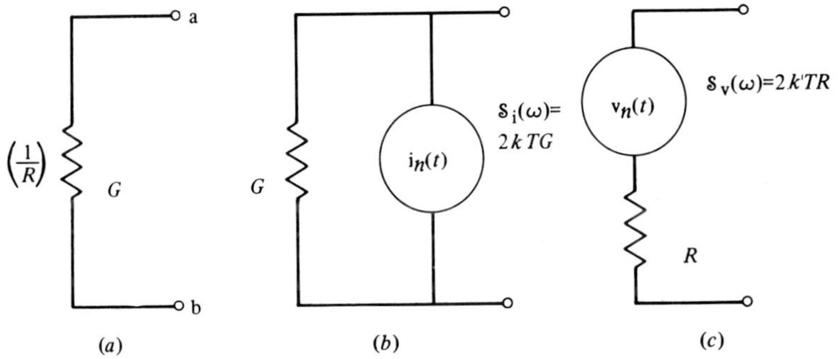

FIG. A3-3.

The order of the magnitude of α is about 10^{14}. Hence for all practical purposes (for $\omega < 10^{13}$) $\mathcal{S}_i(\omega)$ may be considered uniform with a value of

$$\mathcal{S}_i(\omega) \simeq 2kTG$$

A noisy conductor can therefore be represented by a noiseless conductor in parallel with a noise-current source of uniform power density $2kTG$ as shown in Fig. A3-3b. Alternatively, one may construct a Thévenin equivalent of this model as shown in Fig. A3-3b. The resistor $R(R = 1/G)$ can be represented by a noiseless resistance in series with a voltage source $v_n(t)$ of uniform power-density spectrum $2kTR$. This follows from the fact that

$$v_n(t) = iR$$

Hence

$$\begin{aligned}\mathcal{S}_v(\omega) &= R^2\mathcal{S}_i(\omega)\\ &= R^2(2kTG)\\ &= 2kTR\end{aligned}$$

APPENDIX 3-2. CALCULATION OF THE MEAN-SQUARE VALUE FOR RATIONAL POWER SPECTRA

The mean square value of a random process $x(t)$ is given by

$$\overline{x^2} = \frac{1}{2\pi}\int_{-\infty}^{\infty} \mathcal{S}_x(\omega)\,d\omega \tag{A3-7}$$

In the special case when $\mathcal{S}_x(\omega)$ is rational, the integral in Eq. A3-7 can be conveniently evaluated by using the *residue theorem.* To use this theorem, we express the power density spectrum as a function of the complex frequency s, where $s = j\omega$.

Let

$$\mathcal{S}_x(s) = \mathcal{S}_x(\omega)\big|_{\omega = s/j}$$

This in essence is a change of variable from ω to s. Thus $\mathcal{S}_x(s)$ is obtained from $\mathcal{S}_x(\omega)$ by replacing $j\omega$ by s.[15] The reader will recognize that $\mathcal{S}_x(s)$ is the (bilateral) Laplace transform of $R_x(\tau)$. With the change in variable Equation A3-7 becomes

$$x^2 = \frac{1}{2\pi j}\int_{-j\infty}^{j\infty} \mathcal{S}_x(s)\,ds \tag{A3-8}$$

The path of integration is along the entire $j\omega$ axis. We shall assume that $\mathcal{S}_x(s) \to 0$ as s^{-2} at $s = \infty$ (as is true of all physical signals). Under these

[15]Mathematically $\mathcal{S}_x(s)$ represents $\mathcal{S}_x(\omega)$ where ω is replaced by s. Strictly speaking, use of notation $\mathcal{S}_x(s)$ for $\mathcal{S}_x(\omega)$ where $j\omega$ is replaced by s is incorrect. Nevertheless, we use this notation to avoid increasing the number of symbols used and thus reduce the possibility of confusion.

conditions we can (without changing the value of the integral) add the integral along a semicircle of radius $R (R \to \infty)$ in the LHP[16] to that in Eq. A3-8. The path of integration is thus a closed contour, but the value of this integral will be the same as that in Eq. A3-7. The residue theorem states that if $\mathcal{S}_x(s)$ has no poles (singularities) on a closed contour, then the integral taken along the contour (in the counterclockwise direction) is equal to the sum of the residues of $\mathcal{S}_x(s)$ at the poles of $\mathcal{S}_x(s)$ inside the contour (in this case the entire LHP). Thus

$$\overline{x^2} = \frac{1}{2\pi j}\int_{-\infty}^{\infty} \mathcal{S}_x(s)\,ds = \sum \text{residues at LHP poles}$$

We could just as well have taken the infinite semicircle enclosing the RHP. In that case, we must find the sum of residues of the poles of $\mathcal{S}_x(s)$ in the RHP and change the sign of the sum. This is because the integration (Eq. A3-8) will be in the counterclockwise direction when the semicircle encloses the RHP. Either method will yield the same answer. Thus

$$\overline{x^2} = \sum \text{residues at LHP poles} = -\sum \text{residues at RHP poles} \qquad \text{(A3-9)}$$

It is important to note that there should be no poles of $\mathcal{S}_x(s)$ on the path of integration. This implies that $\mathcal{S}_x(s)$ has no poles along the $j\omega$ axis. In addition, the order of the denominator polynomial of $\mathcal{S}_x(s)$ must be at least two orders higher than that of the numerator polynomial.

Example 3-11. As an example, consider the case

$$\mathcal{S}_x(\omega) = \frac{\omega^2 + 9}{\omega^4 + 5\omega^2 + 4}$$

Substituting $\omega = s/j$, we have

$$\begin{aligned}\mathcal{S}_x(s) &= \frac{9 - s^2}{s^4 - 5s^2 + 4}\\ &= \frac{9 - s^2}{(s+1)(s-1)(s+2)(s-2)}\\ &= \frac{4/3}{s+1} - \frac{4/3}{s-1} - \frac{5/12}{s+2} + \frac{5/12}{s-2}\end{aligned}$$

from Eq. A3-9, we have (taking the residues at LHP poles),

$$\overline{x^2} = \frac{4}{3} - \frac{5}{12} = \frac{11}{12}$$

We obtain the same answer by considering the residues at RHP poles. Alternatively, the integral of the form A3-7 may be obtained from certain

[16]This is because along this path the integrand varies as s^{-2}, whereas ds varies as s. Hence the integral along this path varies as s^{-1} and is evidently zero along a radius $R \to \infty$.

standard tables. Since the power density spectrum is an even function of ω, the poles (and also zeros) of $\mathcal{S}_x(s)$ are symmetrically placed with respect to the $j\omega$ axis. In other words, the poles (as well as zeros) of $\mathcal{S}_x(s)$ in the LHP are mirror images about $j\omega$ axis of the poles (and zeros) of $\mathcal{S}_x(s)$ in the RHP. Hence $\mathcal{S}_x(s)$ can be expressed as

$$\mathcal{S}_x(s) = \frac{N(s)}{\mathfrak{D}(s)\,\mathfrak{D}(-s)}$$

where $N(s)$ and $\mathfrak{D}(s)$ are rational polynomials of the form

$$N(s) = a_0 s^{2n-2} + a_1 s^{2n-4} + \cdots + a_{n-1}$$
$$\mathfrak{D}(s) = b_0 s^n + b_1 s^{n-1} + \cdots + b_n$$

where $\mathfrak{D}(s)$ has all poles and zero in the LHP and $\mathfrak{D}(-s)$ has all poles and zeros in the RHP. Let us define an integral I_n as

$$I_n = \overline{x^2} = \frac{1}{2\pi j}\int_{-j\infty}^{j\infty} \frac{N(s)}{\mathfrak{D}(s)\mathfrak{D}(-s)}\,ds$$

The values of I_n in terms of the coefficients a_i and b_i are available (Ref. 3-3) for all values of $n \leq 10$. The values of I_n for $n = 1, 2$, and 3 are

$$I_1 = \frac{a_0}{2b_0b_1}$$

$$I_2 = \frac{-a_0 + (b_0a_1/b_2)}{2b_0b_1}$$

$$I_3 = \frac{-a_0b_2 + a_1b_0 - (a_2b_0b_1/b_3)}{2b_0(b_0b_3 - b_1b_2)}$$

Example 3-12. Consider the power-density function of Ex. 3-11:

$$\mathcal{S}_x(s) = \frac{9 - s^2}{s^4 - 5s^2 + 4}$$
$$= \frac{9 - s^2}{(s^2 + 3s + 2)(s^2 - 3s + 2)}$$

We have in this case

$$n = 2$$
$$a_0 = -1,\ a_1 = 9$$
$$b_0 = 1,\ b_1 = 3 \text{ and } b_2 = 2$$

Substituting in the expression for I_n with $n = 2$, we have

$$\overline{x^2} = I_2 = \frac{1 + (9/2)}{2(3)}$$
$$= \frac{11}{12}$$

Which is the same as the result obtained in Ex. 3-11.

The second method (of using tables) is essentially the same as the residue method except that it is more mechanized. It should be realized that both the first and the second methods are applicable under the same constraints. (In other words, the power density vanishes at $\omega = \infty$ as ω^{-2}, and $S_x(s)$ has no poles on the $j\omega$ axis.)

REFERENCES

1. Y. W. Lee, *Statistical Theory of Communication*, Wiley, New York, 1960.
2. B. P. Lathi, *Signals, Systems and Communication*, Wiley, New York, 1965.
3. H. M. James, N. G. Nichols, and R. J. Phillips, *Theory of Servomechanism*, McGraw-Hill, New York, 1947.
4. J. M. Wozencraft and I. M. Jacobs, *Principles of Communication Engineering*, Wiley, New York, 1965.
5. A. Papoulis, *Probability, Random Variables and Stochastic Processes*, McGraw-Hill, New York, 1965.
6. W. B. Davenport and W. L. Root, *Random Signals and Noise,* McGraw-Hill, New York, 1958.
7. J. S. Bendat and A. Piersol, *Measurement and Analysis of Random Data*, Wiley, New York, 1966.
8. J. Otterman, "Properties and Methods of Computation of Exponentially Mapped Past Statistical Variables," *IRE Trans. on Automatic Control*, Vol. AC-6 (January 1960).
9. J. J. Freeman, *Principles of Noise*, Wiley, New York, 1958.
10. R. B. Blackman and J. W. Tuckey, *The Measurement of Power Spectra*, Dover Publications, New York, 1958.

PROBLEMS

3-1. Sketch a few sample functions of the random process defined by

$$x(t) = A\cos(\omega t + \phi)$$

(a) If A is the random variable distributed uniformly in the range $(-1, 1)$. (b) If ω is the random variable distributed uniformly in the range $(0, 10)$. (c) If ϕ is the random variable distributed uniformly over the range $(-\pi, \pi)$.

3-2. Repeat Prob. 3-1 if

$$x(t) = A + \omega t + \phi t^2$$

3-3. Find the first-order probability-density functions for processes in Prob. 3-1 and 3-2. Determine the mean and mean-square values for these processes.

3-4. Find the autocorrelation function $R_x(t_1, t_2)$ for processes in Prob. 3-1 and Prob. 3-2. Note that there are 6 processes in all.

3-5. Classify all the 6 processes in Prob. 3-1 and 3-2, into stationary (in wide sense) or nonstationary groups. Justify your conclusions.

3-6. Show that for the random process in Ex. 3-2 in the text, the time autocorrelation function of a sample function is given by

$$\mathcal{R}_x(\tau) = \widetilde{x(t)x(t+\tau)} = \frac{k^2}{2}\cos\omega_0\tau$$

3-7. A random process

$$x(t) = A\cos(\omega_0 t + \phi)$$

where A and ω_0 are constant and ϕ is a random variable uniformly distributed in the interval $(0, \pi/2)$ over the ensemble. Determine the first probability-density function. Is this process wide-sense stationary?

3-8. If $x(t)$ and $y(t)$ are two independent random processes and the two new processes $u(t)$ and $v(t)$ are formed as follows:

$$u(t) = x(t) + y(t) \;;\; v(t) = 2x(t) + 3y(t)$$

Find $R_u(\tau)$, $R_v(\tau)$, $R_{uv}(\tau)$ and $R_{vu}(\tau)$ in terms $R_x(\tau)$, $R_y(\tau)$.

3-9. For each of the following functions state whether it can or cannot be a valid power-density spectrum of a random process; if not, explain why.

(a) $\dfrac{\omega^2}{\omega^2 + 4}$ (b) $\dfrac{1}{\omega^2 - 4}$ (c) $\dfrac{\omega}{\omega^2 + 4}$ (d) $\delta(\omega) + \dfrac{1}{\omega^2 + 4}$

(e) $\delta(\omega + \omega_0) + \delta(\omega - \omega_0)$ (f) $j[\delta(\omega + \omega_0) - \delta(\omega - \omega_0)]$ (g) $\dfrac{j\omega^2}{\omega^2 + 4}$

3-10. Find the mean-square values for processes whose power-density spectra are given below:

(a) $\dfrac{1}{\omega^4 + 10\omega^2 + 9}$ (b) $\dfrac{1}{\omega^4 + 9\omega^2 + 10}$

(c) $\dfrac{\omega^2 + 2}{\omega^4 + 13\omega^2 + 36}$ (d) $\dfrac{\omega^2 + 3}{\omega^4 + 10\omega^2 + 36}$

(e) $\dfrac{\omega^4 + 3\omega^2 + 2}{\omega^6 + 14\omega^4 + 41\omega^2 + 34}$

3-11. Using the properties of autocorrelation function (Sec. 3-11), determine the mean value and the mean-square value of each of the processes in Examples 3-5 through 3-9 in the text.

3-12. Consider a random process

$$x(t) = A\cos(kt + \varphi)$$

where A and k are constants and φ is a random variable with probability density $p_\varphi(\phi)$. If $P_\varphi(\omega)$ is the Fourier transform of $p_\varphi(\phi)$, then show that $x(t)$ is a wide-sense stationary process if and only if

$$P_\varphi(1) = P_\varphi(2) = 0$$

3-13. The autocorrelation function of three processes are given below:

$$R_x(\tau) = Ae^{-b|\tau|} + \cos\omega_0\tau$$
$$R_x(\tau) = Ae^{-b|\tau|} + k$$
$$R_x(\tau) = Ae^{-b|\tau|} + \cos\omega_0\tau + k$$

From this information alone, state what you can find out about these processes (e.g., mean value, mean-square value, periodicity, etc.)

3-14. Three random processes $x(t)$, $y(t)$, and $z(t)$ have cross-correlation functions as follows:

$$R_{xy}(\tau) = 6 + f_1(\tau)$$
$$R_{yz}(\tau) = 15 + f_2(\tau)$$
$$R_{zx}(\tau) = 10 + f_3(\tau)$$

All the three function $f_1(\tau)$, $f_2(\tau)$ and $f_3(\tau)$ decay with time and vanish at $\tau = \infty$. Determine the mean values for the three processes.

3-15. If the average power of a complex signal $f(t)$ is defined as

$$\lim_{T\to\infty} \frac{1}{T} \int_{-T/2}^{T/2} |f^2(t)|\, dt$$

(a) Find the power-density spectrum of a complex random process $x(t)$; where the power-density spectrum of a random process is the ensemble average of the power density spectra of the sample functions. (Derive Eq. 3-40.) (b) If the autocorrelation function is the inverse Fourier transform of the power-density spectrum, find the expression for the autocorrelation function of a complex random process (see Eq. 3-41). (c) Find the expression for the cross-correlation function of two complex random process by extending the result in (b) (See Eq. 3-43.)

3-16. Show that if the power-density spectrum of a random process is band-limited to B Hz; and if the signal values are uncorrelated at $\tau = n/2B$, for all integral values of n, then the process must have a uniform power-density spectrum over the band $(0, B)$. In other words, the process must be a band-limited white process.

3-17. A random process $y(t)$ is related to the random process $x(t)$ as

$$y(t) = x(t)\cos(\omega_0 t + \Theta)$$

where Θ is independent random variable uniformly distributed over the interval $(0, 2\pi)$. Show that

$$R_y(\tau) = \frac{1}{2} R_x(\tau)\cos\omega_0\tau$$

and

$$\mathcal{S}_y(\omega) = \frac{1}{4}[\mathcal{S}_x(\omega + \omega_c) + \mathcal{S}_x(\omega - \omega_c)]$$

This is the extension of modulation theorem to random processes. *Hint:* If two random processes $x(t)$ and $g(t)$ are independent, then

$$\overline{x(t)g(t)x(t+\tau)g(t+\tau)} = \overline{x(t)x(t+\tau)}\;\overline{g(t)g(t+\tau)}$$
$$= R_x(\tau)R_g(\tau)$$

3-18. Two random processes are given by

$$x(t) = A\cos(\omega_1 t + \varphi); \quad y(t) = B\cos(\omega_2 t + \Theta)$$

where A, B, ω_1 and ω_2 are constants. The initial phases φ and Θ are related by equation $\Theta = 2\varphi$ and the random variable φ is distributed uniformly over the interval $(0, 2\pi)$. Show that the cross-correlation function and the cross-power-density spectrum of the two processes are zero.

3-19. Find the autocorrelation function and the power-density spectrum of a periodic random process $x(t)$ shown in Fig. P. 3-19 in terms of its Fourier series coefficients. The distance b in the figure varies randomly over the ensemble and is

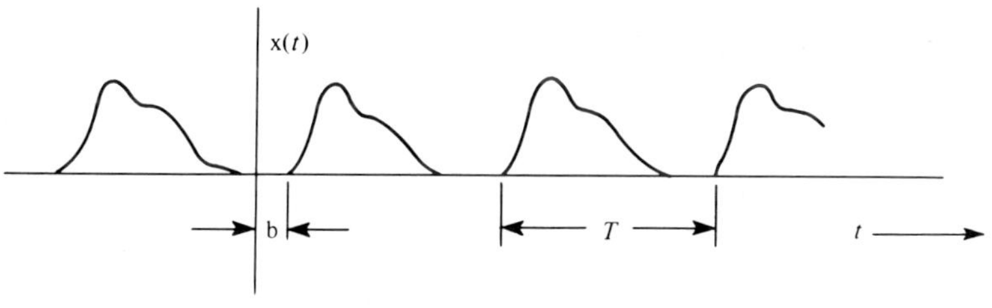

FIG. P. 3-19.

uniformly distributed over the interval $(0, T)$. *Hint:* Express $x(t)$ as a trigonometric Fourier series:

$$x(t) = \sum_{n=0}^{\infty} a_n [\cos n\omega_0(t - b) + \theta_n], \qquad \omega_0 = \frac{2\pi}{T}$$

Here a_n and θ_n are constants and b is a random variable uniformly distributed over $(0, T)$. Now use property 3-81 extended to summation of infinite terms and note that the cross-correlation of any two harmonics will be zero.

3-20. Consider the white-noise process in Ex. 3-7. Find the autocorrelation function and the power-density spectrum for this process if the impulses are not of uniform strength but can assume a random value a where

$$p_a(a) = \frac{1}{\sqrt{2\pi}} e^{-a^2/2}$$

3-21. Consider the process in Ex. 3-8. Find the autocorrelation function and the power-density spectrum for this process if the probability of state transition (from 1 to -1 or vice versa) is not 0.5 but 0.4.

3-22. A random telegraph signal is shown in Fig. P. 3-22. This signal can assume a value A or $-A$ with equal probability (0.5). The transitions from one value to the other are independent and random (this makes the zero crossings Poisson-distributed). Find the autocorrelation function and the power-density spectrum of the process.

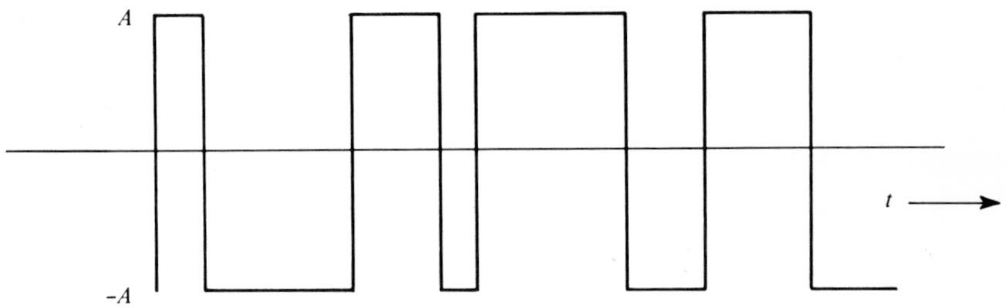

FIG. P. 3-22.

3-23. Rectangular pulses shown in Fig. P. 3-23 alternate sign every T seconds. The heights of pulses in a random variable x with probability-density function given by

$$p_x(x) = \frac{1}{2} |x| e^{-|x|}$$

The pulse heights are independent of all the past history. Find the autocorrelation function and the power-density spectrum of the process.

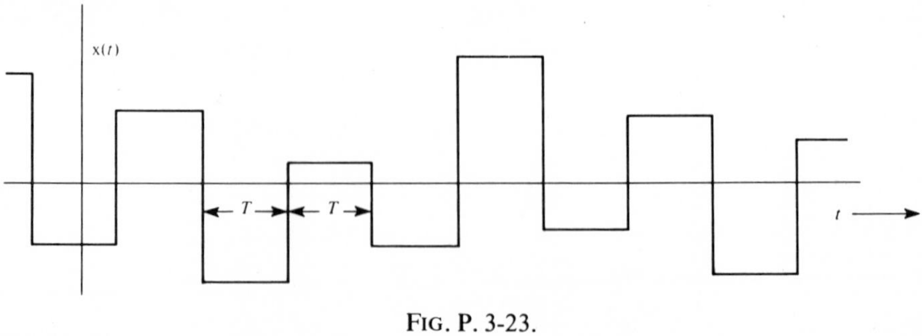

FIG. P. 3-23.

4

Transmission and Filtering of Random Signals

4-1. INTRODUCTION

In the preceding chapter various ways of describing a random process were discussed. The complete statistical description of the process is given by the various joint probability-density functions (of all orders) discussed in Sec. 3-3. It is, however, extremely difficult to obtain from empirical data the probability-density function of order higher than 2. Even the determination of second-order probability-density function is not easy. Fortunately, however, where the transmission of random signals through linear systems is involved, a knowledge of only the first- and second-order probability-density functions is adequate. This is because for linear systems the power-density spectra of the input signal and the response signal are simply related through the transfer function of the system. To determine the power-density spectrum of a random process, we need to know only the autocorrelation function which in turn can be obtained from the second-order probability-density function.

For nonrandom signals we have considered two possible types of representation, viz., the time-domain representation and the frequency-domain representation. Similarly, when linear systems are involved, we may describe a random process either by its autocorrelation function or its power-density spectrum. For nonrandom signals the analysis of signal transmission through linear systems can be carried out in the time domain (convolution analysis) or in the frequency domain. A similar line of attack is followed for random signals. We shall consider both techniques and derive certain useful relationships between the input and the response signals. As usual, we shall assume all the signals to be from stationary or at least wide-sense stationary processes.

4-2. INPUT-OUTPUT RELATION IN THE FREQUENCY DOMAIN

We have shown in Chapter 1 (Eq. 1-122) that for a nonrandom signal the input-signal power density and the response-signal power density are related simply as

$$\mathcal{S}_y(\omega) = |H(\omega)|^2 \mathcal{S}_x(\omega) \tag{4-1}$$

where $\mathcal{S}_y(\omega)$ and $\mathcal{S}_x(\omega)$ represent the response-signal power density and the input-signal power density respectively, and $H(\omega)$ is the transfer function of the system. We may wonder whether such a simple relationship exists for random signals as well. We shall presently show that Eq. 4-1 does indeed apply to random signals as well. To show this consider a system with a transfer function $H(\omega)$. Let the random process applied at the input terminals be $\mathrm{x}(t)$. The output signal is another random process $\mathrm{y}(t)$. A sample function $x(t, \zeta_k)$ of the input process yields a sample function $y(t, \zeta_k)$ at the response. This is shown in Fig. 4-1. The power-density spectrum of $x(t, \xi_k)$ is given by Eq. 1-120.

$$\mathcal{S}_x(\omega, \xi_k) = \lim_{T\to\infty} \frac{|X_T(\omega, \xi_k)|^2}{T} \tag{4-2}$$

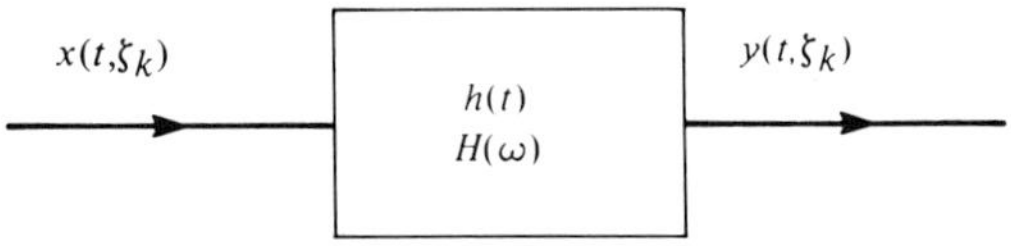

FIG. 4-1.

where $X_T(\omega, \xi_k)$ is the Fourier transform of the truncated signal $x_T(t, \xi_k)$. From Eq. 1-122 we obtain $\mathcal{S}_y(\omega, \xi_k)$, the power density of $y(t, \xi_k)$ as

$$\begin{aligned}\mathcal{S}_y(\omega, \xi_k) &= |H(\omega)|^2 \mathcal{S}_x(\omega, \xi_k) \\ &= |H(\omega)|^2 \left[\lim_{T\to\infty} \frac{|X_T(\omega, \xi_k)|^2}{T}\right]\end{aligned} \tag{4-3}$$

$\mathcal{S}_y(\omega)$, the power density of $\mathrm{y}(t)$, is the ensemble average of power density. In other words, $\mathcal{S}_y(\omega)$ is obtained by averaging $\mathcal{S}_y(\omega, \xi_k)$ over all ξ. Since $|H(\omega)|^2$ is nonrandom, this implies

$$\mathcal{S}_y(\omega) = |H(\omega)|^2 \left[\lim_{T\to\infty} \frac{\overline{|X_T(\omega, \xi_k)|^2}}{T}\right] \tag{4-4}$$

The quantity inside the brackets on the right-hand side is the ensemble average of the power density of input sample function $x(t, \xi_k)$. This by definition is $\mathcal{S}_x(\omega)$, the power density of the input process $\mathrm{x}(t)$. Hence

$$\mathcal{S}_y(\omega) = |H(\omega)|^2 \mathcal{S}_x(\omega) \tag{4-5}$$

This is the result we seek. It follows that the power-density relationship between random input and output processes is the same as that between nonrandom input and output signals.

It is interesting to observe that for an ergodic process, the power-density spectra of all the sample functions are identical. Hence the power-density spectrum of the process is the same as that of any individual sample function. The relationship in Eq. 4-5, however, applies to all stationary processes (at least in the wide sense), ergodic as well as nonergodic.

4-3. INPUT-OUTPUT RELATION IN THE τ-DOMAIN

Corresponding to relationship 4-5 in the frequency domain, there exists a relationship in τ-domain. This relationship can be obtained in several ways. We have (Eq. 4-5)

$$\begin{aligned}\mathcal{S}_y(\omega) &= |H(\omega)|^2 \mathcal{S}_x(\omega)\\ &= H(-\omega)H(\omega)\mathcal{S}_x(\omega)\end{aligned} \tag{4-6}$$

and

$$\begin{aligned}R_y(\tau) &\leftrightarrow \mathcal{S}_y(\omega)\\ R_x(\tau) &\leftrightarrow \mathcal{S}_x(\omega)\\ h(\tau) &\leftrightarrow H(\omega)\\ h(-\tau) &\leftrightarrow H(-\omega)\end{aligned}$$

Hence application of the time convolution theorem to Eq. 4-6 yields

$$R_y(\tau) = h(-\tau) * h(\tau) * R_x(\tau) \tag{4-7}$$

This is the required relationship in the τ-domain. Alternatively, the τ-domain relationship (Eq. 4-7) can be derived as follows. The autocorrelation function $R_y(\tau)$ is the ensemble average of $y(t)y(t+\tau)$. The output $y(t)$ is related to the input $x(t)$ by the convolution relationship

$$\begin{aligned}y(t) &= h(t) * x(t)\\ &= \int_{-\infty}^{\infty} h(\alpha)x(t-\alpha)\,d\alpha\end{aligned} \tag{4-8}$$

and

$$y(t+\tau) = \int_{-\infty}^{\infty} h(\beta)x(t+\tau-\beta)\,d\beta$$

Hence

$$\overline{y(t)y(t+\tau)} = \overline{\int_{-\infty}^{\infty} h(\alpha)x(t-\alpha)\,d\alpha \int_{-\infty}^{\infty} h(\beta)x(t+\tau-\beta)\,d\beta}$$

$$= \overline{\int_{-\infty}^{\infty} \int_{-\infty}^{\infty} h(\alpha) h(\beta) \mathrm{x}(t - \alpha) \mathrm{x}(t + \tau - \beta) \, d\alpha d\beta}$$

Interchanging the operations of integration and averaging, we have

$$R_{\mathrm{y}}(\tau) = \overline{\mathrm{y}(t)\mathrm{y}(t + \tau)} = \int_{-\infty}^{\infty} \int_{-\infty}^{\infty} h(\alpha) h(\beta) \overline{\mathrm{x}(t - \alpha) \mathrm{x}(t + \tau - \beta)} \, d\alpha d\beta$$

Observe that

$$\overline{\mathrm{x}(t - \alpha)\mathrm{x}(t + \tau - \beta)} = R_{\mathrm{x}}(\tau + \alpha - \beta)$$

Hence

$$R_{\mathrm{y}}(\tau) = \int_{-\infty}^{\infty} \int_{-\infty}^{\infty} h(\alpha) h(\beta) R_{\mathrm{x}}(\tau + \alpha - \beta) \, d\alpha d\beta$$

$$= h(-\tau) * h(\tau) * R_{\mathrm{x}}(\tau) \tag{4-9}$$

This, of course is the same result as was previously obtained in Eq. 4-7.

Note that we have derived the output autocorrelation function by two methods: (i) by using the frequency-domain relationship (Eq. 4-5) and then using the time convolution theorem, and (ii) by directly obtaining the result in the τ-domain. The latter method is usually more involved than the frequency-domain approach. Moreover, in the former method the result is displayed in the form of convolution of certain quantities. This is at once compact and meaningful, and can be derived merely by inspection of the frequency-domain relationship. The latter method yields results in an integral form which, unless studied very closely, appears to be a clumsy expression that somehow mechanically relates $R_{\mathrm{y}}(\tau)$ to $R_{\mathrm{x}}(\tau)$.

For physical systems $h(t)$ is causal; that is, $h(t) = 0$ for $t < 0$. Hence Eq. 4-9 may be expressed as

$$R_{\mathrm{y}}(\tau) = \int_{0}^{\infty} \int_{0}^{\infty} h(\alpha) h(\beta) R_{\mathrm{x}}(\tau + \alpha - \beta) \, d\alpha d\beta \tag{4-10}$$

4-4. THE MEAN VALUE AND THE MEAN-SQUARE VALUE OF THE OUTPUT

The ensemble mean of the output signal can easily be determined from Eq. 4-8:

$$\mathrm{y}(t) = \int_{-\infty}^{\infty} h(\alpha) \mathrm{x}(t - \alpha) \, d\alpha$$

Hence

$$\overline{\mathrm{y}} = \overline{\mathrm{y}(t)} = \int_{-\infty}^{\infty} h(\alpha) \overline{\mathrm{x}(t - \alpha)} \, d\alpha$$

$$= \bar{x} \int_{-\infty}^{\infty} h(\alpha)\, d\alpha \tag{4-11a}$$

For physical systems, the lower limit in the above integral may be replaced by zero. Observe that

$$\int_{-\infty}^{\infty} h(\alpha)\, d\alpha = H(0)$$

Therefore

$$\bar{y} = H(0)\,\bar{x} \tag{4-11b}$$

This shows that the mean of the output process is equal to the mean of the input process multiplied by $H(0)$, the system transfer function at d-c. This is an expected result.

The mean-square value of the output process can be obtained either from the autocorrelation function or the output power-density spectrum. According to Eqs. 3-90c and 4-9,

$$\overline{y^2} = R_y(0) = \int_{-\infty}^{\infty} \int_{-\infty}^{\infty} h(\alpha) h(\beta) R_x(\alpha - \beta)\, d\alpha\, d\beta \tag{4-12a}$$

Also from Eqs. 3-90d and 4-5,

$$\overline{y^2} = \frac{1}{2\pi} \int_{-\infty}^{\infty} \mathcal{S}_y(\omega)\, d\omega = \frac{1}{2\pi} \int_{-\infty}^{\infty} |H(\omega)|^2 \mathcal{S}_x(\omega)\, d\omega \tag{4-12b}$$

Again the lower limits of the integral in Eq. 4-12a may be replaced by zero for physical systems.

Stationary of the Output. From the above discussion it is evident that if $x(t)$ is wide-sense stationary, then the response process $y(t)$ is also wide-sense stationary. This follows from Eqs. 4-9 and 4-11. It can be seen from these equations that $\bar{y}$ (the mean) is a constant and the autocorrelation function is a function of τ alone. Note, however, that the impulse response $h(t)$ must be integrable (implying that the system is stable) for these results to be valid.

It can also be shown that if $x(t)$ is strictly stationary, then $y(t)$ is also strictly stationary. This follows from the fact that

$$y(t + t_0) = \int_{-\infty}^{\infty} h(\alpha)\, x(t + t_0 - \alpha)\, d\alpha$$

If processes $x(t)$ and $x(t + t_0)$ have the same statistical properties, it is obvious that processes $y(t)$ and $y(t + t_0)$ must also have identical statistical properties. Hence $y(t)$ is strictly stationary if $x(t)$ is strictly stationary.

Example 4-1. A random voltage signal in the form of Poisson-distributed unit impulses (shown in Fig. 4-2a) is applied at the input terminals of an *R-C*

(a)

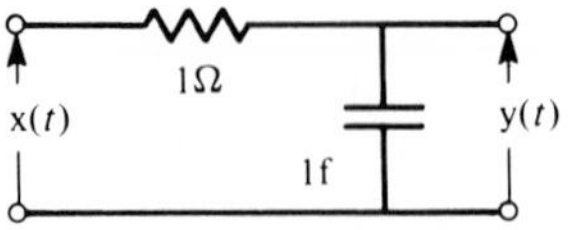

(b)

FIG. 4-2.

circuit shown in Fig. 4-2b. Determine the power-density spectrum and the autocorrelation function of the output voltage. Also determine the mean value and mean-square value of the output voltage.

We have already determined the autocorrelation function and the power-density spectrum for Poisson-distributed impulses in Ex. 3-6. It was observed that the autocorrelation function and the power-density spectrum for this wave is given by

$$R_x(\tau) = k\delta(\tau) + k^2$$

and

$$\mathcal{S}_x(\omega) = k + 2\pi k^2\delta(\omega)$$

where k is the average number of impulses per second.

The transfer function $H(\omega)$ for the R-C circuit in Fig. 4-2b is given by

$$H(\omega) = \frac{1}{j\omega + 1}$$

Hence

$$\begin{aligned}\mathcal{S}_y(\omega) &= |H(\omega)|^2\mathcal{S}_x(\omega)\\ &= \frac{1}{\omega^2 + 1}[k + 2\pi k^2\delta(\omega)]\end{aligned} \tag{4-13}$$

and

$$R_y(\tau) = \mathfrak{F}^{-1}[\mathcal{S}_y(\omega)] = \frac{k}{2}e^{-|\tau|} + k^2 \tag{4-14}$$

This result could also be obtained by using the convolution relationship given in Eq. 4-7.

The mean value of the process can be obtained from Eq. 4-11b.

$$\overline{y} = H(0)\overline{x}$$

From Eq. 3-83,

$$\begin{aligned}\overline{x}^2 &= \lim_{\tau \to \infty} R_x(\tau)\\ &= k^2\end{aligned}$$

Therefore

$$\overline{x} = k$$

also

$$H(0) = 1$$

Hence

$$\overline{y} = k$$

The mean-square value of the output voltage can be obtained by using Eqs. 4-12a and 4-14:

$$\overline{y^2} = R_y(0) = \frac{k}{2} + k^2$$

Alternatively, we could use Eqs. 4-12b and 4-13:

$$\overline{y^2} = \frac{1}{2\pi} \int_{-\infty}^{\infty} \frac{1}{\omega^2 + 1} [k + 2\pi k^2 \delta(\omega)]\, d\omega$$

$$= \frac{k}{2} + k^2$$

Example 4-2. Shot Noise. In this example we shall use the results derived earlier to obtain a statistical description of shot noise. In a vacuum tube shot noise is generated by the random emission of electrons from the cathode. Each electron induces a triangular current pulse $i_e(t)$, as shown in Fig. 4-3a (Ref. 4-1). The instants of electron emission are Poisson-distributed. We now wish to calculate the power-density spectrum and the autocorrelation function of the shot noise.

Let us consider a linear system whose unit impulse response is the $i_e(t)$ shown in Fig. 4-3a:

$$h(t) = i_e(t)$$

Obviously if we apply a unit impulse to this system, the output will be $i_e(t)$. If we apply α number of Poisson-distributed impulses at the input, the output will consist of Poisson-distributed pulses $i_e(t)$. This, precisely, is shot-noise current. If we denote the input process by x(t) and the output process by y(t), then from Eq. 3-53b (Ex. 3-6)

$$\mathcal{S}_x(\omega) = \alpha + 2\pi\alpha^2\delta(\omega)$$

Also,

$$h(t) = \frac{2q}{\tau_a^2} [tu(t) - \tau_a u(t - \tau_a) - (t - \tau_a) u(t - \tau_a)]$$

and

$$H(\omega) = \frac{2q}{-\omega^2\tau_a^2} [1 - e^{-j\omega\tau_a} - j\omega\tau_a e^{-j\omega\tau_a}]$$

Hence

$$\mathcal{S}_y(\omega) = |H(\omega)|^2 \mathcal{S}_x(\omega)$$

$$= \underbrace{\frac{4qI_0}{(\omega\tau_a)^4} [(\omega\tau_a)^4 + 2(1 - \cos\omega\tau_a - \omega\tau_a \sin\omega\tau_a)]}_{\text{shot-noise component}} + \underbrace{2\pi I_0^2 \delta(\omega)}_{\text{d-c component}}$$

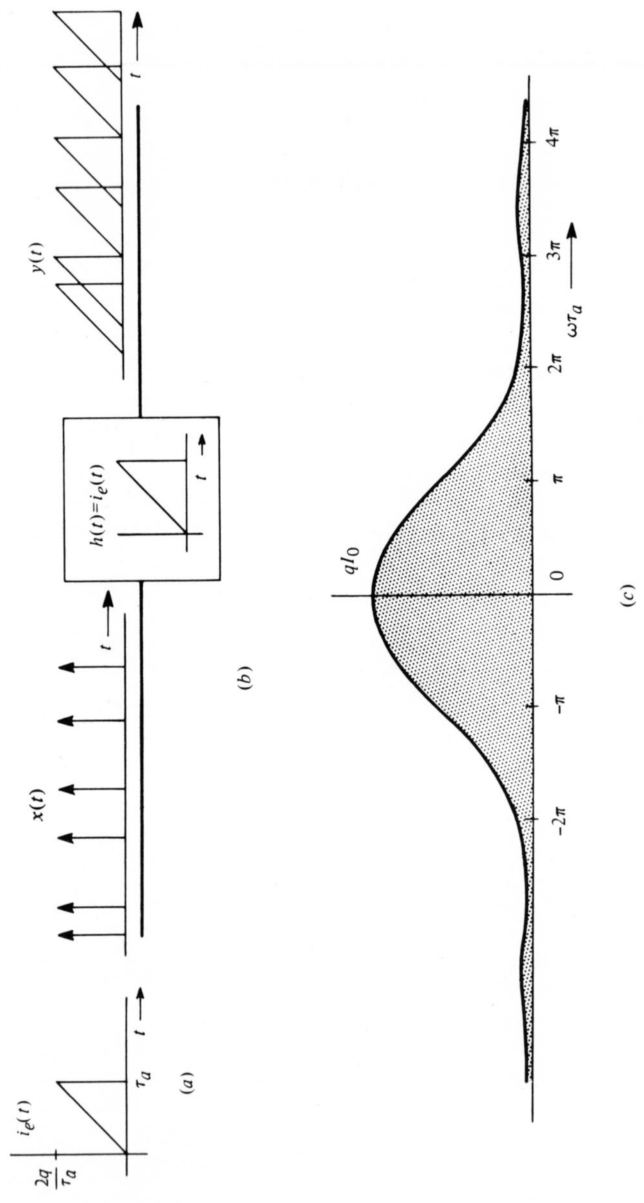

$i_e(t)$
$\frac{2q}{\tau_a}$
τ_a
t
(a)
$x(t)$
t
$h(t)=i_e(t)$
t
$y(t)$
t
(b)
qI_0
-2π
$-\pi$
0
π
2π
3π
4π
$\omega\tau_a$
(c)

FIG. 4-3.

where $I_0 = \alpha q = \overline{y}$ is the mean value of y(t) and

$$R_y(\tau) = \mathfrak{F}^{-1}[\mathcal{S}_y(\omega)]$$

$$= \begin{cases} 4qI_0\left[1 - \dfrac{3}{2}\dfrac{|\tau|}{\tau_a} + \dfrac{1}{2}\dfrac{|\tau|^3}{\tau_a^3}\right] + I_0 & |\tau| < \tau_a \\ 0 & \text{otherwise} \end{cases}$$

The shot-noise component of the power-density spectrum $\mathcal{S}_y(\omega)$ as a function of $\omega\tau_a$ is shown in Fig. 4-3c. Observe that for $\omega\tau_a < 0.5$, the spectrum can be assumed to be flat. Typically τ_a is of the order 10^{-9}. Hence for frequencies below 100 MHz shot noise may be considered to have a flat power-density spectrum of magnitude qI_0.

Shot noise results from a large number of independent perturbations. Hence from the multidimensional central limit theorem it can be shown that shot noise is a gaussian random process (Ref. 3-2). The autocorrelation function determined above, therefore, gives complete statistical information about the process.

4-5. INPUT-OUTPUT CROSS-CORRELATION

We shall now show that

$$R_{xy}(\tau) = h(\tau) * R_x(\tau) \tag{4-15a}$$

$$= \int_{-\infty}^{\infty} h(\alpha)\, R_x(\tau - \alpha)\, d\alpha \tag{4-15b}$$

By definition,

$$R_{xy}(\tau) = \overline{x(t)\,y(t + \tau)}$$

and

$$y(t + \tau) = h(t) * x(t + \tau) = \int_{-\infty}^{\infty} h(\alpha)\, x(t + \tau - \alpha)\, d\alpha$$

Hence

$$R_{xy}(\tau) = \overline{x(t) \int_{-\infty}^{\infty} h(\alpha)\, x(t + \tau - \alpha)\, d\alpha}$$

$$= \int_{-\infty}^{\infty} h(\alpha)\, \overline{x(t)\, x(t + \tau - \alpha)}\, d\alpha$$

$$= \int_{-\infty}^{\infty} h(\tau)\, R_x(\tau - \alpha)\, d\alpha$$

$$= h(\tau) * R_x(\tau)$$

which is the desired result. Application of the time convolution theorem to Eq. 4-15a now yields

$$\mathcal{S}_{xy}(\omega) = H(\omega)\, \mathcal{S}_x(\omega)$$

To summarize,

$$R_{xy}(\tau) = h(\tau) * R_x(\tau) \tag{4-16a}$$

$$\mathcal{S}_{xy}(\omega) = H(\omega)\,\mathcal{S}_x(\omega) \tag{4-16b}$$

For physical systems $h(t)$ is causal and the lower limit in the integra in Eq. 4-15b may be replaced by zero.

The result in Eq. 4-16 has an interesting application in determining the impulse response $h(t)$ of a system. We shall now show that if a white-noise signal is applied to a system, then the input-output cross correlation is proportional to the impulse response of the system (Fig 4-4).

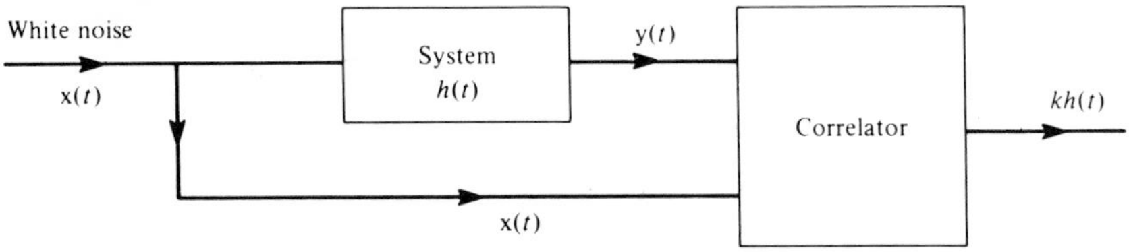

FIG. 4-4.

To show this, we observe that white noise has a constant power density spectrum

$$\mathcal{S}_x(\omega) = k$$

Hence from Eq. 4-16,

$$\mathcal{S}_{xy}(\omega) = kH(\omega)$$

and

$$\begin{aligned} R_{xy}(\tau) &= \mathcal{F}^{-1}[\mathcal{S}_{xy}(\omega)] \\ &= k\,\mathcal{F}^{-1}[H(\omega)] \\ &= kh(\tau) \end{aligned} \tag{4-17}$$

4-6. THE TRANSIENT ANALYSIS

So far we have considered input processes which are wide-sens stationary. Signals from such processes must necessarily exist over th entire interval $(-\infty, \infty)$, since their mean value is constant for all value of t in the interval $(-\infty, \infty)$. So when we talk about applying a wide sense stationary process at the input of a linear system, we imply tha the signal has been applied at $t = -\infty$. The response y(t) has two com ponents; the transient and the steady state. The statistical properties o the transient component vary with time. But for a stable system, th transient decays with time. If the input signal is applied at $t = -\infty$ (a in the ideal model of a wide-sense stationary process), the transient component at any finite time has become negligible and we have only a steady-state component which is wide-sense stationary.

All signals in practice must start at some finite time. Hence strictly speaking, in practice, all random processes are nonstationary. However, a long time after a process has started, one may assume a stationary model if the behavior of the process has stationary characteristics over the finite-time duration of its existence.

Even if the input process is stationary in the above sense, the output process is not stationary because of the transient (source-free) component. However, for a stable system this component decays with time and eventually becomes negligible. After the transient component becomes negligible, the output process may be assumed to be stationary. The various relationships between the input and the response developed so far apply to the input process and response process over the period where both may be assumed to be stationary in the sense mentioned earlier. If the system is unstable, the source-free component grows with time and the response can never be considered stationary.

In this section we shall assume the input signals applied to a system at some finite time $t = 0$ and observe the response. We shall, in particular, determine the mean and the mean-square value (ensemble means) of the output process at some instant t without ignoring the source-free component. No restriction regarding stability will be placed on the system.

Let a sample function $\mathrm{x}(t)$ be applied at the instant $t = 0$ to the input terminals of a system whose unit impulse response is $h(t)$. The output is given by

$$\begin{aligned} \mathrm{y}(t) &= h(t) * \mathrm{x}(t) \\ &= \int_{-\infty}^{\infty} h(\alpha)\mathrm{x}(t - \alpha)\, d\alpha \end{aligned}$$

Since the signal $\mathrm{x}(t)$ starts at $t = 0$, $\mathrm{x}(t - \alpha)$ is zero for $\alpha > t$. Hence the upper limit of integration can be replaced by t and $h(\alpha) = 0$ for $\alpha < 0$, inasmuch as a physically realizable system must have a causal impulse response. Hence the lower limit can be replaced by 0. Thus

$$\mathrm{y}(t) = \int_0^t h(\alpha)\mathrm{x}(t - \alpha)\, d\alpha \tag{4-18}$$

The ensemble average of the output process is given by

$$\begin{aligned} \overline{\mathrm{y}} = \overline{\mathrm{y}(t)} &= \int_0^t h(\alpha)\overline{\mathrm{x}(t - \alpha)}\, d\alpha \\ &= \overline{\mathrm{x}} \int_0^t h(\alpha)\, d\alpha \end{aligned} \tag{4-19}$$

This expression is identical with that for the steady-state case (Eq. 4-11) except for the limits of integration.

To find $\overline{\mathrm{y}^2}$, we follow a similar procedure. From Eq. 4-18 we have

$$\mathrm{y}^2(t) = \int_0^t h(\alpha)\mathrm{x}(t-\alpha)\,d\alpha \int_0^t h(\beta)\mathrm{x}(t-\beta)\,d\beta$$

$$= \int_0^t \int_0^t h(\alpha)h(\beta)\mathrm{x}(t-\alpha)\mathrm{x}(t-\beta)\,d\alpha\,d\beta$$

Hence

$$\overline{\mathrm{y}^2} = \overline{\mathrm{y}^2(t)} = \int_0^t \int_0^t h(\alpha)h(\beta)\overline{\mathrm{x}(t-\alpha)\mathrm{x}(t-\beta)}\,d\alpha\,d\beta$$

$$= \int_0^t \int_0^t h(\alpha)h(\beta)R_{\mathrm{x}}(\alpha-\beta)\,d\alpha\,d\beta \tag{4-20}$$

Again we observe that this expression is identical with that for the steady-state case (Eq. 4-12a) except for the limits of integration.

Example 4-3. A white-noise voltage signal is applied at $t = 0$ at the input terminals of the *R-C* circuit shown in Fig. 4-5. Find the mean value and the mean-square value of the output voltage at any instant t.

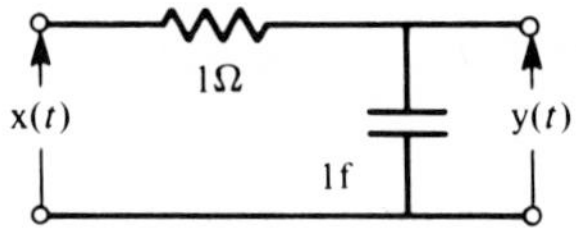

FIG. 4-5.

We have

$$\mathcal{S}_{\mathrm{x}}(\omega) = k$$

and

$$R_{\mathrm{x}}(\tau) = k\delta(\tau)$$

The unit impulse response of the *R-C* circuit is obviously $e^{-t}u(t)$. Hence from Eq. 4-19,

$$\overline{\mathrm{y}} = \overline{\mathrm{x}} \int_0^t h(\alpha)\,d\alpha$$

$$= \overline{\mathrm{x}} \int_0^t e^{-\alpha}\,d\alpha$$

$$= \overline{\mathrm{x}}(1 - e^{-t})$$

The white-noise source has a constant power-density spectrum and no impulses at $\omega = 0$. This means the d-c component is zero, implying that $\overline{\mathrm{x}} = 0$. Hence

$$\overline{\mathrm{y}} = 0$$

'rom Eq. 4-20,

$$\begin{aligned}\overline{y^2} &= \int_0^t \int_0^t e^{-\alpha}e^{-\beta}k\delta(\alpha - \beta)\,d\alpha\,d\beta \\ &= k\int_0^t e^{-\beta} \int_0^t e^{-\alpha}\delta(\alpha - \beta)\,d\alpha\,d\beta \\ &= k\int_0^t e^{-\beta}(e^{-\beta})\,d\beta \\ &= k\int_0^t e^{-2\beta}\,d\beta \\ &= \frac{k}{2}(1 - e^{-2t})\end{aligned}$$

Note that as t becomes large, the term e^{-2t} becomes negligible and

$$\overline{y^2} \rightarrow \frac{k}{2}$$

'his is the mean-square value in the steady state. The reader may verify that he use of Eq. 4-12a instead of 4-20 does indeed yield

$$y^2 = \frac{k}{2}$$

4-7. MULTIPLE TERMINALS

So far we have discussed the case of only one source applied to a ystem. In practice there may be multiple sources applied at various ocations in the system. The response or the output contains the comonents due to each source. In this section we shall consider the relaionship between the output signal and the various input sources. For he sake of simplicity, we shall first discuss the case of two separate ources and later generalize the results to any number of sources. Let us onsider a system with two separate input terminals to which processes $x_1(t)$ and $x_2(t)$ are applied as shown in Fig. 4-6a. Let $H_1(\omega)$ and $H_2(\omega)$ e the transfer functions of the system relating the output to the two nput signals respectively. The output signal $y(t)$ consists of two componets $y_1(t)$ and $y_2(t)$ which are the responses to inputs $x_1(t)$ and $x_2(t)$ espectively.

We have

$$y_1(t) = \int_{-\infty}^{\infty} h_1(\alpha)x_1(t - \alpha)\,d\alpha \tag{4-21}$$

and

$$y_2(t) = \int_{-\infty}^{\infty} h_2(\beta)x_2(t - \beta)\,d\beta \tag{4-22}$$

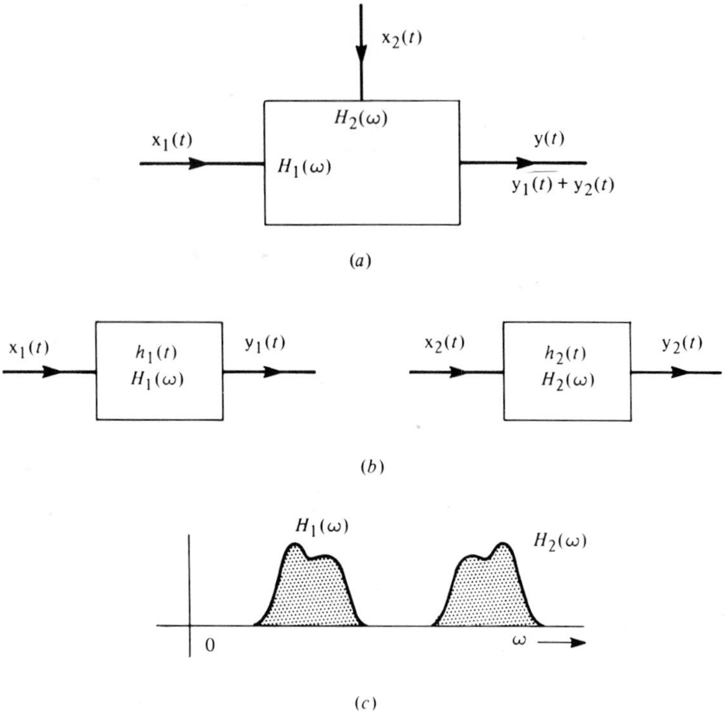

FIG. 4-6.

The cross-correlation between $y_1(t)$ and $y_2(t)$ is given by

$$
\begin{aligned}
R_{y_1y_2}(\tau) &= \overline{y_1(t)\,y_2(t+\tau)} \\
&= \overline{\int_{-\infty}^{\infty} h_1(\alpha)\,x_1(t-\alpha)\,d\alpha \int_{-\infty}^{\infty} h_2(\beta)\,x_2(t+\tau-\beta)\,d\beta} \\
&= \int_{-\infty}^{\infty}\int_{-\infty}^{\infty} h_1(\alpha)\,h_2(\beta)\,\overline{x_1(t-\alpha)\,x_2(t+\tau-\beta)}\,d\alpha\,d\beta \\
&= \int_{-\infty}^{\infty}\int_{-\infty}^{\infty} h_1(\alpha)\,h_2(\beta)\,R_{x_1x_2}(\tau+\alpha-\beta)\,d\alpha\,d\beta \qquad (4\text{-}23)
\end{aligned}
$$

where $R_{x_1x_2}(\tau)$ is the cross-correlation of $x_1(t)$ and $x_2(t)$. Note that this expression is analogous to Eq. 4-9 and hence can be expressed as

$$R_{y_1y_2}(\tau) = h_1(-\tau) * h_2(\tau) * R_{x_1x_2}(\tau) \qquad (4\text{-}24a)$$

Using the time convolution theorem, we obtain

$$\mathcal{S}_{y_1y_2}(\omega) = H_1(-\omega)\,H_2(\omega)\,\mathcal{S}_{x_1x_2}(\omega) \qquad (4\text{-}24b)$$

If the processes $x_1(t)$ and $x_2(t)$ are uncorrelated, then

$$R_{x_1x_2}(\tau) = \overline{x}_1\overline{x}_2$$

where $\overline{x}_1$ and $\overline{x}_2$ are the means of the random processes $x_1(t)$ and $x_2(t)$ respectively. Substituting this result in Eq. 4-23, we have

$$R_{y_1y_2}(\tau) = \overline{x}_1\overline{x}_2 \int_{-\infty}^{\infty} \int_{-\infty}^{\infty} h_1(\alpha) h_2(\beta)\, d\alpha\, d\beta$$

(In all these integrals, the lower limits of integration may be replaced by zero for physical systems.)

$$R_{y_1y_2}(\tau) = \left[\overline{x}_1 \int_{-\infty}^{\infty} h_1(\alpha)\, d\alpha\right]\left[\overline{x}_2 \int_{-\infty}^{\infty} h_2(\beta)\, d\beta\right]$$

From Eq. 4-19, it now follows that

$$R_{y_1y_2}(\tau) = \overline{y_1(t)\, y_2(t+\tau)} = \overline{y}_1\,\overline{y}_2 \tag{4-26}$$

It is obvious from Eq. 4-26 that variables $y_1(t)$ and $y_2(t + \tau)$ are uncorrelated. This is the result we seek. To recapitulate, if $x_1(t)$ and $x_2(t)$ are uncorrelated, then the output components due to $x_1(t)$ and $x_2(t)$ are also uncorrelated. This result can be extended to any number of separate inputs. If $x_1(t)$ and $x_2(t)$ are not uncorrelated, then the appropriate relationship is given by Eq. 4-23 (or Eq. 4-24a).

We observe here that the results derived for systems with multiple inputs are very general. The two responses $y_1(t)$ and $y_2(t)$ need not be observed at the same terminal as shown in Fig. 4-6a. It applies also to the case where $y_1(t)$, the response due to $x_1(t)$, appears at one terminal and $y_2(t)$, the response due to $x_2(t)$ appears at some other terminal. In fact, it also applies to two different systems with impulse responses $h_1(t)$, $h_2(t)$ and inputs $x_1(t)$, $x_2(t)$ respectively, as shown in Fig. 4-6b. The respective outputs are $y_1(t)$ and $y_2(t)$.

At this point we make a very interesting observation. In Fig. 4-6b, if $H_1(\omega)$ and $H_2(\omega)$ have nonoverlapping spectra (disjoint spectra as shown in Fig. 4-6c), then $H_1(-\omega) H_2(\omega) = 0$ and from Eq. 4-24b, we have

$$\mathcal{S}_{y_1y_2}(\omega) = 0$$

or

$$R_{y_1y_2}(\tau) = 0$$

Hence $y_1(t)$ and $y_2(t)$ are orthogonal. Thus for any $x_1(t)$ and $x_2(t)$, $y_1(t)$ and $y_2(t)$ are orthogonal if $H_1(\omega)$ and $H_2(\omega)$ are disjoint. Note that if $H_1(\omega)$ and $H_2(\omega)$ are disjoint as shown in Fig. 4-6c, then $\mathcal{S}_{y_1}(\omega)$ and $\mathcal{S}_{y_2}(\omega)$ are also disjoint. This result is valid for any $x_1(t)$ and $x_2(t)$. This suggests a method of obtaining two orthogonal processes from a given process $x(t)$. If we transmit $x(t)$ through two systems whose transfer functions are disjoint, the output processes of the two systems will be orthogonal. In particular, if a process $x(t)$ is transmitted through an ideal filter (low-pass, bandpass, or high-pass), the output signal and the signal suppressed by the filter will be orthogonal. If in addition, $\overline{x}$, the mean of the input process is zero, then $\overline{y}_1$ and $\overline{y}_2$ are also zero (Eq. 4-11b). In this case orthogonality of $y_1(t)$ and $y_2(t)$ implies that the processes are uncorrelated. We shall use this result in later chapters. It also follows from this that any two random processes with disjoint power-density spectra are orthogonal.

Superposition of Power Spectra and Autocorrelation Functions for Uncorrelated Processes. We shall now derive the relationship between output power-density and input power-density spectra for the case of multiple inputs. The output $y(t)$ (Fig. 4-6a) is given by

$$y(t) = y_1(t) + y_2(t)$$

The autocorrelation function $R_y(\tau)$ of the output is given by (see Eq. 3-81):

$$R_y(\tau) = R_{y_1}(\tau) + R_{y_2}(\tau) + R_{y_1y_2}(\tau) + R_{y_2y_1}(\tau) \tag{4-27}$$

where $R_{y_1}(\tau)$ and $R_{y_2}(\tau)$ are the autocorrelation functions of $y_1(t)$ and $y_2(t)$ respectively, and $R_{y_1y_2}(\tau)$ and $R_{y_2y_1}(\tau)$ are their cross-correlation functions. If $x_1(t)$ and $x_2(t)$ are uncorrelated, then $y_1(t)$ and $y_2(t)$ are also uncorrelated. Hence

$$R_{y_1y_2}(\tau) = R_{y_2y_1}(\tau) = \overline{y}_1\overline{y}_2 \tag{4-28}$$

and

$$R_y(\tau) = R_{y_1}(\tau) + R_{y_2}(\tau) + 2\overline{y}_1\overline{y}_2 \tag{4-29}$$

Let us first consider the case when the random processes have zero mean values. Let $\overline{x}_1$, $\overline{x}_2$, the means of the input processes $x_1(t)$ and $x_2(t)$ respectively be zero. This will cause $\overline{y}_1$ and $\overline{y}_2$ to be zero and Eq. 4-29 reduces to

$$R_y(\tau) = R_{y_1}(\tau) + R_{y_2}(\tau) \tag{4-30}$$

Where $R_{y_1}(\tau)$ is the autocorrelation function of the output if only the random process $x_1(t)$ is applied at the input (and letting $x_2(t) = 0$) and $R_{y_2}(\tau)$ is the autocorrelation function of the output if only the process $x_2(t)$ is applied at the input (and letting $x_1(t) = 0$). This is the principle of superposition as applied to autocorrelation function. This result can be extended to any number of input signals. We may state this result as follows. If there are n uncorrelated random signal inputs, the autocorrelation function of the output of a linear system is given by the sum of the n autocorrelation functions of the output signal observed when each of the n signals is applied individually (and the remaining $n - 1$ signals suppressed). This principle of superposition also applies to power density spectra. The Fourier transform of Eq. 4-30 yields

$$\mathcal{S}_y(\omega) = \mathcal{S}_{y_1}(\omega) + \mathcal{S}_{y_2}(\omega) \tag{4-31}$$

where $\mathcal{S}_{y_1}(\omega)$ and $\mathcal{S}_{y_2}(\omega)$ are the power density spectra of $y_1(t)$ and $y_2(t)$ respectively. If $\mathcal{S}_{x_1}(\omega)$ and $\mathcal{S}_{x_2}(\omega)$ represent the power density spectra of the input processes $x_1(t)$ and $x_2(t)$ respectively, then

$$\mathcal{S}_{y_1}(\omega) = |H_1(\omega)|^2 \mathcal{S}_{x_1}(\omega)$$

and (4-32)

$$\mathcal{S}_{y_2}(\omega) = |H_2(\omega)|^2 \mathcal{S}_{x_2}(\omega)$$

and

$$\mathcal{S}_y(\omega) = |H_1(\omega)|^2 \mathcal{S}_{x_1}(\omega) + |H_2(\omega)|^2 \mathcal{S}_{x_2}(\omega) \tag{4-33}$$

This result can be extended to any number of input processes.

In the above discussion, we have restricted the argument to input random processes with zero mean. If the processes have nonzero mean values, then the expression 4-29 holds true. The output power density in this case is given by

$$\mathcal{S}_y(\omega) = \mathcal{S}_{y_1}(\omega) + \mathcal{S}_{y_2}(\omega) + 4\pi \overline{y}_1 \overline{y}_2 \delta(\omega) \tag{4-34}$$

$$= |H_1(\omega)|^2 \mathcal{S}_{x_1}(\omega) + |H_2(\omega)|^2 \mathcal{S}_{x_2}(\omega) + 4\pi \overline{y}_1 \overline{y}_2 \delta(\omega) \tag{4-35}$$

Use of Eq. 4-11b in Eq. 4-34 yields

$$\mathcal{S}_y(\omega) = |H_1(\omega)|^2 \mathcal{S}_{x_1}(\omega) + |H_2(\omega)|^2 \mathcal{S}_{x_2}(\omega) + 4\pi H_1(0) H_2(0) \overline{x}_1 \overline{x}_2 \delta(\omega) \tag{4-36}$$

(Note that for the principle of superposition to hold, it is not necessary that both $\overline{x}_1$ and $\overline{x}_2$ be zero. It is sufficient that either $\overline{x}_1$ or $\overline{x}_2$ or both be zero.)

In addition, it should also be noted that the principle of superposition also holds with respect to the means.

Example 4-4. Uncorrelated Sources. Two random-voltage processes $x_1(t)$ and $x_2(t)$ are applied at the terminals of an *R-C* network as shown in Fig. 4-7. The two sources are independent. One source generates the random process

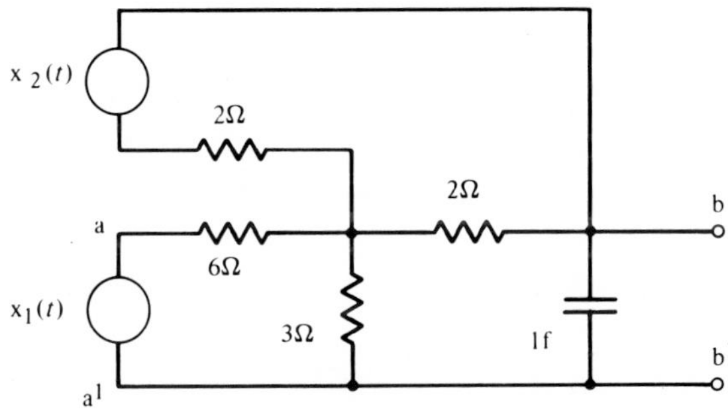

FIG. 4-7.

$x_2(t)$ shown in Fig. 3-22 with a mean-square value of unity and zero mean, and the other source generates a white-noise process $x_1(t)$ with flat (white) power-density spectrum of magnitude k. Find the power-density spectrum and the autocorrelation function of the signal appearing across output terminals.

We shall first find the two transfer functions $H_1(\omega)$ and $H_2(\omega)$. The transfer function $H_1(\omega)$ relates the source voltage across the terminals aa' to that across cc' and is given by

$$H_1(\omega) = \frac{1}{3(3j\omega + 1)} \tag{4-37}$$

Similarly,

$$H_2(\omega) = \frac{1}{2(1 + j3\omega)} \tag{4-38}$$

The autocorrelation and the power-density spectrum of the process $x_2(t)$ are given by (see Eqs. 3-70a and 3-70b):

$$R_{x_2}(\tau) = e^{-\alpha|\tau|} \tag{4-39}$$

since $\overline{x_2^2} = 1$ and $\overline{x}_2$ is assumed zero, and

$$\mathcal{S}_{x_2}(\omega) = \frac{2\alpha}{\alpha^2 + \omega^2} \tag{4-40}$$

The autocorrelation and the power-density spectrum of a white-noise source is given by

$$R_{x_1}(\tau) = k\delta(\tau)$$
$$\mathcal{S}_{x_1}(\omega) = k$$

Observe that both the processes $x_1(t)$ and $x_2(t)$ have a zero mean. Since the two sources are independent (and hence uncorrelated), we have, from Eq. 4-33,

$$\mathcal{S}_y(\omega) = |H_1(\omega)|^2\mathcal{S}_{x_1}(\omega) + |H_2(\omega)|^2\mathcal{S}_{x_2}(\omega)$$
$$= \frac{k}{9(1 + 9\omega^2)} + \frac{\alpha}{2}\frac{1}{(1 + 9\omega^2)(\alpha^2 + \omega^2)} \tag{4-41}$$

The autocorrelation function of the output is given by

$$R_y(\tau) = \mathcal{F}^{-1}[\mathcal{S}_y(\omega)]$$
$$= \frac{3\alpha}{4(9\alpha^2 - 1)} + \frac{k}{54}e^{-|\tau|/3} - \frac{1}{4(9\alpha^2 - 1)}e^{-\alpha|\tau|}$$

Alternatively, we could have obtained $R_y(\tau)$ using Eqs. 4-30 since $\overline{x}_1$ and $\overline{x}_2$ are zero:

$$R_y(\tau) = R_{y_1}(\tau) + R_{y_2}(\tau)$$

where (from Eq. 4-7)

$$R_{y_1}(\tau) = [\tfrac{1}{9}e^{\tau/3}u(-\tau)] * [\tfrac{1}{9}e^{-\tau/3}u(\tau)] * [k\delta(\tau)]$$

and

$$R_{y_2}(\tau) = \tfrac{1}{6}e^{\tau/3}u(-\tau) * \tfrac{1}{6}e^{-\tau/3}u(\tau) * e^{-\alpha|\tau|}$$

The reader should evaluate this expression and verify the result.

4-8. MULTIPLE RELATED SOURCES

If the input processes $x_1(t)$ and $x_2(t)$ are dependent or related, then one must use the general equation for $R_{y_1y_2}(\tau)$ as given by Eq. 4-24a (or 4-23). In such a case the principle of superposition of power spectra does not apply. The output autocorrelation function is given by Eq. 4-27.

$$R_y(\tau) = R_{y_1}(\tau) + R_{y_2}(\tau) + R_{y_1y_2}(\tau) + R_{y_2y_1}(\tau)$$

Use of Eqs. 4-7 and 4-23 in the above equation yields

$$\begin{aligned} R_y(\tau) = {} & [h_1(-\tau) * h_1(\tau) * R_{x_1}(\tau)] \\ & + [h_2(-\tau) * h_2(\tau) * R_{x_2}(\tau)] \\ & + [h_1(-\tau) * h_2(\tau) * R_{x_1x_2}(\tau)] \\ & + [h_2(-\tau) * h_1(\tau) * R_{x_2x_1}(\tau)] \end{aligned} \tag{4-42a}$$

$$\begin{aligned} = {} & \int_{-\infty}^{\infty}\int_{-\infty}^{\infty} h_1(\alpha)h_1(\beta)R_{x_1}(\tau+\alpha-\beta)\,d\alpha\,d\beta \\ & + \int_{-\infty}^{\infty}\int_{-\infty}^{\infty} h_2(\alpha)h_2(\beta)R_{x_2}(\tau+\alpha-\beta)\,d\alpha\,d\beta \\ & + \int_{-\infty}^{\infty}\int_{-\infty}^{\infty} h_1(\alpha)h_2(\beta)R_{x_1x_2}(\tau+\alpha-\beta)\,d\alpha\,d\beta \\ & + \int_{-\infty}^{\infty}\int_{-\infty}^{\infty} h_2(\alpha)h_1(\beta)R_{x_2x_1}(\tau+\alpha-\beta)\,d\alpha\,d\beta \end{aligned} \tag{4-42b}$$

The corresponding expression in the frequency domain is obtained by taking the Fourier transform of both sides of Eq. 4-42:

$$\begin{aligned} \mathcal{S}_y(\omega) = {} & |H_1(\omega)|^2\mathcal{S}_{x_1}(\omega) + |H_2(\omega)|^2\mathcal{S}_{x_2}(\omega) \\ & + H_1(-\omega)H_2(\omega)\mathcal{S}_{x_1x_2}(\omega) + H_2(-\omega)H_1(\omega)\mathcal{S}_{x_2x_1}(\omega) \end{aligned} \tag{4-43}$$

The results in Eqs. 4-42 and 4-43 can easily be extended to any number of sources.

Example 4-5. Related Sources. In this example the case of two sources which are not independent will be studied. We shall again consider the same circuit as that of Example 4-3 (Fig. 4-7). The two sources, however, are not independent in this example. The source $x_1(t)$ is the same as before—that is, a white-noise source with constant power-density spectrum k. The second source $x_2(t)$ is not an independent source but is derived from $x_1(t)$ as shown in Fig. 4-8. The process $x_1(t)$ is passed through a system with a transfer function $1/(j\omega + 1)$. The output of this system forms the second voltage source $x_2(t)$. It is obvious that the two sources are not uncorrelated. Hence we must use either of Eqs. 4-42 or 4-43.

1. *Frequency-Domain Approach.* We have

$$\mathcal{S}_{x_1}(\omega) = k$$

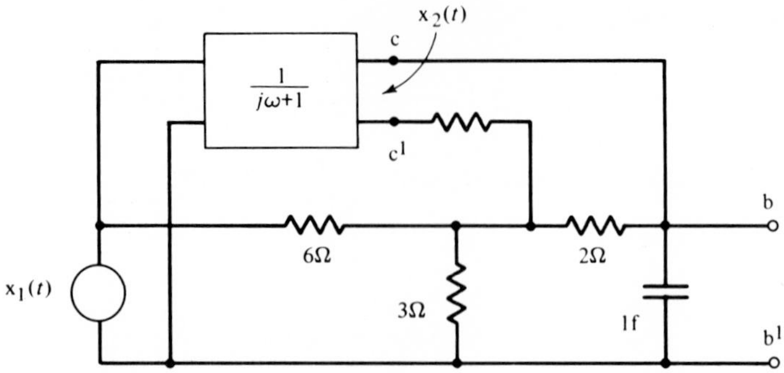

FIG. 4-8.

and

$$\mathcal{S}_{x_2}(\omega) = \frac{k}{|j\omega + 1|^2} = \frac{k}{1 + \omega^2}$$

Hence

$$R_{x_2}(\tau) = \frac{k}{2} e^{-|\tau|}$$

To determine $\mathcal{S}_{x_1x_2}(\omega)$, we observe that $x_2(t)$ is the response to $x_1(t)$ of a system whose transfer function is given by

$$H(\omega) = \frac{1}{j\omega + 1} \quad \text{and} \quad h(t) = e^{-t}u(t)$$

Hence from Eq. 4-16b, we obtain

$$\mathcal{S}_{x_1x_2}(\omega) = \frac{1}{j\omega + 1} \mathcal{S}_{x_1}(\omega) = \frac{k}{j\omega + 1} \tag{4-44}$$

Since

$$\mathcal{S}_{x_1x_2}(\omega) = \mathcal{S}_{x_2x_1}(-\omega),$$

$$\mathcal{S}_{x_2x_1}(\omega) = \frac{k}{1 - j\omega} \tag{4-45}$$

We shall assume that the output impedance of the subsystem with transfer function $1/(j\omega + 1)$ is zero. This assumption yields $H_1(\omega)$ and $H_2(\omega)$ as

$$H_1(\omega) = \frac{1}{3(1 + j3\omega)} \quad \text{and} \quad H_2(\omega) = \frac{1}{2(1 + j3\omega)}$$

Substituting these quantities in Eq. 4-43, we obtain

$$\mathcal{S}_y(\omega) = \frac{k}{1 + 9\omega^2}\left[\frac{1}{9} + \frac{7}{12}\,\frac{1}{1 + \omega^2}\right]$$

The autocorrelation function of the output is given by

$$\begin{aligned} R_y(\tau) &= \mathcal{F}^{-1}[\mathcal{S}_y(\omega)] \\ &= k[0.057e^{-|\tau|/3} - 0.073e^{-|\tau|}] \end{aligned}$$

τ-Domain Approach. The autocorrelation function $R_y(\tau)$ can be obtained from Eq. 4-42. Here we need to determine $R_{x_1x_2}(\tau)$ and $R_{x_2x_1}(\tau)$. This can easily be accomplished by noting that $x_2(t)$ is the response of a system (Fig. 4-8) with transfer function $1/(j\omega + 1)$ to a driving function $x_1(t)$. Hence from Eq. 4-15a we obtain

$$R_{x_1x_2}(\tau) = h(\tau) * R_{x_1}(\tau)$$

The rest is left as an exercise for the reader. Note that

$$R_{x_1x_2}(\tau) = R_{x_2x_1}(-\tau)$$

4-9. TRANSMISSION OF A GAUSSIAN PROCESS THROUGH LINEAR SYSTEMS

It was shown in Sec. 3-14 that the response of a linear system to a gaussian random process is also gaussian. Thus if the input process $x(t)$ is a gaussian process, $y(t)$ the response is also gaussian. In addition, the statistics of the gaussian process are completely specified by its autocorrelation function and the mean value. If we know the mean and the autocorrelation function of the input process, we can determine the mean and the autocorrelation function of the output process using Eqs. 4-7 and 4-11. Hence the statistics of the output process are completely known if the autocorrelation function and the mean of the input process are known. This property is very significant and is not true in general for processes that are not gaussian. If the input process is not gaussian, the output process will not in general be gaussian and hence a knowledge of its mean and autocorrelation function does not give complete statistical information about the process.

4-10. LINEAR MEAN-SQUARE FILTERING: WIENER-HOPF CONDITION

A signal is modified when it is transmitted through a system. This is because the system performs certain operations on the input signal. We may be given a certain random process and we may wish to obtain from this process another desired process. We may, for example, be given a random signal mixed with noise having a power spectrum overlapping that of the signal. One problem of great practical importance is to design a filter (a system) which will filter out the noise as best as possible.

When the signals are deterministic, it is easy to design a system to perform a desired operation. The transfer function of the required system is the ratio of transforms of the desired output and the given input signal. When the signals are random the solution is not so easy. One may argue here that if we are given the input process $x(t)$ and the desired output process $y(t)$, the transfer function $H(\omega)$ of the required system can be obtained from the relationship in Eq. 4-16a (or Eq. 4-16b).

Since x(t) and y(t) are given, we can calculate $\mathcal{S}_{xy}(\omega)$ and $\mathcal{S}_x(\omega)$ and $H(\omega)$, the required transfer function is given by the ratio of these two spectra. This statement appears flawless, yet there is a subtle error in it. If we are given x(t) and $R_{xy}(\tau)$, then we can easily show that y(t) is not unique. Let x(t) and y(t) be two processes with cross-correlation $R_{xy}(\tau)$. We can easily show that the cross-correlation of x(t) with another process y(t) + z(t) is also $R_{xy}(\tau)$ if z(t) is uncorrelated to x(t) and has a zero mean—that is, if $R_{xz}(\tau) = 0$.

$$\overline{x(t)[y(t+\tau) + z(t+\tau)]} = R_{xy}(\tau) + R_{xz}(\tau)$$
$$= R_{xy}(\tau)$$

In addition, it can be seen that the cross-correlation $R_{xy}(\tau)$ remains unchanged even if one adds to x(t) another process which is uncorrelated with y(t).

Hence it follows that the system obtained by using Eq. 4-16 will not necessarily yield the desired process[1] y(t). If we use such a system, the only thing we can be certain of is that the cross-correlation of the given process (input) with the output process of this system is identical with the cross-correlation of the given process with the desired output process. But the two processes are not necessarily identical. Generally speaking, they will be different. We cannot hope to design a system that will yield exactly the desired process. This uncertainty is unavoidable when random signals are involved. One can appreciate this fact from the filtering problem mentioned above. If a random signal and noise with overlapping power-density spectra are mixed, then it is impossible to recover the signal perfectly. There will always be some residual noise left, and the signal may also be distorted somewhat in the process of filtering. We therefore resign ourselves to the fact that there will be a certain amount of error between what we desire and what we can actually obtain. The design problem then reduces to the optimization problem—that is, the problem of minimizing the error. We must, however, make a suitable definition of the error before proceeding to optimize the system. Once we agree upon some criterion for defining the error, we can design a system which will have the least amount of this error. It is therefore evident that for random signals the problem of design implies optimization with respect to some meaningful error criterion.

In any optimization problem we must specify the nature of the signals, the optimization criterion, and the restrictions on the systems to

[1]This suggests that the optimum system should be such that if $\hat{y}(t)$ is the optimum output, then $\hat{y}(t) - y(t)$ should be uncorrelated with the input process x(t). As we shall see, this indeed is the Wiener-Hopf condition required of optimum systems.

be used. In the discussion that follows we shall concern ourselves only with strictly stationary or wide-sense stationary processes—that is, the processes that are at least wide-sense stationary. We shall also restrict our systems to be linear and time-invariant. There are many possible optimization criteria which are meaningful. However, the mean-square-error criterion lends itself easily to quantitative analysis. Under this criterion, the optimum system is one which minimizes the mean square of the error between the desired output and the actual output. We are now ready to make a statement of the problem.

Statement of the Problem. We are given a process $\mathrm{x}(t)$ which is at least wide-sense stationary. It is desired to obtain another process $\mathrm{y}(t)$ from $\mathrm{x}(t)$, using a linear time-invariant system (Fig. 4-9). In general as explained earlier it is not possible to design a system that will yield the desired process $\mathrm{y}(t)$ exactly. If we denote the actual output process by $\hat{\mathrm{y}}(t)$, then the error between what is desired and what is possible to obtain is $\mathrm{y}(t) - \hat{\mathrm{y}}(t)$. We wish to design the system so that the mean-square error is minimized.

We shall denote the error by $\epsilon(t)$. Thus

$$\epsilon(t) = \mathrm{y}(t) - \hat{\mathrm{y}}(t) \tag{4-46}$$

and

$$\hat{\mathrm{y}}(t) = \int_I \mathrm{x}(t - \alpha)h(\alpha)\,d\alpha \tag{4-47}$$

where $h(t)$ is the unit impulse response of the optimum system and I is the domain of integration. If the optimum system is required to be physically realizable, then $h(t)$ is causal and I is $(0, \infty)$. If the system is not constrained to be physically realizable, then $h(t)$ could be noncausal in general and I is $(-\infty, \infty)$.

We now wish to find $h(t)$ for which $\overline{\epsilon^2(t)}$ (see Eq. 4-46) is minimum.

We maintain that $h(t)$ will be optimum if and only if the processes $\epsilon(t)$ and $\mathrm{x}(t)$ are uncorrelated; that is,[2]

[2]Note that when $\epsilon(t)$ and $\mathrm{x}(t)$ are uncorrelated, we have

$$R_{\mathrm{x}\epsilon}(\tau) = \bar{\mathrm{x}}\bar{\epsilon}$$

But if the system is optimum, that is $\overline{\epsilon^2(t)}$ is minimum, then the mean value $\overline{\epsilon(t)}$ must necessarily be zero. This is because if $\overline{\epsilon(t)} \neq 0$, we can always choose $\mathrm{y}(t)$ by subtracting the constant $\overline{\epsilon(t)}$, so that the new error function $\epsilon'(t)$ has a zero mean and consequently will have a lower mean-square value. Hence we conclude that $\overline{\epsilon(t)} = 0$ for an optimum system. Therefore the uncorrelatedness of $\mathrm{x}(t)$ and $\epsilon(t)$ is equivalent to Eq. 4-48.

$$R_{\mathrm{x}\epsilon}(\tau) = 0$$

This implies that the data $\mathrm{x}(t)$ and the error $\epsilon(t)$ are orthogonal.

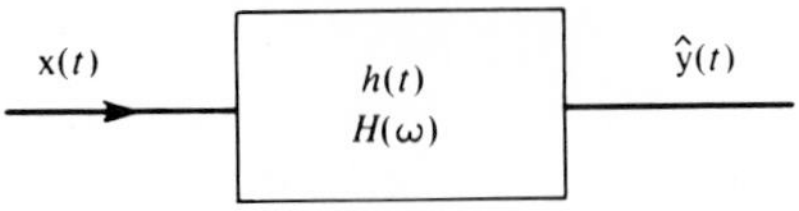

FIG. 4-9.

$$R_{x\epsilon}(\tau) = 0 \qquad \text{for all } \tau \text{ in } I \tag{4-48}$$

or

$$\overline{x(t)\epsilon(t+\tau)} = \overline{x(t-\tau)\epsilon(t)} = 0 \qquad \text{for all } \tau \text{ in } I \tag{4-49}$$

Proof: Suppose $h(t)$ is such that $R_{x\epsilon}(\tau) = 0$ (see Eq. 4-48), but let us assume that a linear system with unit impulse response $g(t)$ is optimum, where $g(t)$ is not necessarily equal to $h(t)$.

Let us denote the output of this system by $\tilde{y}(t)$. It is evident that

$$\tilde{y}(t) = x(t) * g(t) = \int_I x(t-\alpha)g(\alpha)\,d\alpha \tag{4-50a}$$

also

$$\hat{y}(t) = x(t) * h(t) = \int_I x(t-\alpha)h(\alpha)\,d\alpha \tag{4-50b}$$

We now have two systems with impulse response $h(t)$ and $g(t)$ respectively. The corresponding outputs are $\hat{y}(t)$ and $\tilde{y}(t)$. We shall denote the errors by $\epsilon_h(t)$ and $\epsilon_g(t)$ respectively.

$$\begin{aligned} \epsilon_h(t) &= y(t) - \hat{y}(t) \\ \epsilon_g(t) &= y(t) - \tilde{y}(t) \end{aligned} \tag{4-51}$$

where $\epsilon_h(t)$ satisfies the condition in Eq. 4-48 (or 4-49). From Eq. 4-51, we have

$$\begin{aligned} \overline{\epsilon_g^2(t)} &= \overline{[y(t) - \tilde{y}(t)]^2} \\ &= \overline{\{[y(t) - \hat{y}(t)] + [\hat{y}(t) - \tilde{y}(t)]\}^2} \\ &= \overline{\{\epsilon_h(t) + [\hat{y}(t) - \tilde{y}(t)]\}^2} \\ &= \overline{\epsilon_h^2(t)} + \overline{2\epsilon_h(t)[\hat{y}(t) - \tilde{y}(t)]} + \overline{[\hat{y}(t) - \tilde{y}(t)]^2} \end{aligned} \tag{4-52}$$

We shall now show that the second term on the right-hand side of Eq. 4-52 is zero. Using Eqs. 4-50a and 4-50b, we get

$$\begin{aligned} \overline{\epsilon_h(t)[\hat{y}(t) - \tilde{y}(t)]} &= \overline{\epsilon_h(t)\int_I x(t-\tau)[h(\tau) - g(\tau)]\,d\tau} \\ &= \overline{\int_I \epsilon_h(t)x(t-\tau)[h(\tau) - g(\tau)]\,d\tau} \\ &= \int_I \overline{\epsilon_h(t)x(t-\tau)}[h(\tau) - g(\tau)]\,d\tau \\ &= \int_I R_{x\epsilon_h}(\tau)[h(\tau) - g(\tau)]\,d\tau \end{aligned}$$

But by assumption $\epsilon_h(t)$ satisfies condition 4-48—that is, $R_{x\epsilon_h}(\tau) = 0$. Hence the integral on the right-hand side is zero. Substitution of this result in Eq. 4-52 yields

$$\overline{\epsilon_g^2(t)} = \overline{\epsilon_h^2(t)} + \overline{[\hat{y}(t) - \tilde{y}(t)]^2}$$

The second term on the right-hand side is a nonnegative quantity. It therefore follows that $\overline{\epsilon_g^2(t)}$ can never be less than $\overline{\epsilon_h^2(t)}$. It may at best be equal to $\overline{\epsilon_h^2(t)}$ but will never be less than $\overline{\epsilon_h^2(t)}$. The two errors will be equal if and only if

$$g(t) = h(t)$$

Hence our assumption that $g(t)$ is optimum is incorrect unless $g(t) = h(t)$. Therefore $h(t)$ is the impulse response of the optimum system. Thus for the optimum system, the error $\epsilon(t)$ *must* satisfy condition 4-48 (or 4-49).

Under optimum conditions,

$$\overline{\epsilon(t)\,x(t - \tau)} = \overline{[y(t) - \hat{y}(t)]\,x(t - \tau)} = 0 \qquad \text{for } \tau \text{ in } I$$

This yields

$$\overline{y(t)\,x(t - \tau)} = \overline{\hat{y}(t)\,x(t - \tau)} \qquad \text{for all } \tau \text{ in } I$$

or

$$R_{xy}(\tau) = R_{x\hat{y}}(\tau) \qquad \text{for all } \tau \text{ in } I \tag{4-53}$$

Thus for an optimum system the cross-correlation of the input signal and the desired output signal should be equal to the cross-correlation of the input signal and the actual output signal for all τ in I. This is known as the *Wiener-Hopf condition.* We can express the Wiener-Hopf condition in terms of $h(t)$ by observing that

$$\begin{aligned} R_{x\hat{y}}(\tau) &= \overline{\hat{y}(t)\,x(t - \tau)} \\ &= \overline{x(t - \tau)\int_I x(t - \alpha)h(\alpha)\,d\alpha} \\ &= \int_I \overline{x(t - \tau)\,x(t - \alpha)}\,h(\alpha)\,d\alpha \\ &= \int_I h(\alpha)\,R_x(\tau - \alpha)\,d\alpha \end{aligned}$$

Hence the Wiener-Hopf condition may be expressed as

$$R_{xy}(\tau) = \int_I h(\alpha)\,R_x(\tau - \alpha)\,d\alpha \qquad \text{for all } \tau \text{ in } I \tag{4-54}$$

The Mean-Square Error. The mean-square error in the optimum system is given by

$$\overline{\epsilon^2(t)} = \overline{\epsilon(t)[y(t) - \hat{y}(t)]}$$
$$= \overline{\epsilon(t)y(t)} - \overline{\epsilon(t)\hat{y}(t)}$$

We can easily show that the second term on the right-hand side is zero:

$$\overline{\epsilon(t)\hat{y}(t)} = \overline{\epsilon(t)\int_I x(t-\tau)h(\tau)\,d\tau}$$
$$= \int_I \overline{\epsilon(t)x(t-\tau)}\,h(\tau)\,d\tau$$

From Eq. 4-49 it follows that the integrand is zero. Hence

$$\overline{\epsilon(t)\hat{y}(t)} = R_{e\hat{y}}(0) = 0$$

and

$$\overline{\epsilon^2(t)} = \overline{\epsilon(t)y(t)}$$
$$= \overline{[y(t) - \hat{y}(t)]y(t)}$$
$$= \overline{y(t)y(t)} - \overline{y(t)\hat{y}(t)}$$
$$= R_y(0) - \int_I \overline{y(t)x(t-\tau)}\,h(\tau)\,d\tau$$
$$= R_y(0) - \int_I R_{xy}(\tau)h(\tau)\,d\tau \tag{4-55}$$

To recapitulate, for the optimum system (in the sense of minimum mean-square error), the error $\epsilon(t)$ and the data $x(t)$ are uncorrelated. Since $\overline{\epsilon(t)}$ is zero, the uncorrelatedness implies orthogonality. Thus for the optimum system, the process $x(t)$ (the data) and the process $\epsilon(t)$ (the error) are orthogonal. This condition is equivalent to the Wiener-Hopf condition in Eq. 4-53 or 4-54 which states that the input, desired output correlation should be equal to the input, actual output correlation. The unit impulse response $h(t)$ of the optimum system can be obtained by solving the Wiener-Hopf condition (Eq. 4-54). The (optimum) mean-square error is given by Eq. 4-55.

We shall now consider the two cases (i) no constraint of physical realizability on the system; and (ii) system constrained to be physically realizable.

4-11. OPTIMUM SYSTEMS WITHOUT THE CONSTRAINT OF PHYSICAL REALIZABILITY

When the system is not constrained to be physically realizable, the impulse response $h(t)$ could be noncausal in general; that is, $h(t)$ could exist over the entire time interval $(-\infty < t < \infty)$, and the domain of integration I in this case is $(-\infty, \infty)$. The Wiener-Hopf condition becomes

$$R_{xy}(\tau) = R_{x\hat{y}}(\tau) \qquad -\infty < \tau < \infty$$

or

$$R_{xy}(\tau) = \int_{-\infty}^{\infty} h(\alpha) R_x(\tau - \alpha)\, d\alpha$$
$$= h(\tau) * R_x(\tau) \tag{4-56}$$

We wish to solve this equation to obtain the optimum $h(t)$. This can easily be done by taking the Fourier transform of both sides of Eq. 4-56:

$$\mathcal{S}_{xy}(\omega) = H(\omega)\mathcal{S}_x(\omega)$$

where $H(\omega)$ is the transfer function of the desired optimum system. We now have

$$H(\omega) = \frac{\mathcal{S}_{xy}(\omega)}{\mathcal{S}_x(\omega)} \tag{4-57}$$

This is the desired solution.[3] Note that $H(\omega)$ is physically unrealizable. This follows from the fact that $\mathcal{S}_x(\omega)$ is an even function of ω. Hence the corresponding complex frequency form $\mathcal{S}_x(s)$ will have zeros in the LHP as well as the RHP.[4] Thus $H(s)$ will have poles in the RHP. This implies two possibilities: (i) the system is stable but physically unrealizable (impulse response existing for $t < 0$), or (ii) the system is physically realizable but unstable. In this case, however, the RHP poles of $H(s)$ cannot represent physically realizable but unstable system because we have assumed $x(t)$ and $y(t)$ to be at least wide-sense stationary processes. If the system were unstable, $y(t)$ cannot be stationary. Hence the possibility of the system being physically realizable but unstable is ruled out. The only remaining alternative is that the RHP poles represent a stable but physically unrealizable system.

The Mean-Square Error. The mean-square error in this case is given by (see Eq. 4-55):

$$\overline{\epsilon^2(t)} = R_y(0) - \int_{-\infty}^{\infty} R_{xy}(\tau) h(\tau)\, d\tau \tag{4-58}$$

If we let

$$\phi(\tau) = R_{yx}(\tau) * h(\tau)$$
$$= \int_{-\infty}^{\infty} R_{yx}(\tau - \alpha) h(\alpha)\, d\alpha$$

[3]We had anticipated this result earlier (see page 252). At that time, however, it was not clear whether it would yield optimum results or not. We have now shown that it does yield an optimum system when the performance criterion is minimum mean-square error.

[4]See Appendix at end of chapter.

$$= \int_{-\infty}^{\infty} R_{xy}(\alpha - \tau) h(\alpha) \, d\alpha \tag{4-59}$$

Hence

$$\phi(0) = \int_{-\infty}^{\infty} R_{xy}(\alpha) h(\alpha) \, d\alpha \tag{4-60}$$

Substituting Eq. 4-60 in Eq. 4-58, we obtain

$$\overline{\epsilon^2(t)} = R_y(0) - \phi(0) \tag{4-61}$$

Note that if

$$f(t) \leftrightarrow F(\omega)$$

then

$$f(t) = \frac{1}{2\pi} \int_{-\infty}^{\infty} F(\omega) e^{j\omega t} \, d\omega$$

and

$$f(0) = \frac{1}{2\pi} \int_{-\infty}^{\infty} F(\omega) \, d\omega \tag{4-62}$$

Also

$$R_y(\tau) \leftrightarrow \mathcal{S}_y(\omega)$$

$$\phi(\tau) = R_{yx}(\tau) * h(\tau) \leftrightarrow \mathcal{S}_{yx}(\omega) H(\omega)$$

Hence from the results in Eq. 4-62, we have

$$\phi(0) = \frac{1}{2\pi} \int_{-\infty}^{\infty} \mathcal{S}_{yx}(\omega) H(\omega) \, d\omega$$

and

$$R_y(0) = \frac{1}{2\pi} \int_{-\infty}^{\infty} \mathcal{S}_y(\omega) \, d\omega$$

The use of these results in Eq. 4-61 yields

$$\overline{\epsilon^2(t)} = \frac{1}{2\pi} \int_{-\infty}^{\infty} [\mathcal{S}_y(\omega) - H(\omega) \mathcal{S}_{yx}(\omega)] \, d\omega$$

$$= \frac{1}{2\pi} \int_{-\infty}^{\infty} [\mathcal{S}_y(\omega) - \mathcal{S}_{xy}(-\omega) H(\omega)] \, d\omega \tag{4-63a}$$

Substituting for $H(\omega)$ from Eq. 4-57 we get

$$\overline{\epsilon^2(t)} = \frac{1}{2\pi} \int_{-\infty}^{\infty} \left[\mathcal{S}_y(\omega) - \frac{|\mathcal{S}_{xy}(\omega)|^2}{\mathcal{S}_x} \right] d\omega \tag{4-63b}$$

Example 4-6. Filtering. To start with we shall consider the problem of filtering a signal mixed with noise. Let $s(t)$ and $n(t)$ be the signal and noise respectively. The received signal $x(t)$ is given by

$$x(t) = s(t) + n(t)$$

We wish to separate the signal from the noise. The desired output $y(t)$ is given by

$$y(t) = s(t)$$

For convenience we shall assume the means of $s(t)$ and $n(t)$ to be zero. This can be done without loss of generality.

$$\begin{aligned} R_x(\tau) &= \overline{[s(t) + n(t)][s(t+\tau) + n(t+\tau)]} \\ &= R_s(\tau) + R_n(\tau) + R_{sn}(\tau) + R_{ns}(\tau) \end{aligned}$$

since the processes $s(t)$ and $n(t)$ are uncorrelated (and $\overline{s} = \overline{n} = 0$)

$$R_{sn}(\tau) = R_{ns}(\tau) = 0$$

Hence

$$R_x(\tau) = R_s(\tau) + R_x(\tau)$$

and

$$\mathcal{S}_x(\omega) = \mathcal{S}_s(\omega) + \mathcal{S}_n(\omega)$$

Similarly

$$\begin{aligned} R_{xy}(\tau) &= \overline{[s(t) + n(t)]\, s(t+\tau)} \\ &= R_s(\tau) + R_{ns}(\tau) \\ &= R_s(\tau) \end{aligned}$$

and

$$\mathcal{S}_{xy}(\omega) = \mathcal{S}_s(\omega)$$

Also we are given

$$\mathcal{S}_y(\omega) = \mathcal{S}_s(\omega)$$

The transfor function $H(\omega)$ of the optimum linear filter is given by Eq. 4-57:

$$H(\omega) = \frac{\mathcal{S}_{xy}(\omega)}{\mathcal{S}_x(\omega)} = \frac{\mathcal{S}_s(\omega)}{\mathcal{S}_s(\omega) + \mathcal{S}_n(\omega)} \tag{4-64}$$

It can be seen from Eq. 4-64, that the optimum filter accentuates the frequency components in the range where the desired signal is strong and the noise is weak and suppresses the frequency components in the range where the desired signal is weak and the noise is strong. The mean-square error $\overline{\epsilon^2}$ is given by Eq. 4-63:

$$\begin{aligned} \overline{\epsilon^2} &= \frac{1}{2\pi}\int_{-\infty}^{\infty} \left[\mathcal{S}_s(\omega) - \frac{\mathcal{S}_s^2(\omega)}{\mathcal{S}_s(\omega) + \mathcal{S}_n(\omega)}\right] d\omega \\ &= \frac{1}{2\pi}\int_{-\infty}^{\infty} \frac{\mathcal{S}_s(\omega)\mathcal{S}_n(\omega)}{\mathcal{S}_s(\omega) + \mathcal{S}_n(\omega)}\, d\omega \end{aligned} \tag{4-65}$$

Observe that if the spectra of $s(t)$ and $n(t)$ do not overlap as shown in Fig. 4-10a, then

$$\mathcal{S}_s(\omega)\mathcal{S}_n(\omega) = 0$$

and

$$\overline{\epsilon^2} = 0$$

Thus the filtering is perfect. The result is of course trivial in this case when the signal and noise spectra are nonoverlapping.

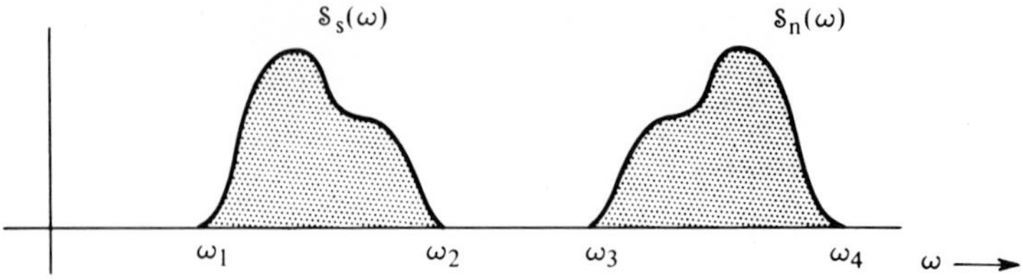

(a)

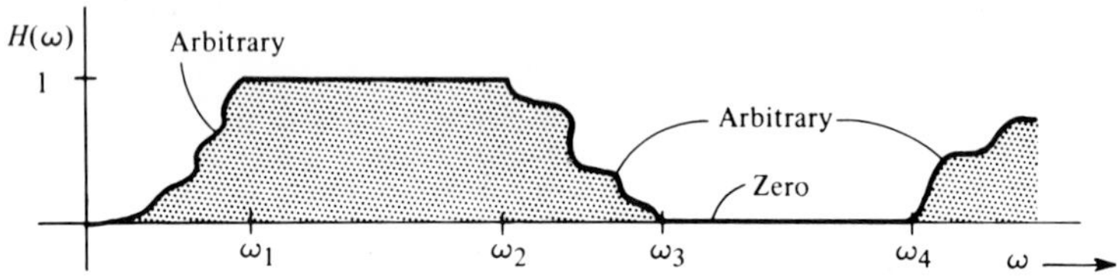

(b)

FIG. 4-10.

The transfer function in this case becomes (see Eq. 4-64)

$$H(\omega) = \begin{cases} 1 & \text{where } \mathcal{S}_s(\omega) \neq 0 \text{ and } \mathcal{S}_n(\omega) = 0 \\ 0 & \text{where } \mathcal{S}_s(\omega) = 0 \text{ and } \mathcal{S}_n(\omega) \neq 0 \\ \text{arbitrary} & \text{where } \mathcal{S}_s(\omega) = \mathcal{S}_n(\omega) = 0 \end{cases}$$

The required transfer function $H(\omega)$ is shown in Fig. 4-10b. Note that $H(\omega)$ vanishes over certain band and hence violates the Paley-Wiener criterion. Therefore $H(\omega)$ is physically unrealizable.

Example 4-7. Filtering. We shall again consider the problem of filtering a signal mixed with noise. The process $s(t)$ is the random process shown in Fig. 3-22 with a mean-square value of unity and the noise signal is a white-noise signal $n(t)$ with a constant power-density spectrum of magnitude $\mathfrak{N}/2$. Thus

$$\mathcal{S}_s(\omega) = \frac{2\alpha}{\alpha^2 + \omega^2}$$

$$\mathcal{S}_n(\omega) = \frac{\mathfrak{N}}{2} \quad \text{also } \mathcal{S}_{sn}(\omega) = \mathcal{S}_{ns}(\omega) = 0$$

Hence from Eq. 4-64

$$H(\omega) = \frac{4\alpha}{4\alpha + \mathfrak{N}(\alpha^2 + \omega^2)}$$

$$= \frac{4\alpha}{\mathfrak{N}\omega^2 + (\mathfrak{N}\alpha^2 + 4\alpha)} \tag{4-66}$$

The mean-square error is given by Eq. 4-63:

$$\overline{\epsilon^2} = \frac{1}{2\pi} \int_{-\infty}^{\infty} \frac{2\mathfrak{N}\alpha}{\mathfrak{N}\omega^2 + (\mathfrak{N}\alpha^2 + 4\alpha)} \, d\omega$$

$$= \frac{\alpha}{\alpha^2 + 4\alpha/\mathfrak{N}} \tag{4-67}$$

The unit impulse response $h(t)$ is given by

$$h(t) = \mathfrak{F}^{-1}\left[\frac{4\alpha}{\mathfrak{N}\omega^2 + (\mathfrak{N}\alpha^2 + 4\alpha)/\mathfrak{N}}\right]$$

$$= \frac{2\alpha}{\beta} e^{-\beta|t|} \tag{4-68}$$

where $\beta = \alpha^2 + 4\alpha/\mathfrak{N}$.

The impulse response is shown in Fig. 4-11a. It can be easily seen that this is a noncausal function and hence the system is physically unrealizable.

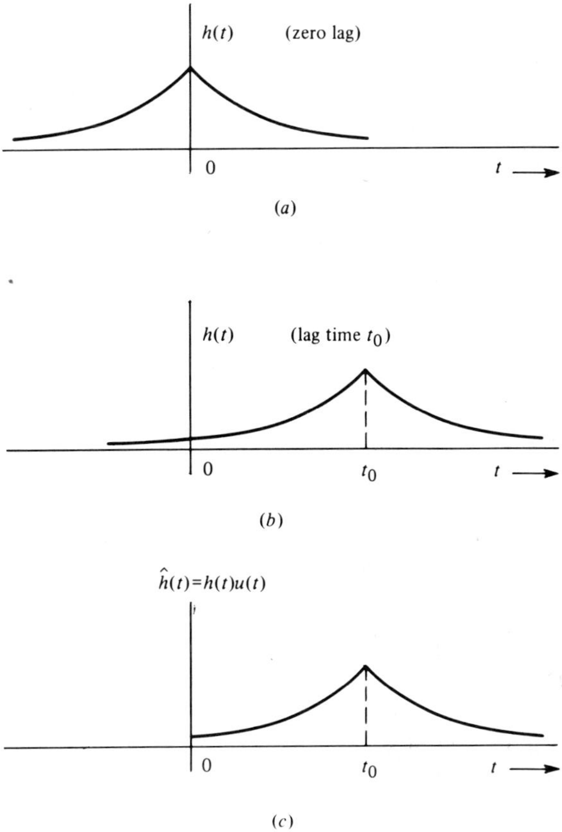

FIG. 4-11.

Example 4-8. Lag Filtering. In this problem we consider the same signal and noise as in Example 4-7. The desired output, however, is not the signal s(t) but the delayed signal s$(t - t_0)$. Thus

$$\mathrm{x}(t) = \mathrm{s}(t) + \mathrm{n}(t)$$

$$\begin{aligned} y(t) &= s(t - t_0) \\ R_{xy}(\tau) &= \overline{[s(t) + n(t)]\, s(t - t_0 + \tau)} \\ &= \overline{s(t)\, s(t - t_0 + \tau)} + \overline{n(t)\, s(t - t_0 + \tau)} \\ &= R_s(\tau - t_0) + \underbrace{R_{ns}(\tau - t_0)}_{0} \\ &= R_s(\tau - t_0) \end{aligned}$$

Hence

$$\mathcal{S}_{xy}(\omega) = \mathcal{S}_s(\omega)\, e^{-j\omega t_0}$$

Thus from Eq. 4-57 we have

$$\begin{aligned} H(\omega) &= \frac{\mathcal{S}_s(\omega)\, e^{-j\omega t_0}}{\mathcal{S}_s(\omega) + \mathcal{S}_n(\omega)} \\ &= \frac{4\alpha e^{-j\omega t_0}}{\mathfrak{N}\omega^2 + (\mathfrak{N}\alpha^2 + 4\alpha)} \end{aligned} \tag{4-69}$$

This mean-square error is identical with that in Example 4-7 since $\mathcal{S}_y(\omega)$, $\mathcal{S}_x(\omega)$ and $|\mathcal{S}_{yx}(\omega)|^2$ are identical in both cases.

The impulse response $h(t)$ is given by

$$h(t) = \mathcal{F}^{-1}[H(\omega)] = \frac{2\alpha}{\mathfrak{N}\beta}\, e^{-\beta|t - t_0|} \tag{4-70}$$

where $\beta = \alpha^2 + 4\alpha/\mathfrak{N}$.

This impulse response is shown in Fig. 4-11b. Observe that the impulse response is identical with that obtained in Example 4-7 (Fig. 4-11a) except that it is shifted by t_0 seconds to the right. This impulse response (Fig. 4-11b) is also noncausal and consequently physically unrealizable. However, for sufficiently large t_0 most of $h(t)$ lies in the region $t > 0$ and we can ignore the tail of $h(t)$ lying in the region $t < 0$. We can construct a new function $\hat{h}(t)$ by letting the negative time part of $h(t)$ go to zero:

$$\hat{h}(t) = h(t)\, u(t)$$

The transfer function $\hat{h}(t)$ can now be realized physically. The performance of this system will approach that of the optimum system as $t_0 \to \infty$. Thus if one is willing to accept a delay, the optimum filter can approximately be realized. Note that t_0 need be only 3 to 4 time constants of the exponential in order for the performance of this physically realizable system closely to approach that of the optimum system (which is physically unrealizable). We therefore conclude that although an optimum filter is physically unrealizable, a close approximation to it can be realized physically if we are willing to accept a delay of t_0 in the response. This conclusion is true in general of any physically unrealizable system if $h(t) \to 0$ for large negative values of t.

4-12. OPTIMUM PHYSICALLY REALIZABLE SYSTEMS

We shall now find an optimum system which is constrained to be physically realizable. In this case $h(t) = 0$ for $t > 0$ and I, the domain of integration is $(0, \infty)$. Hence the Wiener-Hopf condition (Eq. 4-54) becomes

$$R_{xy}(\tau) = \int_0^\infty R_x(\tau - \alpha) h(\alpha)\, d\alpha \qquad \tau > 0 \tag{4-71}$$

or

$$R_{xy}(\tau) = R_{x\hat{y}}(\tau) \qquad \text{for } \tau > 0 \tag{4-72}$$

The corresponding mean-square error is given by (Eq. 4-55)

$$\overline{\epsilon^2(t)} = R_y(0) - \int_0^\infty R_{xy}(\tau) h(\tau)\, d\tau \tag{4-73}$$

To obtain $h(t)$, we must solve Eq. 4-71 (or 4-72):

$$R_{xy}(\tau) = R_{x\hat{y}}(\tau) \qquad \text{for } \tau > 0 \tag{4-74}$$

Consider the equation

$$R_{xy}(\tau) - R_{x\hat{y}}(\tau) = q(\tau) \tag{4-75}$$

In view of Eq. 4-74, the function $q(\tau)$ must be zero for $\tau > 0$. This fact is of crucial importance in finding the solution to Eq. 4-74. Taking the Fourier transform of Eq. 4-75, we obtain

$$\mathcal{S}_{xy}(\omega) - \mathcal{S}_{x\hat{y}}(\omega) = Q(\omega) \tag{4-76a}$$

Note that $\hat{y}(t)$ is the output of the system to the input $x(t)$. Therefore application of Eq. 4-16b yields

$$\mathcal{S}_{xy}(\omega) - H(\omega)\mathcal{S}_x(\omega) = Q(\omega) \tag{4-76b}$$

In the above equation when $j\omega$ is replaced by a complex variable s, we obtain the representation in terms of the (bilateral) Laplace transform.[5]

$$\mathcal{S}_{xy}(\omega)\big|_{\omega = s/j} = \mathcal{S}_{xy}(s)$$

Eq. 4-76 now becomes

$$\mathcal{S}_{xy}(s) - H(s)\mathcal{S}_x(s) = Q(s) \tag{4-77}$$

At this point we note (see appendix at the end of the chapter) that if $F(s)$ is the bilateral Laplace transform of an absolutely integrable function $f(t)$, the positive time component of $f(t)$ contributes to the poles of $F(s)$ in the LHP and the negative time component of $f(t)$ contributes to the poles of $F(s)$ in the RHP. Since $q(\tau)$ is zero for $\tau > 0$, $Q(s)$ cannot have poles in the LHP. We shall use this fact in Eq. 4-77 to obtain $H(s)$. Since $h(t) = 0$ for $t < 0$, $H(s)$ has no poles in the RHP. Also $\mathcal{S}_x(s)$ is an even function of s. Hence it can be expressed in the form

$$\mathcal{S}_x(s) = \mathcal{S}_x(s)\,\mathcal{S}_x(-s) \tag{4-78}$$

[5]Mathematically $\mathcal{S}_x(s)$ represents $\mathcal{S}_x(\omega)$ where ω is replaced by s. Strictly speaking, use of notation $\mathcal{S}_x(s)$ for $\mathcal{S}_x(\omega)$ where $j\omega$ is replaced by s is incorrect. Nevertheless, we use this notation to avoid increasing the number of symbols used and thus reduce the possibility of confusion.

where $\mathcal{S}_x(s)$ has all its poles and zeros in the LHP. It is obvious that $\mathcal{S}_x(-s)$ has all the poles and zeros in the RHP. The poles of $\mathcal{S}_x(-s)$ are mirror images of the poles of $\mathcal{S}_x(s)$ about the vertical axis. A similar conclusion applies to zeros. Equation 4-77 can now be expressed as

$$\mathcal{S}_{xy}(s) - H(s)\,\mathcal{S}_x(s)\,\mathcal{S}_x(-s) = Q(s)$$

or

$$H(s)\,\mathcal{S}_x(s) = \frac{\mathcal{S}_{xy}(s)}{\mathcal{S}_x(-s)} - \frac{Q(s)}{\mathcal{S}_x(-s)} \tag{4-79}$$

Observe that the term $H(s)\,\mathcal{S}_x(s)$ has all its poles in the LHP and hence its inverse transform is exclusively a positive time function. On the other hand, the term $Q(s)/\mathcal{S}_x(-s)$ has all its poles in the RHP and hence its inverse transform is exclusively a negative time function. The third term $\mathcal{S}_{xy}(s)/\mathcal{S}_x(-s)$ has poles in the LHP as well as the RHP and hence its inverse transform has both positive and negative time components. It is evident from Eq. 4-79 that the positive time component of this term must be equal to the inverse transform of $H(s)\,\mathcal{S}_x(s)$ and its negative time component must equal the inverse transform of $Q(s)/\mathcal{S}_x(-s)$. For convenience we shall designate the term $\mathcal{S}_{xy}(s)/\mathcal{S}_x(-s)$ by $F(s)$.

$$F(s) = \frac{\mathcal{S}_{xy}(s)}{\mathcal{S}_x(-s)} \tag{4-80}$$

By partial-fraction expansion, the function $F(s)$ can be split into two terms $F^+(s)$ and $F^-(s)$, the first with all the poles in the LHP and the second with all its poles in the RHP. The former term is obviously equal to $H(s)\,\mathcal{S}_x(s)$ and the latter is equal to $Q(s)/\mathcal{S}_x(-s)$. If these two terms are designated as $F^+(s)$ and $F^-(s)$—that is,

$$F(s) = \frac{\mathcal{S}_{xy}(s)}{\mathcal{S}_x(-s)} = F^+(s) + F^-(s) \tag{4-81}$$

then

$$H(s)\,\mathcal{S}_x(s) = F^+(s)$$

and

$$H(s) = \frac{F^+(s)}{\mathcal{S}_x(s)} \tag{4-82}$$

This is the desired solution of the Wiener-Hopf equation. We can also obtain this solution in the time domain. Let the inverse transform of $\mathcal{S}_{xy}(s)/\mathcal{S}_x(-s)$ be $f(t)$. Then it follows from the above discussion that $f(t)\,u(t)$ is the inverse transform of $F^+(s)$. The term $\mathcal{S}_x(s)$ has all its poles and zeros in the LHP. Hence its reciprocal, $1/\mathcal{S}_x(s)$ has all its poles and zeros in the LHP and corresponds to a positive time function. Let the inverse transform of $1/\mathcal{S}_x(s)$ be $g(t)$. Then application of time convolution theorem to Eq. 4-82 yields

$$h(t) = f(t)u(t) * g(t)$$
$$= \int_0^t f(\alpha)g(t - \alpha)\,d\alpha$$

The limits of integration are $(0, t)$ because both $f(t)u(t)$ and $g(t)$ are causal.

The Mean-Square Error. The mean-square error $\overline{\epsilon^2}$ is obtained from Eq. 4-73. This expression is similar to that in Eq. 4-58 except that the lower limit of integration in Eq. 4-73 is 0 instead of $-\infty$. This, however, is immaterial, since $h(t)$ is zero for $t < 0$. Hence

$$\overline{\epsilon^2(t)} = R_y(0) - \int_{-\infty}^{\infty} h(\beta)\,R_{xy}(\beta)\,d\beta \qquad \text{(4-83a)}$$

Following the reasoning used in deriving Eq. 4-63a, we obtain

$$\overline{\epsilon^2(t)} = \frac{1}{2\pi}\int_{-\infty}^{\infty} [\mathcal{S}_y(\omega) - H(\omega)\mathcal{S}_{xy}(-\omega)]\,d\omega \qquad \text{(4-83b)}$$

Alternatively we observe that

$$R_{xy}(\tau) = R_{x\hat{y}}(\tau) \qquad \tau \geq 0$$

Equation 4-83a can therefore be expressed as

$$\overline{\epsilon^2(t)} = R_y(0) - \int_{-\infty}^{\infty} h(\beta)\,R_{x\hat{y}}(\beta)\,d\beta \qquad \text{(4-83c)}$$

Note that since $h(\beta) = 0$ for $\beta < 0$, equating $R_{xy}(\beta)$, to $R_{x\hat{y}}(\beta)$ for all β in the above integral is permissible. Now following the reasoning used in deriving Eq. 4-63a, we obtain[6]

$$\overline{\epsilon^2(t)} = \frac{1}{2\pi}\int_{-\infty}^{\infty} [\mathcal{S}_y(\omega) - H(\omega)\mathcal{S}_{x\hat{y}}(-\omega)]\,d\omega$$

The use of Eq. 4-16b in this equation and remembering the fact that $\mathcal{S}_x(\omega)$ is an even function of ω yields

$$\overline{\epsilon^2(t)} = \frac{1}{2\pi}\int_{-\infty}^{\infty} [\mathcal{S}_y(\omega) - |H(\omega)|^2\mathcal{S}_x(\omega)]\,d\omega \qquad \text{(4-83d)}$$

In the case of problems involving filtering,

$$x(t) = s(t) + n(t)$$
$$y(t) = s(t)$$

and (when signal and noise are uncorrelated)

$$R_{xy}(\tau) = R_s(\tau)$$

[6]Note that the output of the system is $\hat{y}(t)$ and hence Eq. 4-16b holds for $\hat{y}(t)$ as the output to $x(t)$.

Hence Eqs. 4-83a and 4-83d reduce to

$$\overline{\epsilon^2(t)} = R_s(0) - \int_{-\infty}^{\infty} h(\beta)\, R_s(\beta)\, d\beta \tag{4-83e}$$

$$= \frac{1}{2\pi} \int_{-\infty}^{\infty} [\mathcal{S}_s(\omega) - |H(\omega)|^2 \mathcal{S}_x(\omega)]\, d\omega \tag{4-83f}$$

In filtering problems if the noise is a white noise of power density $\mathfrak{N}/2$, the expression for the mean-square error is further simplified to

$$\overline{\epsilon^2(t)} = \frac{\mathfrak{N}}{2} h(0) \tag{4-83g}$$

This can be shown as follows. For a filtering problem with white noise (noise uncorrelated with the signal),

$$R_{xy}(\tau) = R_s(\tau)$$

$$R_x(\tau) = R_s(\tau) + R_n(\tau) = R_s(\tau) + \frac{\mathfrak{N}}{2}\delta(\tau)$$

Hence the Wiener-Hopf Eq. 4-71 reduces to

$$R_s(\tau) = \int_0^{\infty} \left[\frac{\mathfrak{N}}{2}\delta(\tau - \alpha) + R_s(\tau - \alpha)\right] h(\alpha)\, d\alpha$$

Since $h(\alpha) = 0$ for $\alpha < 0$, the lower limit of the above integral may be taken as $-\infty$.

$$R_s(\tau) = \int_{-\infty}^{\infty} \left[\frac{\mathfrak{N}}{2}\delta(\tau - \alpha) + R_s(\tau - \alpha)\right] h(\alpha)\, d\alpha$$

$$= \frac{\mathfrak{N}}{2} h(\tau) + \int_{-\infty}^{\infty} R_s(\tau - \alpha)\, h(\alpha)\, d\alpha$$

Therefore

$$R_s(\tau) - \int_{-\infty}^{\infty} R_s(\tau - \alpha)\, h(\alpha)\, d\alpha = \frac{\mathfrak{N}}{2} h(\tau)$$

From Eq. 4-83e, it now follows that

$$\overline{\epsilon^2(t)} = \frac{\mathfrak{N}}{2} h(0)$$

because $R_s(\tau)$ is an even function of τ.

Example 4-9. Optimum Prediction. We are given a process $x(t)$ and wish to predict what it will be at some future instant $t + t_0$. Thus

$$y(t) = x(t + t_0)$$

It is evident that

$$R_{xy}(\tau) = R_x(\tau + t_0)$$

and

$$\mathcal{S}_{xy}(\omega) = \mathcal{S}_x(\omega)\, e^{j\omega t_0} \tag{4-84}$$

As a specific example, consider the random process in Fig. 3-22, with a mean-square value of unity. For this process,

$$\mathcal{S}_x(\omega) = \frac{2\alpha}{\alpha^2 + \omega^2}$$

$$\mathcal{S}_x(s) = \frac{2\alpha}{\alpha^2 - s^2} = \frac{2\alpha}{(\alpha + s)(\alpha - s)} = \mathcal{S}_x(s)\, \mathcal{S}_x(-s)$$

Hence

$$\mathcal{S}_x(s) = \frac{\sqrt{2\alpha}}{\alpha + s} \quad \text{and} \quad \mathcal{S}_x(-s) = \frac{\sqrt{2\alpha}}{\alpha - s}$$

Also from Eq. 4-84,

$$\mathcal{S}_{xy}(s) = \mathcal{S}_x(s) e^{st_0} = \frac{2\alpha e^{st_0}}{(\alpha + s)(\alpha - s)}$$

Hence

$$F(s) = \frac{\mathcal{S}_{xy}(s)}{\mathcal{S}_x(-s)} = \frac{\sqrt{2\alpha}\, e^{st_0}}{\alpha + s} \tag{4-85}$$

We are now required to split $\mathcal{S}_{xy}(s)/\mathcal{S}_x(-s)$ into two terms corresponding to positive and negative time functions respectively. This can easily be done as follows. Let

$$F(s) = \frac{\mathcal{S}_{xy}(s)}{\mathcal{S}_x(-s)} = F^+(s) + F^-(s)$$

and

$$f(t) \leftrightarrow F(s)$$

Then

$$f(t)\, u(t) \leftrightarrow F^+(s)$$
$$f(t)\, u(-t) \leftrightarrow F^-(s)$$

The function $f(t)$ is the inverse transform of $F(s)$ in Eq. 4-85. This is obviously

$$f(t) = \sqrt{2\alpha}\, e^{-\alpha(t + t_0)} u(t + t_0)$$

This function is shown in Fig. 4-12. It is evident that

$$f(t)\, u(t) = \sqrt{2\alpha}\, e^{-\alpha(t + t_0)} u(t)$$

Hence

$$F^+(s) = \frac{\sqrt{2\alpha}\, e^{-\alpha t_0}}{s + \alpha}$$

The desired transfer function $H(s)$ is now given by Eq. 4-82:

$$H(s) = \frac{F^+(s)}{\mathcal{S}_x(s)} = \frac{\sqrt{2\alpha}\, e^{-\alpha t_0}}{s + \alpha} \frac{s + \alpha}{\sqrt{2\alpha}} = e^{-\alpha t_0} \tag{4-86}$$

The transfer function $H(s)$ in this case is a constant $e^{-\alpha t_0}$ as given in Eq. 4-86, This implies that the best prediction that can be made at any instant of the signal value in the future is its value at the instant of prediction attenuated by

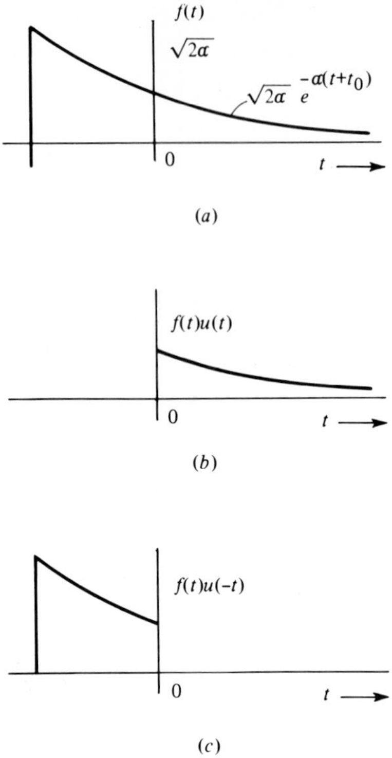

FIG. 4-12.

a constant. As the prediction is made for a more distant future, (t_0 increasing), the attenuation increases. This is the best possible prediction in the sense of minimum mean-square error that can be obtained using a linear system. The mean-square error is obtained from Eq. 4-83d.

In this problem

$$y(t) = x(t + t_0)$$

$$R_y(\tau) = \overline{x(t + t_0)\, x(t + t_0 + \tau)} = R_x(\tau)$$

and

$$\mathcal{S}_y(\omega) = \mathcal{S}_x(\omega) = \frac{2\alpha}{\alpha^2 + \omega^2}$$

Also

$$H(\omega) = e^{-\alpha t_0}$$

Hence from Eq. 4-83d,

$$\overline{\epsilon^2(t)} = \frac{1 - e^{-2\alpha t_0}}{2\pi} \int_{-\infty}^{\infty} \frac{2\alpha}{\alpha^2 + \omega^2}\, d\omega$$

$$= 1 - e^{-2\alpha t_0}$$

Observe that the maximum error is 1 and occurs for $t_0 = \infty$.

It can easily be seen from this equation that the error increases and the prediction becomes less effective as the prediction is made for a more distant future (t_0 increasing). In the limit as $t_0 \to \infty$, the transfer function $H(\omega) = 0$ and the best prediction is zero. The error in this case is maximum.

Example 4-10. Optimum Filtering and Prediction. We shall now consider a combined operation of filtering and prediction on a signal $s(t)$ contaminated with noise $n(t)$:

$$x(t) = s(t) + n(t)$$

We desire the output $y(t)$ to be

$$y(t) = s(t + t_0)$$

Since $s(t)$ and $n(t)$ come from independent sources, they are uncorrelated, and

$$\mathcal{S}_x(\omega) = \mathcal{S}_s(\omega) + \mathcal{S}_n(\omega)$$

and

$$\begin{aligned} R_{xy}(\tau) &= \overline{x(t)\,y(t+\tau)} \\ &= \overline{[s(t) + n(t)][s(t + t_0 + \tau)]} \\ &= R_s(\tau + t_0) \end{aligned}$$

Hence

$$\mathcal{S}_{xy}(\omega) = \mathcal{S}_s(\omega)\, e^{j\omega t_0}$$

As a specific example, let the signal $s(t)$ be the random process shown in Fig. 3-22 with a mean-square value of 1 and $n(t)$ be white noise with a flat power-density spectrum of magnitude $\mathfrak{N}/2$. We then have

$$\mathcal{S}_x(\omega) = \frac{2\alpha}{\alpha^2 + \omega^2} + \frac{\mathfrak{N}}{2}$$

$$\mathcal{S}_{xy}(\omega) = \frac{2\alpha}{\alpha^2 + \omega^2}\, e^{j\omega t_0}$$

It follows from these relationships that

$$\mathcal{S}_x(s) = \frac{2\alpha}{\alpha^2 - s^2} + \frac{\mathfrak{N}}{2} = \frac{4\alpha + \mathfrak{N}\alpha^2 - \mathfrak{N}s^2}{2(\alpha^2 - s^2)} = \mathcal{S}_x(s)\,\mathcal{S}_x(-s)$$

Hence

$$\mathcal{S}_x(-s) = \sqrt{\frac{\mathfrak{N}}{2}}\left(\frac{a - s}{\alpha - s}\right) \quad \text{and} \quad \mathcal{S}_x(s) = \sqrt{\frac{\mathfrak{N}}{2}}\left(\frac{a + s}{\alpha + s}\right)$$

where

$$a^2 = \frac{4\alpha}{\mathfrak{N}} + \alpha^2$$

Also

$$\mathcal{S}_{xy}(s) = \frac{2\alpha}{\alpha^2 - s^2}\, e^{st_0}$$

Hence

$$F(s) = \frac{\mathcal{S}_{xy}(s)}{\mathcal{S}_x(-s)} = \sqrt{\frac{2}{\mathfrak{N}}}\, \frac{2\alpha e^{st_0}}{(\alpha + s)(a - s)}$$

We must split $F(s)$ into $F^+(s)$ and $F^-(s)$ corresponding to positive and negative time components. The inverse transform of $F(s)$ is $f(t)$ and can be obtained by expanding $F(s)$ into partial fractions.

$$F(s) = \frac{2\alpha\sqrt{2}}{\sqrt{\mathfrak{N}}(a + \alpha)}\left[\frac{1}{s + \alpha} + \frac{1}{a - s}\right]e^{st_0}$$

Hence

$$f(t) = \frac{2\alpha\sqrt{2}}{\sqrt{\mathfrak{N}}(a + \alpha)}\left[e^{-\alpha(t+t_0)}u(t + t_0) + e^{a(t+t_0)}u(-t - t_0)\right]$$

The function $f(t)$ is shown in Fig. 4-13. It is evident from this figure that

$$f(t)u(t) = \frac{2\alpha\sqrt{2}}{\sqrt{\mathfrak{N}}(a + \alpha)}e^{-\alpha(t+t_0)}u(t)$$

and

$$\begin{aligned} F^+(s) &= \mathcal{L}\left[f(t)u(t)\right] \\ &= \frac{2\alpha\sqrt{2}}{\sqrt{\mathfrak{N}}(a + \alpha)}\left(\frac{e^{-\alpha t_0}}{s + \alpha}\right) \end{aligned}$$

But

$$\begin{aligned} H(s) &= \frac{F^+(s)}{\mathcal{S}_x(\omega)} \\ &= \frac{4\alpha e^{-\alpha t_0}}{\mathfrak{N}(a + \alpha)}\left(\frac{1}{s + a}\right) \end{aligned}$$

Since $4\alpha/\mathfrak{N} = a^2 - \alpha^2$, we have

$$H(s) = \frac{(a - \alpha)e^{-\alpha t_0}}{s + a}$$

and

$$h(t) = (a - \alpha)e^{-(\alpha t_0 + at)}u(t)$$

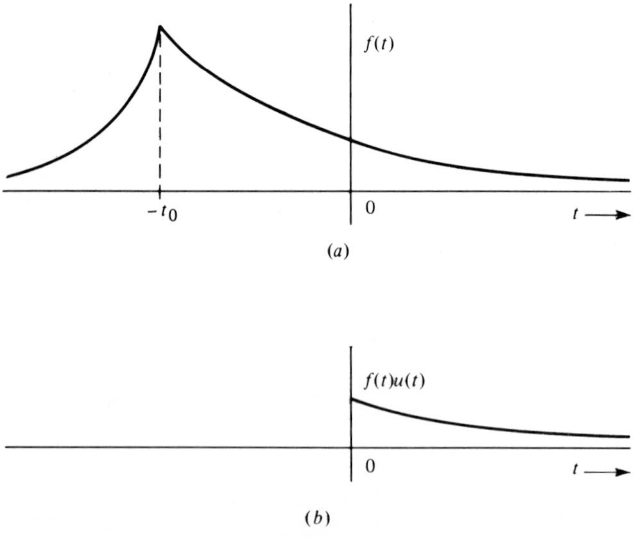

FIG. 4-13.

If in this problem, we desire only to filter the signal x(t), we let $t_0 = 0$. This yields

$$H(s) = \frac{a - \alpha}{s + a}$$

Evaluation of the mean-square error for this problem is left as an exercise for the reader (use Eq. 4-83d). In the special case of white noise mixed with the desired signal s(t) when the power density spectrum of s(t) is rational, the expression for the Wiener-Hopf (realizable) filter is greatly simplified.

For such a case it is shown in Appendix 4-2, that $H(s)$, the transfer function of the optimum filter, is given by

$$H(s) = 1 - \frac{\sqrt{\mathfrak{N}/2}}{\mathcal{S}_x(s)} \tag{4-87a}$$

and

$$\overline{\epsilon^2} = \frac{\mathfrak{N}}{\pi} \int_{-\infty}^{\infty} 1 - \frac{\sqrt{\mathfrak{N}/2}}{\mathcal{S}_x(\omega)}\, d\omega \tag{4-87b}$$

$$= \frac{\mathfrak{N}}{4\pi} \int_{-\infty}^{\infty} \log\left[\frac{2}{\mathfrak{N}}\, \mathcal{S}_x(\omega)\right] d\omega \tag{4-87c}$$

4-13. INFINITE LAG FILTER: IRREDUCIBLE ERROR

In the design of optimum systems, when no constraint is placed on the system (except for linearity), the resulting system was found to be physically unrealizable. It is therefore obvious that the theoretical performance of the unconstrained (physically unrealizable) system will be better than that of the constrained (physically realizable) system. In other words, the mean-square error of the physically unrealizable system will be smaller than that of the physically realizable system. It was shown in Sec. 4-11 (Fig. 4-11) that a physically unrealizable system can in practice be approximated very closely if we are willing to accept a certain delay in arriving at the outcome. Thus instead of directly designing a physically realizable system, if we accept a certain time delay in obtaining the outcome, we can approach superior performance (lower mean-square error) of the physically unrealizable system. The procedure used here is first to design a physically unrealizable system and then to realize an approximation to it by using a time delay. We can achieve the same result by approaching the problem from the other direction. We first design a physically realizable system which has a built-in delay in the desired outcome. In the limit if the delay is made infinite, we expect the performance to approach that of the unrealizable filter with infinite delay. We shall now show that this indeed is the case.

Let x(t) be the input process and y(t) the desired output process. We shall now accept a delay of t_0 in the desired output. The problem then reduces to designing a physically realizable system with input $x(t)$ and output $y(t - t_0)$ instead of $y(t)$. Let

$$y(t - t_0) = z(t)$$

This yields

$$\begin{aligned} R_{xz}(\tau) &= \overline{x(t)z(t + \tau)} \\ &= \overline{x(t)y(t - t_0 + \tau)} \\ &= R_{xy}(\tau - t_0) \end{aligned} \tag{4-88}$$

Similarly,

$$\begin{aligned} R_z(\tau) &= \overline{z(t)z(t + \tau)} \\ &= \overline{y(t - t_0)y(t - t_0 + \tau)} \\ &= R_y(\tau) \end{aligned}$$

Therefore

$$\mathcal{S}_{xz}(s) = \mathcal{S}_{xy}(s)e^{-st_0} \tag{4-89}$$

and

$$\mathcal{S}_z(s) = \mathcal{S}_y(s)$$

To recapitulate, the input process is $x(t)$ and the desired process is $z(t)(= y(t - t_0))$. We wish to design a linear optimum physically realizable system to accomplish this. From Eqs. 4-80 and 4-89,

$$F(s) = \frac{\mathcal{S}_{xz}(s)}{\mathcal{S}_x(-s)} = \frac{\mathcal{S}_{xy}(s)}{\mathcal{S}_x(-s)}e^{-st_0}$$

We must now separate $F(s)$ into positive time and negative time components. Observe that the inverse transform of $\mathcal{S}_{xy}(s)/\mathcal{S}_x(-s)$ will exist for $t < 0$. However, the factor e^{-st_0} is a delay factor and if $t_0 \to \infty$, the entire inverse transform of $F(s)$ will be pushed into the positive time region. Hence

$$\lim_{t_0 \to \infty} F(s) = F^+(s)$$

and from Eq. 4-82, the desired transfer function $H(s)$ is given by

$$\begin{aligned} \lim_{t_0 \to \infty} H(s) &= \frac{F^+(s)}{\mathcal{S}_x(s)} = \frac{\mathcal{S}_{xy}(s)}{\mathcal{S}_x(-s)\,\mathcal{S}_x(s)}e^{-st_0} \\ &= \frac{\mathcal{S}_{xy}(s)}{\mathcal{S}_x(s)}e^{-st_0} \end{aligned}$$

Note that this transfer function is identical with the transfer function of the corresponding physically unrealizable system (Eq. 4-57) except for the delay factor e^{-st_0}. This is precisely the result we wanted to prove.

The corresponding mean-square error is given by Eq. 4-83b:

$$\overline{\epsilon^2} = \frac{1}{2\pi}\int_{-\infty}^{\infty} \mathcal{S}_z(\omega) - \mathcal{S}_{xz}(-\omega)H(\omega)]\,d\omega$$

$$= \frac{1}{2\pi} \int_{-\infty}^{\infty} \left[S_y(\omega) - S_{xy}(-\omega) e^{j\omega t_0} \left(\frac{S_{xy}(\omega)}{S_x(\omega)} e^{-j\omega t_0} \right) \right] d\omega$$

$$= \frac{1}{2\pi} \int_{-\infty}^{\infty} \left[S_y(\omega) - \frac{|S_{xy}(\omega)|^2}{S_x(\omega)} \right] d\omega$$

This result is identical with the mean-square error obtained for the physically unrealizable system (Eq. 4-63b).

The foregoing expression represents the lower bound on the mean-square error and is called *irreducible error.* It can be achieved by a physically unrealizable system with no lag or a physically realizable system with infinite lag. In practice one can approach it very closely with finite lag (see the discussion in Sec. 4-11).

4-14. WIENER FILTERING: ANOTHER POINT OF VIEW

So far we have considered Wiener filtering as a problem in the design of linear systems. There is, however, another very instructive point of view. We can consider Wiener filtering as a solution to the problem of statistical estimation. (For more details refer to Papoulis, Ref. 4-3.)

When two random variables x and y are related (or dependent), then a knowledge of one gives certain information about the other. Hence it is possible to estimate the value of y from a knowledge of the value of x. The estimate of y will be another random variable $\hat{\mathrm{y}}$. The estimated value $\hat{y}$ will in general be different from the actual value y. One may choose various criteria of goodness for estimation. Minimum mean-square error is one possible criterion. The optimum estimate should then be such that the mean-square error $\overline{\epsilon^2}$ is minimized. $\overline{\epsilon^2}$ is of course given by

$$\overline{\epsilon^2} = \overline{(\mathrm{y} - \hat{\mathrm{y}})^2}$$

It can be shown that under this criterion if it is given that $\mathrm{x} = x$, then the best estimate of y is $\hat{y}$ given by (see Ref. 3-3)

$$\hat{y} = E[\mathrm{y} \mid \mathrm{x} = x]$$

The best estimate of y is the conditional mean of y when $\mathrm{x} = x$. In general, the conditional mean $E[\mathrm{y} \mid \mathrm{x} = x]$ is a nonlinear function of x and is difficult to evaluate. We try to simplify the problem by constraining the estimate $\hat{\mathrm{y}}$ to be a linear function of x of the form[7]

$$\hat{\mathrm{y}} = a\mathrm{x}$$

[7]Throughout the discussion, the variables x, y, . . . , etc., will be assumed to have zero mean. This can be done without loss of generality. If the variables have nonzero means, we can form new variables $\mathrm{x}' = \mathrm{x} - \overline{\mathrm{x}}$ and $\mathrm{y}' = \mathrm{y} - \overline{\mathrm{y}}$, etc. The new variables obviously have zero-mean values.

In this case

$$\overline{\epsilon^2} = \overline{(y - \hat{y})^2} = \overline{(y - ax)^2}$$
$$= \overline{y^2} + a^2\overline{x^2} - 2a\overline{xy}$$

To minimize $\overline{\epsilon^2}$, we have

$$\frac{\partial \overline{\epsilon^2}}{\partial a} = 2a\overline{x^2} - 2\overline{xy} = 0$$

Hence

$$a = \frac{\overline{xy}}{\overline{x^2}} = \frac{R_{xy}}{R_{xx}} \tag{4-90}$$

Note that for this value of a,

$$\epsilon = y - ax = y - \frac{R_{xy}}{R_{xx}} x$$

Hence

$$\overline{x\epsilon} = \overline{x\left(y - \frac{R_{xy}}{R_{xx}} x\right)}$$
$$= \overline{xy} - \frac{R_{xy}}{R_{xx}} \overline{x^2}$$

Since by definition $\overline{xy} = R_{xy}$ and $\overline{xx} = \overline{x^2} = R_{xx}$, we have

$$\overline{x\epsilon} = R_{xy} - R_{xy} = 0 \tag{4-91}$$

Hence the data (x) and the error (ϵ) should be orthogonal (implying uncorrelatedness in this case) if the mean-square error is to be minimum.

The mean-square error is given by

$$\overline{\epsilon^2} = \overline{(y - ax)^2}$$
$$= \overline{y^2} - 2a\overline{xy} + a^2\overline{x^2}$$
$$= R_{yy} - \frac{2R_{xy}^2}{R_{xx}} + \frac{R_{xy}^2}{R_{xx}}$$
$$= R_{yy} - \frac{R_{xy}^2}{R_{xx}} = R_{yy} - aR_{xy} \tag{4-92}$$

Estimation of a Random Variable Using *n* Random Variables. If a random variable x_0 is related to n random variables, $x_1, x_2, \ldots, x_n$, then we can estimate x_0 using a linear combination[8] of $x_1, x_2, \ldots, x_n$:

$$\hat{x}_0 = a_1 x_1 + a_2 x_2 + \cdots + a_n x_n$$
$$= \sum_{i=1}^{n} a_i x_i \tag{4-93}$$

[8]Here as before we assume that all the random variables have zero-mean values. This can be done without loss of generality.

The mean-square error is given by

$$\overline{\epsilon^2} = \overline{[x_0 - (a_1x_1 + a_2x_2 + \cdots + a_nx_n)]^2}$$

To minimize $\overline{\epsilon^2}$, we must set

$$\frac{\partial\overline{\epsilon^2}}{\partial a_1} = \frac{\partial\overline{\epsilon^2}}{\partial a_2} = \cdots = \frac{\partial\overline{\epsilon^2}}{\partial a_n} = 0$$

Consider

$$\frac{\partial\overline{\epsilon^2}}{\partial a_i} = \frac{\partial}{\partial a_i}\overline{[x_0 - (a_1x_1 + a_2x_2 + \cdots + a_nx_n)]^2} = 0$$

Interchanging the order of differentiation and averaging, we have

$$\frac{\partial\overline{\epsilon^2}}{\partial a_i} = -2\overline{[x_0 - (a_1x_1 + a_2x_2 + \cdots + a_nx_n)]\,x_i} = 0 \tag{4-94}$$

or

$$R_{0i} = a_1R_{i1} + a_2R_{i2} + \cdots + a_nR_{in} \tag{4-95}$$

where

$$R_{ij} = \overline{x_ix_j}$$

differentiating $\overline{\epsilon^2}$ with respect to $a_1, a_2, \ldots, a_n$ and equating to zero, we obtain n simultaneous equations of the form shown in Eq. 4-95. The desired constants $a_1, a_2, \ldots, a_n$ can be found from these equations by using Cramer's rule.

At this point we make an important observation about Eq. 4-94. The quantity inside the brackets on the right-hand side is the error ϵ.

Hence under optimum conditions,

$$\overline{\epsilon x_i} = 0 \qquad i = 1, 2, \ldots, n$$

or

$$\overline{\epsilon x_1} = \overline{\epsilon x_2} = \cdots = \overline{\epsilon x_n} = 0 \tag{4-96}$$

Thus the random variable ϵ is orthogonal to each variable $x_1, x_2, \ldots, x_n$. This is the principle of orthogonality in mean-square estimation. The error and the data are orthogonal (implying uncorrelatedness in this case). Thus we conclude that the constants a_i that minimize $\overline{\epsilon^2}$ are such that it renders the resulting error and the data orthogonal (uncorrelated).

Estimation of a Random Variable from Continuous Data. So far, we have considered the problem of estimating a random variables with a discrete set of random variables. We can easily extend these results to the problem of estimation of a random variable from continuous data. Consider two random processes $x(t)$ and $y(t)$ which are related in some manner. The process $y(t)$ may be some signal, and $x(t)$ may for example be $y(t) + n(t)$ where $n(t)$ is undesired noise. We wish to estimate $y(t)$ from a knowledge of $x(t)$. The process $x(t)$ is available over the interval $a < t < b$. Hence we can use all this data to estimate $y(t)$. In short, our problem reduces to estimating $y(t)$ from continuous data over the interval (a, b). The data is in the form of the random variable $x(t)$ for all values of t in the given interval. We shall use a limiting process in attacking this problem. Let us first assume that $x(t)$ is known at discrete instants uniformly spaced $\Delta\alpha$ seconds apart in the interval (a, b). The best linear estimate of the random variable y(t) is given by

$$\hat{y}(t) = \sum_{\alpha=a}^{b} h(\alpha)\,\Delta\alpha\, x(\alpha) \tag{4-97}$$

where the quantity $h(\alpha)\,\Delta\alpha$ is the weighting constant analogous to a_i in Eq. 4-93. It should be observed that the constant $h(\alpha)$ will be different for different values of t. It is therefore a function of variables α and t and should really be represented by $h(t, \alpha)$. The parameter t is the instant at which the estimate is desired, using the data available at the instant α to make this estimate. As we are dealing with stationary processes only, it is obvious that the estimation will be a function of $t - \alpha$. As long as the difference $t - \alpha$ is unchanged the actual values of α and t are immaterial. Thus the parameters α and t will appear only in the form $t - \alpha$. Hence we may replace $h(t, \alpha)$ with $h(t - \alpha)$. Equation 4-97 can therefore be expressed as

$$\hat{y}(t) = \sum_{\alpha=a}^{b} h(t - \alpha)\, x(\alpha)\,\Delta\alpha \tag{4-98}$$

The error ϵ is given by

$$\epsilon(t) = y(t) - \hat{y}(t) = y(t) - \sum_{\alpha=a}^{b} h(t - \alpha)\,\Delta\alpha\, x(\alpha) \tag{4-99}$$

According to orthogonality principle, the mean-square error is minimized if the error $\epsilon(t)$ and the entire data over the interval (a, b) are orthogonal (or uncorrelated). Thus we must choose $h(\alpha)$ such that

$$\overline{\epsilon(t)\, x(\lambda)} = 0 \qquad a < \lambda < b$$

Substitution of Eq. 4-99 in this equation yields

$$\overline{y(t)\, x(\lambda)} = \sum_{\alpha=a}^{b} h(t - \alpha)\,\Delta\alpha\, \overline{x(\alpha)\, x(\lambda)} \qquad a < \lambda < b \tag{4-100}$$

Note that $\overline{y(t)\, x(\lambda)} = R_{xy}(t - \lambda)$ and $\overline{x(\alpha)\, x(\lambda)} = R_x(\alpha - \lambda)$. In the limit as $\Delta\alpha \to 0$, Eq. 4-100 becomes

$$R_{xy}(t - \lambda) = \int_a^b R_x(\alpha - \lambda)\, h(t - \alpha)\, d\alpha \qquad a < \lambda < b \tag{4-101}$$

and Eq. 4-98 becomes

$$\hat{y}(t) = \int_a^b h(t - \alpha)\, x(\alpha)\, d\alpha \tag{4-102}$$

If the data $x(t)$ is available over the entire interval $(-\infty, \infty)$, Eqs. 4-102 and 4-101 become

$$\hat{y}(t) = \int_{-\infty}^{\infty} h(t - \alpha)x(\alpha)\,d\alpha = h(t) * x(t) \tag{4-103}$$

and $h(t)$ is the solution of

$$R_{xy}(t - \lambda) = \int_{-\infty}^{\infty} R_x(\alpha - \lambda)h(t - \alpha)\,d\alpha \qquad -\infty < \lambda < \infty$$

Changing the variable so that $t - \alpha = \beta$, we have

$$R_{xy}(t - \lambda) = \int_{-\infty}^{\infty} R_x(t - \beta - \lambda)h(\beta)\,d\beta \qquad -\infty < \lambda < \infty$$

Again changing the variable so that $t - \lambda = \tau$, we have

$$R_{xy}(\tau) = \int_{-\infty}^{\infty} R_x(\tau - \beta)h(\beta)\,d\beta \qquad -\infty < \tau < \infty$$

$$= h(\tau) * R_x(\tau) \qquad -\infty < \tau < \infty \tag{4-104}$$

This is the Wiener-Hopf equation to estimate $y(t)$ from $x(t)$ over the interval $(-\infty, \infty)$. This is the same equation (Eq. 4-56) that was obtained earlier for the physically unrealizable filter. It is evident that the best linear (minimum mean square) estimate of $y(t)$, when $x(t)$ is available over the entire interval $(-\infty, \infty)$ can be obtained by feeding $x(t)$ to a particular physically unrealizable linear system. The impulse response of this system satisfies Eq. 4-104.

If $x(t)$ is available only over the interval $(-\infty, t)$, then Eqs. 4-102 and 4-101 become

$$\hat{y}(t) = \int_{-\infty}^{t} h(t - \alpha)x(\alpha)\,d\alpha$$

and

$$R_{xy}(t - \lambda) = \int_{-\infty}^{t} R_x(\alpha - \lambda)h(t - \alpha)\,d\alpha \qquad \lambda < t$$

Changing the variable $t - \alpha = \beta$ first and $t - \lambda = \tau$ later, we have

$$R_{xy}(\tau) = \int_{0}^{\infty} R_x(\tau - \beta)h(\beta)\,d\beta \tag{4-105}$$

This is the Wiener-Hopf equation used to estimate $y(t)$ from $x(t)$ over the interval $(-\infty, t)$. This is also the same equation (Eq. 4-71) that was obtained earlier for a filter constrained to be physically realizable. We therefore conclude that the best linear mean-square estimate of $y(t)$, when $x(t)$ is available in the interval $(-\infty, t)$ can be obtained by feeding $x(t)$ to a particular physically realizable linear system. The impulse response of such a system satifies Eq. 4-105.

To recapitulate, we have considered two cases: (i) the best estimate of $y(t)$ from the data $x(t)$ available over the entire interval $(-\infty, \infty)$, and (ii) the best estimate of $y(t)$ from the data $x(t)$ available only over the interval $(-\infty, t)$. The former estimate is necessarily superior to the latter because more information is available for estimation. Hence the performance of the former (physically unrealizable filter) is inherently superior to that of the latter (physically realizable filter).

Some Insight into Realizable and Unrealizable Optimum Systems. In estimating $y(t)$ from $x(t)$, we have seen that the optimum system is physically unrealizable if the data is available over the entire interval $(-\infty, \infty)$ and is physically realizable if the data is available only over the interval $(-\infty, t)$. It is very illuminating to reflect upon this behavior. The discussion in the previous section has paved the way to an understanding of the reason for this behavior. We estimate $y(t)$ at some instant t from information about another process $x(t)$ available over the entire interval $-\infty < t < \infty$. If we apply a driving signal $x(t)$ to any physical system and observe its output at the instant t, we will be using the information about $x(t)$ only over the interval $-\infty$ to t. If we are to use the information contained in $x(t)$ over the entire interval $-\infty < t < \infty$ we have two choices:

(a) Use a physical system but wait until $t = \infty$ when all the information in $x(t)$ is fed to the system.

(b) If we want the estimate of $y(t)$ at the very instant t, we must use a nonrealizable system which can use the information of $x(t)$ in the interval t to ∞ even before it is fed in. It is a curious fact that physically unrealizable systems can respond to signals even before these signals are applied. This is indeed the reason why such systems are physically unrealizable. The impulse response $h(t)$ of these systems exist (for negative values of t) even before the impulse is applied to the input. If we feed a signal $x(t)$ to a physically unrealizable system with impulse response $h(t)$, the output of $y(t)$ of this system is given by

$$y(t) = \int_{-\infty}^{\infty} h(\alpha)x(t - \alpha)\,d\alpha$$

Observe that the limits of α are $-\infty$ to ∞, and in this interval $x(t - \alpha)$ ranges from $x(\infty)$ to $x(-\infty)$. Thus in order to obtain the response at some instant t, we are utilizing the signal $x(t)$ in the entire interval $-\infty$ to ∞. We therefore conclude that if we insist upon obtaining the estimate of $y(t)$ at the instant t from the information contained in $x(t)$ in the interval $-\infty$ to ∞, applied to the input of a linear system, the system must necessarily be physically unrealizable. The second alternative is to use a physically realizable system, and to wait for the estimate until the complete information of x(t) in the interval ($-\infty$, ∞) is fed to the system. This amounts to an infinite lag system discussed in Sec. 4-13. We have already seen that the performance of such a system is indeed equal to that of the optimum physically unrealizable system.

4-15. THE REVERSIBILITY THEOREM

Suppose that we are given a process x(t) and it is required to design an optimum linear system which will yield y(t) when the input is x(t) with the least mean-square error possible. Let the optimum system transfer function be $H(s)$. This is shown in Fig. 4-14a. Next we consider the process of optimization in two steps. First x(t) is converted to z(t) by a system with transfer function $G(\omega)$. Next we design the optimum system which will give the minimum mean-square-error approximation of y(t), with z(t) as its input. The transfer function of this system is $H'(\omega)$ (see Fig. 4-14b). We now ask ourselves whether the two systems will yield identical results. The answer is provided by the theorem on

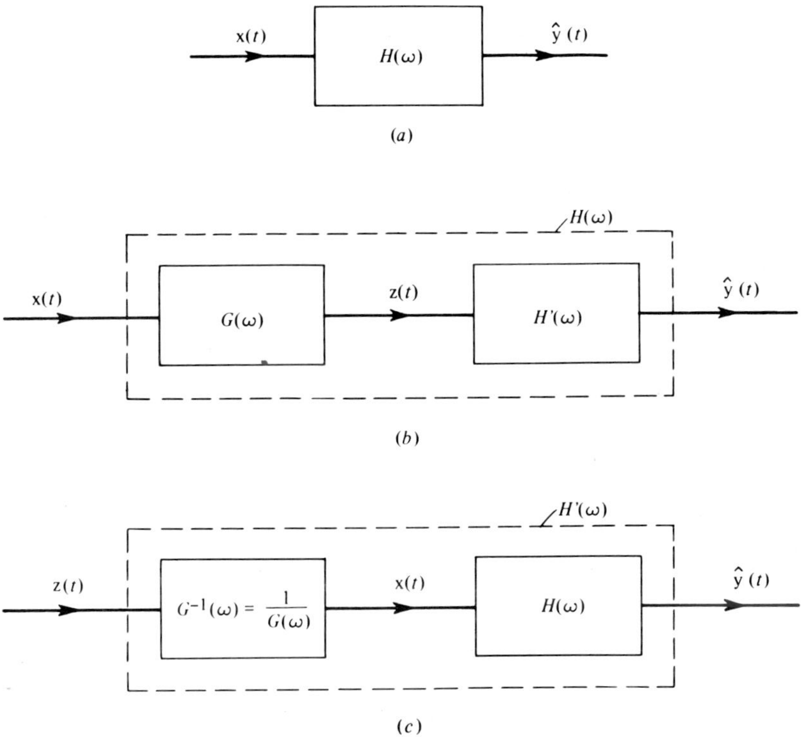

FIG. 4-14.

reversibility, which states that if the intermediate operation is reversible, that is, if one can reconstruct $x(t)$ from $z(t)$, then both systems will yield identical results. In this problem reversibility will mean that the transfer function $1/G(\omega)$ is physically realizable. If $z(t)$ is transmitted through a system with transfer function $1/G(\omega)$ we shall obtain $x(t)$ at the output.

The proof of the theorem on reversibility is very simple. We observe that the two-step system (in Fig. 4-14b) can never be better than the one-step system (in Fig. 4-14a) since the one-step system (in Fig. 4-14a) is by assumption the optimum system. Next we show that the two-step system (Fig. 4-14b) must perform at least as good as the one-step system (Fig. 4-14a). However, let us assume to the contrary, that the one-step system gives better results than the two-step system. We can always construct $H'(\omega) = [1/G(\omega)]\,H(\omega)$ as shown in Fig. 4-14c. The first system $1/G(\omega)$ yields $x(t)$. When $x(t)$ is transmitted through $H(\omega)$, we get the same result as in one-step system. Hence it is always possible to design $H'(\omega)$ which will give at least as good result as the one-step system. Hence the assumption is wrong and obviously when $H'(\omega)$ is optimum, the performance of the two systems must be identical.

The theorem on reversibility is very general. Although here we assumed the least mean-square optimization criterion, and considered only linear systems, it applies to any general form of optimization, and regardless of whether the system is linear or nonlinear. The only condition is that the intermediate operation should be reversible. Following the lines used above, the reader can easily generalize the theorem on reversibility to any optimization criterion and any type of system.

The theorem of reversibility proves very useful when one is dealing with *colored noise* (noise with nonuniform power-density spectrum) mixed with a signal. In such a case one may design an optimum filter

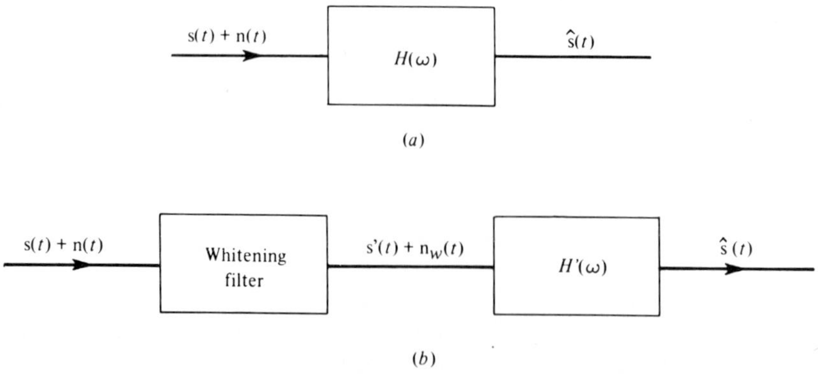

FIG. 4-15.

by a straight-forward method to obtain the minimum mean square error estimate $\hat{s}(t)$ as discussed in earlier sections. Alternatively, one may transmit the received signal $[s(t) + n(t)]$ through a whitening filter (filter which converts the colored noise into white noise). The output of the whitening filter yields a modified signal $s'(t)$ mixed with white noise. One can now design an optimum filter to obtain the minimum mean square estimate $\hat{s}(t)$ from the modified signal $s'(t)$ mixed with white noise. In this case $x(t) = s'(t) + n_w(t)$ and $y(t) = s(t)$. The procedure is shown in Fig. 4-15. We shall demonstrate the use of reversibility theorem by an example.[9]

Example 4-11. Design an optimum filter to filter $s(t)$ from $s(t) + n(t)$ if $s(t)$ is the process shown in Fig. 3-22 with $\alpha = 1$ and mean-square value 2:

$$\mathcal{S}_s(\omega) = \frac{4}{\omega^2 + 1}$$

The noise power-density spectrum is given by $\mathcal{S}_n(\omega) = \omega^2 + 4/\omega^2 + 1$.

This problem can be solved directly by techniques discussed earlier. Here, however, we shall demonstrate the two step approach using an intermediate whitening filter.

$$\mathcal{S}_n(\omega) = \frac{\omega^2 + 4}{\omega^2 + 1} = \frac{(j\omega + 2)(j\omega - 2)}{(j\omega + 1)(j\omega - 1)}$$

We shall first whiten the incoming signal $x(t)$. We have

$$\begin{aligned}\mathcal{S}_x(\omega) &= \mathcal{S}_s(\omega) + \mathcal{S}_n(\omega)\\ &= \frac{4}{\omega^2 + 1} + \frac{\omega^2 + 4}{\omega^2 + 1} = \frac{\omega^2 + 8}{\omega^2 + 1} = \left(\frac{j\omega + \sqrt{8}}{j\omega + 1}\right)\left(\frac{-j\omega + \sqrt{8}}{-j\omega + 1}\right)\end{aligned}$$

To whiten $x(t)$ the transfer function $H_w(\omega)$ of the whitening filter must satisfy the condition

$$|H_w(\omega)|^2 = \frac{1}{\mathcal{S}_x(\omega)}$$

[9]The theorem on reversibility provides an alternative method of solving the Wiener-Hopf equation when the system is constrained to be physically realizable. If the input process $x(t)$ has a uniform power density spectrum $\mathfrak{N}/2$, then $R_x(\tau) = (\mathfrak{N}/2)\,\delta(\tau)$ and Eq. 4-71 becomes

$$\begin{aligned}R_{xy}(\tau) &= \int_0^\infty \frac{\mathfrak{N}}{2}\,\delta(\tau)\,h(\alpha)\,d\alpha \qquad \tau \ge 0\\ &= \frac{\mathfrak{N}}{2}\,h(\tau) \qquad \tau \ge 0\end{aligned}$$

Hence the solution in this case is immediate. The impulse response of the optimum system is $R_{xy}(\tau)$ for $\tau > 0$ and is zero for $\tau < 0$. If the input process does not have a uniform power density, we use a whitening filter to convert it into white noise and obtain the optimum system's impulse response as indicated above. From the theorem on reversibility, it is obvious that the desired optimum system is the whitening filter in cascade with the second system obtained above. This, of course, yields the same solution as obtained in Eq. 4-82. See Prob. 4-28 at end of chapter.

If $\mathcal{S}_x(\omega)$ is expressed as

$$\mathcal{S}_x(\omega) = S_x(\omega)\, S_x(-\omega)$$

then

$$|\, S_x(\omega)\,|^2 = |\, S_x(-\omega)\,|^2 = \mathcal{S}_x(\omega)$$

We may therefore choose

$$H_w(\omega) = \frac{1}{S_x(\omega)} = \frac{j\omega + 1}{j\omega + \sqrt{8}}$$

Let the output of the whitening filter (Fig. 4-16) be z(t). It follows that

$$\mathcal{S}_z(\omega) = 1 \quad \text{and} \quad R_z(\tau) = \delta(\tau)$$

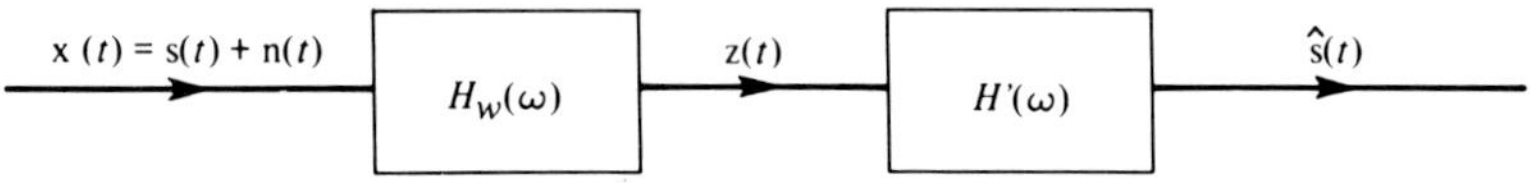

FIG. 4-16.

Our problem is now to obtain s(t) from z(t). If $h'(t)$ is the impulse response of this filter (Fig. 4-16), then from Eq. 4-71, we have

$$\begin{aligned} R_{zs}(\tau) &= \int_0^{\infty} R_z(\tau - \alpha)\, h'(\alpha)\, d\alpha \qquad && \tau > 0 \\ &= \int_0^{\infty} \delta(\tau - \alpha)\, h'(\alpha)\, d\alpha && \tau > 0 \\ &= h'(\tau) && \tau > 0 \end{aligned}$$

Thus

$$h'(t) = R_{zs}(t)\, u(t)$$

and

$$H'(s) = \mathcal{S}^{+}_{zs}(s)$$

To obtain $\mathcal{S}^{+}_{zs}(s)$, we observe that

$$\begin{aligned} R_{zs}(\tau) &= \overline{z(t)\, s(t + \tau)} \\ &= \overline{s(t + \tau) \int_{-\infty}^{\infty} h_w(\alpha)\, x(t - \alpha)\, d\alpha} \\ &= \int_{-\infty}^{\infty} h_w(\alpha)\, \overline{x(t - \alpha)\, s(t + \tau)}\, d\alpha \\ &= \int_{-\infty}^{\infty} h_w(\alpha)\, R_{xs}(\tau + \alpha)\, d\alpha \\ &= \int_{-\infty}^{\infty} h_w(-\beta)\, R_{xs}(\tau - \beta)\, d\beta \\ &= h_w(-\tau) * R_{xs}(\tau) \end{aligned}$$

Therefore

$$\mathcal{S}_{zs}(\omega) = H_w(-\omega)\, \mathcal{S}_{xs}(\omega)$$

But

$$\mathcal{S}_{xs}(\omega) = \mathcal{S}_s(\omega) = \frac{4}{\omega^2 + 1}$$

and

$$\begin{aligned}\mathcal{S}_{zs}(\omega) &= \left(\frac{-j\omega + 1}{-j\omega + \sqrt{8}}\right)\left(\frac{4}{\omega^2 + 1}\right) \\ &= \frac{4}{(-j\omega + \sqrt{8})(j\omega + 1)} \\ &= \frac{4/1 + \sqrt{8}}{j\omega + 1} + \frac{4/1 + \sqrt{8}}{-j\omega + 8}\end{aligned}$$

and

$$\mathcal{S}_{zs}^{+}(s) = \frac{4}{1 + \sqrt{8}}\left(\frac{1}{s + 1}\right) = H'(s)$$

The transfer function $H(s)$ of the overall (optimum) system is given by (Fig. 4-16)

$$H(s) = H_w(s)\,H'(s) = \frac{s + 1}{s + \sqrt{8}}\,\frac{4/(1 + \sqrt{8})}{s + 1} = \frac{4}{1 + \sqrt{8}}\left(\frac{1}{s + \sqrt{8}}\right)$$

The reader can verify the theorem on reversibility by working this problem by one step operation.

Some Comments on the Process of Optimization. It should be understood that the optimization discussed in this chapter has been made with certain qualifications and under certain limitations. First of all, the estimation (approximation) obtained by various optimum systems discussed is not always the best possible estimate, but it is the best under the restriction of linear estimation. It is possible to obtain a better estimate (approximation) using nonlinear operations on the given data. Such methods, however, are extremely complex and the solution of the general problem is very difficult. The second qualification made about the estimation is that it is to be optimized with respect to the mean-square error. The errors are weighted as their squares. Thus larger errors weigh much more heavily than smaller errors. Therefore the mean-square optimization concentrates its attention primarily on large errors and is relatively insensitive to small errors. In many cases this criterion is adequate, but in a number of others it may be of little value.

In some problems it is important to make as many very accurate predictions (or estimates) as possible even if this means that we make an occasional gross error as a result. In the case of a gunnery fire-control problem, on the other hand, all errors within a certain limit are equally acceptable and all errors outside this limit are equally serious, that is, whether we miss the target by an inch or by a mile, it is still a miss.

In the optimization techniques discussed above the optimization is with respect to all sample functions in the ensemble and not for any one particular sample function. In other words, the mean-square error is minimized over the whole ensemble. The optimum system thus obtained may not be optimum for any one specific sample function of the process but is optimum for the whole process. This is quite reasonable, since there is no way of ascertaining which sample function of the process we are dealing with.

4-16. THE MATCHED FILTER

So far we have studied the problem of filtering random signals. In these problems the merit criterion was the fidelity of reproduction of the waveform as reflected by the mean-square-error performance. In some problems, however, the waveform is of no consequence. This is true in digital data communication where there are finite number of messages to be transmitted. Consider, for example, transmission of an English text by using some code such as a Morse code. Here the messages are transmitted only by two symbols; mark and space. A pulse $s(t)$ of durating T seconds may be transmitted for a mark and no pulse (no signal) transmitted over the duration of T seconds for space. The transmitted signal is masked by noise over the channel. The receiver must detect the presence or absence of pulse over each interval of T seconds. The detection of waveform of $s(t)$ itself is of no consequence. The optimum receiver designed on the basis of mean-square error in such cases will, therefore, be inappropriate. The important problem here is to detect whether the pulse is present or absent. In such cases, it will appear that an optimum receiver should be such that it peaks out the signal component at some instant and suppresses the noise amplitude at the same time. This will give a sharp contrast between the signal and noise. If the pulse is present, the output will appear to have a large peak at some instant, and if the pulse is absent, no such peak will appear. This arrangement reduces the probability of error in the decision as to whether the pulse is present or absent. A filter which accomplishes this is known as the *matched filter*. The purpose of this filter is to increase the signal component and decrease the noise component at the same time. This is obviously equivalent to maximizing the ratio of signal amplitude to the noise amplitude at some instant at the output. Since the noise is a random signal it is more appropriate to talk in terms of mean-square value of the noise amplitude. So in a matched filter (Fig. 4-17) we wish to maximize the ratio

$$\rho = \frac{s_0^2(t)}{\overline{n_0^2(t)}} \tag{4-106}$$

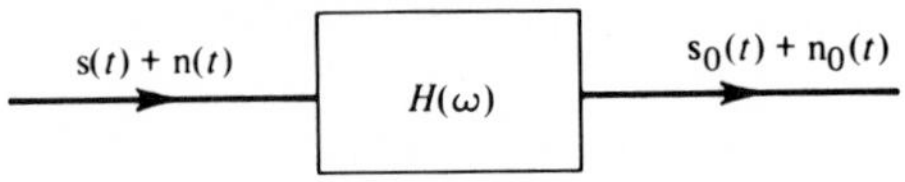

FIG. 4-17.

where $s_0(t)$ is the output of the filter due to desired signal $s(t)$ at the input, and $n_0(t)$ is the output of the filter due to noise input to the filter.

Let us assume that the signal-to-noise ratio in Eq. 4-106 is maximized at some instant t_m:

$$\rho = \frac{s_0^2(t_m)}{\overline{n_0^2(t_m)}} \tag{4-107}$$

Here it must be pointed out that the signal $s(t)$ is not a random signal but a deterministic signal whose waveform is known. Moreover, this signal is of finite duration and has a Fourier transform. Let $S(\omega)$ be the Fourier transform of $s(t)$ and $H(\omega)$ be the transfer function of the desired optimum filter. Then

$$s_0(t) = \mathfrak{F}^{-1}[S(\omega)H(\omega)]$$
$$= \frac{1}{2\pi}\int_{-\infty}^{\infty} H(\omega)S(\omega)e^{j\omega t}\,d\omega$$

and

$$s_0(t_m) = \frac{1}{2\pi}\int_{-\infty}^{\infty} H(\omega)S(\omega)e^{j\omega t_m}\,d\omega \tag{4-108}$$

The mean-square value of the noise signal can be expressed in terms of the noise power-density spectrum at the output.

If $\mathcal{S}_n(\omega)$ is the power-density spectrum of the input noise signal $n(t)$, then $|H(\omega)|^2\mathcal{S}_n(\omega)$ is the power-density spectrum of $n_0(t)$. Hence

$$\overline{n_0^2(t)} = \frac{1}{2\pi}\int_{-\infty}^{\infty} \mathcal{S}_n(\omega)|H(\omega)|^2\,d\omega$$

Note that $n_0(t)$ being a stationary process its mean-square value is independent of t and

$$\overline{n_0^2(t_m)} = \frac{1}{2\pi}\int_{-\infty}^{\infty} \mathcal{S}_n(\omega)|H(\omega)|^2\,d\omega$$

Let us assume that the input noise $n(t)$ is white. The results can later be extended to colored noise by invoking the theorem on reversibility.

Let

$$\mathcal{S}_n(\omega) = \frac{\mathfrak{N}}{2}$$

and

$$\overline{n_0^2(t_m)} = \frac{\mathfrak{N}}{4\pi} \int_{-\infty}^{\infty} |H(\omega)|^2 \, d\omega \tag{4-109}$$

Substituting Eqs. 4-108 and 4-109 in Eq. 4-107, we get

$$\rho = \frac{s_0^2(t_m)}{\overline{n_0^2(t_m)}} = \frac{\left| \int_{-\infty}^{\infty} H(\omega)\, S(\omega)\, e^{j\omega t_m}\, d\omega \right|^2}{\pi \mathfrak{N} \int_{-\infty}^{\infty} |H(\omega)|^2 \, d\omega} \tag{4-110}$$

Note that $s_0(t)$ being a real number, $s_0^2(t) = |s_0(t)|^2$. At this point we use the Schwarz inequality. One form of this inequality states that if $F_1(\omega)$ and $F_2(\omega)$ are complex functions, then

$$\left| \int_{-\infty}^{\infty} F_1(\omega)\, F_2(\omega)\, d\omega \right|^2 \le \int_{-\infty}^{\infty} |F_1(\omega)|^2 \, d\omega \int_{-\infty}^{\infty} |F_2(\omega)|^2 \, d\omega$$

The equality holds only if (see Ref. 4-6)

$$F_1(\omega) = kF_2^*(\omega) \tag{4-111}$$

where k is an arbitrary constant. If we let

$$F_1(\omega) = H(\omega)$$
$$F_2(\omega) = S(\omega)\, e^{j\omega t_m}$$

then

$$\left| \int_{-\infty}^{\infty} H(\omega)\, S(\omega)\, e^{j\omega t}\, d\omega \right|^2 \le \int_{-\infty}^{\infty} |H(\omega)|^2 \, d\omega \int_{-\infty}^{\infty} |S(\omega)|^2 \, d\omega \tag{4-112}$$

Substitution of inequality 4-112 in Eq. 4-110 yields

$$\rho = \frac{s_0^2(t_m)}{\overline{n_0^2(t_m)}} \le \frac{1}{\pi \mathfrak{N}} \int_{-\infty}^{\infty} |S(\omega)|^2 \, d\omega$$

Hence

$$\rho_{\max} = \left. \frac{s_0^2(t_m)}{\overline{n_0^2(t_m)}} \right|_{\max} = \frac{1}{\pi \mathfrak{N}} \int_{-\infty}^{\infty} |S(\omega)|^2 \, d\omega \tag{4-113}$$

and occurs when the inequality in Eq. 4-112 becomes an equality. This is possible only if (see Eq. 4-111)

$$\begin{aligned} H(\omega) &= kS^*(\omega)\, e^{-j\omega t_m} \\ &= kS(-\omega)\, e^{-j\omega t_m} \end{aligned} \tag{4-114}$$

where k is arbitrary constant. The impulse response $h(t)$ of the optimum system is then given by

$$\begin{aligned} h(t) &= \mathfrak{F}^{-1}[H(\omega)] \\ &= \mathfrak{F}^{-1}[kS(-\omega)e^{-j\omega t_m}] \end{aligned}$$

Note that the inverse Fourier transform of $S(-\omega)$ is $s(-t)$ and the term $e^{-j\omega t_m}$ represents a time shift of t_m seconds. Hence

$$h(t) = ks(t_m - t) \tag{4-115}$$

For the sake of convenience we shall assume $k = 1$.

As mentioned earlier, the message signal $s(t)$ is of finite duration. Let $s(t)$ be zero outside the interval $(0, T)$ as shown in Fig. 4-18a. The signal $s(t_m - t)$ can be obtained by flipping $s(t)$ over and about the vertical axis and shifting to the right by t_m seconds. Three cases where $t_m < T$, $t_m = T$, and $t_m > T$ are shown in Fig. 4-18. Observe that the impulse response $h(t)$ is noncausal for $t_m < T$ (Fig. 4-18c). This represents a physically unrealizable system. For physical realizability $t_m \geq T$ as shown in Figs. 4-18d and 4-18e. Both these systems will yield the desired result. However, it is desirable to make the observation time t_m as small as possible in order to arrive at the decision as quickly as possible. For larger values of t_m, one must wait a correspondingly longer time to make the decision. Hence $t_m = T$ is preferable to $t_m > T$. We therefore conclude that the impulse response of the optimum system is the mirror image of the desired message signal $s(t)$ about the vertical axis but shifted to the right by T seconds. Such a receiver is called the *matched filter* or the matched receiver. In this filter the signal-to-noise ratio (Eq. 4-106) becomes maximum at the instant T which is also the instant at which all of the signal $s(t)$ has entered the receiver (Fig. 4-19). In Fig. 4-19b, the input and the output of the filter are shown without noise. Fig. 4-19c shows the same signals with noise. It can be easily seen from this figure, that the desired signal's amplitude peaks at the instant (T) of decision. This allows one to make decision whether $s(t)$ is present or absent with less error.

It should be realized that the matched filter is *the* optimum linear filter. In general a better signal-to-noise ratio can be obtained if the constraint of linearity on the filter is removed. This will of course yield a nonlinear filter. However, if the white noise $n(t)$ is gaussian (a reasonable model in most cases in practice), then the matched filter is also the absolute optimum; that is, no other filter, linear or otherwise can improve the efficiency of signal detection attained by a matched filter. This fact will be proved later in Chapter 6.

To recapitulate, we have obtained a linear filter which maximizes the signal-to-noise ratio at the instant $t = T$. From an observation of $r(T)$, the filter output at $t = T$, the decision as to whether the signal

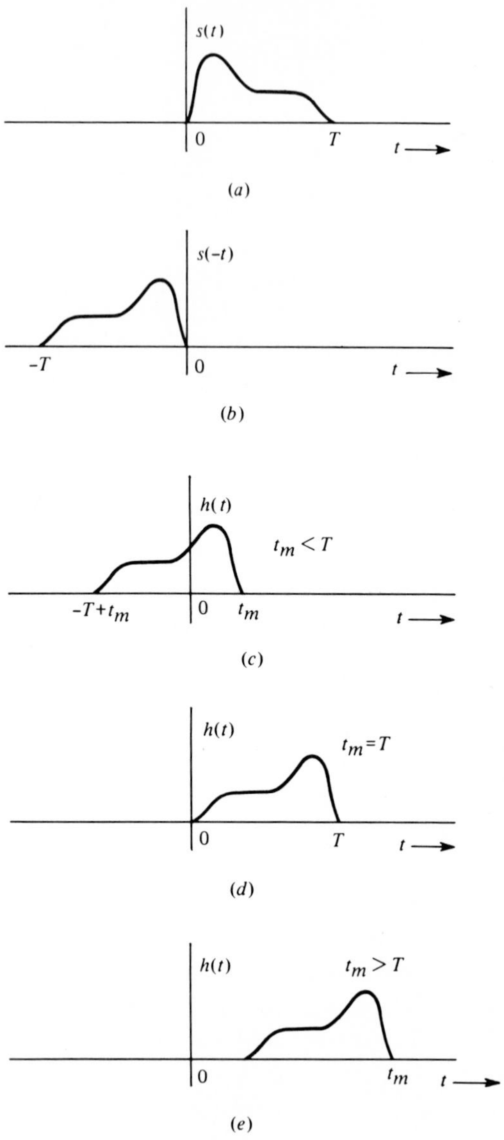

FIG. 4-18.

$s(t)$ is present or absent can be made with more confidence (less error probability), than would be possible by direct examination of the incoming signal waveform.

The maximum value of the signal-to-noise ratio attained by the matched filter is given by Eq. 4-113. Note that E, the energy of the signal $s(t)$ is given by

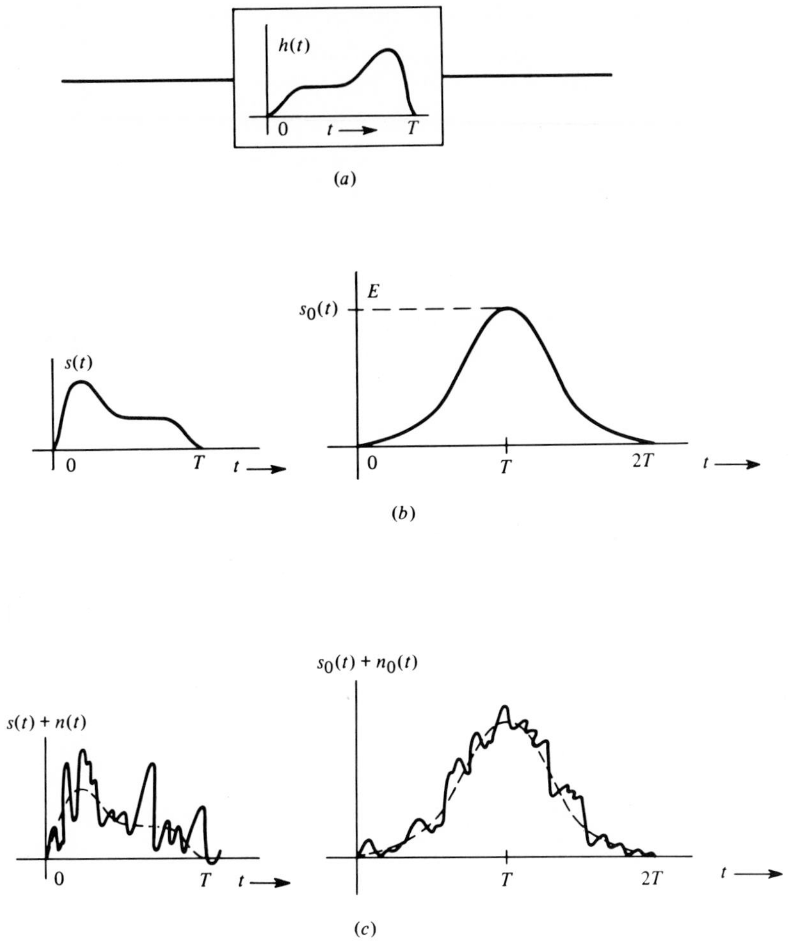

FIG. 4-19.

$$E = \int_{-\infty}^{\infty} s^2(t)\, dt = \frac{1}{2\pi} \int_{-\infty}^{\infty} |S(\omega)|^2 d\omega$$

Hence

$$\rho_{\max} = \frac{s_0^2(T)}{\overline{n_0^2(T)}} = \frac{E}{\mathfrak{N}/2} = \frac{2E}{\mathfrak{N}} \tag{4-116}$$

$$\rho_{\max} = \frac{\text{Energy of signal } s(t)}{\text{Power-density spectrum of input noise signal}}$$

The signal amplitude $s_0(T)$ is obtained by substituting Eq. 4-114 in Eq. 4-108:

$$s_0(T) = \frac{1}{2\pi} \int_{-\infty}^{\infty} |S(\omega)|^2 d\omega = E \tag{4-117}$$

Hence the maximum amplitude of the signal component at the output that occurs at $t = T$ and has the magnitude E, the energy of the signal $s(t)$. This is a remarkable result. The maximum amplitude is independent of the waveform $s(t)$ and depends only upon its energy.

The mean-square value of the noise signal at the output can be obtained by substituting Eq. 4-117 in Eq. 4-116:

$$\overline{n_0^2(T)} = \frac{\mathfrak{N}E}{2} \tag{4-118}$$

Matched Filter as a Correlator. The matched filter may also be viewed as a form of time correlator. If the input to the matched filter is $f(t)$, then $r(t)$, its output is given by

$$r(t) = f(t) * h(t) = \int_{-\infty}^{\infty} f(\alpha)h(t - \alpha)\,d\alpha$$

The impulse response $h(t)$ is $s(T - t)$. Hence

$$r(t) = \int_{-\infty}^{\infty} f(\alpha)s(T - t + \alpha)\,d\alpha$$

The limits of integration for this integral may be replaced by 0 and t. This is because the impulse response $h(t)$ and $f(t)$ are both causal so the signal $f(\alpha)$ may be assumed to start at $\alpha = 0$. Hence

$$r(t) = \int_0^t f(\alpha)s(T - t + \alpha)\,d\alpha$$

The response at $t = T$ is given by

$$r(T) = \int_0^T f(\alpha)s(\alpha)\,d\alpha \tag{4-119}$$

We can obtain $r(T)$ with the time correlator shown in Fig. 4-20. This arrangement is therefore equivalent to the matched filter. The output of the time correlator at $t = T$ enables one to make a decision as to whether $s(t)$ is present or absent.

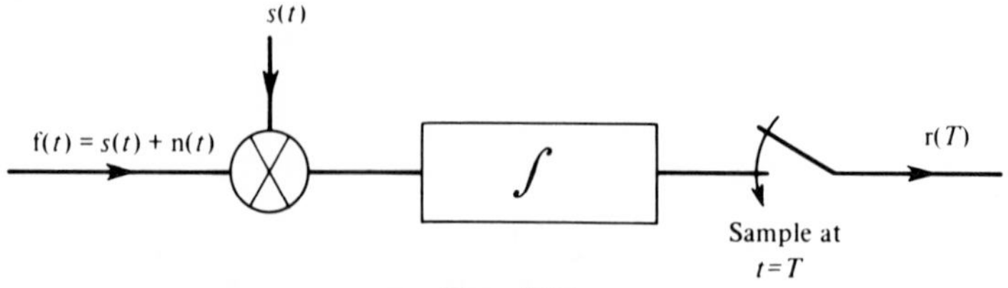

FIG. 4-20.

4-17. MATCHED FILTER FOR COLORED NOISE

In our discussion so far we have assumed the noise to be white. Following exactly similar lines, one can obtain the matched filter for nonwhite noise (colored noise). We shall, however, invoke the theorem on reversibility to obtain the matched filter for colored noise.

The incoming signal (message plus noise) is first transmitted through a whitening filter. As a result of this, the message $s(t)$ is modified to $s'(t)$ at the output of the whitening filter, and the noise is rendered white. We can now design a second subsystem which is a filter matched to $s'(t)$, as shown in Fig. 4-21a.

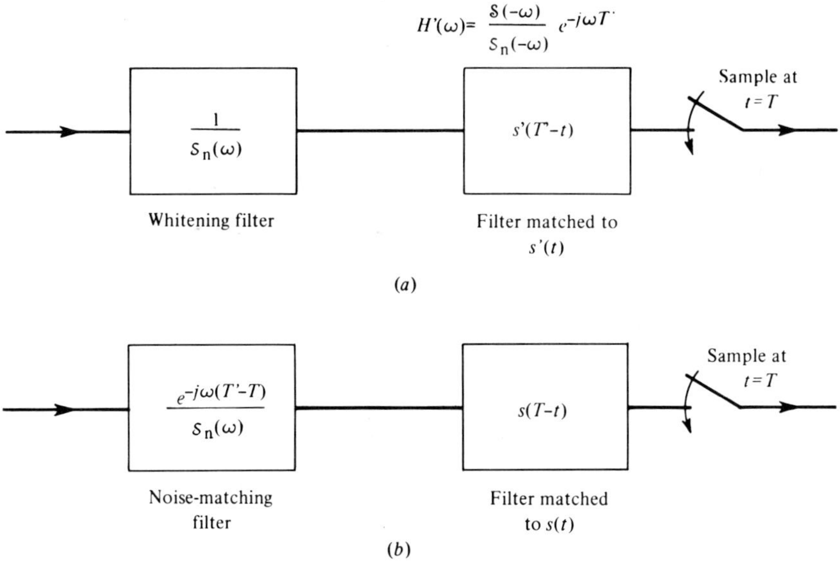

FIG. 4-21.

If $\mathcal{S}_n(\omega)$ is the power-density spectrum of the noise, then we can express $\mathcal{S}_n(\omega)$ as

$$\mathcal{S}_n(\omega) = S_n(\omega)\, S_n(-\omega)$$

where $S_n(\omega)$ has all its poles and zeros in the left half of the s plane and $S_n(-\omega)$ has all its poles and zeros in the right half of the s plane. It is evident that

$$|\, S_n(\omega)\,|^2 = \mathcal{S}_n(\omega)$$

and the whitening filter's transfer function is given by[10]

$$H_w(\omega) = \frac{1}{S_n(\omega)} \tag{4-120}$$

[10]If $\mathcal{S}_n(\omega) \to 0$ as $\omega \to \infty$, the whitening filter transfer function $1/\mathcal{S}_n(\omega) \to \infty$ as $\omega \to \infty$. In practice all circuits ultimately become capacitive at high frequencies, and true whitening at very high frequencies cannot be achieved. This difficulty, however, is not serious since we need to whiten the noise only over the finite band which contains most of the energy of $s(t)$.

The message signal at the output of the whitening filter is $s'(t)$ and its Fourier transform is given by

$$S'(\omega) = \frac{S(\omega)}{\mathcal{S}_n(\omega)}$$

It is evident that the filter matched to $s'(t)$ will have the transfer function (see Eq. 4-114):

$$\begin{aligned} H'(\omega) &= S'(-\omega)\,e^{-j\omega T'} \\ &= \frac{S(-\omega)}{\mathcal{S}_n(-\omega)}\,e^{-j\omega T'} \end{aligned} \tag{4-121}$$

where T' is the duration of $s'(t)$. The overall transfer function $H_c(\omega)$ of the filter for the colored noise is obtained from Eqs. 4-120 and 4-121:

$$\begin{aligned} H_c(\omega) &= H_w(\omega)\,H'(\omega) \\ &= \frac{S(-\omega)}{\mathcal{S}_n(\omega)\;\mathcal{S}_n(-\omega)}\,e^{-j\omega T'} \\ &= \frac{S(-\omega)}{\mathcal{S}_n(\omega)}\,e^{-j\omega T'} \end{aligned} \tag{4-122}$$

We shall now discuss the question relating to the physical realizability of these transfer functions. In general, the whitening filter is physically realizable, at least over the finite band of interest. The matched filter for $s'(t)$ has an impulse response $s'(T' - t)$ and is obviously physically realizable if T' the duration of $s'(t)$ is finite. Under this condition the combined transfer function $H_c(\omega)$ in Eq. 4-122 is physically realizable. This arrangement is shown in Fig. 4-21a. The signal $s'(t)$ is the response of the whitening filter to $s(t)$. Although $s(t)$ is of finite duration (T seconds), $s'(t)$ in general may have infinite duration; that is, $T' = \infty$. But in practice $s'(t)$ decays with time and one may ignore its tail beyond some finite time $t = T'$. Alternatively $H_c(\omega)$ in Eq. 4-122 may be realized by $1/\mathcal{S}_n(\omega)$ in cascade with $S(-\omega)e^{-j\omega T'}$. But $1/\mathcal{S}_n(\omega)$ is not physically realizable because it has poles in the RHP. However, by adding some delay, we can closely approximate it with a physical system. Consider, for example,

$$\mathcal{S}_n(\omega) = \frac{\omega^2 + 1}{\omega^2 + 3}$$

and

$$H(\omega) = \frac{1}{\mathcal{S}_n(\omega)} = 1 + \frac{2}{\omega^2 + 1}$$

The impulse response of this system is given by

$$h(t) = \mathcal{F}^{-1}\left[1 + \frac{2}{\omega^2 + 1}\right]$$

$$= \delta(t) + e^{-|t|}$$

This impulse response (Fig. 4-22a) is obviously physically unrealizable. But if we consider

$$H_d(\omega) = H(\omega)e^{-j\omega t_0}$$
$$= 1 + \frac{2}{\omega^2 + 1} e^{-j\omega t_0}$$

then

$$h_d(t) = \delta(t - t_0) + e^{-|t-t_0|}$$

This system can be realized to a very close approximation by choosing t_0 large enough, as seen from Fig. 4-22b. In this case t_0 may be chosen to be about 4. Thus a physically unrealizable filter can always be realized to a very close approximation by introducing a delay factor.[11]

Returning to our original problem, we can make $1/\mathcal{S}_n(\omega)$ physically realizable by adding a delay factor. We may split $H_c(\omega)$ in Eq. 4-122 as follows:

$$H_c(\omega) = \left[\frac{1}{\mathcal{S}_n(\omega)} e^{-j\omega(T'-T)}\right]\left[S(-\omega)e^{-j\omega T}\right]$$

where T is the duration of $s(t)$. We recognize that the second term $S(-\omega)e^{-j\omega T}$ is a filter matched to $s(t)$ and is physically realizable. Since $H_c(\omega)$ as a whole is physically realizable, the first term must also

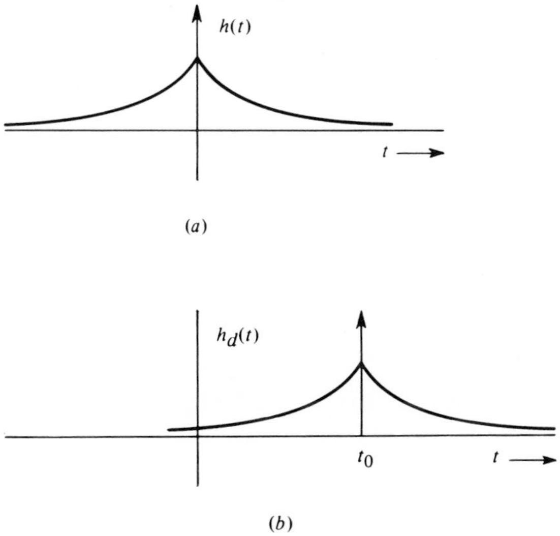

FIG. 4-22.

[11]It is implicitly assumed here that $h(t) \to 0$ as $t \to -\infty$. In almost all practical cases, this is true.

be realizable; that is, the delay of $T' - T$ seconds should render $1/\mathcal{S}_n(\omega)$ physically realizable. This system is shown in Fig. 4-21b. The first filter $[1/\mathcal{S}_n(\omega)]\,e^{-j\omega(T'-T)}$ is known as the *noise-matching filter*.

4-18. DECISION THRESHOLD IN A MATCHED FILTER

Let us consider the case of a message signal mixed with white gaussian noise of zero mean value. The output of the matched filter at the instant T of observation is given by

$$\mathrm{r}(T) = s_0(T) + \mathrm{n}_0(T)$$

Substituting Eq. 4-117 in this equation, we obtain

$$\mathrm{r}(T) = E + \mathrm{n}_0(T) \tag{4-123}$$

Since the input noise is assumed to be gaussian, the output noise of the linear filter must also be gaussian. Hence $\mathrm{n}_0(T)$ is a gaussian random variable with mean-square value $\mathfrak{N}E/2$ (see Eq. 4-118). Since the input noise process has a zero mean, the output noise must also have a zero mean. Hence $\mathrm{n}_0(T)$ has a variance of $\mathfrak{N}E/2$. Therefore, the random variable $\mathrm{r}(T)$, as seen from Eq. 4-123, is a gaussian variable with a mean E and variance $\mathfrak{N}E/2$. Hence

$$p_{\mathrm{r}}(r) = \frac{1}{\sqrt{\pi\mathfrak{N}E}}\,e^{-(r-E)^2/\mathfrak{N}E} \tag{4-124}$$

This is the probability distribution of $\mathrm{r}(T)$, the output signal amplitude at T, when the message signal $s(t)$ is present at the input. This is shown in Fig. 4-23. When the message $s(t)$ is absent, the output is exclusively given by $\mathrm{n}_0(T)$ which is gaussian with a zero mean and variance $\mathfrak{N}E/2$. In this case the probability density $p_{\mathrm{r}}(r)$ of the output signal amplitude is given by

$$p_{\mathrm{r}}(r) = \frac{1}{\sqrt{\pi\mathfrak{N}E}}\,e^{-r^2/\mathfrak{N}E} \tag{4-125}$$

This distribution is also shown in Fig. 4-23. We must now choose some fixed decision threshold a (Fig. 4-23) such that when we observe the signal $\mathrm{r} > a$, the decision made is "$s(t)$ present" at the input. If $\mathrm{r} < a$ then the decision made is "no $s(t)$." How shall we choose the threshold a? We observe from Fig. 4-23 that no matter how we choose a, there will be instances when we shall observe output $\mathrm{r} > a$, even if the message $s(t)$ is absent. In this case, we shall commit a *false alarm* (FA) type of error. On the other hand, there will be instances when $\mathrm{r} < a$ even if $s(t)$ is present. In this case the error committed will be of the *false-dismissal* (FD) type. In any practical situation, we must pay some price for each

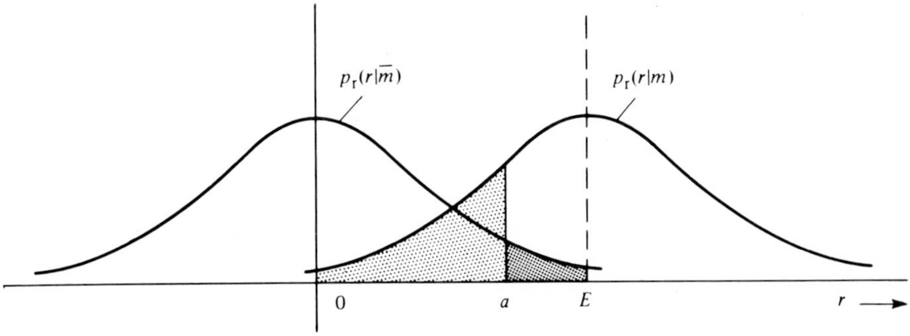

FIG. 4-23.

kind of error. We must choose the decision threshold a, so that the average cost of these errors is minimized.

For the sake of convenience, the events that "the signal $s(t)$ is present" and that "it is not present" will be denoted by m and $\overline{m}$ respectively. Let $P(m)$ and $P(\overline{m})$ be the probabilities that the signal is present or absent respectively ($P(m) + P(\overline{m}) = 1$). Let us also define the following conditional probability densities.

$P(m \mid \mathrm{r} = r)$ = probability that $s(t)$ is present when the output amplitude is equal to r.

$p_{\mathrm{r}}(r \mid m)$ = probability density of the output amplitude when it is given that the signal is present and is given by (Eq. 4-124).

$$p_{\mathrm{r}}(r \mid m) = \frac{1}{\sqrt{\pi \mathfrak{N} E}} e^{-(r-E)^2/\mathfrak{N} E} \tag{4-126}$$

We can similarly define $P(\overline{\mathrm{m}} \mid \mathrm{r} = r)$ and $p_{\mathrm{r}}(r \mid \overline{m})$. Note that

$$p_{\mathrm{r}}(r \mid \overline{m}) = \frac{1}{\sqrt{\pi \mathfrak{N} E}} e^{-r^2/\mathfrak{N} E} \tag{4-127}$$

Let C_a and C_d be the costs of the false-alarm error and the false-dismissal error. Suppose the output amplitude $\mathrm{r} = r$. If we make the decision that m (message present) is true, then we are *liable* to commit an error of the false-alarm type. A false alarm will occur if the event is actually $\overline{m}$, given that $\mathrm{r} = r$. Hence the probability of a false alarm is $P(\overline{m} \mid \mathrm{r} = r)$. The average cost of this decision is e_1 given by

$$e_1 = C_a P(\overline{m} \mid \mathrm{r} = r)$$

On the other hand, if we make a decision that $\overline{m}$ (no message) is true, then we are liable to commit the error of false dismissal with probability $P(m \mid \mathrm{r} = r)$. The average cost of this decision is e_2:

$$e_2 = C_d P(m \mid \mathrm{r} = r)$$

To minimize the cost, we must make the decision that m (signal present) is true if

$$e_2 > e_1$$

or

$$C_d P(m \mid \mathrm{r} = r) > C_a P(\overline{m} \mid \mathrm{r} = r) \tag{4-128}$$

At this point we use Bayes' mixed rule (Eq. 2-43c):

$$P(m \mid \mathrm{r} = r) = \frac{p_{\mathrm{r}}(r \mid m)\, P(m)}{p_{\mathrm{r}}(r)}$$

Similarly, (4-129)

$$p(\overline{m} \mid \mathrm{r} = r) = \frac{p_{\mathrm{r}}(r \mid \overline{m})\, P(\overline{m})}{p_{\mathrm{r}}(r)}$$

Substituting Eq. 4-129 in Eq. 4-128, we should make the decision that m (signal present) is true if

$$C_d p_{\mathrm{r}}(r \mid m)\, P(m) > C_a p_{\mathrm{r}}(r \mid \overline{m})\, P(\overline{m})$$

Substituting Eqs. 4-126 and 4-127 in the above equation, we make the decision that m is true if

$$e^{(2r-E)/\mathfrak{N}} > \frac{C_a P(\overline{m})}{C_d P(m)} \tag{4-130}$$

The decision threshold $\mathrm{r} = a$ is the point where both losses are equal and 4-130 becomes an equality

$$e^{(2a-E)/\mathfrak{N}} = \frac{C_a P(\overline{m})}{C_d P(m)} \tag{4-131}$$

Hence the decision threshold a is given by

$$a = \frac{E}{2} + \frac{\mathfrak{N}}{2} \ln \frac{C_a P(\overline{m})}{C_d P(m)}$$

$$= \frac{1}{2}\left[E + \mathfrak{N} \ln \frac{C_a}{C_d}\right] + \frac{\mathfrak{N}}{2} \ln \frac{P(\overline{m})}{P(m)} \tag{4-132}$$

Note that if the message is equally likely to be present or absent $P(m) = P(\overline{m}) = 0.5$, and if in addition we assume that costs of either types of errors to be equal $C_d = C_a$, then

$$a = \frac{E}{2} \tag{4-133}$$

Figure 4-24 shows the filter with the decision-making arrangement.

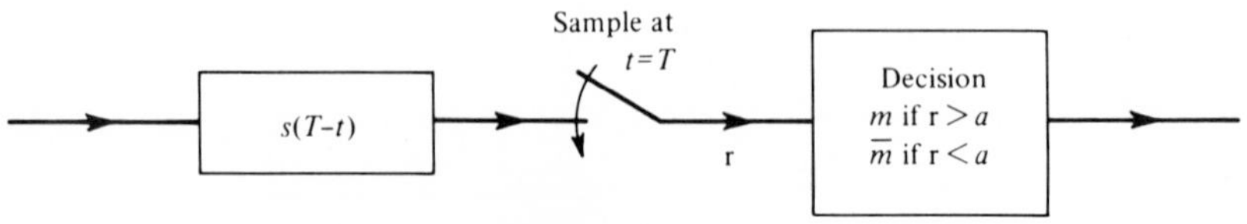

FIG. 4-24.

4-19. ERROR PROBABILITY OF A MATCHED RECEIVER

We shall now calculate the probability of error in our decision when the threshold of decision is chosen to be a. If we denote the conditional probability of making the decision that m is true when the message $s(t)$ is indeed present, by $P(C \mid m)$, then

$$\begin{aligned} P(C \mid m) &= \text{probability that r} > a \text{ when } m \text{ is true} \\ &= \int_a^\infty p_r(r \mid m)\, dr \\ &= \frac{1}{\pi \mathfrak{N} E} \int_a^\infty e^{-(r-E)^2/\mathfrak{N}E}\, dr \end{aligned} \tag{4-134}$$

Here we make use of the error function erf (α) and the complementary error function erfc (α) defined as[12]

$$\operatorname{erf}(\alpha) = \frac{1}{\sqrt{2\pi}} \int_{-\infty}^{\alpha} e^{-x^2/2}\, dx \tag{4-135a}$$

$$\operatorname{erfc}(\alpha) = \frac{1}{\sqrt{2\pi}} \int_{\alpha}^{\infty} e^{-x^2/2}\, dx \tag{4-135b}$$

It is obvious that

$$\operatorname{erf}(\alpha) + \operatorname{erfc}(\alpha) = 1 \tag{4-135c}$$

Appropriate change of variables in Eqs. 4-135a and 4-135b yields

$$\operatorname{erf}\left(\frac{\alpha}{\sigma}\right) = \frac{1}{\sigma\sqrt{2\pi}} \int_{-\infty}^{\alpha} e^{-x^2/2\sigma^2}\, dx \tag{4-135d}$$

$$\operatorname{erfc}\left(\frac{\alpha}{\sigma}\right) = \frac{1}{\sigma\sqrt{2\pi}} \int_{\alpha}^{\infty} e^{-x^2/2\sigma^2}\, dx \tag{4-135e}$$

It also follows from these definitions that

$$\operatorname{erfc}(\alpha) + \operatorname{erfc}(-\alpha) = 1 \tag{4-136}$$

A useful approximation for erfc (α) is given by (see Ref. 4-5):

$$\operatorname{erfc}(\alpha) \simeq \frac{1}{\alpha\sqrt{2\pi}} \left(1 - \frac{1}{\alpha^2}\right) e^{-\alpha^2/2} \quad \text{for } \alpha > 2 \tag{4-137}$$

The error in this approximation is about 10 percent for $\alpha = 2$ and about 1 percent for $\alpha = 3$.

We can now express Eq. 4-134 as

$$P(C \mid m) = \operatorname{erfc}\left(\frac{a - E}{\sqrt{\mathfrak{N}E/2}}\right) \tag{4-138}$$

[12]At present there exist in the literature, several definitions of erf (α) and erfc (α) which are essentially equivalent but for minor differences.

Similarly,

$$
\begin{aligned}
P(C \mid \overline{m}) &= \text{probability that r} < a \text{ when no message present} \\
&= \int_{-\infty}^{a} p_r(r \mid \overline{m})\, dr \\
&= \frac{1}{\sqrt{\pi \mathfrak{N} E}} \int_{-\infty}^{a} e^{-r^2/\mathfrak{N}E}\, dr \\
&= 1 - \operatorname{erfc} \frac{a}{\sqrt{\mathfrak{N}E/2}}
\end{aligned}
$$

The probability of making correct decision $P(C)$ is given by

$$
\begin{aligned}
P(C) &= P(C,m) + P(C,\overline{m}) \\
&= P(m)\,P(C \mid m) + P(\overline{m})\,P(C \mid \overline{m}) \\
&= P(m) \operatorname{erfc}\left(\frac{a-E}{\sqrt{\mathfrak{N}E/2}}\right) + P(\overline{m})\left[1 - \operatorname{erfc}\left(\frac{a}{\sqrt{\mathfrak{N}E/2}}\right)\right]
\end{aligned}
$$

The error probability $P(\epsilon)$ is given by

$$
\begin{aligned}
P(\epsilon) &= 1 - P(C) \\
&= P(m) + P(\overline{m}) - P(C) \\
&= P(m)\left[1 - \operatorname{erfc}\left(\frac{a-E}{\sqrt{\mathfrak{N}E/2}}\right)\right] + P(\overline{m}) \operatorname{erfc}\left(\frac{a}{\sqrt{\mathfrak{N}E/2}}\right)
\end{aligned}
$$

Using Eq. 4-136, we obtain

$$
P(\epsilon) = P(m) \operatorname{erfc}\left(\frac{E-a}{\sqrt{\mathfrak{N}E/2}}\right) + P(\overline{m}) \operatorname{erfc}\left(\frac{a}{\sqrt{\mathfrak{N}E/2}}\right) \tag{4-139}
$$

In the case,

$$P(m) = P(\overline{m}) = 0.5$$

and

$$C_d = C_a$$

Then

$$a = E/2$$

and

$$
\begin{aligned}
P(\epsilon) &= \operatorname{erfc}\left(\frac{E/2}{\sqrt{\mathfrak{N}E/2}}\right) \\
&= \operatorname{erfc}\left(\sqrt{\frac{E}{2\mathfrak{N}}}\right) && \text{(4-140a)} \\
&= \operatorname{erfc}\left(\frac{1}{2}\sqrt{\frac{E}{\mathfrak{N}/2}}\right) \\
&= \operatorname{erfc}\left(\frac{\sqrt{\rho}}{2}\right) && \text{(4-140b)}
\end{aligned}
$$

Where ρ is the signal-to-noise power ratio as defined in Eq. 4-107 (see Eq. 4-116). Thus Eq. 4-140b gives the explicit relationship between the signal-to-noise power ratio of the matched filter and its probability of error.

The error probability as a function of $E/\mathfrak{N}$ is plotted in Fig. 4-25.

Note that the error probability depends upon the message only through its energy E. The waveform is of no consequence. This is a rather significant result. The energy of a message emerges as one of the most significant parameters in the process of communication of the type (digital) discussed in this section.

Example 4-12. A message m is transmitted by means of a sinusoidal waveform A sin $\omega_c t$ lasting over the interval $(0, T)$ as shown in Fig. 4-26a.

$$s(t) = \begin{cases} A \sin \omega_c t & 0 \leq t < T \\ 0 & \text{Otherwise} \end{cases}$$

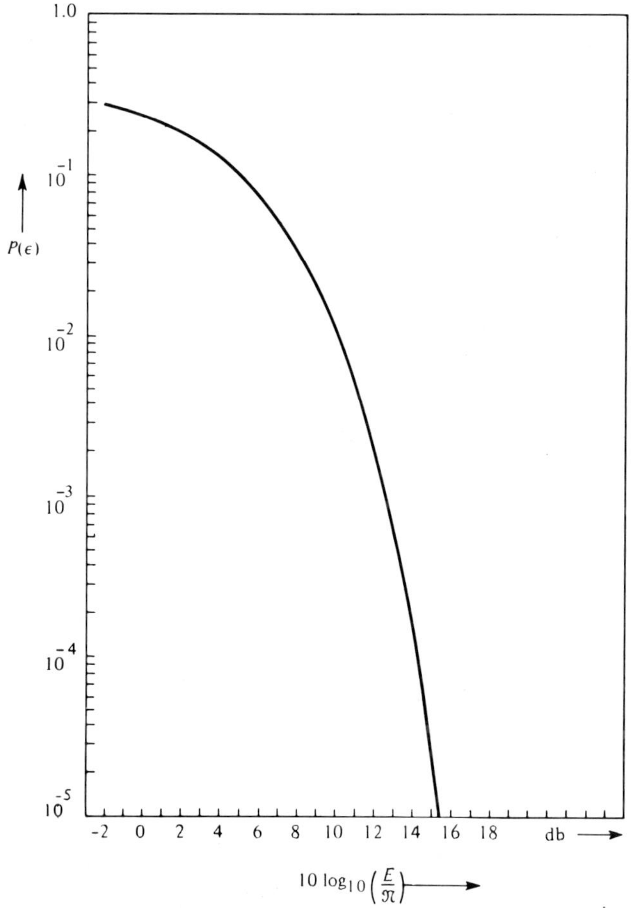

FIG. 4-25.

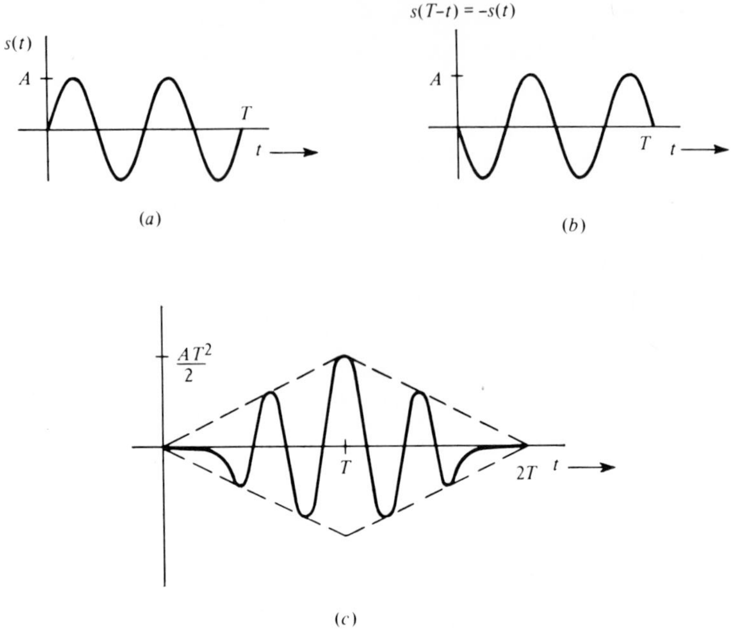

FIG. 4-26.

Assume

$$P(m) = P(\overline{m}) = 0.5$$

and

$$C_a = C_d$$

Design the matched receiver and find the error probability of this receiver.

The impulse response of the matched receiver is given by $h(t) = s(T - t)$.

Observe that $s(T - t)$ is obtained by taking the mirror image of $s(t)$ about the vertical axis and shifting it to the right by T seconds. This is obviously $-s(t)$ (Fig. 4-26b). Hence the matched receiver is a system whose impulse response is $-s(t)$.

The output waveform when $s(t)$ is present is given by the convolution of $s(t)$ with $h(t)$; that is, the convolution of $s(t)$ with $-s(t)$.

This is shown in Fig. 4-26c. The output is maximum at $t = T$ as expected and has a magnitude equal to the energy E of the signal (Eq. 4-117). In this case,

$$E = \frac{A^2T}{2}$$

When the signal $s(t)$ is present, the output will be as shown in Fig. 4-26c plus gaussian noise with a variance of $\mathfrak{N}E/2$ (Eq. 4-118). The decision threshold in this case is $E/2$. The error probability $P(\epsilon)$ is given by Eq. 4-140a:

$$P(\epsilon) = \operatorname{erfc}\left(\sqrt{\frac{E}{2\mathfrak{N}}}\right)$$

$$= \operatorname{erfc}\left(\sqrt{\frac{A^2T}{4\mathfrak{N}}}\right)$$

$$= \operatorname{erfc}\left(\frac{A}{2}\sqrt{\frac{T}{\mathfrak{N}}}\right)$$

4-20. BINARY COMMUNICATION USING TWO WAVEFORMS (BIPOLAR SCHEME)

In our discussion of binary communication, we have basically used only one type of pulse shape $s(t)$. The two symbols transmitted were mark [$s(t)$ present] and space [$s(t)$ absent]. We shall now investigate the case where the two messages m_1 and m_2 are transmitted by means of two pulses $(1/2)s(t)$ and $-(1/2)s(t)$ respectively. In this case we are using a bipolar pulse to transmit binary signals. We shall show that the latter scheme (bipolar) has the same error probability as the former (unipolar) but has the advantage that it requires less power than the former.

The received signal f(t) now consists of sequences of the pulses $(1/2)s(t)$ and $-(1/2)s(t)$ in some order as shown in Fig. 4-27b. To this signal we shall add a periodic signal $\phi(t)$ each individual cycle of which has the waveform $(1/2)s(t)$ as shown in Fig. 4-27c. The periodic signal is properly synchronized with the incoming signal. Note that this addition being a reversible operation, does not affect the optimality of the performance. The signal $f_1(t)$, the sum of the two signals $f(t)$ and $\phi(t)$ is now made up of sequences of only one basic type of pulse, viz., $s(t)$. In essence we have converted bipolar scheme into a unipolar. The optimum detection of such a signal can be performed by a matched filter as usual. This is shown in Fig. 4-27e. The decision threshold a (assuming $C_d = C_a$) is given by

$$a = \frac{1}{2}\left[E + \frac{\mathfrak{N}}{2}\ln\frac{P(m_2)}{P(m_1)}\right] \tag{4-141}$$

where E is the energy of pulse $s(t)$.

The error probability is given by Eq. 4-139. For convenience we shall consider the case of equiprobable messages. In this case (see Eq. 4-140),

$$P(\epsilon) = \operatorname{erfc}\left(\sqrt{\frac{E}{2\mathfrak{N}}}\right) \tag{4-142}$$

where E is the energy of $s(t)$.

It is evident that the error performance of the bipolar scheme [using $s(t)/2$ and $-s(t)/2$] is identical to that of the unipolar scheme [using $s(t)$ and no $s(t)$]. Let us now consider the average power required in

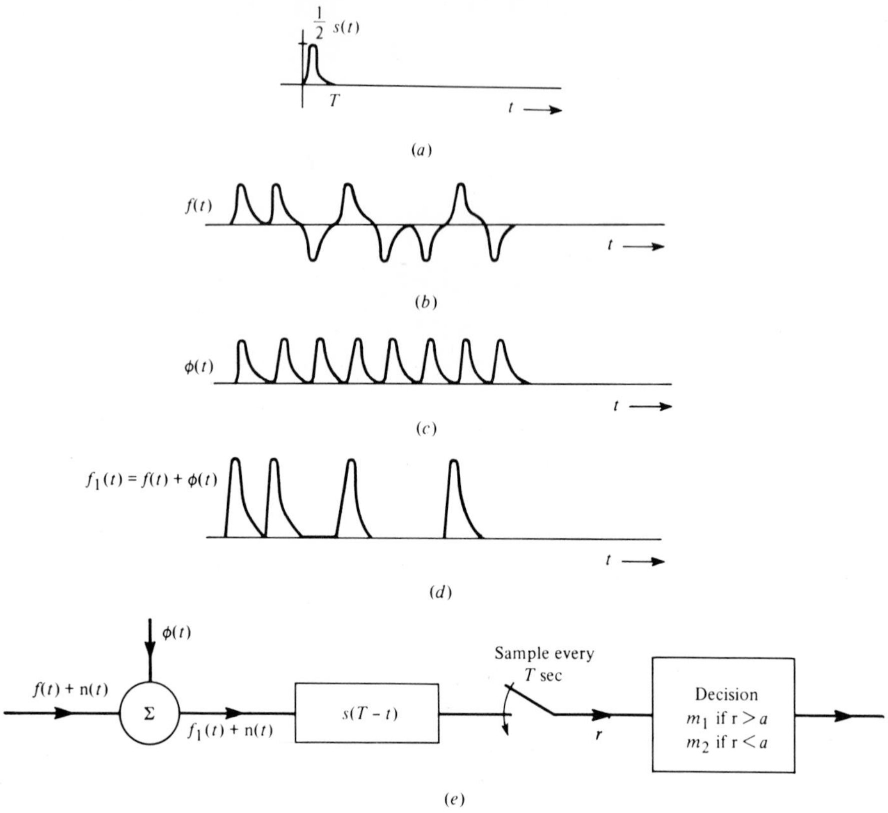

FIG. 4-27.

each scheme. For convenience, we shall consider the case where m_1 and m_2 are equally likely. In the unipolar scheme the signal $s(t)$ is present half the time and absent half the time. The duration of $s(t)$ is T seconds. Hence the average power S_1 is given by

$$S_1 = \frac{1}{T}\left[\frac{1}{2}(E) + \frac{1}{2}(0)\right] = \frac{E}{2T} \tag{4-143a}$$

Hence

$$P(\epsilon) = \text{erfc}\left(\sqrt{\frac{S_1 T}{\mathfrak{N}}}\right) \tag{4-143b}$$

Now consider the bipolar case (of two pulses $s(t)/2$ and $-s(t)/2$). In this case the energy of each of the two waveforms is $E/4$. Hence the average power S_2 is given by

$$S_2 = \frac{1}{T}\left[\frac{1}{2}\left(\frac{E}{4}\right) + \frac{1}{2}\left(\frac{E}{4}\right)\right] = \frac{E}{4T} = \frac{S_1}{2} \tag{4-144a}$$

and

$$P(\epsilon) = \text{erfc}\left(\sqrt{\frac{2S_2T}{\mathfrak{N}}}\right) \tag{4-144b}$$

It is evident that to attain a given error probability, the bipolar scheme needs only half the power as that required for the unipolar scheme. This result will again be proved in Chapter 6 from a more fundamental point of view.

In terms of signal energy Eq. 4-144b may be expressed as

$$P(\epsilon) = \text{erfc}\left(\sqrt{\frac{2E_s}{\mathfrak{N}}}\right) \tag{4-145}$$

where E_s is the energy of waveforms $s(t)/2$ or $-s(t)/2$. Also,

$$E_s = \frac{E}{4}$$

Substituting this equation in Eq. 4-142, we obtain the same $P(\epsilon)$ as in Eq. 4-145. Hence the error performance of the bipolar scheme is the same as that of the unipolar scheme, but requires only half as much power.

Implementation of the Receiver of Two-Waveform Case. The optimum receiver for the bipolar case is shown in Fig. 4-27. This method, requires the generation of $\phi(t)$ at the receiver and is therefore inconvenient. We shall now show that Figs. 4-28a and 4-28b represent alternative equivalents. This can be easily shown by observing that the output of the matched receiver at $t = T$ due to $\phi(t)$ is $E/2$ because $\phi(t)$ consists of pulses $(1/2)s(t)$. Hence we may add $E/2$ to r and eliminate $\phi(t)$ entirely. Adding $E/2$ to r is equivalent to changing the threshold from a to $a - E/2$. The arrangement in Fig. 4-28b is equivalent to that in Eq. 4-28a. This follows from the fact that

$$\alpha_1 = \frac{\mathrm{r}}{2} + \frac{\mathfrak{N}}{2} \ln P(m_1)$$

$$\alpha_2 = -\frac{\mathrm{r}}{2} + \frac{\mathfrak{N}}{2} \ln P(m_2)$$

and

$$\alpha_1 - \alpha_2 = \mathrm{r} - \frac{\mathfrak{N}}{2} \ln \frac{P(m_2)}{P(m_1)}$$

Therefore if

$$\alpha_1 > \alpha_2, \mathrm{r} > \frac{\mathfrak{N}}{2} \ln \frac{P(m_2)}{P(m_1)} \text{ and the decision is "}m_1\text{ true."}$$

Similarly, if $\alpha_1 < \alpha_2$, the decision is "m_2 is true."

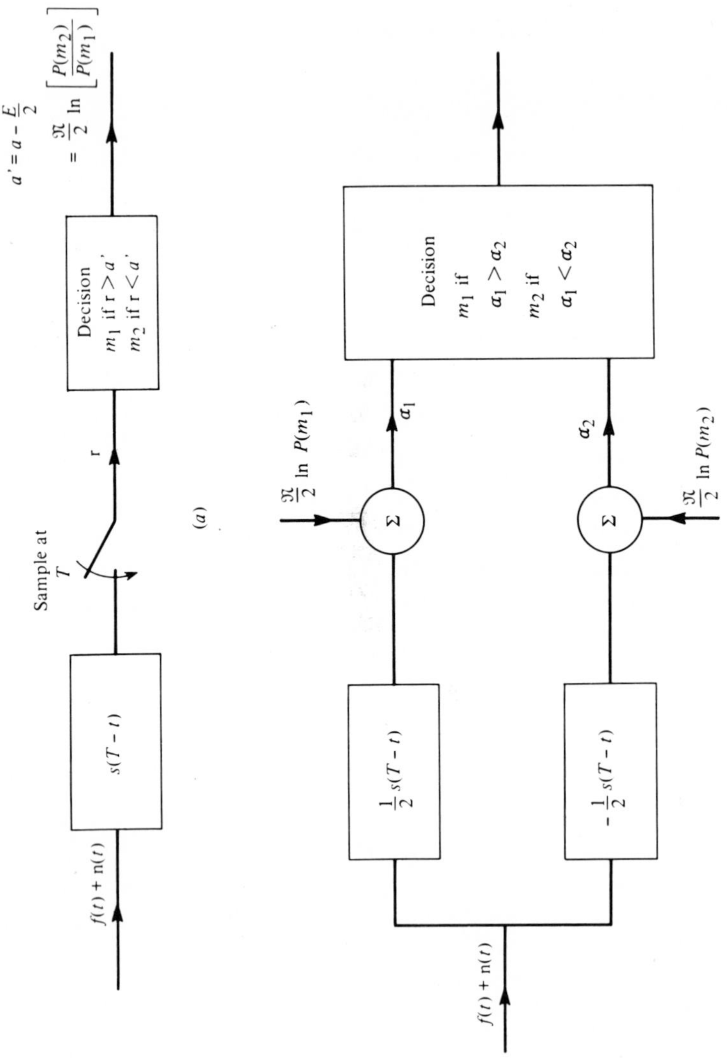

$a' = a - \frac{E}{2}$
$= \frac{\mathfrak{N}}{2} \ln \left[\frac{P(m_2)}{P(m_1)}\right]$
Decision
m_1 if r > a'
m_2 if r < a'
r
Sample at T
$s(T - t)$
$f(t) + n(t)$
(a)
$\frac{\mathfrak{N}}{2} \ln P(m_1)$
α_1
Σ
$\frac{1}{2} s(T - t)$
Decision
m_1 if
$\alpha_1 > \alpha_2$
m_2 if
$\alpha_1 < \alpha_2$
α_2
Σ
$\frac{\mathfrak{N}}{2} \ln P(m_2)$
$-\frac{1}{2} s(T - t)$
$f(t) + n(t)$
(b)

FIG. 4-28.

Fig. 4-28b represents two filters matched to the two signals $s(t)/2$ nd $-s(t)/2$ respectively. If the signal $s(t)/2$ is present α_1, the output f the corresponding matched filter is higher and α_2 the output of the ther matched filter is lower. The decision is therefore "m_1 true" if $_1 > \alpha_2$. The constants $(\mathfrak{N}/2)\ln P(m_1)$ and $(\mathfrak{N}/2)\ln P(m_2)$ in the two ircuits add the desired bias required for minimizing the error when the vo signals are not equiprobable.

PPENDIX 4-1. THE BILATERAL LAPLACE TRANSFORM

The Fourier transform is a tool used to represent an arbitrary unction $f(t)$ as a continuous sum of exponential functions of the form $^{j\omega t}$. The frequencies of these exponential functions are therefore re-tricted to the $j\omega$ axis in the complex frequency plane. In general, how-ver, it is desirable to represent a function $f(t)$ by a continuous sum of xponentials of the form e^{st} where $s = \sigma + j\omega$. Therefore we need to xtend the results obtained for the special case when $s = j\omega$ to the more eneral one when $s = \sigma + j\omega$. It is indeed possible to express a function (t) as a continuous sum of exponential functions with complex fre-uencies (Ref. 4-1):

$$f(t) = \frac{1}{2\pi j}\int_{\sigma - j\omega}^{\sigma + j\omega} F(s)\, e^{st}\, ds \tag{A4-1}$$

vhere

$$F(s) = \int_{-\infty}^{\infty} f(t)\, e^{-st}\, dt \tag{A4-2}$$

nd $s = \sigma + j\omega$.

The *complex Fourier transform* is also known as the *bilateral Laplace ransform* (or the two-sided Laplace transform). The complex Fourier ransform or the bilateral Laplace transform will be denoted symboli-ally as

$$F(s) = \mathfrak{F}_c[f(t)] \quad \text{and} \quad f(t) = \mathfrak{F}_c^{-1}[F(s)]$$

It is evident from Eq. A4-2 that the function $F(s)$ can be obtained rom $F(\omega)$, the Fourier transform of $f(t)$, merely by replacing[13] $j\omega$ y s, i.e.,

$$F(s) = F(\omega)\big|_{\omega = s/j} \tag{A4-3}$$

[13]A note of caution is in order here. The complex Fourier transform of a function an be found from an ordinary Fourier transform by replacing $j\omega$ by s only for ab-olutely integrable functions. This procedure cannot be applied for those functions vhose Fourier transform exists in the limit. For such functions (e.g., $u(t)$), one should valuate the transform directly from Eq. A4-2.

As an example consider the function $e^{-at}u(t)$. The Fourier transform of this function is given by

$$\mathfrak{F}[e^{-at}u(t)] = \frac{1}{a + j\omega} \tag{A4-4}$$

The bilateral Laplace transform of $e^{-at}u(t)$ may be evaluated directly from Eq. A4-2, or may be found by replacing $j\omega$ by s in Eq. A4-4. Thus

$$\mathfrak{F}_c[e^{-at}u(t)] = \frac{1}{s + a}$$

Existence of the Bilateral Laplace Transform. From Eq. A4-2 it is evident that the bilateral Laplace transform of $f(t)$ exists if

$$\int_{-\infty}^{\infty} f(t)\, e^{-st}\, dt$$

is finite, since

$$|f(t)\, e^{-st}| = |f(t)|\, e^{-\sigma t}$$

the existence of $F(s)$ is guaranteed if

$$\int_{-\infty}^{\infty} |f(t)|\, e^{-\sigma t}\, dt \qquad \text{is finite} \tag{A4-5}$$

If there exists a real finite numbers M, σ and β such that

$$f(t) \leq \begin{cases} Me^{\alpha t} & \text{for } t > 0 \\ Me^{\beta t} & \text{for } t < 0 \end{cases} \tag{A4-6}$$

then the condition A4-5 is satisfied for any value of σ greater than α but less than β, i.e., the integral $\int f(t)\, e^{-st}\, dt$ is absolutely convergent for values of s given by

$$\beta > \sigma > \alpha \tag{A4-7}$$

This may easily be shown by breaking up the integral into two parts:

$$F(s) = \int_{-\infty}^{0} f(t)\, e^{-st}\, dt + \int_{0}^{\infty} f(t)\, e^{-st}\, dt$$

Use of inequality A4-6 yields

$$|F(s)| \leq \int_{-\infty}^{0} Me^{(\beta - s)t}\, dt + \int_{0}^{\infty} Me^{(\alpha - s)}\, dt$$

$$\leq M \left\{ \frac{1}{\beta - s}\, e^{(\beta - s)t} \Big|_{-\infty}^{0} + \frac{1}{\alpha - s}\, e^{(\alpha - s)t} \Big|_{0}^{\infty} \right\}$$

It is obvious that the first integral will converge for Re $s < \beta$ and the second integral will converge for Re $s > \alpha$. The two regions are shown in Fig. A4-1. The region where both integrals converge is the common

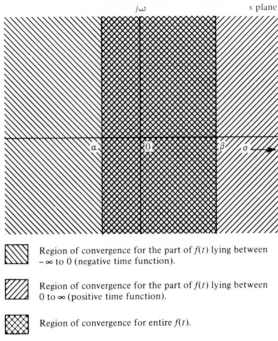

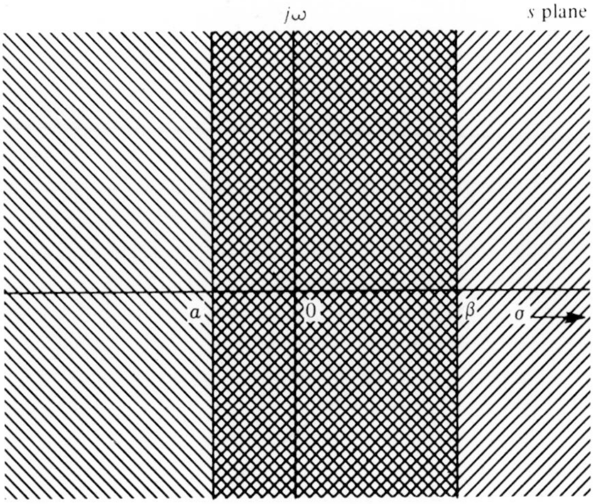

Region of convergence for the part of $f(t)$ lying between $-\infty$ to 0 (negative time function).

Region of convergence for the part of $f(t)$ lying between 0 to ∞ (positive time function).

Region of convergence for entire $f(t)$.

FIG. A4-1.

egion given by

$$\beta > \sigma > \alpha$$

The region of convergence for various time functions is shown in 'ig. A4-2 by shaded areas.

It is obvious that for all values of s lying in the region of convergence '(s) is finite. So any singularities of $F(s)$ [poles of $F(s)$] must lie out-ide the region of convergence. If $f(t)$ is entirely a positive-time func-ion, i.e., $f(t)$ exists in the interval $(0, \infty)$ only, then it is obvious from 'ig. A4-2 that the poles of $F(s)$ must lie to the left of the region of con-ergence. Similarly if $f(t)$ is entirely a negative-time function, i.e., $f(t)$ xists in the interval $(-\infty, 0)$ only, then the poles of $F(s)$ must lie to the ight of the region of convergence. For a general $f(t)$ which exists in the nterval $(-\infty, \infty)$, $F(s)$ may have several poles, some lying on the left nd some lying to the right of the region of convergence. It is now evi-ent that the poles to the left of the region arise due to the positive-time art of $f(t)$ and those to the right of the region arise due to the negative-ime part of $f(t)$. This fact is of crucial importance in finding the inverse ransform.

The shaded regions in Fig. A4-2 are the regions of convergence of the omplex Fourier transform. The ordinary Fourier transform is a special

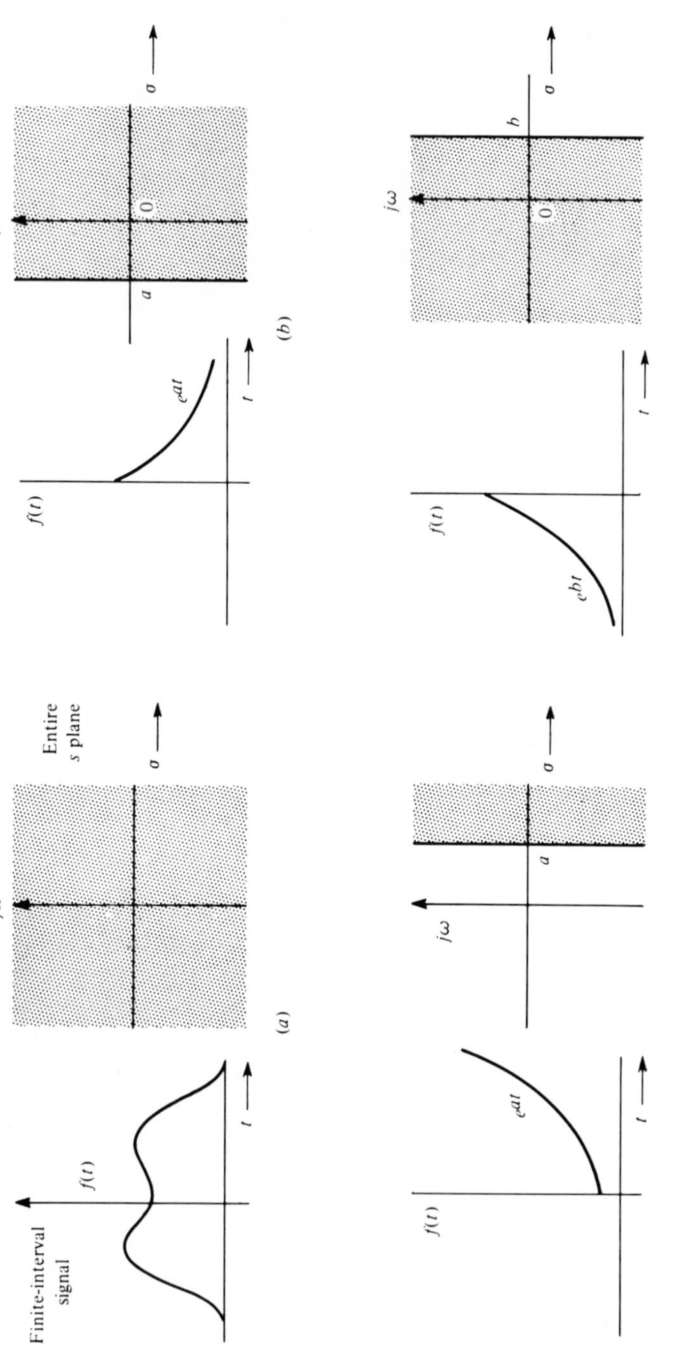
Finite-interval signal
f(t)
t
Entire s plane
jω
σ
(a)
f(t)
e^{at}
t
jω
a
0
σ
(b)
f(t)
e^{at}
t
jω
a
σ
(c)
f(t)
e^{bt}
t
jω
0
b
σ
(d)

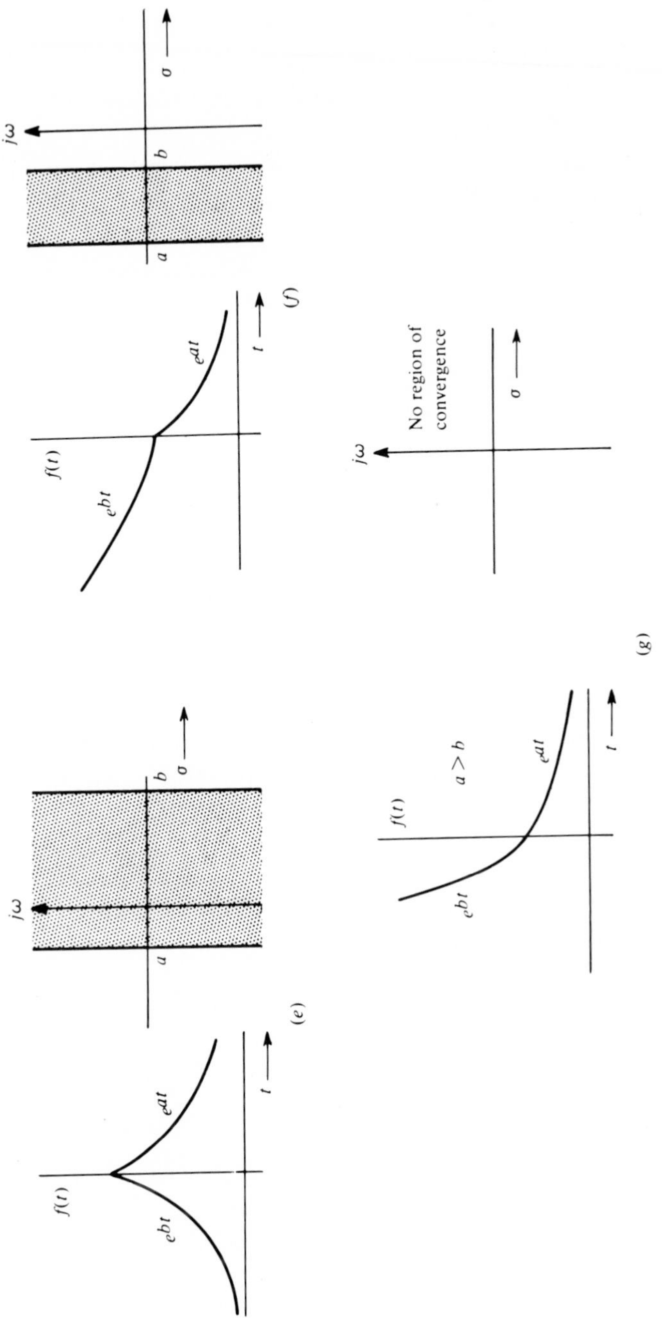

FIG. A4-2.

case for which the region of convergence that is utilized is confined t the imaginary axis $s = j\omega$. If the region of convergence of the bilatera Laplace transform of a function $f(t)$ includes the $j\omega$ axis, then th ordinary Fourier transform of $f(t)$ exists and can be obtained by sub stituting $j\omega$ for s in $F(s)$. This is the case for the functions in Fig. A4-2a b, d, and e. If, however, the region of convergence of $F(s)$ does no include the imaginary axis, the function fails to satisfy the condition o absolute integrability and does not possess the ordinary Fourier trans form. Functions shown in Fig. A4-2c and f represent such cases.

At this point we make an important observation. If a function $f(t$ is absolutely integrable, its ordinary Fourier transform exists and henc the region of convergence of $F(s)$ must include the $j\omega$ axis. In such case the poles of $F(s)$ lying to the left of the region of convergence must al lie in the LHP. Similarly all the poles of $F(s)$ lying to the right of th region must lie in the RHP. From the previous discussion with regar to positive- and negative-time functions it follows that if $f(t)$ is ab solutely integrable—that is,

$$\int_{-\infty}^{\infty} |f(t)|\,dt \qquad \text{is finite}$$

then all the terms of $F(s)$ represented by LHP poles correspond to th positive-time component. Similarly all the terms of $F(s)$ represented by RHP poles correspond to the negative-time component. Alternatively if $f(t)$ is absolutely integrable and is a positive-time function, all th poles of $F(s)$ must lie in the LHP. Similarly if $f(t)$ is absolutely integrable and is a negative-time function, all the poles of $F(s)$ must lie in the RHP.

EVALUATION OF THE BILATERAL LAPLACE TRANSFORM FROM UNILATERAL TRANSFORMS

We shall now show that any bilateral Laplace transform can be expressed as a sum of two unilateral Laplace transforms. It is, therefore, possible to evaluate bilateral transforms from the tables of unilateral transforms.

Consider a function $f(t)$ shown in Fig. A4-3a. We can separate $f(t)$ into two component functions $f_1(t)$ and $f_2(t)$ representing respectively the positive- and negative-time components as shown in Fig. A4-3b and c. In other words,

$$f_1(t) = f(t)\,u(t)$$

and

$$f_2(t) = f(t)\,u(-t)$$

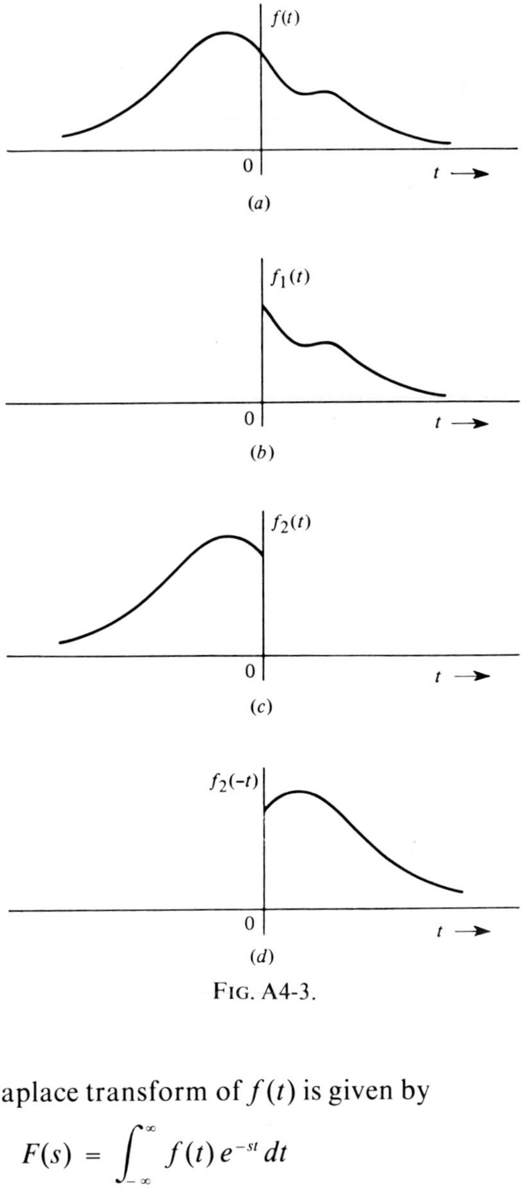

FIG. A4-3.

The bilateral Laplace transform of $f(t)$ is given by

$$F(s) = \int_{-\infty}^{\infty} f(t)\,e^{-st}\,dt$$

$$= \int_{-\infty}^{0} f_2(t)\,e^{-st}\,dt + \int_{0}^{\infty} f_1(t)\,e^{-st}\,dt$$

Changing the dummy variable t to $-t$ in the first integral, we have

$$F(s) = \int_{0}^{\infty} f_2(-t)e^{st}\,dt + \int_{0}^{\infty} f_1(t)\,e^{-st}\,dt$$

$$= F_2(-s) + F_1(s)$$

where $F_1(s)$ and $F_2(s)$ are both unilateral Laplace transforms:

$$F_1(s) = \mathcal{L}[f_1(t)]$$

and

$$F_2(s) = \mathcal{L}[f_2(-t)]$$

Function $f_2(-t)$ is a mirror image of $f_2(t)$ about the vertical axis $t = 0$ as shown in Fig. A4-3d. Hence the contribution due to the negative-time function may be found as follows:

1. Take the reflection (mirror image) of $f_2(t)$ about the vertical axis $t = 0$ and find its Laplace transform.
2. Replace s by $-s$ in the transform obtained in (1). This gives the contribution to the transform due to negative-time component of $f(t)$.

The region of convergence of $F(s)$ is given by the region of convergence common to both $F_1(s)$ and $F_2(-s)$.

The procedure will now be illustrated by an example.

Example A4-1. Find the bilateral Laplace transform of $f(t)$ given by (Fig. A4-4):

$$f_1(t) = e^{-2t}u(t)$$
$$f_2(t) = e^{t}u(-t)$$

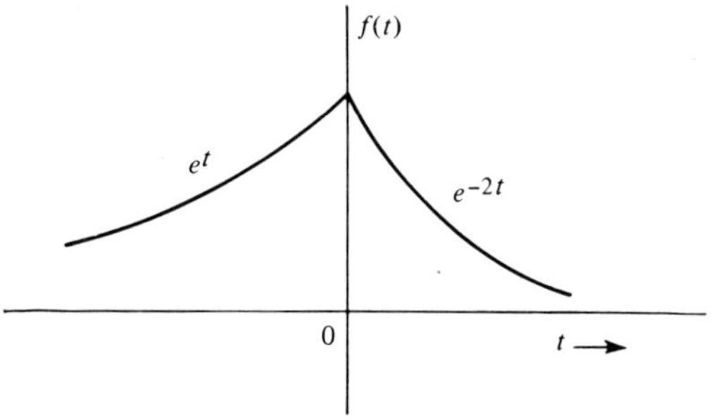

FIG. A4-4.

We have

$$f_2(-t) = e^{-t}u(t)$$

Hence

$$F_1(s) = \frac{1}{s+2}$$

$$F_2(s) = \frac{1}{s+1}$$

Therefore

$$\mathfrak{F}_c[f(t)] = F_1(s) + F_2(-s)$$
$$= \frac{1}{s+2} + \frac{1}{-s+1}$$

$$= \frac{-3}{(s+2)(s-1)}$$

Note that the region of convergence is given by $-2 < \sigma < 1$.

Example A4-2. Find the inverse transform of $F(s)$ given by

$$F(s) = \frac{s}{(s+2)(s+1)} \qquad -2 < \sigma < -1$$

Expansion of $F(s)$ by partial fraction yields

$$F(s) = \frac{2}{s+2} - \frac{1}{s+1}$$

Since the pole -2 lies to the left of the region of convergence, the term $2/(s+2)$ represents positive-time function. The term $1/(s+1)$ represents negative-time function since the pole at $s = -1$ lies to the right of the region of convergence. Hence

$$F_1(s) = \frac{2}{s+2} \quad \text{and} \quad f_1(t) = 2e^{-2t}u(t)$$

$$F_2(-s) = \frac{-1}{s+1} \quad \text{and} \quad F_2(s) = \frac{1}{s-1}$$

Therefore

$$f_2(-t) = \mathcal{L}^{-1}[F_2(s)] = e^{t}u(t)$$

and

$$f_2(t) = e^{-t}u(-t)$$

Thus

$$f(t) = 2e^{-2t}u(t) + e^{-t}u(-t)$$

The function $f(t)$ is shown in Fig. A4-5.

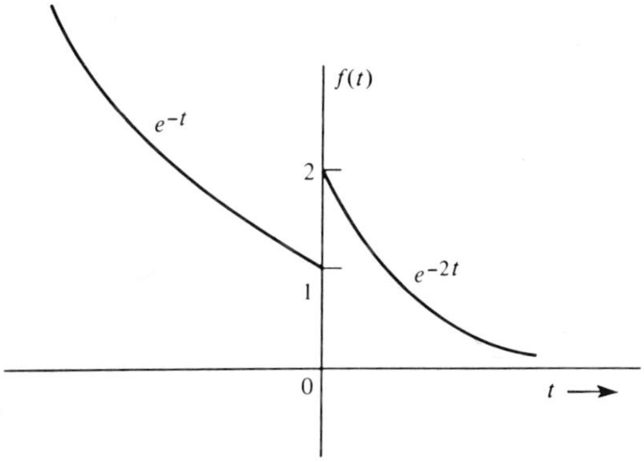

FIG. A4-5.

APPENDIX 4-2. FILTERING OF SIGNALS WITH RATIONAL SPECTRA FROM WHITE NOISE

In the special case of white noise mixed with the desired signal $s(t)$ when the power-density spectrum of $s(t)$ is rational with its order of the denominator greater than that of the numerator, the expressions for the Wiener-Hopf (realizable) filter transfer function and the mean-square error are greatly simplified. The results, originally due to Yovits and Jackson, Ref. 4-4 (also see Helstrom, Ref. 4-5) are given by:

$$H(s) = 1 - \frac{\sqrt{\mathfrak{N}/2}}{\mathcal{S}_x(s)} \tag{A4-8}$$

The corresponding mean-square error is given by

$$\overline{\epsilon^2} = \frac{\mathfrak{N}}{\pi} \int_{-\infty}^{\infty} \left[1 - \frac{\sqrt{\mathfrak{N}/2}}{\mathcal{S}_x(\omega)}\right] d\omega \tag{A4-9}$$

The mean-square error can alternately be expressed as

$$\overline{\epsilon^2} = \frac{\mathfrak{N}}{4\pi} \int_{-\infty}^{\infty} \log \left[\frac{2}{\mathfrak{N}} \mathcal{S}_x(\omega)\right] d\omega \tag{A4-10}$$

This result can be proved as follows. We have for the case of filtering without delay (see Ex. 4-6),

$$\begin{aligned} \mathcal{S}_{xy}(s) &= \mathcal{S}_s(s) \\ &= \mathcal{S}_s(s) + \frac{\mathfrak{N}}{2} - \frac{\mathfrak{N}}{2} = \mathcal{S}_x(s) - \frac{\mathfrak{N}}{2} \end{aligned}$$

and

$$\begin{aligned} F(s) &= \frac{\mathcal{S}_{xy}(s)}{\mathcal{S}_x(-s)} \\ &= \frac{\mathcal{S}_x(s) - \mathfrak{N}/2}{\mathcal{S}_x(-s)} \\ &= \frac{\mathcal{S}_x(s)\,\mathcal{S}_x(-s) - \mathfrak{N}/2}{\mathcal{S}_x(-s)} \\ &= \mathcal{S}_x(s) - \frac{\mathfrak{N}/2}{\mathcal{S}_x(-s)} \qquad \text{(A4-11)} \\ &= [\mathcal{S}_x(s) - \sqrt{\mathfrak{N}/2}] + \left[\sqrt{\mathfrak{N}/2} - \frac{\mathfrak{N}/2}{\mathcal{S}_x(-s)}\right] \qquad \text{(A4-12)} \end{aligned}$$

Since the poles of $\mathcal{S}_x(s)$ are in the LHP, the first term of Eq. A4-12 represents $F^+(s)$, the positive time component of $F(s)$.[14] Similarly, it

[14]The reader may wonder about the addition and subtraction of $\sqrt{\mathfrak{N}/2}$ in Eq. A4-12. The reason for this is as follows.

Since by assumption the order of numerator polynomial of $\mathcal{S}_s(s)$ is lower than the order of the denominator polynomial, the order of numerator and denominator of $\mathcal{S}_x(s)$ are the same. A similar conclusion is arrived at for $\mathcal{S}_x(s)$ and $\mathcal{S}_x(-s)$. It also follows that the denominator polynomial of $F(s)$ is of higher order than its numerator polynomial, implying that $f(t)$ has no impulse (or its derivatives) at the origin. Obviously, the order of the numerator of $F^+(s)$ [and also $F^-(s)$] must be smaller than that

can be seen that the second term of Eq. A4-12 represents $F^-(s)$, the negative time component of $F(s)$. Hence

$$F^+(s) = \mathcal{S}_x(s) - \sqrt{\mathfrak{N}/2}$$

and

$$H(s) = \frac{F^+(s)}{\mathcal{S}_x(s)}$$

$$= 1 - \frac{\sqrt{\mathfrak{N}/2}}{\mathcal{S}_x(s)} \qquad \text{Q.E.D.}$$

The reader can verify the results of Ex. 4-10 (for zero prediction time) by using Eq. A4-8. To calculate $\overline{\epsilon^2(t)}$, we use Eq. 4-83g.

$$\overline{\epsilon^2(t)} = \frac{\mathfrak{N}}{2} h(0)$$

In the present case

$$h(t) = \mathcal{L}^{-1}\left[1 - \frac{\sqrt{\mathfrak{N}/2}}{\mathcal{S}_x(s)}\right]$$

$$= \mathfrak{F}^{-1}\left[1 - \frac{\sqrt{\mathfrak{N}/2}}{\mathcal{S}_x(\omega)}\right]$$

$$= \frac{1}{2\pi}\int_{-\infty}^{\infty}\left(1 - \frac{\sqrt{\mathfrak{N}/2}}{\mathcal{S}_x(\omega)}\right)e^{j\omega t}\,d\omega$$

Hence

$$\overline{\epsilon^2(t)} = \frac{\mathfrak{N}}{2} h(0) = \frac{\mathfrak{N}}{4\pi}\int_{-\infty}^{\infty}\left(1 - \frac{\sqrt{\mathfrak{N}/2}}{\mathcal{S}_x(\omega)}\right)d\omega$$

In terms of the complex variable $s = j\omega$, $\overline{\epsilon^2}$ can be expressed as

$$\overline{\epsilon^2(t)} = \frac{\mathfrak{N}}{2}\frac{1}{2\pi j}\int_{-j\infty}^{j\infty}\left[1 - \frac{\sqrt{\mathfrak{N}/2}}{\mathcal{S}_x(s)}\right]ds$$

$\mathcal{S}_x(s)$ is rational, with the order of the numerator polynomial equal to the order of the denominator polynomial. Moreover as $s \to \infty$, $\mathcal{S}_x(s) \to \mathfrak{N}/2$. Hence

$$\mathcal{S}_x(s) = \mathcal{S}_x(s)\,\mathcal{S}_x(-s) = \frac{\mathfrak{N}}{2}\left(\prod_{k=1}^{n}\frac{s+\alpha_k}{s+\beta_k}\right)\left(\prod_{k=1}^{n}\frac{s-\alpha_k}{s-\beta_k}\right)$$

---

of its denominator. Therefore, $\mathcal{S}_x(s)$ in Eq. A4-11 cannot be $F^+(s)$ because (the inverse transform of) $\mathcal{S}_x(s)$ has an impulse of strength $\sqrt{\mathfrak{N}/2}$ at the origin [following from the fact that $\mathcal{S}_x(s) \to \sqrt{\mathfrak{N}/2}$ as $s \to \infty$ because of the white noise present]. It is, therefore, evident that $F^+(s) = \mathcal{S}_x(s) - \sqrt{\mathfrak{N}/2}$. Note that $F^-(s)$ as given by the second term of Eq. A4-12 also has no impulse at the origin. This is because as $s \to \infty$, $\mathcal{S}_x(-s) \to \sqrt{\mathfrak{N}/2}$ and hence $F^-(s) \to 0$.

where α_k's and β_k's are positive real numbers. Evidently

$$\mathcal{S}_x(s) = \sqrt{\frac{\mathfrak{N}}{2}} \prod_{k=1}^{n} \frac{s + \alpha_k}{s + \beta_k}$$

and

$$1 - \frac{\sqrt{\mathfrak{N}/2}}{\mathcal{S}_x(s)} = 1 - \prod_{k=1}^{n} \left(\frac{s + \beta_k}{s + \alpha_k}\right)$$

$$= 1 - \prod_{k=1}^{n} \left(1 - \frac{\alpha_k - \beta_k}{s + \alpha_k}\right)$$

$$= \sum_{k=1}^{n} \frac{\alpha_k - \beta_k}{s + \alpha_k} + \phi(s)$$

where $\phi(s)$ tends to zero at least as s^{-2} as $s \to \infty$. Hence the integral of $\phi(s)$ along a semicircle of radius $R(R \to \infty)$ will be zero. Note that $\phi(s)$ has no poles in the RHP. Hence the contour integral of $\phi(s)$ along the entire $j\omega$ axis and the semicircle in the RHP of radius $R(R \to \infty)$ is zero. Since the integral along the semicircle is zero, the integral along the $j\omega$ axis must also be zero. Hence

$$\overline{\epsilon^2} = \frac{\mathfrak{N}}{2} \sum_{k=1}^{n} \frac{1}{2\pi j} \int_{-j\infty}^{j\infty} \frac{\alpha_k - \beta_k}{s + \alpha_k} \, ds$$

Using the residue theorem, this can be expressed as

$$\overline{\epsilon^2} = \frac{\mathfrak{N}}{2} \sum_{k=1}^{n} \alpha_k - \beta_k$$

To express this in a more meaningful form, we observe that

$$\frac{1}{2\pi} \int_{-\infty}^{\infty} \log \left[\frac{2}{\mathfrak{N}} \mathcal{S}_x(\omega)\right] d\omega = \frac{1}{2\pi} \int_{-\infty}^{\infty} \log \left[\prod_{k=1}^{n} \left(\frac{\omega^2 + \alpha_k^2}{\omega^2 + \beta_k^2}\right)\right] d\omega$$

$$= \frac{1}{2\pi} \sum_{k=1}^{n} \int_{-\infty}^{\infty} \log \left(\frac{\omega^2 + \alpha_k^2}{\omega^2 + \beta_k^2}\right) d\omega$$

$$= \sum_{k=1}^{n} \alpha_k - \beta_k$$

Hence

$$\overline{\epsilon^2} = \frac{\mathfrak{N}}{4\pi} \int_{-\infty}^{\infty} \log \left[\frac{2}{\mathfrak{N}} \mathcal{S}_x(\omega)\right] d\omega$$

REFERENCES

1. Lathi, B. P., *Signals, Systems and Communication*, Wiley, New York, 1965.

2. Davenport, W. B., and W. L. Root, *An Introduction to the Theory of Random Signals and Noise*, McGraw-Hill, New York, 1958.

3. Papoulis, A., *Probability, Random Variables and Stochastic Processes*, McGraw-Hill, New York, 1965.
4. Yovits, M. C., and J. L. Jackson, "Linear Filter Optimization with Game Theory Considerations," *IRE Nat. Conv. Record*, Part 4 (March 1955), pp. 193–199.
5. Helstrom, C. W., "Topics in the Transmission of Continuous Information," Westinghouse Research Labs., Report 64-8C3-522-R1 (Aug. 27, 1964).
6. Lathi, B. P., *Communication Systems*, Wiley, New York, 1968.
7. Wozencraft, J. M., and I. M. Jacobs, *Principles of Communication Engineering*, Wiley, New York, 1965.

PROBLEMS

4-1. A white-noise source in Fig. 3-18 (with $\alpha = 10^{-3}$) is applied at the input terminals of an *R-C* network shown in Fig. P. 4-1. (a) Sketch the output signal (express output signal as a superposition of responses to impulses in the input signal). (b) Find the power-density spectrum of the output and determine the output autocorrelation function from the power-density spectrum. (c) Determine the mean value and the mean-square value of the output voltage.

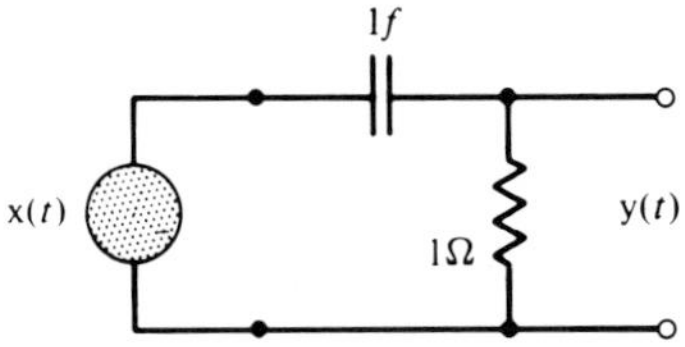

FIG. P. 4-1.

4-2. In Prob. 4-1, instead of determining the output power-density spectrum first, obtain the output autocorrelation function first and from this obtain the power-density spectrum.

4-3. In Prob. 4-1, determine the input-output cross-correlation function and the input-output cross-power-density spectrum.

4-4. In Prob. 4-1, if the input signal is applied at $t = 0$, determine the mean value (ensemble mean, of course) and the mean-square value of the output voltage at some instant t $(t > 0)$.

4-5. Repeat Prob. 4-1, 4-2, 4-3 and 4-4 if the input voltage source is a random voltage source in Fig. 3-22 with mean value and mean-square values given by

$$\bar{x} = 0, \qquad \overline{x^2} = 1$$

and $\alpha = 10^3$.

4-6. Two independent random voltage sources $x_1(t)$ and $x_2(t)$ are connected to a circuit as shown in Fig. P. 4-6. The source $x_1(t)$ is a white-noise source in Fig. 3-18

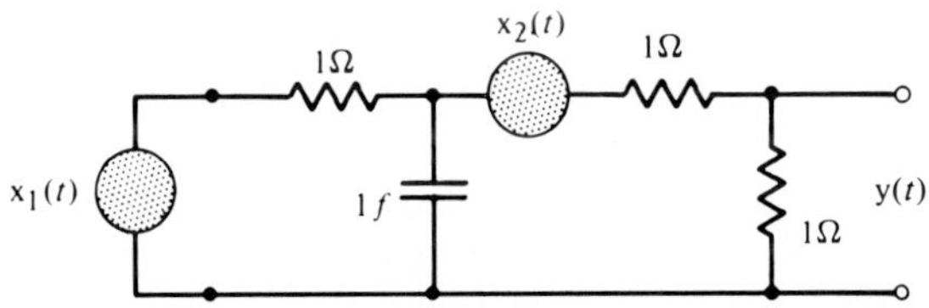

FIG. P. 4-6.

(with $\alpha = 10^{-3}$) and the source $x_2(t)$ is a random source in Fig. 3-22 with $\alpha = 10^3$, $\overline{x_2} = 0$ and $\overline{x_2^2} = 1$. (a) Find the power-density spectrum of the output voltage and hence determine the autocorrelation function of the output voltage $y(t)$. (b) Determine first the output voltage autocorrelation function and from this find the output voltage power-density spectrum.

4-7. In Fig. P. 4-6, the input signal $x_1(t)$ is a random signal shown in Fig. 3-22 (with $\alpha = 10^3$, $\overline{x_1} = 0$, $\overline{x_1^2} = 1$), and the signal $x_2(t)$ is given by

$$x_2(t) = x_1(t - 1)$$

Determine the power-density spectrum and the autocorrelation function of the output voltage $y(t)$.

4-8. In Fig. P. 4-6, the input voltage signal $x_1(t)$ is a white-noise source shown in Fig. 3-18 (with $\alpha = 10^{-3}$), and the voltage $x_2(t)$ is given by

$$x_2(t) = x_1(t - 1)$$

Determine the power-density spectrum and the autocorrelation function of the output voltage $y(t)$.

4-9. In Fig. P. 4-6, the input voltage signal $x_1(t)$ is a random signal in Fig. 3-22 ($\alpha = 10^3$, $\overline{x_1} = 0$ and $\overline{x_1^2} = 1$) and $x_2(t)$ is given by

$$x_2(t) = x_1(t) \cos(\omega_c t + \theta)$$

where θ is independent random variable over the ensemble, uniformly distributed over the range $(0, 2\pi)$. Determine the power-density spectrum and the autocorrelation function of the output signal $y(t)$.

4-10. Extend the transient analysis in Sec. 4-6 to the case of two sources applied at different input terminals. Consider both cases (i) source uncorrelated, (ii) sources dependent. Determine the mean and the mean-square value of the output signal at some instant t.

4-11. Determine the mean and the mean-square value of the output voltage $y(t)$ at $t = 1$ in Prob. 4-6 if the sources $x_1(t)$ and $x_2(t)$ were turned on at $t = 0$.

4-12. Determine the mean and the mean-square value of the output voltage $y(t)$ at $t = 1$ in Prob. 4-7 if $x_1(t)$ and $x_2(t)$ were turned on at $t = 0$.

4-13. A system whose output at any instant t depends upon the input of the system at that instant t alone, is called a memoryless system. Such a system may be linear or nonlinear. Let us consider a general time-invariant memoryless system whose input and output signals $x(t)$ and $y(t)$ are related by

$$y = \phi(x)$$

or

$$y(t) = \phi[x(t)]$$

(a) Show that

$$R_y(\tau) = \int_{-\infty}^{\infty} \int_{-\infty}^{\infty} \phi(x_1)\,\phi(x_2)\,p_{x_1 x_2}(x_1, x_2; \tau)\,dx_1 dx_2$$

where

$$x_1 = x(t) \quad \text{and} \quad x_2 = x(t + \tau)$$

(b) Consider a square-law device

$$y = x^2$$

The input process $x(t)$ is a stationary gaussian process with autocorrelation function $R_x(\tau)$. Determine $p_y(y)$ the first-order probability density of $y(t)$. Determine the autocorrelation function of $y(t)$.

4-14. A stationary white gaussian random process whose autocorrelation function $R_x(\tau)$ is given by

$$R_x(\tau) = \frac{\mathfrak{N}}{2}\,\delta(\tau)$$

is applied at the input terminals of a system shown in Fig. P. 4-1. Find the nth-order probability-density function of the output process.

4-15. If $\mathcal{S}_x(\omega)$ is a real and even function of ω, show that one can always find a process $x(t)$ whose power-density spectrum is $\mathcal{S}_x(\omega)$. This is known as the existence theorem. *Hint:* Apply a white-noise signal at the input of a linear system whose transfer function magnitude is given by

$$|H(\omega)| = \sqrt{\mathcal{S}_x(\omega)}$$

4-16. A random-voltage signal $s(t)$ has a power-density spectrum

$$\mathcal{S}_s(\omega) = \frac{1}{1+\omega^2}$$

This signal is passed through an *R-C* circuit shown in Fig. P. 4-16. The output signal $s_1(t)$ is transmitted over the channel where it is additively disturbed by a channel noise $n(t)$ with power-density spectrum

$$\mathcal{S}_n(\omega) = \frac{1}{16+\omega^2}$$

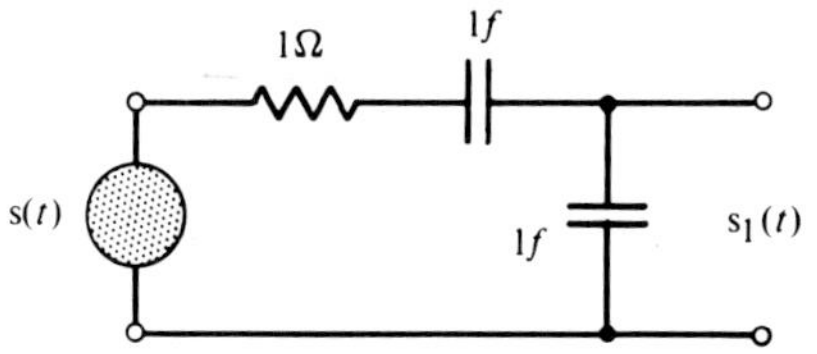

FIG. P. 4-16.

At the receiver we have $s_1(t) + n(t)$ available. Design the optimum filter (physically unrealizable and physically realizable), to obtain $s(t)$. Find the mean-square errors in each case.

4-17. Signal $s(t)$ is mixed with (independent) noise $n(t)$. It is given that

$$\mathcal{S}_s(\omega) = \frac{1}{1+\omega^2} \quad \text{and} \quad \mathcal{S}_n(\omega) = \frac{8}{16+\omega^2}$$

Design the optimum (Wiener) physically realizable filter to filter $s(t)$ from $s(t) + n(t)$. Find your answer by two methods: (a) Straightforward as given in Sec. 4-12. (b) Using a two-step filter, the first stage being a noise whitener.

4-18. Signal $s(t)$ is mixed with (independent) noise $n(t)$. It is given that

$$\mathcal{S}_s(\omega) = \frac{2}{1+\omega^2}, \quad \mathcal{S}_n(\omega) = \frac{8}{16+\omega^2}$$

(a) Design the optimum (Wiener) filter to obtain $s(t)$ from $s(t) + n(t)$ without the constraint of physical realizability. Sketch the filter unit impulse response and find the mean-square error. (b) Design the physically realizable optimum (Wiener) filter and find the mean-square error. Sketch the filter unit-impulse response. (c) Design the optimum-lag filter (physically realizable of course) with lag time 8 sec. Sketch the filter unit-impulse response and find the corresponding mean-square error. Compare this error with that obtained in parts (a) and (b). Compare the impulse response in (c) with that in (a). Comment.

4-19. A random process shown in Fig. 3-22 has $\alpha = 2$, $\overline{x} = 0$ and $\overline{x^2} = 1$. The waveform has a value of 5 volts at some instant t_0. Find the best (linear minimum mean square) estimate of the waveform at $t = t_0 + 1$. Find the mean-square error in the estimate. *Hint*: Use Eq. 4-90. Here $x = x(t_0)$ and $y = x(t_0 + 1)$. Also note that $R_{xy} = \overline{xy} = \overline{x(t_0)\,x(t_0 + 1)} = R_x(1)$ and $R_{xx} = \overline{x(t_0)\,x(t_0)} = R_x(0)$.

4-20. A random process $s(t)$ is mixed with an (independent) noise process $n(t)$. Obtain the best (linear minimum mean square) estimate of $s(t)$ from the knowledge of $s(t) + n(t)$ at the instant t alone. *Hint*: Use Eq. 4-90. Here $x = s(t) + n(t)$ and $y = s(t)$. Also

$$R_{xy} = \overline{[s(t) + n(t)]\,s(t)} = R_s(0) + R_{ns}(0) = R_s(0)$$

4-21. For a binary random process (Fig. 3-20) the parameter $b = 3$ sec. It is given that the signal value at $t = -2$ is 1. (a) Find the best estimate (linear minimum mean square) for the signal at $t = 0$. (b) Find the best estimate for signal at $t = 0$ when it is given that signal values at $t = 1$ and $t = 2$ are -1 and 1 respectively. (c) Find the mean-square errors in (a) and (b).

4-22. In the text, the matched filter analysis was carried out in frequency domain. Obtain the unit impulse response of the matched filter (for white noise) using time-domain analysis. *Hint*:

$$s_0^2(t_m) = \left[\int_{-\infty}^{\infty} h(\alpha)\,s(t_m - t)\,d\alpha\right]^2$$

and from Eq. 4-12a and the fact $R_n(\tau) = (\mathfrak{N}/2)\,\delta(\tau)$, we have

$$\overline{n_0^2(t_m)} = \frac{\mathfrak{N}}{2}\int_{-\infty}^{\infty} h^2(\alpha)\,d\alpha$$

Now use the Schwarz inequality which states that for real functions $x(t)$ and $g(t)$

$$\left[\int_{-\infty}^{\infty} x(t)\,g(t)\,dt\right]^2 \le \left[\int_{-\infty}^{\infty} x^2(t)\,dt\right]\left[\int_{-\infty}^{\infty} g^2(t)\,dt\right]$$

4-23. Obtain the transfer function of a matched filter for colored noise without using the theorem on reversibility. *Hint*: In this case

$$\overline{n_0^2(t_m)} = \frac{1}{2\pi}\int_{-\infty}^{\infty} \mathcal{S}_n(\omega)\,|H(\omega)|^2\,d\omega$$

Now use the Schwarz inequality for complex functions using

$$F_1(\omega) = \mathcal{S}_n(\omega)\,H(\omega) \quad \text{and} \quad F_2(\omega) = \frac{S(\omega)}{\mathcal{S}_n(\omega)}\,e^{j\omega t_m}$$

where

$$\mathcal{S}_n(\omega) = \mathcal{S}_n(\omega)_{\searrow}\mathcal{S}_n(-\omega)$$

4-24. Compare the performance of a simple *R-C* filter (Fig. P. 4-24a) with that of the matched filter when the binary communication is effected by a rectangular pulse (Fig. P. 4-24b). In particular calculate the maximum signal-to-noise power ratio $s_0^2(t)/\overline{n_0^2(t)}$ that can be obtained by this type of filter and compare it with that obtained by using the matched filter. *Hint*: Observe that $s_0(t)$ in this case is maximum at $t = T$. The SNR is a function of the time constant RC, and is maximum for some value of RC.

4-25. Find the decision threshold of a matched filter when the channel noise is colored. In particular, assume

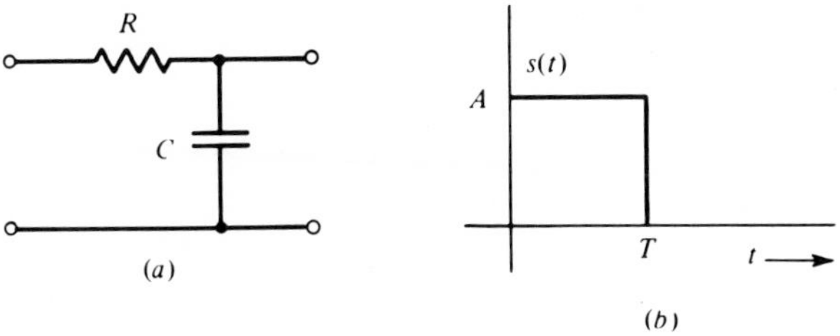

FIG. P. 4-24.

$$s(t) = u(t) - u(t-1)$$

$$\mathcal{S}_n(\omega) = \frac{\omega^2 + 1}{\omega^2 + 4}$$

$$C_a = 1 \quad \text{and} \quad C_d = 5$$

$$P(m) = 0.1$$

Find the average cost of decision.

4-26. Give the schematic of an optimum receiver for colored noise using correlator and whitening filter(s) only. The message signal $s(t)$ is available at the receiver, if desired.

4-27. A signal with power-density spectrum

$$\mathcal{S}_s(\omega) = \frac{k \,\mathrm{Sa}(\pi/2n)}{1 + (\omega/2\pi B)^{2n}}$$

is mixed with white noise with power density $\mathfrak{N}/2$. Find the transfer function of the physically realizable Wiener filter and show that the corresponding mean-square error is given by

$$\overline{\epsilon^2} = \frac{2n\mathfrak{N}B}{\mathrm{Sa}(\pi/2n)}\left\{\left[1 + \frac{2k}{\mathfrak{N}}\,\mathrm{Sa}(\pi/2n)\right]^{1/2n} - 1\right\}$$

You may use the fact that

$$\log\,[1 + k/(1 + x)^{2n}] = \int_0^k (1/x^{2n} + 1 + y)\,dy$$

4-28. If $\mathrm{x}(t)$ and $\mathrm{y}(t)$ represent the input and the desired output processes respectively, the solution to Wiener-Hopf equation (for physically realizable systems) is given by Eq. 4-71. If $\mathrm{x}(t)$ has a flat power density of magnitude k, then the solution to this equation is trivial because $R_{\mathrm{x}}(\tau) = k\delta(\tau)$ and Eq. 4-71 yields

$$R_{\mathrm{xy}}(\tau) = kh(\tau)$$

Use this result to obtain the solution of Wiener-Hopf equation (physically realizable case) when $\mathrm{x}(t)$ does not have a flat power-density spectrum. *Hint:* This can be done by the help of the reversibility theorem. First use a filter to whiten the incoming signal $\mathrm{x}(t)$. Let the output of this filter be $\mathrm{z}(t)$ with a power density of unity ($k = 1$). From the above discusion, it follows that if $h(t)$ is the impulse response of the second subsystem (which yields $\mathrm{y}(t)$ from $\mathrm{z}(t)$), then

$$R_{\mathrm{zy}}(\tau) = h(\tau) \qquad \tau \geq 0$$

Also note that if $h_{\mathrm{w}}(t)$ is the impulse response of the whitening filter, then

$$R_{\mathrm{zy}}(\tau) = \overline{\mathrm{z}(t)\,\mathrm{y}(t + \tau)}$$

$$= \int_{-\infty}^{\infty} h_w(\alpha) \overline{x(t - \alpha) y(t + \tau)} \, d\alpha$$

$$= \int_{-\infty}^{\infty} h_w(\alpha) R_{xy}(\tau - \alpha) \, d\alpha$$

$$= h_w(\tau) * R_{xy}(\tau)$$

Therefore

$$\mathcal{S}_{zy}(s) = H_w(s) \mathcal{S}_{xy}(s)$$

The impulse response of the second subsystem is evidently given by

$$h(t) = R_{zy}(\tau) = \mathcal{L}^{-1}[\mathcal{S}_{zy}^{+}(s)] \qquad \tau \geq 0$$

So

$$H(s) = \mathcal{S}_{zy}^{+}(s) = [H_w(s) \mathcal{S}_{xy}(s)]^{+}$$

5

Analog Data Communication: Modulation

5-1. ANALOG AND DIGITAL MESSAGES

The central problem in communication engineering is the efficient transmission of messages from one point and reception at the other. Broadly speaking, there are two distinct classes of messages (or data): the continuous (or analog) and the discrete (or digital). A continuous message is, as the name indicates, a continuous function of time. A voice signal is a good example of this. Other examples are atmospheric temperature and pressure, and rate of reaction in a chemical process. Obviously the possible number of messages are infinite in the case of continuous messages. In contrast to this we have digital messages which are in the form of a sequence of digital data. A good example is provided by any written message that is expressed in terms of finite (and discrete) alphabetical symbols of the language. A message in the English language, for example, can be transmitted by 27 symbols (26 letters and a space). Thus discrete messages can be represented by a finite number of symbols. The problem of communicating discrete messages therefore reduces to transmitting and receiving a finite number of symbols or messages. This is in contrast to continuous communication, where the number of messages may be infinite. There is the possibility of converting a continuous message into a digital message by using an approximation technique known as *quantization.* In this chapter we shall discuss some important techniques of communicating continuous messages (or analog data). The case of discrete messages is treated in Chapter 6.

5-2. MODULATION

At the transmitter the messages are rarely produced in a form suitable for direct transmission over a channel. Considerations of transmission efficiency requires that these messages be processed in some manner before transmitting over the channel. Consider for example the case of transmission of voice signals. The frequency spectrum of a voice signal, generally speaking, occupies the low-frequency band 0 to 5 kHz. There are at least two difficulties involved with transmitting these signals directly. If a number of voice signals are transmitted directly over a channel, they will interfere with each other and it will be impossible to recover the original signals separately at the receiver. Hence these signals must be so transformed that if after transformation they are mixed together, they can still be separated at the receiver. This is usually done either by translating their spectra (frequency-division multiplexing) or by sampling (time-division multiplexing). Secondly, the spectra of these signals consist of low-frequencies, thus making it difficult to transmit them by radiation. This is because for effective radiation the radiating-system dimensions must be of the order of a wavelength of the signal to be transmitted. Hence low-frequency signals need impracticably large antennas. For this reason it is desirable to transform the original signal so that it's spectrum is translated to a higher-frequency range. In addition, it is possible to transform a signal in such a way as to make it more immune to contamination by channel noise. As we shall see later, the increased immunity to noise contamination must be paid for in terms of an increase in the required channel bandwidth. All these transformations fall under the general term *modulation.* Thus modulation is the transformation of messages into signals suitable for transmission over the medium. Mathematically we may consider modulation as a one to one mapping from message space into signal space. At the receiver inverse transformation is performed to recover the original message (transformation from signal space to message space).

We shall study some important modulation techniques used to transmit continuous messages and compare their performance. The immediate question is: What is the basis of comparison? The answer will depend upon our communication objective. In most cases the communication objective is fidelity of reproduction of the original message. The message is distorted by the noise in the process of transmission. The ability of the listener to interpret the message depends upon the signal-to-noise ratio of the message. Therefore the signal-to-noise ratio of a message at the receiver may be taken as a measure of fidelity or goodness of a communication system. Thus the signal-to-noise power ratio (SNR) at the receiver is our figure of merit.[1] The power of a signal is defined in the usual way, viz., its mean-square value.

If $m(t)$ is the message signal and $\hat{m}(t)$ the final output of the demodulator, then the noise signal is equal to $m(t) - \hat{m}(t)$, and Λ, the signal-to-

[1]It is possible to use other merit criteria such as maximum likelihood or maximum a posteriori probability (MAP). For more discussion of these criteria, see Refs. 1, 5, and 9.

noise ratio of the output signal is given by

$$\Lambda = \frac{\overline{m^2(t)}}{\overline{[m(t) - \hat{m}(t)]^2}} \tag{5-1}$$

For a given m(t), it is evident that maximizing Λ is equivalent to minimizing the mean-square error $\overline{[m(t) - \hat{m}(t)]^2}$.

The message signals m(t) which represent the original information are called *baseband signals*. The word baseband is used as a generic term to include, for example, audio signals (radio), the video signal (television), and pulse trains representing a data source.

There are various ways of transmitting analog data. Here we shall discuss analog modulation[2] and pulse-code modulation, the latter being essentially a digital modulation used to transmit analog data. Other modulation techniques such as pulse-amplitude modulation (PAM), pulse-position modulation (PPM), and pulse-frequency modulation, will not be covered here. For further discussion on these techniques the reader is referred to Ref. 1.

When a signal is modulated, its spectrum occupies a certain finite passband. This signal is transmitted over the channel. In the process of transmission the modulated signal is corrupted by the omnipresent noise (channel noise). Generally this channel noise is additive—that is, it adds on to the desired signal by way of addition as shown in Fig. 5-1.

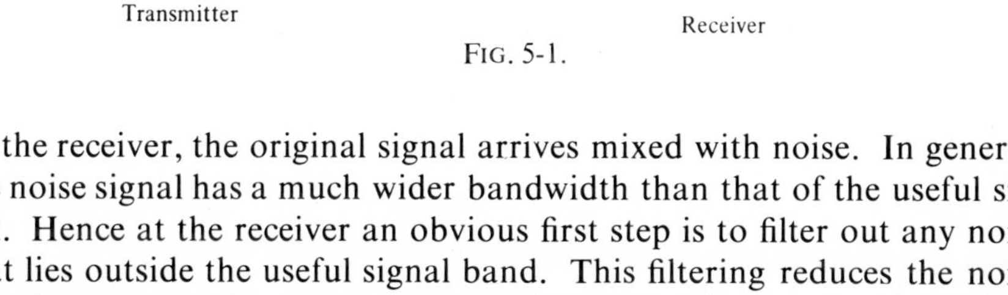

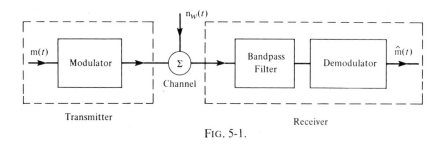

FIG. 5-1.

At the receiver, the original signal arrives mixed with noise. In general, the noise signal has a much wider bandwidth than that of the useful signal. Hence at the receiver an obvious first step is to filter out any noise that lies outside the useful signal band. This filtering reduces the noise signal but leaves the useful signal unchanged. The output of the filter contains the useful signal plus bandpass-noise. We shall now develop a representation of bandpass noise, that will enable us to deal with it effectively.

[2]Analog modulation is the generic term used for communication systems employing amplitude or angle modulation of the carrier.

5-3. REPRESENTATION OF RANDOM BANDPASS SIGNALS

A signal whose power-density spectrum exists only over a band centered at some frequency ω_c ($\omega_c \neq 0$) is called a *bandpass signal.* Figure 5-2 shows an example of the power-density spectrum of a wide-

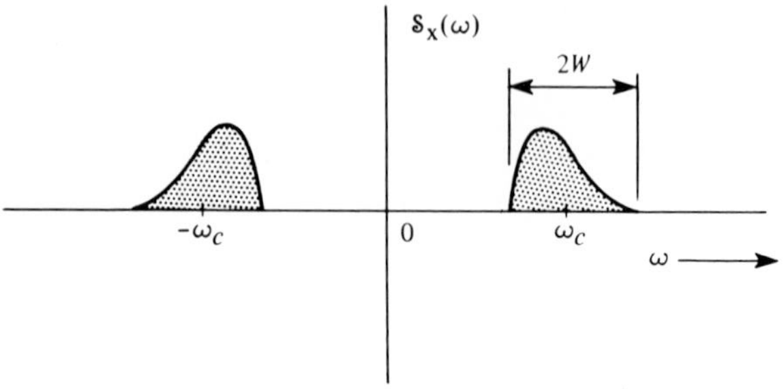

FIG. 5-2.

sense stationary bandpass process $x(t)$ of bandwidth $2W$ centered at $\omega = \omega_c$. We shall now show that such a process can be expressed as a sum of two component processes.

$$x(t) = x_c(t)\cos\omega_c t + x_s(t)\sin\omega_c t \tag{5-2}$$

where $x_c(t)$ and $x_s(t)$ are wide-sense stationary processes band-limited to W rps. Both $x_c(t)$ and $x_s(t)$ have identical power-density spectra which are related to $\mathcal{S}_x(\omega)$, the power density of $x(t)$ by the relation

$$\mathcal{S}_{x_c}(\omega) = \mathcal{S}_{x_s}(\omega) = \begin{cases} \mathcal{S}_x(\omega + \omega_c) + \mathcal{S}_x(\omega - \omega_c) & |\omega| < W \\ 0 & |\omega| > W \end{cases} \tag{5-3a}$$

It will also be shown that $\mathcal{S}_{x_c x_s}(\omega)$, the cross-power-density spectrum of $x_c(t)$ and $x_s(t)$ is given by

$$\mathcal{S}_{x_c x_s}(\omega) = -\mathcal{S}_{x_s x_c}(\omega) = \begin{cases} j\,[\mathcal{S}_x(\omega - \omega_c) - \mathcal{S}_x(\omega + \omega_c)] & |\omega| < W \\ 0 & |\omega| > W \end{cases} \tag{5-3b}$$

It can also be shown easily from Eq. 5-3 that the mean-square values (power) of $x(t)$, $x_c(t)$ and $x_s(t)$ are identical.

$$\overline{x^2} = \overline{x_c^2} = \overline{x_s^2} \tag{5-3c}$$

These results can be proved as follows:

Consider a random process $x(t)$ with power density spectrum $\mathcal{S}_x(\omega)$ as shown in Fig. 5-3a. We shall find the random process $z(t)$ whose power density spectrum is $4\mathcal{S}_x(\omega)$ for $\omega > 0$ and zero for $\omega < 0$ (Fig.

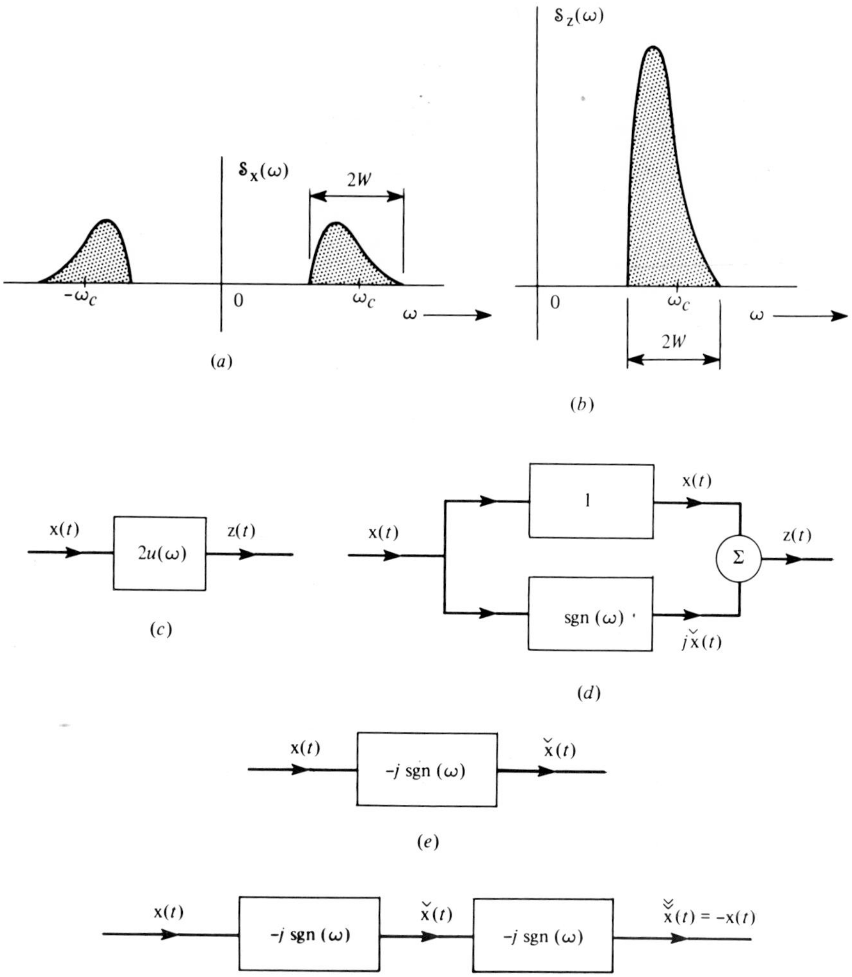

FIG. 5-3.

5-3b). This relationship may be expressed as

$$S_z(\omega) = 4\, S_x(\omega)\, u(\omega) \tag{5-4}$$

From this equation, it is evident that z(t) is the response to x(t) of a system whose transfer function $H(\omega)$ is given by (Fig. 5-3c)

$$H(\omega) = 2u(\omega)$$

This is because

$$|H(\omega)|^2 = 4u(\omega)$$

Note that

$$2u(\omega) = 1 + \text{sgn}(\omega)$$

Hence z(t) may be obtained from x(t) *by an alternate arrangement* shown in Fig. 5-3d. From this figure, we have

$$z(t) = x(t) + j\check{x}(t) \tag{5-5}$$

where $j\check{x}(t)$ is the response to x(t) of a system with transfer function sgn(ω). In other words $\check{x}(t)$ is the response to x(t) of a system with transfer function (Fig. 5-3e):

$$H_h(\omega) = -j\ \text{sgn}(\omega)$$

From Eq. 1-85b, we have

$$\frac{1}{\pi t} \leftrightarrow -j\ \text{sgn}(\omega)$$

Hence the unit impulse response of the system in Fig. 5-3e is $\frac{1}{\pi t}$ and

$$\check{x}(t) = \frac{1}{\pi t} * x(t) \tag{5-6a}$$

$$= \frac{1}{\pi} \int_{-\infty}^{\infty} \frac{x(\tau)}{t - \tau}\, d\tau \tag{5-6b}$$

$\check{x}(t)$ is called the *Hilbert transform* of x(t). The process z(t) (Eq. 5-5) is called *analytic signal* associated with process x(t). The power density of the analytic signal z(t) represents 4 times the unilateral (for $\omega > 0$) power density spectrum of x(t). Note that since

$$[H_h(\omega)]^2 = [-j\ \text{sgn}(\omega)]^2 = -1 \tag{5-7}$$

the response to $\check{x}(t)$ of $H_h(\omega)$ will be $-x(t)$ as shown in Fig. 5-3f. Thus[3]

$$\check{\check{x}}(t) = -x(t) \tag{5-8a}$$

Also, since

$$|H_h(\omega)|^2 = 1$$

$$S_{\check{x}}(\omega) = |H_h(\omega)|^2 S_x(\omega) = S_x(\omega) \tag{5-8b}$$

and

$$R_{\check{x}}(\tau) = R_x(\tau) \tag{5-8c}$$

Thus for a system with transfer function $-j$ sgn(ω), $\check{x}(t)$ is the response to input x(t) and $-x(t)$ is the response to input $\check{x}(t)$. Using relation 4-16b, we can express

$$S_{x\check{x}}(\omega) = -j\ \text{sgn}(\omega)\, S_x(\omega) \tag{5-9a}$$

$$S_{\check{x}x}(\omega) = j\ \text{sgn}(\omega)\, S_x(\omega) \tag{5-9b}$$

Therefore

[3]Throughout our discussion on Hilbert transform, it will be assumed that the process x(t) has a zero mean. If this condition is not satisfied, then Eq. 5-8a is invalid. This is because finite mean of x(t) implies an impulse at $\omega = 0$. Since $H_h(0) = 0$, this implies $\check{x}(t)$ must have a zero mean and consequently $\check{\check{x}}(t)$ must also have a zero mean. This obviously violates Eq. 5-8a.

$$\mathcal{S}_{x\check{x}}(\omega) = -\mathcal{S}_{\check{x}x}(\omega) \tag{5-10a}$$

and

$$R_{x\check{x}}(\tau) = -R_{\check{x}x}(\tau) \tag{5-10b}$$

Next we consider the process $z(t)e^{-j\omega_c t}$. Let

$$z(t)e^{-j\omega_c t} = x_c(t) - j\,x_s(t) \tag{5-11}$$

Then

$$x_c(t) = \text{Re}\,[z(t)\,e^{-j\omega_c t}] = x(t)\cos\omega_c t + \check{x}(t)\sin\omega_c t \tag{5-12a}$$

$$x_s(t) = \text{Im}\,[z(t)e^{-j\omega_c t}] = x(t)\sin\omega_c t - \check{x}(t)\cos\omega_c t \tag{5-12b}$$

From Eqs. 5-12a and 5-12b, it follows that

$$x(t) = x_c(t)\cos\omega_c t + \check{x}_s(t)\sin\omega_c t \tag{5-13a}$$

$$\check{x}(t) = x_c(t)\sin\omega_c t - x_s(t)\cos\omega_c t \tag{5-13b}$$

Equation 5-13a is the desired representation of the bandpass process. We shall now show that $x_c(t)$ and $x_s(t)$ are stationary low pass processes and derive their power density spectra. From Eq. 5-12a we have

$$\begin{aligned} R_{x_c}(\tau) &= \overline{[x(t)\cos\omega_c t + \check{x}(t)\sin\omega_c t)(x(t+\tau)\cos\omega_c(t+\tau)} \\ &\quad \overline{+\,\check{x}(t+\tau)\sin\omega_c(t+\tau)]} \\ &= \frac{1}{2}\,\{[R_x(\tau) + R_{\check{x}}(\tau)]\cos\omega_c\tau + [R_{x\check{x}}(\tau) - R_{\check{x}x}(\tau)]\sin\omega_c\tau \\ &\quad + [R_x(\tau) - R_{\check{x}}(\tau)]\cos\omega_c(2t+\tau) \\ &\quad + [R_{x\check{x}}(\tau) + R_{\check{x}x}(\tau)]\sin\omega_c(2t+\tau)\} \end{aligned}$$

Using Eqs. 5-8c and 5-10b, we have

$$R_{x_c}(\tau) = R_x\cos\omega_c\tau + R_{x\check{x}}(\tau)\sin\omega_c\tau \tag{5-14a}$$

Similarly it can be shown that

$$R_{x_s}(\tau) = R_x\cos\omega_c\tau + R_{x\check{x}}(\tau)\sin\omega_c\tau \tag{5-14b}$$

Note that

$$R_{x_c}(\tau) = R_{x_s}(\tau) \tag{5-14c}$$

Both $R_{x_c}(\tau)$ and $R_{x_s}(\tau)$ are independent of t. Therefore $x_c(t)$ and $x_s(t)$ are stationary (at least in the wide sense) processes.

To find the power density spectra of $x_c(t)$ and $x_s(t)$, we use Eqs. 5-14 and 5-9a and Eqs. 1-88a and 1-88b.

$$\begin{aligned} \mathcal{S}_{x_c}(\omega) &= \frac{1}{2}\,[\mathcal{S}_x(\omega+\omega_c) + \mathcal{S}_x(\omega-\omega_c)] \\ &\quad + \frac{1}{2}\,[\mathcal{S}_x(\omega+\omega_c)\,\text{sgn}\,(\omega+\omega_c) \\ &\quad - \mathcal{S}_x(\omega-\omega_c)\,\text{sgn}\,(\omega-\omega_c)] \end{aligned} \tag{5-15a}$$

In order to simplify this expression, let us construct a low frequency power spectrum $\mathcal{S}_\ell(\omega)$ as shown in Fig. 5-4b. It is evident from Figs.

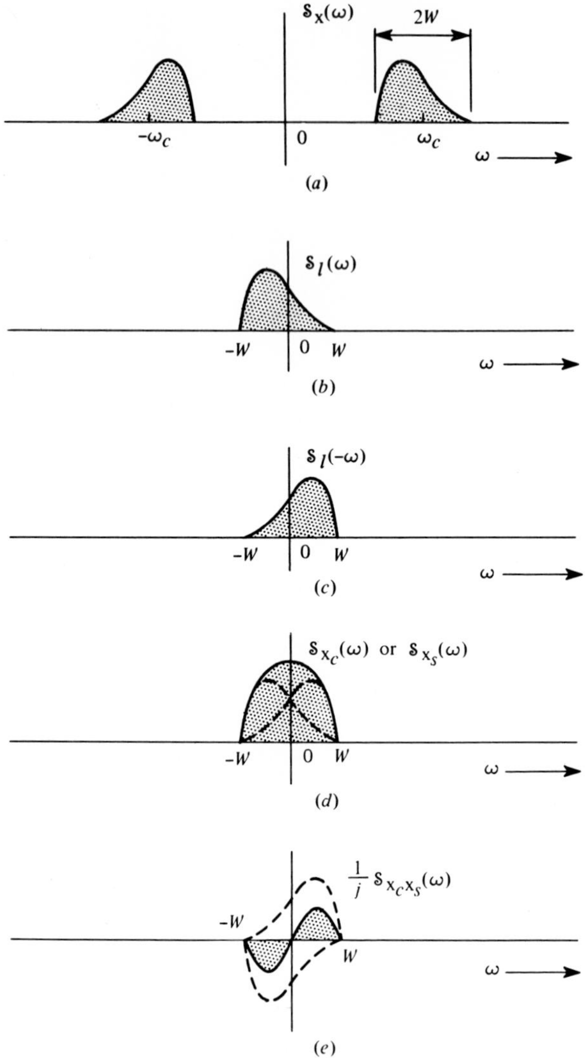

FIG. 5-4.

5-4a, 5-4b, and 5-4c that

$$\begin{aligned}
\mathcal{S}_x(\omega + \omega_c) &= \mathcal{S}_\ell(\omega) + \mathcal{S}_\ell(-\omega - 2\omega_c) \\
\mathcal{S}_x(\omega - \omega_c) &= \mathcal{S}_\ell(-\omega) + \mathcal{S}_\ell(\omega - 2\omega_c) \\
\mathcal{S}_x(\omega + \omega_c)\,\mathrm{sgn}\,(\omega + \omega_c) &= \mathcal{S}_\ell(\omega) - \mathcal{S}_\ell(-\omega - 2\omega_c) \\
\mathcal{S}_x(\omega - \omega_c)\,\mathrm{sgn}\,(\omega - \omega_c) &= -\mathcal{S}_\ell(-\omega) + \mathcal{S}_\ell(\omega + 2\omega_c)
\end{aligned} \tag{5-15b}$$

Substituting Eqs. 5-15b in Eq. 5-15a, we obtain

$$\mathcal{S}_{x_c}(\omega) = \mathcal{S}_\ell(\omega) + \mathcal{S}_\ell(-\omega) \tag{5-16a}$$

Note that

$$\mathcal{S}_{\ell}(\omega) + \mathcal{S}_{\ell}(-\omega) = \begin{cases} \mathcal{S}_{x}(\omega + \omega_c) + \mathcal{S}_{x}(\omega - \omega_c) & |\omega| < W \\ 0 & |\omega| > W \end{cases} \qquad (5\text{-}16b)$$

Also from Eq. 5-14c, we have

$$\mathcal{S}_{x_s} = \mathcal{S}_{x_c} \qquad (5\text{-}16c)$$

Obviously $x_c(t)$ and $x_s(t)$ are lowpass processes bandlimited to W rps. To find the cross power density spectrum $\mathcal{S}_{x_c x_s}(\omega)$, we follow a similar procedure. From Eqs. 5-12a and 5-12b, we can show that

$$R_{x_c x_s}(\tau) = -R_{x_s x_c}(\tau) = R_x(\tau)\sin\omega_c\tau - R_{x\hat{x}}(\tau)\cos\omega_c\tau \qquad (5\text{-}17a)$$

and

$$\mathcal{S}_{x_c x_s}(\omega) = -\mathcal{S}_{x_s x_c}(\omega) = j[\mathcal{S}_{\ell}(-\omega) - \mathcal{S}_{\ell}(\omega)] \qquad (5\text{-}17b)$$

$$= \begin{cases} j[\mathcal{S}_x(\omega - \omega_c) - \mathcal{S}_x(\omega + \omega_c)] & |\omega| < W \\ 0 & |\omega| > W \end{cases} \qquad (5\text{-}17c)$$

Note that since $\mathcal{S}_{x_c}(\omega) = \mathcal{S}_{x_s}(\omega)$, the mean square values of $x_c(t)$ and $x_s(t)$ are identical.

$$\overline{x_c^2} = \overline{x_s^2}$$

Furthermore, the power spectrum of $x_c(t)$ is identical in shape to that of $x(t)$ except that the two components (positive and negative) are shifted to the origin (Eq. 5-16a, Fig. 5-4d). Since the mean square value is proportional to the area under the power density spectrum, we must have

$$\overline{x^2} = \overline{x_c^2} = \overline{x_s^2} \qquad (5\text{-}17d)$$

The representation of a process in the form 5-13 is also known as the representation by *quadrature components.*

In passing, we note that $\mathcal{S}_{x_c x_s}(\omega)$ is an imaginary function of ω (Eq. 5-17b or Fig. 5-4e). Therefore, $R_{x_c x_s}(\tau)$ is an odd function of τ (see Probs. 1-7and 1-8). Therefore,

$$R_{x_c x_s}(0) = \overline{x_c(t)x_s(t)} = 0 \qquad (5\text{-}18a)$$

It follows that the random variables $x_c(t)$ and $x_s(t)$ are uncorrelated. If in addition, the spectrum $\mathcal{S}_{\ell}(\omega)$ is an even function of ω, then (Eq. 5-17b)

$$\mathcal{S}_{x_c x_s}(\omega) = 0$$

and

$$R_{x_c x_s}(\tau) = 0 \qquad (5\text{-}18b)$$

We therefore conclude, that if the spectrum $\mathcal{S}_{\ell}(\omega)$ is symmetrical about $\omega = 0$ (this is equivalent to saying that the positive spectrum $\mathcal{S}_x(\omega)$ is symmetrical about $\omega = \omega_c$), then the quadrature processes $x_c(t)$ and $x_s(t)$ are uncorrelated.

Bandpass White Noise Representation. For a special case of bandpass white noise process $n(t)$, the power density spectrum $\mathcal{S}_n(\omega)$ is shown in Fig. 5-5a. We can represent this process on the quadrature com-

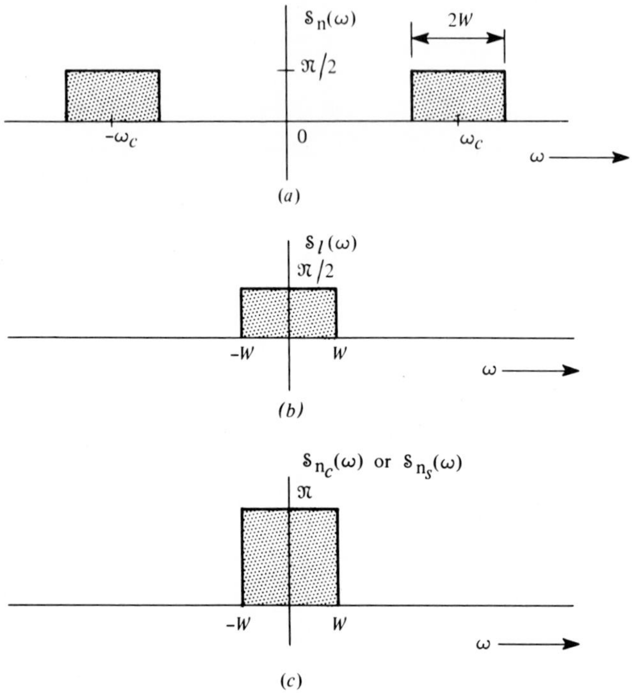

FIG. 5-5.

ponents as follows

$$n(t) = n_c(t) \cos \omega_c t + n_s(t) \sin \omega_c t \tag{5-19a}$$

where

$$\overline{n_c^2} = \overline{n_s^2} = \overline{n^2} = \left(\frac{2W}{2\pi}\right) = \frac{W}{\pi} \tag{5-19b}$$

and $\mathcal{S}_\ell(\omega)$ for this case as shown in Fig. 5-5b. Note that

$$\mathcal{S}_\ell(\omega) = \mathcal{S}_\ell(-\omega) = \begin{cases} \dfrac{\mathcal{N}}{2} & |\omega| > W/2 \\ 0 & \text{Otherwise} \end{cases}$$

and the power density spectrum $\mathcal{S}_{n_c}$ (and $\mathcal{S}_{n_s}$) is given by

$$\mathcal{S}_{n_c}(\omega) = \mathcal{S}_{n_s}(\omega) = \begin{cases} \mathcal{N} & |\omega| > W \\ 0 & |\omega| < W \end{cases} \tag{5-19c}$$

Also note that since $\mathcal{S}_{\ell}(\omega)$ is an even function of ω, the processes $n_c(t)$ and $n_s(t)$ are uncorrelated (Eq. 5-18b)

$$R_{n_c n_s}(\tau) = 0 \tag{5-19d}$$

Envelope of a Bandpass Process. Note that the bandpass process $x(t)$ in Eq. 5-13a may be expressed in a slightly different way:

$$x(t) = \mathcal{E}(t)\cos[\omega_c t + \psi(t)] \tag{5-20a}$$

where

$$\mathcal{E}(t) = \sqrt{x_c^2(t) + x_s^2(t)} \tag{5-20b}$$

$$\psi(t) = -\tan^{-1}\frac{x_s(t)}{x_c(t)} \tag{5-20c}$$

From Eq. 5-19a we observe that $\mathcal{E}(t)$ appears as the envelope of the bandpass process $x(t)$. In the literature $\mathcal{E}(t)$ is defined as the envelope of $x(t)$. From Eq. 5-11, we have

$$|z(t)| = \sqrt{x_c^2(t) + x_s^2(t)}$$

Therefore

$$\mathcal{E}(t) = |z(t)|$$

and

$$x(t) = |z(t)|\cos[\omega_c t + \psi(t)] \tag{5-20d}$$

We, therefore, conclude that the envelope of a bandpass process is given by the magnitude of $z(t)$, the analytic signal associated with $x(t)$.

It should be realized that Eq. 5-2 expresses $x(t)$ in terms of quadrature components. Therefore the addition of an arbitrary phase angle θ to both the sine and cosine terms in Eq. 5-2 should be immaterial. The reader can easily prove this result by replacing the term $\omega_c t$ by $\omega_c t + \theta$ in Eqs. 5-11, 5-12a, and 5-12b. Thus a more general representation of a bandpass signal $x(t)$ in terms of its quadrature components is given by

$$x(t) = x_c(t)\cos(\omega_c t + \theta) + x_s(t)\sin(\omega_c t + \theta) \tag{5-21}$$

where θ is arbitrary. Later we shall find it convenient to let θ be a random variable with a uniform distribution over the range $(0, 2\pi)$. Obviously Eqs. 5-2, 5-3a, 5-3b, and 5-3c are valid for any value of θ in Eq. 5-21.

5-4. AMPLITUDE MODULATION

1. DSB-SC (Double Sideband–Suppressed Carrier). In amplitude modulation, the spectrum of the message signal $m(t)$ is translated to a higher frequency ω_c by multiplying $m(t)$ by a sinusoidal signal $\cos(\omega_c t + \Theta)$. The message signal $m(t)$ is assumed to be band-limited to W rps

($W < \omega_c$). Let the modulated signal be denoted by $\phi(t)$. Then

$$\phi(t) = \mathrm{m}(t)\cos(\omega_c t + \Theta)$$

where Θ is a random variable distributed uniformly over the range $(0, 2\pi)$. To find the power-density spectrum of the modulated signal, we shall find the autocorrelation function of $\phi(t)$:

$$R_\phi(\tau) = \overline{\phi(t)\,\phi(t+\tau)}$$

$$= \overline{\mathrm{m}(t)\,\mathrm{m}(t+\tau)\cos(\omega_c t + \Theta)\cos(\omega_c t + \omega_c\tau + \Theta)} \qquad (5\text{-}22)$$

Note that both the random variables m(t) and m$(t + \tau)$ are independent (and hence uncorrelated) of random variable Θ. Hence Eq. 5-22 can be expressed as the product of two means:

$$R_\phi(\tau) = \overline{\mathrm{m}(t)\,\mathrm{m}(t+\tau)}\;\overline{\cos(\omega_c t + \Theta)\cos[\omega_c(t+\tau) + \Theta]}$$

The first term on the right-hand side of the above equation is obviously $R_{\mathrm{m}}(\tau)$, and the second term as evaluated in Ex. 3-5 is $\frac{1}{2}\cos\omega_c\tau$. Hence

$$R_\phi(\tau) = \tfrac{1}{2}R_{\mathrm{m}}(\tau)\cos\omega_c\tau \qquad (5\text{-}23a)$$

Using the modulation theorem, we obtain

$$\mathcal{S}_\phi(\omega) = \tfrac{1}{4}[\mathcal{S}_{\mathrm{m}}(\omega + \omega_c) + \mathcal{S}_{\mathrm{m}}(\omega - \omega_c)] \qquad (5\text{-}23b)$$

Equation 5-23b is the extension of the modulation theorem to random processes. The power-density spectra $\mathcal{S}_{\mathrm{m}}(\omega)$ and $\mathcal{S}_\phi(\omega)$ are shown in Figs.

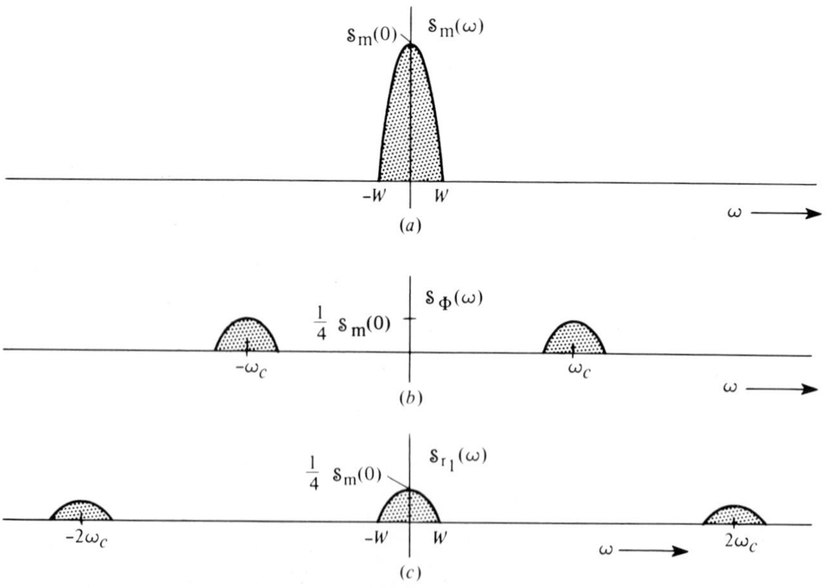

FIG. 5-6.

5-6a and 5-6b respectively. And from Eq. 5-23a note that

$$\overline{\phi^2(t)} = R_\phi(0) = \tfrac{1}{2} R_m(0) = \tfrac{1}{2}\overline{m^2(t)} \tag{5-24}$$

Hence the mean-square value of the amplitude-modulated signal is half that of the modulating signal.

Synchronous (Coherent) Demodulation. The original signal can be recovered from the DSB-SC signal $m(t)\cos(\omega_c t + \Theta)$ by synchronous demodulation. This is accomplished by multiplying the incoming signal by $\cos(\omega_c t + \Theta)$. The resultant product $r_1(t)$ is given (Fig. 5-7) by

$$\begin{aligned} r_1(t) &= [m(t)\cos(\omega_c t + \Theta)]\cos(\omega_c t + \Theta) \\ &= m(t)\cos^2(\omega_c t + \Theta) \\ &= \tfrac{1}{2}[m(t) + m(t)\cos(2\omega_c t + \Theta)] \end{aligned}$$

The first term on the right-hand side is the desired signal. The power spectrum of the second term is centered at $2\omega_c$ and can be filtered out by a low-pass filter with a cutoff frequency W rps (or B Hz, $W = 2\pi B$). The output of this filter is obviously $\tfrac{1}{2} m(t)$.

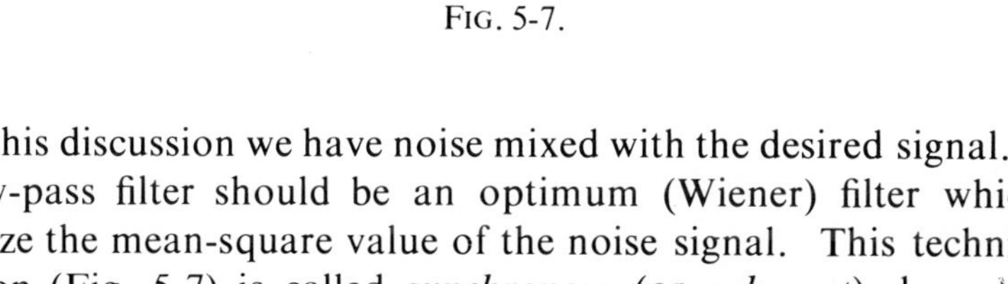

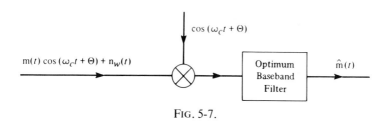

FIG. 5-7.

In this discussion we have noise mixed with the desired signal. Hence the low-pass filter should be an optimum (Wiener) filter which will minimize the mean-square value of the noise signal. This technique of detection (Fig. 5-7) is called *synchronous* (or *coherent*) demodulation because it uses a local carrier at the receiver of exactly the same frequency and phase as that of the incoming signal.

So far we have outlined a convenient method of demodulation without any regard for optimality. It will be shown in the next section that in the case of white gaussian channel noise, synchronous demodulation in Fig. 5-7 is indeed the optimum technique.

Optimality of Synchronous Demodulation. We shall now show that when we have white gaussian channel noise, the synchronous demodulation in Fig. 5-7 is optimum—that is, no other type of demodulator will yield a better signal-to-noise ratio (SNR) than that of the synchronous demodulator followed by an optimum baseband filter.[4] To begin with, in our development, we shall place no restrictions of physical realizability on our systems. The received signal $r(t)$ is given by

$$r(t) = m(t) \cos(\omega_c t + \Theta) + n_w(t)$$

where $n_w(t)$ is white gaussian channel noise of power density $\mathfrak{N}/2$. The message signal $m(t)$ and the channel noise $n_w(t)$ are assumed to be independent.

The message signal $m(t)$ is band-limited to W rps (B Hz). Hence the signal $m(t) \cos(\omega_c t + \Theta)$ is a bandpass signal with its spectrum in the frequency range $\omega_c \pm W$. The noise spectrum, however, exists over the entire frequency range $(-\infty, \infty)$. This noise can be separated into two components by using an ideal bandpass filter with passband $\omega_c \pm W$, (i) the out-of-band noise component $n_0(t)$, and (ii) the in-band noise component $n(t)$:

$$n_w(t) = n_0(t) + n(t)$$

The signal $n(t)$ is the output of a bandpass filter as shown in Fig. 5-8.

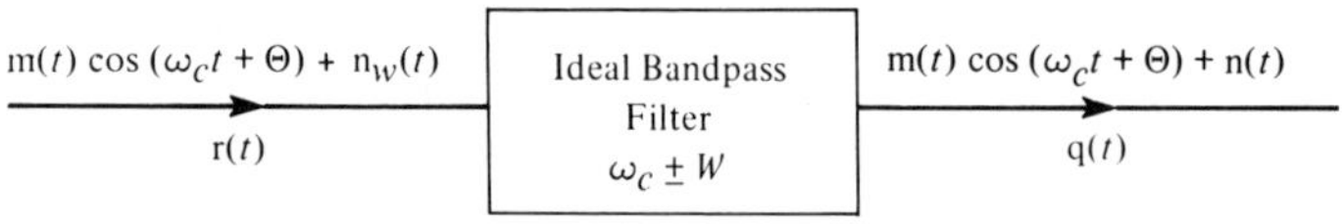

FIG. 5-8

Note that it is possible to discard the component $n_0(t)$, without disturbing the useful signal, by using an ideal bandpass filter. Therefore, as a first step to improve the SNR at the receiver it will appear reasonable to filter out the component $n_0(t)$. This conclusion, however, is not necessarily true. We cannot discard $n_0(t)$ unless we are certain that it is in no way related to the useful signal $m(t)$ as well as the remaining noise component $n(t)$. If, for example, it is related to the in-band noise $n(t)$, it will be possible to obtain information about $n(t)$ from $n_0(t)$ and thus reduce $n(t)$ itself. Therefore if the process $n_0(t)$ is independent of the process $n(t)$ we can reject $n_0(t)$ without degrading the optimum performance. It is also evident that $n_0(t)$ cannot be discarded unless it is independent of the process $m(t)$. Otherwise it will carry some information about $m(t)$. Therefore we can discard the process $n_0(t)$ if it is inde-

[4]The channel noise should be gaussian but not necessarily white. We shall show that it is only necessary for the noise to be gaussian and a power-density spectrum symmetrical about $\omega = \omega_c$ for the synchronous demodulation to be optimum.

pendent of both processes $m(t)$ and $n(t)$. The noise component $n_0(t)$ when independent of both $m(t)$ and $n(t)$, constitutes irrelevant data[5] in improving the SNR. Since by assumption, the message signal $m(t)$ is independent of the channel noise $n_w(t)$, it is also independent of the channel-noise component $n_0(t)$. Also the power-density spectra of $n_0(t)$ and $n(t)$ are disjoint (Fig. 5-9a and b). Hence these processes are

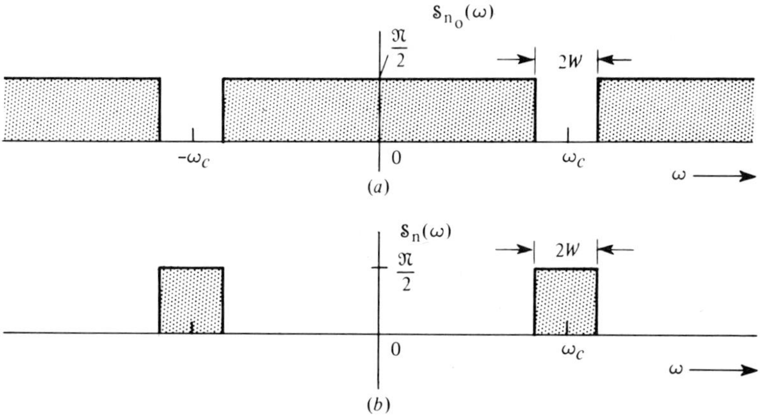

FIG. 5-9.

orthogonal (uncorrelated) as shown in Sec. 4-7. Note that both $n(t)$ and $n_0(t)$ are obtained from a gaussian process through linear operations. Hence $n(t)$ and $n_0(t)$ are both gaussian processes. So if they are uncorrelated they must necessarily be independent. Hence $n_0(t)$ is independent of both $m(t)$ and $n(t)$ and can therefore be discarded without any loss of information.

The signal that remains after discarding $n_0(t)$ will be denoted by $q(t)$, where

$$q(t) = m(t)\cos(\omega_c t + \theta) + n(t)$$

Our problem now reduces to that of finding the optimum system (operation) which will maximize the SNR.

Now $n(t)$ is a bandpass noise and can be represented in terms of quadrature components as in Eq. 5-21. Hence

$$\begin{aligned} q(t) &= m(t)\cos(\omega_c t + \theta) + n_c(t)\cos(\omega_c t + \theta) + n_s(t)\sin(\omega_c t + \theta) \\ &= [m(t) + n_c(t)]\cos(\omega_c t + \theta) + n_s(t)\sin(\omega_c t + \theta) \qquad (5\text{-}25) \end{aligned}$$

[5]This is a sufficient condition for irrelevance but not a necessary one. For more discussion on irrelevant data and a general theorem on irrelevance, see Ref. 1.

Note that $n_s(t)$ and $m(t)$ are independent processes. Also we have shown that if $\mathcal{S}_n(\omega)$ is symmetrical about $\omega = \omega_c$, then processes $n_c(t)$ and $n_s(t)$ are uncorrelated (Eq. 5-19d). Since $n(t)$ has a uniform power density it is symmetrical about $\omega = \omega_c$ (Fig. 5-9b). Hence $n_c(t)$ and $n_s(t)$ are uncorrelated random processes. Now $n_c(t)$ and $n_s(t)$ are also gaussian processes since they are obtained from a gaussian process $n(t)$ through linear operations. Thus the uncorrelatedness of processes $n_c(t)$ and $n_s(t)$ implies that they are independent. We have accordingly shown that the process $n_s(t)$ is independent of both the processes $m(t)$ and $n_c(t)$. Therefore the noise component $n_s(t) \sin(\omega_c t + \Theta)$ in Eq. 5-25 is irrelevant and may be discarded without any loss of information. Let

$$q_1(t) = [m(t) + n_c(t)] \cos(\omega_c t + \Theta)$$

Hence the data $q_1(t)$ is sufficient to obtain the optimum performance.

Next we observe that multiplying the data q_1 by $\cos(\omega_c t + \Theta)$ is a reversible operation. This is because q_1 can be obtained from $q_1 \cos(\omega_c t + \Theta)$ by remultiplying $q_1 \cos(\omega_c t + \Theta)$ by $\cos(\omega_c t + \Theta)$ and using a bandpass filter:

$$\begin{aligned}[q_1 \cos(\omega_c t + \Theta)] \cos(\omega_c t + \Theta) &= [m(t) + n_c(t)] \cos^3(\omega_c t + \Theta) \\ &= [m(t) + n_c(t)] \{\tfrac{1}{4}[3 \cos(\omega_c t + \Theta) \\ &\quad + \cos(3\omega_c t + 3\Theta)]\} \\ &= \tfrac{3}{4} q_1(t) \\ &\quad + \tfrac{1}{4}[m(t) + n_c(t)] \cos(3\omega_c t + 3\Theta)\end{aligned}$$

Note that $q_1(t)$ is a bandpass signal with spectrum in the frequency range $\omega_c \pm W$. The spectrum of the signal $[m(t) + n_c(t)] \cos(3\omega_c t + 3\Theta)$ lies in the frequency range $3\omega_c \pm W$ and can be filtered out by a bandpass filter with passband $\omega_c \pm W$. Thus $q_1(t)$ can be reconstructed from $q_1(t) \cos(\omega_c t + \Theta)$. Hence multiplying $q_1(t)$ by $\cos(\omega_c t + \Theta)$ is a reversible operation and so will not degrade the optimum performance. For this reason, new data $q_1 \cos(\omega_c t + \Theta)$ is sufficient to obtain the optimum performance. We have

$$\begin{aligned}q_1 \cos(\omega_c t + \Theta) &= [m(t) + n_c(t)] \cos^2(\omega_c t + \Theta) \\ &= \tfrac{1}{2}[m(t) + n_c(t)][1 + \cos(2\omega_c t + 2\Theta)]\end{aligned}$$

We also note that the signal $[m(t) + n_c(t)] \cos(2\omega_c t + 2\Theta)$ can be obtained from the signal $m(t) + n_c(t)$. Hence the signal $[m(t) + n_c(t)]$ carries all the information regarding the signal $q_1 \cos(\omega_c t + \Theta)$. As a result we may discard the signal $[m(t) + n_c(t)] \cos(2\omega_c t + 2\Theta)$ without losing any information. This can be done by using a low-pass filter of cutoff frequency W. (Note that both $m(t)$ and $n_c(t)$ are low-frequency signals band-limited to W.)

The power-density spectrum of $n_c(t)$ is given by Eq. 5-19c.

$$\mathcal{S}_{n_c}(\omega) = \begin{cases} \mathfrak{N} & |\omega| < W \\ 0 & \text{Otherwise} \end{cases}$$

The output signal $[m(t) + n_c(t)]/2$ contains the useful signal $m(t)/2$ with power-density spectrum $\mathcal{S}_m(\omega)/4$ and the noise signal $n_c(t)/2$ with power-density spectrum $(1/4)\,\mathfrak{N} = \mathfrak{N}/4$ (see Eq. 5-19c), over the band $(0, W)$. For convenience we shall multiply this signal by a factor 2. This will yield the signal $m(t) + n_c(t)$. The power-density spectra $\mathcal{S}_m(\omega)$ and $\mathcal{S}_{n_c}(\omega)$ of the signal and noise components, $m(t)$ and $n_c(t)$ respectively are shown in Fig. 5-10.

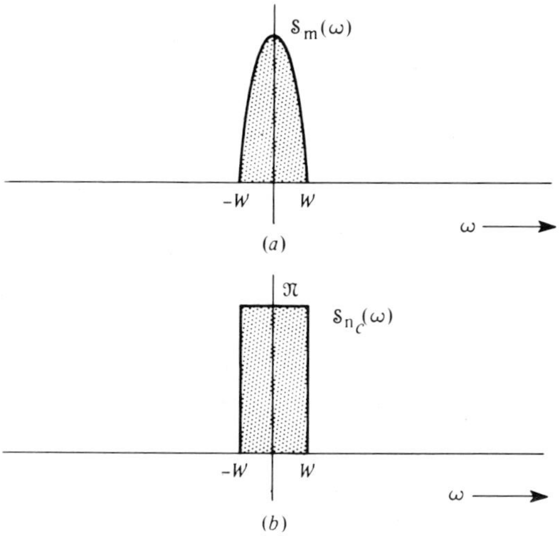

FIG. 5-10.

We must use the data $m(t) + n_c(t)$ to obtain the optimum output. The problem now reduces to the extraction of $m(t)$ from $m(t) + n_c(t)$. From Eq. 5-1, it follows that for a given $m(t)$, maximizing Λ, the signal-to-noise ratio, is equivalent to minimizing the mean-square error. Thus the problem of obtaining the optimum $m(t)$ from $m(t) + n_c(t)$ reduces to one of designing an optimum (Wiener) filter when $m(t) + n_c(t)$ is the input process and $m(t)$ is the desired output. When the system is not constrained to be physically realizable $H(\omega)$, the transfer function of the optimum system is given by Eq. 4-57:

$$H(\omega) = \frac{\mathcal{S}_{xy}(\omega)}{\mathcal{S}_x(\omega)}$$

In this case

$$\mathrm{x}(t) = \mathrm{m}(t) + \mathrm{n}_c(t) \qquad \mathrm{y}(t) = \mathrm{m}(t)$$

Since $\mathrm{m}(t)$ and $\mathrm{n}_c(t)$ are independent processes,

$$\mathcal{S}_{\mathrm{xy}}(\omega) = \mathcal{S}_{\mathrm{m}}(\omega)$$

where $\mathcal{S}_{\mathrm{m}}(\omega)$ is the power-density spectrum of $\mathrm{m}(t)$ and

$$\mathcal{S}_{\mathrm{x}}(\omega) = \mathcal{S}_{\mathrm{m}}(\omega) + \mathcal{S}_{\mathrm{n}c}(\omega)$$

where $\mathcal{S}_{\mathrm{n}_c}(\omega)$ is the power-density spectrum of $\mathrm{n}_c(t)$(Eq. 5-19c). Hence

$$H(\omega) = \frac{\mathcal{S}_{\mathrm{m}}(\omega)}{\mathcal{S}_{\mathrm{m}}(\omega) + \mathfrak{N}}$$

Note that since $\mathcal{S}_{\mathrm{m}}(\omega)$ is band-limited to W,

$$H(\omega) = 0 \qquad |\omega| > W \tag{5-26}$$

The mean-square error is given by Eq. 4-65:

$$\overline{[\mathrm{m}(t) - \hat{\mathrm{m}}(t)]^2} = \frac{1}{\pi}\int_0^W \frac{\mathfrak{N}\,\mathcal{S}_{\mathrm{m}}(\omega)}{\mathfrak{N} + \mathcal{S}_{\mathrm{m}}(\omega)}\,d\omega \tag{5-27a}$$

and

$$\Lambda = \frac{\overline{\mathrm{m}^2(t)}}{\overline{[\mathrm{m}(t) - \hat{\mathrm{m}}(t)]^2}} = \frac{\displaystyle\int_0^W \mathcal{S}_{\mathrm{m}}(\omega)\,d\omega}{\displaystyle\mathfrak{N}\int_0^W \frac{\mathcal{S}_{\mathrm{m}}(\omega)}{\mathfrak{N} + \mathcal{S}_{\mathrm{m}}(\omega)}\,d\omega} \tag{5-27b}$$

As an example, consider the case

$$\mathcal{S}_{\mathrm{m}}(\omega) = \begin{cases} \dfrac{\sigma_{\mathrm{m}}^2}{2B} & |\omega| < W \\ 0 & \text{Otherwise} \end{cases} \tag{5-28}$$

where $B(B = W/2\pi)$ is the bandwidth of $\mathrm{m}(t)$ in Hertz. Substitution of Eq. 5-28 in Eq. 5-27b yields

$$\Lambda = 1 + \frac{\sigma_{\mathrm{m}}^2}{2\mathfrak{N}B} \tag{5-29a}$$

If $\sigma_{\mathrm{m}}^2 \gg 2\mathfrak{N}B$ (small-noise case),

$$\Lambda \simeq \frac{\sigma_{\mathrm{m}}^2}{2\mathfrak{N}B} \tag{5-29b}$$

Note that in this example $\overline{\mathrm{m}^2(t)}$, the message signal power is $(\sigma_{\mathrm{m}}^2/2B)(2B) = \sigma_{\mathrm{m}}^2$. The modulated signal power is obviously $\sigma_{\mathrm{m}}^2/2$ (see Eq. 5-24). Thus in this case for a transmitted signal power $\sigma_{\mathrm{m}}^2/2$, the output SNR is given by Eq. 5-29a (or Eq. 5-29b). The output signal-to-noise ratio Λ increases as the transmitted power is increased. This is of course expected. We can plot Λ (Eq. 5-29a) as a function of the transmitted power $\sigma_m^2/2$. For convenience, however, we shall normalize the transmitted power by dividing it by the noise power in the baseband $(0, B)$. This is obviously $\mathfrak{N}B$. Thus $\sigma_m^2/2\mathfrak{N}B$ represents channel SNR within the modulating-process bandwidth. A plot Λ as a function of channel SNR $\sigma_m^2/2\mathfrak{N}B$ is shown in Fig. 5-11.

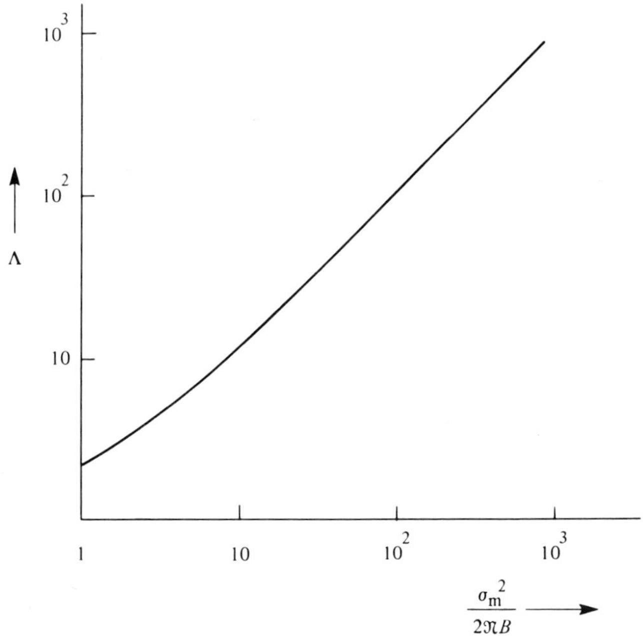

FIG. 5-11.

1. Implementation of the Optimum Demodulator. We shall now discuss the technique of implementing the procedure for optimum demodulation. This, as we shall see, reduces to the synchronous demodulation shown in Fig. 5-7.

To recapitulate, the steps in optimum demodulation are as follows. First we remove the out-of-band noise component $n_0(t)$ with an ideal bandpass filter. Next we discard the quadrature noise component $n_s(t)$. The resulting signal $q_1(t)$ is multiplied by $\cos(\omega_c t + \Theta)$ and the product is passed through an ideal low-pass filter in cascade with an optimum baseband filter. Note that the optimum filter suppresses all frequency components beyond $\omega > W$ (see Eq. 5-26). Hence the first ideal baseband filter is redundant. Therefore, we need retain the optimum (Wiener) filter alone. The procedure is shown in Fig. 5-12a. We observe

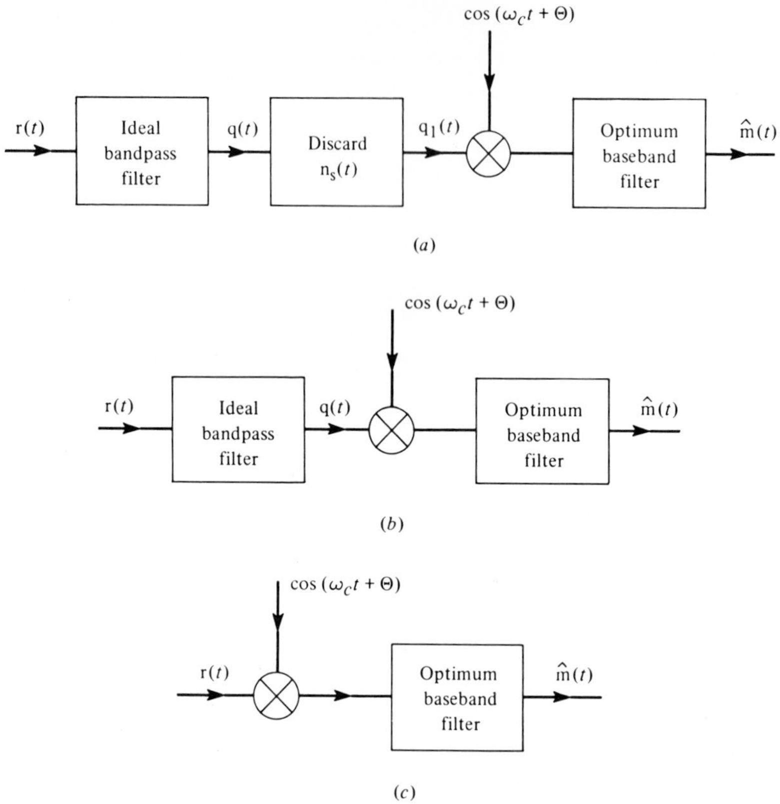

FIG. 5-12.

here that if $q(t)$ is multiplied by $\cos(\omega_c t + \Theta)$ we have

$$q(t)\cos(\omega_c t + \Theta) = [m(t)\cos(\omega_c t + \Theta) + n(t)\cos(\omega_c t + \Theta)]$$
$$= [m(t) + n_c(t)]\cos^2(\omega_c t + \Theta) + n_s(t)\sin(2\omega_c t + 2\Theta)$$

If this signal is passed through optimum baseband filter $H(\omega)$, then the double-frequency term $n_s(t)\sin(2\omega_c t + 2\Theta)$ is suppressed. Hence the operation of multiplication of $q(t)$ by $\cos(\omega_c t + \Theta)$ and transmission through the optimum baseband filter automatically removes $n_s(t)$. Therefore the optimum system reduces to that shown in Fig. 5-12b.

Next we observe that the first bandpass filter in Fig. 5-12b is not essential. The operation of multiplication of $q(t)$ by $\cos(\omega_c t + \Theta)$ and subsequent transmission through the optimum baseband filter passes only the components of $q(t)$ in the range $\omega_c \pm W$ and suppresses the remaining components. Hence the first bandpass filter is redundant and may be removed. The resulting demodulator is shown in Fig. 5-12c.

Note that the final optimum filter is physically unrealizable but can be made realizable by using delay (see discussion in Sec. 4-11). Ideally the delay must be infinite to make the filter realizable, but in practice the use of finite delay is sufficient to realize a very close approximation of the optimum filter.

2. SSB-SC (Single Sideband–Suppressed Carrier). In the DSB-SC system discussed earlier, both the sidebands of the modulated signal (Fig. 5-6b) are transmitted. The two sidebands, however, carry the same information and one of the sidebands is therefore redundant. In the SSB-SC system, one of the sidebands is suppressed and the other transmitted. The power density spectrum of the SSB-SC (upper sideband) signal corresponding to the message signal m(t) is shown in Fig. 5-13b.

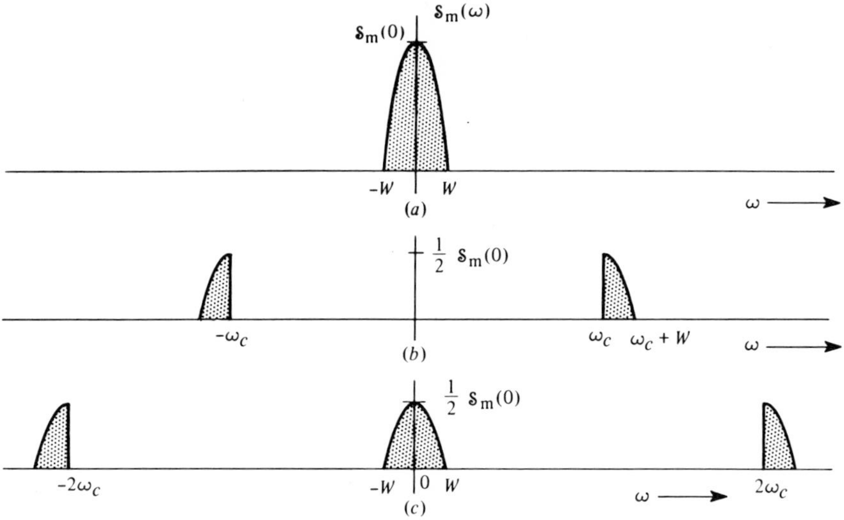

FIG. 5-13.

The SSB signal can be demodulated by synchronous detector. It can be seen from Eq. 5-23b, that if the SSB signal (Fig. 5-13b) is multiplied by a signal cos $(\omega_c t + \Theta)$, the power density spectrum of the resulting signal will be as shown in Fig. 5-13c (see Prob. 5-9 at the end of this chapter). If this signal is transmitted through a baseband filter, the output will be the desired signal.

In order to compare the performance of SSB-SC with DSB-SC, the power density is so normalized that the received power in both the SSB-SC (Fig. 5-13b) and the DSB-SC (Fig. 5-6b) systems is the same. This will facilitate making a comparison between the two systems. Following along the lines similar to those used to analyze the DSB-SC system, we can easily show that for white gaussian channel noise the syn-

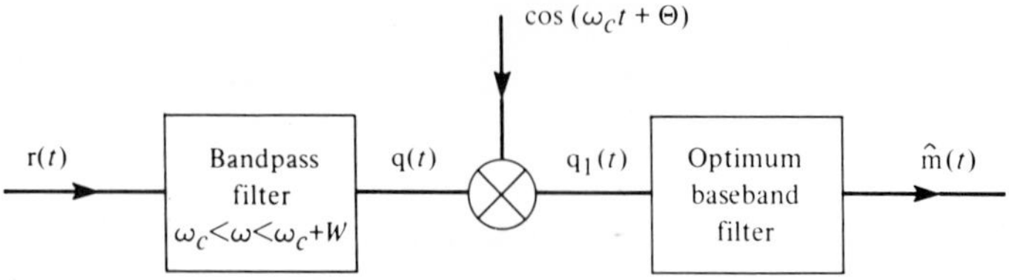

FIG. 5-14. Optimum demodulator for SSB-SC.

chronous detector shown in Fig. 5-14 is the optimum detector for SSB-SC. To prove this, we observe that the first bandpass filter in Fig. 5-14 removes the out-of-band noise component. We have seen that this noise component is irrelevant to the data required for optimum operation and can be discarded. Next it can very easily be shown that the multiplication of q(t) in Fig. 5-14 by cos $(\omega_c t + \Theta)$ is a reversible operation. It is left as an exercise for the reader to show that q(t) can be re-

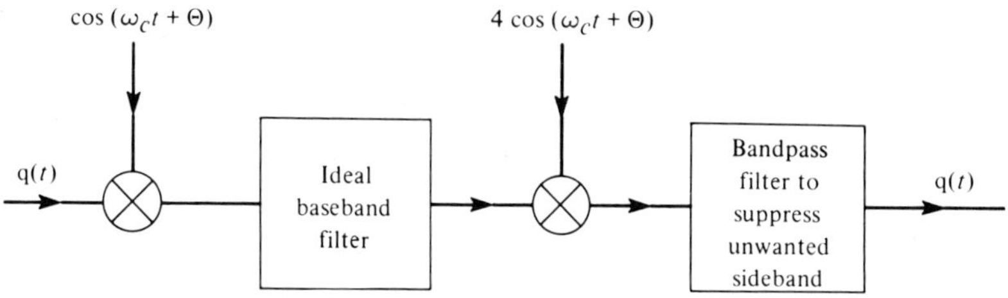

FIG. 5-15.

covered from q(t) cos $(\omega_c t + \Theta)$ by following the procedure shown in Fig. 5-15. (*Hint*: Use Eq. 5-23b). The available data now reduces to $q_1(t)$, where

$$q_1(t) = q(t) \cos (\omega_c t + \Theta)$$

We shall now find Λ, the signal-to-noise ratio at the output. For this purpose we shall first find the power-density spectrum of signal $q_1(t)$ at the input of the optimum filter (Fig. 5-14). The input signal r(t) consists of signal and noise, and similarly $q_1(t)$ consists of signal and noise components. Observe that the demodulator in Fig. 5-14 is linear. Hence we can consider the signal and noise independently. Let us first consider the signal component in r(t). The power-density spectrum of this component is shown in Fig. 5-13b. This signal is unaffected by the first bandpass filter. The signal is then multiplied by cos $(\omega_c t + \Theta)$. Using the results in Eq. 5-23b, we can find $\mathcal{S}_{s_1}(\omega)$, the power-density spectrum of the signal component of $q_1(t)$ from Fig. 5-13b. This is shown in Fig.

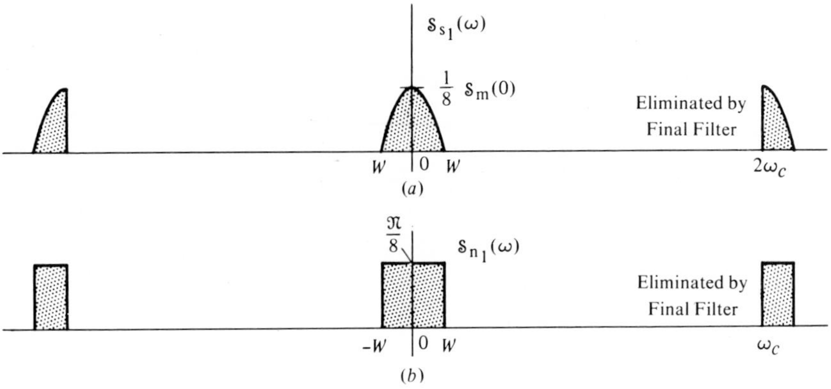

FIG. 5-16.

5-16a. The noise power density at the output of the first bandpass filter is $\mathfrak{N}/2$ over the transmitted band (of width W). This signal is multiplied by $\cos(\omega_c t + \theta)$. Hence from Eq. 5-23b, the power spectrum $\mathcal{S}_{n_1}(\omega)$ of the noise component of $q_1(t)$ can be found. This is shown in Fig. 5-16b. Note that the signal and noise components at $2\omega_c$ will eventually be eliminated by the final filter. Hence the problem is to design an optimum filter when the power spectrum of the input signal is $\mathcal{S}_{q_1}(\omega)$ given by

$$\mathcal{S}_{q_1}(\omega) = \begin{cases} \mathcal{S}_{s_1}(\omega) + \mathcal{S}_{n_1}(\omega) & |\omega| < W \\ 0 & \text{Otherwise} \end{cases}$$

A glance at Figs. 5-10 and 5-16 shows that the power density spectrum of $q_1(t)$ is identical to that of the corresponding signal for DSB-SC (except for a multiplying factor of $1/8$). Hence the filter that was optimum for DSB-SC will be the optimum in this case too. The signal-to-noise ratio will also be identical to that obtained for DSB-SC. Hence we conclude that for identical signal powers transmitted, the SNR Λ is the same for both DSB-SC and SSB-SC. The latter, however, has the advantage of requiring a bandwidth half that of DSB-SC.[6]

There are some minor differences between DSB and SSB demodulators. The first bandpass filter in the case of SSB is not redundant as was true for DSB. This is because the final optimum filter passes all the

[6]Quadrature multiplexing in DSB-SC can be used to transmit two messages over the same DSB-SC channel. In this mode the two carriers used have the same frequency but are in phase quadrature. At the receiver synchronous demodulation by two carriers of the same frequency but in phase quadrature can separate the two messages. Hence the capabilities of DSB-SC are identical to those of SSB-SC. However, from practical point of view, frequency multiplexing of two SSB signals is preferred to quadrature multiplexing of two signals over one DSB channel.

components of the incoming signal in the range $\omega_c \pm W$ (upper and lower sidebands) whereas we require the passage of only one sideband in SSB system. Hence the first bandpass filter which allows only one sideband to pass is essential for SSB-SC system.

3. AM with Large Carrier (Envelope Detection). For suppressed-carrier systems, it is necessary to generate a local carrier at the receiver to be used for synchronous demodulation. This necessitates a rather expensive receiver. This feature is undesirable in public broadcast systems where it is required to have inexpensive receivers. For this reason in such systems a large amount of carrier is transmitted along with sidebands. The modulated signal $\phi(t)$ in this case is given by

$$\begin{aligned}\phi(t) &= A\cos(\omega_c t + \Theta) + \mathrm{m}(t)\cos(\omega_c t + \Theta)\\ &= [A + \mathrm{m}(t)]\cos(\omega_c t + \Theta)\end{aligned} \tag{5-30}$$

If A is made sufficiently large so that

$$A + \mathrm{m}(t) > 0 \tag{5-31}$$

it is then evident from Eq. 5-30 that $A + \mathrm{m}(t)$ will be the envelope of $\phi(t)$. The desired signal $\mathrm{m}(t)$ can be recovered from $\phi(t)$ by using an envelope detector. The channel noise is assumed to be white gaussian. As in the case of suppressed-carrier systems, we suppress the out-of-band noise at the receiver with an ideal bandpass filter of passband $\omega_c \pm W$. The output of this filter is fed to the envelope detector. The output of the latter may in turn be fed to optimum baseband filter, but in conventional receivers the optimum filter is omitted. The improvement obtained by the optimum baseband filter is generally not worth the additional expense in the commercial receivers.

The conventional demodulator is shown in Fig. 5-17. In the absence

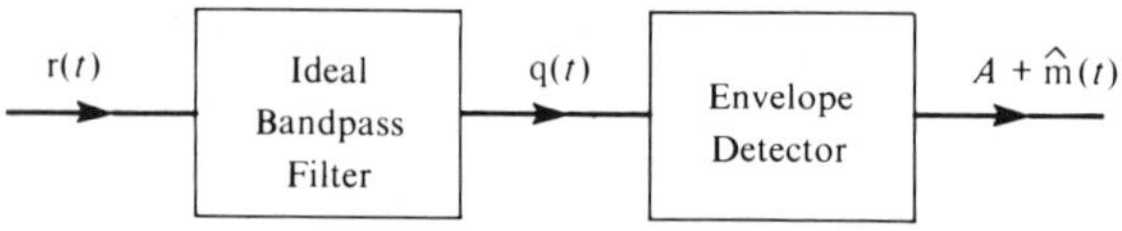

FIG. 5-17.

of noise, the output of the envelope detector is $A + \mathrm{m}(t)$ as seen from Eq. 5-30, provided $A + \mathrm{m}(t) > 0$. The constant A can be subtracted (or suppressed by a capacitor) so the final output is $\mathrm{m}(t)$. In studying the output noise, we observe that

$$\mathrm{r}(t) = [A + \mathrm{m}(t)]\cos(\omega_c t + \Theta) + \mathrm{n_w}(t)$$

The bandpass filter suppresses the out-of-band noise component and the output q(t) is given by

$$q(t) = [A + m(t)] \cos(\omega_c t + \Theta) + n(t)$$

where n(t) is bandpass noise and can be expressed in terms of quadrature components:

$$q(t) = [A + m(t)] \cos(\omega_c t + \Theta)$$
$$+ n_c(t) \cos(\omega_c t + \Theta) + n_s(t) \sin(\omega_c t + \Theta)$$
$$= [A + m(t) + n_c(t)] \cos(\omega_c t + \Theta) + n_s(t) \sin(\omega_c t + \Theta)$$
$$= \mathcal{E}(t) \cos[\omega_c t + \Theta + \psi(t)] \tag{5-32}$$

where the envelope $\mathcal{E}(t)$ is given by

$$\mathcal{E}(t) = \sqrt{[A + m(t) + n_c(t)]^2 + n_s^2(t)} \tag{5-33a}$$

and

$$\psi(t) = -\tan^{-1}\left[\frac{n_s(t)}{A + m(t) + n_s(t)}\right] \tag{5-33b}$$

We shall now separately consider the effect of small and large amounts of noise.

(a) *Small-Noise Case.* If the noise signal n(t) is much smaller than the signal $A + m(t)$, then

$$A + m(t) \gg n_c(t) \quad \text{and} \quad n_s(t)$$

Note that $n_c(t)$ and $n_s(t)$ being gaussian random variables, there will be times when they assume very large values so that $A + m(t)$ may be smaller than $n_c(t)$ or $n_s(t)$ at these times. However, if A is made large enough such occasions will be rare. Hence we say

$$A + m(t) \gg n_c(t) \text{ and } n_s(t) \quad \text{for almost all } t$$

so the term $n_s^2(t)$ in Eq. 5-33a can be ignored. Therefore

$$\mathcal{E}(t) \cong A + m(t) + n_c(t) \quad \text{for almost all } t$$

The constant A on the right-hand side can be subtracted (blocked by a capacitor) to yield the signal:

$$\mathcal{E}_1(t) = m(t) + n_c(t)$$

The problem now reduces to filtering m(t) from $m(t) + n_c(t)$. This is identical to the situation in DSB-SC (or SSB-SC). Hence the final optimum Λ in this case must be the same as that for DSB-SC (or SSB-SC) and Λ, the output SNR will also be the same (Eq. 5-29):

$$\Lambda = 1 + \frac{\sigma_m^2}{2\mathfrak{N}B} \tag{5-34a}$$

$$\cong \frac{\sigma_m^2}{2\mathfrak{N}B} \quad \text{for small noise} \tag{5-34b}$$

The sideband signal in the modulated signal $\phi(t)$ (Eq. 5-27) is $m(t) \cos(\omega_c t + \Theta)$ and has the power $(1/2)\overline{m^2(t)}$. Thus in all the three modes —DSB-SC, SSB-SC, and conventional AM (for the small-noise cases)— we have shown that Λ, the output SNR is the same when the modulation sidebands carry the same transmitted power. Hence all these systems perform identically. Conventional AM, however, requires that extra power be carried by the free carrier. This does not contribute in any way to improving the SNR. The carrier amplitude A must be large enough, so that $A + m(t) > 0$ almost all the time. This may require that the carrier power be anywhere from 3 to 10 db or higher than the power of the modulated sidebands, depending upon the modulation index and the wave shape of the signal. This is the price paid for avoiding the synchronous demodulation. Thus while we avoid an expensive receiver to generate a coherent local carrier, we need a more powerful transmitter.

(b) *Large-Noise Case.* In this case, we shall find convenient to express the bandpass noise $n(t)$ in the form of Eq. 5-20a

$$n(t) = R(t)\cos(\omega_c t + \psi)$$

where $R(t)$ is the envelope of the noise given by

$$R(t) = \sqrt{n_c^2(t) + n_s^2(t)} \quad \text{and} \quad \psi(t) = -\tan^{-1}\frac{n_s(t)}{n_c(t)} \tag{5-35}$$

Equation 5-33a can now be expressed as

$$\mathcal{E}(t) = \sqrt{[A + m(t)]^2 + R^2(t) + 2n_c(t)[A + m(t)]}$$

For a large-noise case, $A + m(t) \ll R(t)$ for almost all t. Hence

$$\mathcal{E}(t) \simeq R(t)\left\{1 + \frac{2[A + m(t)]}{R(t)}\cos\psi(t)\right\}^{1/2}$$

Using binomial approximation, we now have

$$\mathcal{E}(t) \simeq R(t)\left[1 + \frac{A + m(t)}{R(t)}\cos\psi(t)\right]$$

$$= R(t) + [A + m(t)]\cos\psi(t) \tag{5-36}$$

A glance at Eq. 5-36 shows that the envelope $\mathcal{E}(t)$ contains no term proportional to $m(t)$. The signal $m(t)\cos\psi(t)$ represents the message signal $m(t)$ multiplied by a random time varying signal $\psi(t)$ (Eq. 5-35), so this cannot be used to recover $m(t)$.

It is evident from this discussion that for a large-noise case the signal is mutilated by the envelope detector. This behavior accounts for the so-called threshold effect in envelope detectors. The threshold is the value of the input signal-to-noise ratio below which the output signal-to-noise ratio deteriorates much more rapidly than the input signal-to-noise ratio. The threshold effect begins to appear in the region where the carrier-power-to-noise-power ratio approaches unity. For more discussion on the low-SNR case, see Refs. 5-2, 5-3, and 5-11.

It should be stressed that the threshold effect is a property of envelope detectors. No such effect is observed when using synchronous demodulators. The SNR derived for the synchronous case was independent of the relative strengths of the signal and noise. When large noise is present one can avoid the threshold effect in conventional AM by using synchronous demodulation instead of an envelope detection.

Comparison of Various AM Systems. We have discussed the performance of AM-SC (DSB-SC and SSB-SC) and AM (with large carrier) systems. It is interesting to compare these systems with various points of views.

The AM has advantage over AM-SC at the receiver end. The detectors required for AM are relatively simpler (rectifier or envelope detectors) than those required for suppressed carrier systems. For this reason all public communication systems use AM. In addition, AM signals are easier to generate at high power levels compared to suppressed carrier signal. Balanced modulators required in the latter are somewhat difficult to design.

The suppressed carrier systems have the advantage over AM in that they require much less power to transmit the same information. This makes the transmitter less expensive than those required for AM. But the receivers in these systems are more complex since they must be capable of generating a local carrier of the right phase and frequency. For a point-to-point communication systems where there are only a few receivers per transmitter, this complexity in the receiver may be justified.

The effect of *selective fading* (due to multipath propagation), however, is much more disastrous on AM than AM-SC. The fading occurs because of the arrival of the signal at the receiver over more than one path of propagation, each of a different length. This causes the phases of signal arriving by different paths to differ. The resultant signal is the sum of all the signals arriving by various paths. The signal waves along their path are reflected by the ionosphere and the ground. The ionospheric conditions change randomly with time, thus causing the path length to vary randomly. The phases of signals arriving by various paths at the receiver therefore change randomly. Hence the strength of the resultant signal varies randomly. This phenomenon is called *fading*. The fading is also sensitive to frequencies, and because of this its effects become even more serious. Thus the carrier and each of the sidebands undergo different amounts of fading. This effect is called selective fading. The selective fading disturbs different sidebands by different amounts and causes distortion. It also disturbs the relationship between magnitudes of the carrier and the sidebands. It may cause more severe fading of the carrier than the sidebands to the point where the condition (5-31) is no longer fulfilled. Such a waveform, when detected by an envelope detector (or a rectifier detector), will be heavily distorted. The effect of selective fading becomes severe at high frequencies. Therefore at high frequencies, suppressed carrier systems are preferred.

Next we compare DSB-SC with SSB-SC. Here we find that SSB is almost always preferable. The following are the advantages of SSB over DSB.

1. The bandwidth required for SSB signals is half that required for DSB signals.[7]

2. The selective fading effect discussed earlier disturbs the phase relationships of two sidebands in DSB. This gives rise to the distortion in the demodulator output. In SSB since there is only one sideband, this possibility does not exist. Under long-range propagation conditions, the effects of selective fading prove more deleterious to DSB than to SSB.

For these reasons SSB-SC is preferred to DSB-SC; SSB-SC is used in long-range high frequency communications, particularly in audio range where phase distortions are relatively unimportant. Amateur-operated radios use SSB signals.

SSB, however, has a disadvantage in comparison to DSB in one respect. The generation of high level SSB signals is much more difficult than that of DSB signals because of the difficulty in constructing an ideal filter required to suppress one sideband and transmit the other. This disadvantage is overcome in what is called vestigial sideband transmission. This mode of transmission is, in reality, a compromise between DSB and SSB; it combines the advantages of both systems and eliminates their disadvantages at a very little cost. In this mode, instead of rejecting one sideband completely (as in SSB), a gradual cutoff of one sideband is allowed (Ref. 5-10). This system demands a slightly larger bandwidth than that of SSB, and is used in transmitting television signals.

5-5. ANGLE MODULATION

In angle modulation the information in the message signal is contained in the phase angle of a sinusoidal carrier. The angle-modulated

[7]Use of quadrature multiplexing can balance out this difference. This method, however, introduces more cross talk (co-channel interference) due to nonidealities in the bandpass channel in DSB than that attainable by frequency multiplexing in SSB.

carrier is of the form $A \cos [\omega_c t + \Theta + s(t)]$, where $s(t)$ contains the information in $m(t)$. There are two forms of angle modulation: (i) phase modulation, in which the phase of the carrier is varied in proportion to $m(t)$, and (ii) frequency modulation, in which the instantaneous frequency is varied in proportion to $m(t)$. The instantaneous frequency of a carrier is given by the time rate of change of the phase. Thus the instantaneous frequency of $A \cos [\psi(t)]$ is $d\psi/dt$. Note that for a pure sinusoidal $A \cos \omega_c t$, the instantaneous frequency is $\omega_i = d\psi/dt = \omega_c$. This is constant for all t.

For phase modulation (PM), the carrier $\phi_{pm}(t)$ is given by

$$\phi_{pm}(t) = A \cos [\omega_c t + \Theta + k_p m(t)] \tag{5-37}$$

where Θ is a random variable uniformly distributed over the interval $(0, 2\pi)$, and k_p is a constant. For frequency modulation (FM), the carrier $\phi_{fm}(t)$ is given by

$$\phi_{fm}(t) = A \cos \left[\omega_c t + \Theta + k_f \int_{-\infty}^{t} m(\alpha)\, d\alpha\right] \tag{5-38}$$

where Θ is a random variable uniformly distributed over the range $(0, 2\pi)$ and k_f is a constant. Note that for FM carrier ω_i, the instantaneous frequency is given by

$$\begin{aligned} \omega_i &= \frac{d}{dt}\left[\omega_c t + \Theta + k_f \int_{-\infty}^{t} m(\alpha)\, d\alpha\right] \\ &= \omega_c + k_f m(t) \end{aligned} \tag{5-39}$$

Hence ω_i is varied about ω_c in proportion to $m(t)$. From Eqs. 5-37 and 5-38, it is evident that PM and FM are not essentially different. If we integrate $m(t)$ and use the integral to phase modulate a carrier we obtain a frequency-modulated carrier for $m(t)$. Similarly, if a carrier is frequency modulated by dm/dt, we obtain a phase-modulated carrier for $m(t)$. Thus a general expression

$$\phi(t) = A \cos [\omega_c t + \Theta + s(t)] \tag{5-40}$$

can represent both PM and FM by an appropriate choice of $s(t)$. For phase modulation, we let

$$s(t) = k_p m(t)$$

and for frequency modulation, we let

$$\dot{s}(t) = k_f m(t)$$

Autocorrelation Function of Angle-Modulated Wave [4]. We shall now find $R_\phi(\tau)$, the autocorrelation function of $\phi(t)$ in Eq. 5-40. In this case Θ is a random variable uniformly distributed over the interval $(0, 2\pi)$, and $s(t)$ is a process which carries the information of $m(t)$.

$$\begin{aligned}R_\phi(t_1,t_2) &= \overline{\phi(t_1)\phi(t_2)}\\ &= A^2\overline{\cos[\omega_c t_1 + \Theta + s(t_1)]\cos[\omega_c t_2 + \Theta + s(t_2)]}\\ &= \frac{A^2}{2}\{\overline{\cos[\omega_c(t_2-t_1) + s(t_2) - s(t_1)]}\\ &\qquad + \overline{\cos[\omega_c(t_2+t_1) + 2\Theta + s(t_2) + s(t_1)]}\}\end{aligned}$$

Note that the second term contains 2Θ as one of the random variable with uniform distribution of Θ over $(0,2\pi)$. Obviously this term must vanish when averaged (integrated) over the range $(0,2\pi)$ with respect to the variable Θ. Hence

$$\begin{aligned}R_\phi(\tau) &= \frac{A^2}{2}\overline{\cos[\omega_c(t_2-t_1) + s(t_2) - s(t_1)]}\\ &= \frac{A^2}{2}\overline{\cos[\omega_c(t_2-t_1) + s_2 - s_1]}\end{aligned} \tag{5-41}$$

where $s_1 = s(t_1)$ and $s_2 = s(t_2)$.

The right-hand side of Eq. 5-41 contains two random variables s_1 and s_2. Hence averaging this quantity with respect to s_1 and s_2 will eliminate s_1 and s_2, so the final result depends solely on t_1 and t_2 only through the difference $t_2 - t_1$.[8] Hence $\phi(t)$ is wide-sense stationary. From Eq. 5-41, we have

$$\begin{aligned}R_\phi(\tau) &= \frac{A^2}{2}\overline{\cos[\omega_c\tau + s_2 - s_1]}\\ &= \frac{A^2}{2}\operatorname{Re}[e^{j\omega_c\tau}\overline{e^{j(s_2-s_1)}}]\\ &= \frac{A^2}{2}\operatorname{Re}\left[e^{j\omega_c\tau}\int_{-\infty}^{\infty}\int_{-\infty}^{\infty} p(s_1,s_2)e^{-j(s_1-s_2)}\,ds_1\,ds_2\right]\end{aligned} \tag{5-42}$$

Let us consider the case where $s(t)$ is a stationary gaussian process with zero mean and autocorrelation function $R_s(\tau)$. The covariance matrix (see Eq. 3-91) is given by

$$K = \begin{bmatrix} R_s(0) & R_s(\tau)\\ R_s(\tau) & R_s(0)\end{bmatrix}$$

and

$$\begin{aligned}|K| &= R_s^2(0) - R_s^2(\tau)\\ \Delta_{11} &= \Delta_{22} = R_s(0)\\ \Delta_{12} &= \Delta_{21} = -R_s(\tau)\end{aligned} \tag{5-43}$$

Hence

$$p(s_1,s_2) = \frac{1}{2\pi\sqrt{|K|}}\,e^{-[R_s(0)(s_1^2+s_2^2) - 2R_s(\tau)s_1s_2]/2|K|} \tag{5-44}$$

[8]This is because $s(t)$ is a wide sense stationary process.

where

$$|K| = R_s^2(0) - R_s^2(\tau)$$

Substituting Eq. 5-44 in Eq. 5-42 and integrating, we obtain

$$R_\phi(\tau) = \frac{A^2}{2} \text{Re}\,[e^{j\omega_c\tau} e^{R_s(\tau) - R_s(0)}]$$

$$= \frac{A^2}{2} e^{[R_s(\tau) - R_s(0)]} \cos \omega_c \tau \tag{5-45}$$

The power-density spectrum $\mathcal{S}_\phi(\omega)$ is the Fourier transform of $R_\phi(\tau)$. In general, this transform cannot be obtained in a closed form.

For a phase-modulated (PM) carrier,

$$s(t) = k_p m(t)$$

and

$$R_s(\tau) = k_p^2 R_m(\tau)$$

Also,

$$R_s(0) = k_p^2 R_m(0)$$

Hence for PM,

$$R_\phi(\tau) = \frac{A^2}{2} e^{k_p^2 [R_m(\tau) - R_m(0)]} \cos \omega \tau \tag{5-46}$$

and for frequency modulated (FM) carrier,

$$s(t) = k_f \int_{-\infty}^{t} m(\alpha)\, d\alpha$$

If we let

$$\int_{-\infty}^{t} m(\alpha)\, d\alpha = g(t)$$

then

$$s(t) = k_f g(t)$$

and

$$R_\phi(\tau) = \frac{A^2}{2} e^{k_f^2 [R_g(\tau) - R_g(0)]} \cos \omega_c \tau \tag{5-47}$$

Note that Eqs. 5-46 and 5-47 are valid only for a gaussian process $m(t)$.

Bandwidth Estimation for Angle-Modulated Carrier. Since the power-density spectrum of an angle-modulated wave usually cannot be obtained in a closed form, it is not possible to find a general expression for the bandwidth. One can, however, obtain a fair estimate of the bandwidth by studying the variation in the instantaneous frequency ω_i. Consider the FM carrier

$$\phi_{fm}(t) = A \cos [\omega_c t + \Theta + k_f \int_{-\infty}^{t} m(\alpha)\, d\alpha]$$

and

$$\omega_i = \omega_c + k_f \mathrm{m}(t) \tag{5-48}$$

Thus ω_i varies about ω_c. The carrier-frequency deviation is given by

$$\text{Carrier frequency deviation} = \omega_i - \omega_c = k_f \mathrm{m}(t) \tag{5-49}$$

If $\Delta\omega$ is the maximum (one sided) deviation in the carrier frequency, then

$$\Delta\omega = k_f \,|\, \mathrm{m}(t) \,|_{\max} \tag{5-50}$$

Thus the carrier frequency deviates from its quiescent value ω_c at most by $\pm\Delta\omega$. Hence it is reasonable to assume that the frequency spectrum of $\phi_{fm}(t)$ lies within the range $(\omega_c - \Delta\omega,\ \omega_c + \Delta\omega)$, and the bandwidth W_f of $\phi_{fm}(t)$ is given by

$$W_f \simeq 2\Delta\omega$$

$$= 2k_f \,|\, \mathrm{m}(t) \,|_{\max} \tag{5-51}$$

We can also obtain this result by using a quasi-static model. Since $\mathrm{m}(t)$ is band-limited to W rps or B Hz ($B = W/2\pi$), the signal $\mathrm{m}(t)$ does not change appreciably over the Nyquist interval $1/2B$. Hence we can approximate $\mathrm{m}(t)$ by a staircase function as shown in Fig. 5-18a. The FM carrier corresponding to this approximated $\mathrm{m}(t)$ will consist of sinusoidal pulses of width $1/2B$ seconds and constant frequency. The frequency changes abruptly at every sampling instant. Consider a

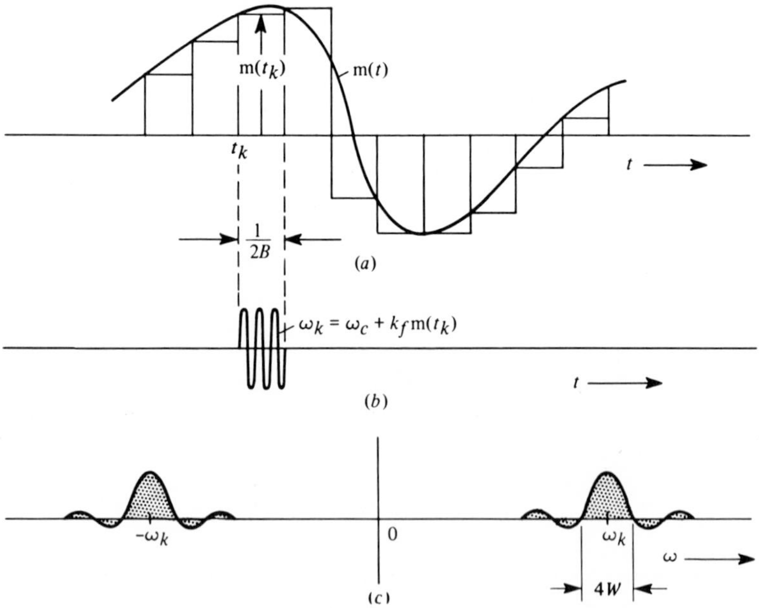

FIG. 5-18.

typical pulse of width $1/2B$ seconds as shown in Fig. 5-18b. The frequency of this pulse is ω_k given by $\omega_k = \omega_c + k_f \mathrm{m}(t_k)$. The frequency spectrum of this pulse can be obtained from pair 13 (Table 1-1) and the modulation theorem. The spectrum is a sampling function shifted to $\pm\omega_k$ as shown in Fig. 5-18c. It is evident from this figure that most of the spectrum of the pulse in Fig. 5-18b lies in the range $\omega_k \pm 2W$ where $\omega_k = \omega_c + k_f \mathrm{m}(t_k)$. The frequency ω_k varies from $\omega_c - k_f |\mathrm{m}(t)|_{\max}$ to $\omega_c + k_f |\mathrm{m}(t)|_{\max}$ or from $\omega_c - \Delta\omega$ to $\omega_c + \Delta\omega$. Hence the spectrum of the entire $\mathrm{m}(t)$ will lie in the range $\omega_c - \Delta\omega - 2W$ to $\omega_c + \Delta\omega + 2W$. This represents a bandwidth $2\Delta\omega + 4W$. Thus

$$\begin{aligned} W_f &\simeq 2\Delta\omega + 4W \\ &= 2(\Delta\omega + 2W) \end{aligned} \tag{5-52}$$

For wideband FM $\Delta\omega \gg W$ and

$$W_f \simeq 2\Delta\omega$$

which is the same expression obtained earlier in Eq. 5-51. The expression 5-52 is, however, a more accurate estimate than that of 5-51. It should be noted that the spectrum of the FM carrier exists over the entire frequency range $(-\infty, \infty)$ in general. However, most of the power is concentrated over the band W_f in Eq. 5-52, and hence this is the bandwidth.

Mean-Square Bandwidth of Angle-Modulated Signals. Whenever a function is distributed over the entire argument over the range $(-\infty, \infty)$, the width of the distribution is obviously infinity. However, most of the function is concentrated over a finite range. In such cases the concept of mean-square width of the function proves quite useful. For a random variable x, the mean-square value $\overline{\mathrm{x}^2}$ is defined as

$$\overline{\mathrm{x}^2} = \int_{-\infty}^{\infty} x^2 p(x)\,dx$$

The mean-square value $\overline{\mathrm{x}^2}$ conveys some information as to the spread of the function $p(x)$ (when the function $p(x)$ is centered at the origin) and is therefore a measure of the width of $p(x)$. Note that the function $p(x)$ is so normalized that area under it is unity. For a function $g(t)$ (centered at the origin), we can define its width (mean-square width) along similar lines, making sure that $g(t)$ has been normalized to unit area. If normalization is not done, then we will find that the same signal multiplied by a factor k will have k times as much width as the original signal. This is obviously incorrect because multiplying a signal $g(t)$ by a constant does not change its width. Hence for a meaningful definition of width, normalization is important. We define the width (*mean-square width*) of $g(t)$ as

$$\overline{W_g^2} = \frac{1}{A_g} \int_{-\infty}^{\infty} t^2 g(t)\, dt \tag{5-53}$$

where

$$A_g = \int_{-\infty}^{\infty} g(t)\, dt \tag{5-54}$$

Thus for a power-density spectrum $\mathcal{S}_x(\omega)$, the mean-square width $\overline{W_x^2}$ is defined as[9]

$$\overline{W_x^2} = \frac{1}{2\pi R_x(0)} \int_{-\infty}^{\infty} \omega^2 \mathcal{S}_x(\omega)\, d\omega \tag{5-55}$$

provided the integral exists.

Note that $2\pi R_x(0)$ is the normalizing factor, since

$$\int_{-\infty}^{\infty} \mathcal{S}_x(\omega)\, d\omega = 2\pi R_x(0)$$

To find $\overline{W_x^2}$, we observe that

$$R_x(\tau) = \frac{1}{2\pi} \int_{-\infty}^{\infty} \mathcal{S}_x(\omega)\, e^{j\omega\tau}\, d\omega$$

upon differentiating this equation twice with respect to τ and setting $\tau = 0$, we obtain

$$\begin{aligned} R_x''(0) &= -\frac{1}{2\pi} \int_{-\infty}^{\infty} \omega^2 \mathcal{S}_x(\omega)\, d\omega \\ &= -R_x(0)\, \overline{W_x^2} \end{aligned}$$

Hence

$$\overline{W_x^2} = \frac{-R_x''(0)}{R_x(0)} \tag{5-56}$$

Thus the mean-square bandwidth as defined exists for $x(t)$ provided the second derivative of its autocorrelation function exists and is continuous at the origin.

We therefore conclude that for a signal $x(t)$, the mean-square width

[9]It should be noted that the mean square bandwidth is one of the possible measure of a bandwidth. There exist other definitions of bandwidth. In another widely used definition of bandwidth, the normalizing factor A_g in Eq. 5-53 is replaced by E_g, the energy of $g(t)$ given by $\int_{-\infty}^{\infty} g^2(t)\, dt$. For angle modulated carriers, the definition in Eq. 5-53 proves convenient.

of its power-density spectrum is given by Eq. 5-56. For an angle-modulated wave $\phi(t)$,

$$\phi(t) = A \cos [\omega_c t + \Theta + s(t)]$$

The spectrum of $\phi(t)$ is actually centered about ω_c as seen from Eq. 5-45. The autocorrelation function has a multiplying factor $\cos \omega_c \tau$. The autocorrelation function of the equivalent low-pass signal $\phi_\ell(t)$ is $e^{[R_s(\tau) - R_s(0)]}$. The spectrum of $\phi(t)$ is the same as that of $\phi_\ell(t)$ but shifted by ω_c. Hence the bandwidth of $\phi(t)$ will be twice that of[10] $\phi_\ell(t)$.

$$R_{\phi\ell}(\tau) = e^{[R_s(\tau) - R_s(0)]}$$

Hence following Eq. 5-56, we have [since $R_{\phi\ell}(0) = 1$]

$$\overline{W_\phi^2} = -2 \frac{\partial^2}{\partial \tau^2} e^{[R_s(\tau) - R_s(0)]}\Big|_{\tau=0}$$

$$= -2 \{R_s''(0) + [R_s'(0)]^2\}$$

Note that the autocorrelation function $R_s(\tau)$ being an even function of τ, $R_s'(0) = 0$. Hence

$$\overline{W_\phi^2} = -2R_s''(0) \tag{5-57}$$

Also from Eq. 5-56, we have

$$\overline{W_s^2} = -\frac{2\,R_s''(0)}{R_s(0)}$$

Hence

$$\frac{\overline{W_\phi^2}}{\overline{W_s^2}} = 2\,R_s(0) \tag{5-58}$$

For phase modulation (PM),

$$s(t) = k_p\, m(t)$$

Therefore

$$\overline{W_s^2} = \overline{W_m^2}$$

and

$$R_s(\tau) = k_p^2 R_m(\tau)$$

Equation 5-58 now becomes

$$\frac{\overline{W_p^2}}{\overline{W_m^2}} = 2k_p^2 R_m(0) \tag{5-59}$$

where $\overline{W_p^2}$ is the mean-square bandwidth of PM carrier, and

$$\overline{W_p^2} = 2k_p^2 \overline{m^2(t)}\, \overline{W_m^2} \tag{5-60}$$

Note that the mean-square bandwidth of a PM carrier is directly proportional to the mean-square bandwidth of the modulating signal.

[10]Multiplying a signal by $\cos \omega_c \tau$ is equivalent to amplitude modulation. Hence the bandwidth is doubled.

For frequency modulation (FM),

$$\dot{s}(t) = k_f m(t) \tag{5-61a}$$

If s(t) is differentiable, the power-density spectrum $s_{\dot{s}}(\omega)$ is given by

$$\mathcal{S}_{\dot{s}}(\omega) = \omega^2 \mathcal{S}_s(\omega) \tag{5-61b}$$

This follows from the fact that $\dot{s}(t)$ can be obtained from s(t) by transmitting it through an ideal differentiator whose transfer function is $j\omega$. Obviously $|j\omega|^2 = \omega^2$. Thus for FM (Eqs. 5-61a and 5-62b)

$$\mathcal{S}_{\dot{s}}(\omega) = k_f^2 \mathcal{S}_m(\omega) = \omega^2 \mathcal{S}_s(\omega)$$

or

$$\mathcal{S}_s(\omega) = \frac{k_f^2}{\omega^2} \mathcal{S}_m(\omega)$$

But by definition,

$$R_m(0) = \frac{1}{2\pi} \int_{-\infty}^{\infty} \mathcal{S}_m(\omega)\, d\omega$$

$$= \frac{1}{2\pi k_f^2} \int_{-\infty}^{\infty} \omega^2 \mathcal{S}_s(\omega)\, d\omega$$

$$= \frac{R_s(0)}{k_f^2} \overline{W_s^2}$$

and

$$R_s(0) = \frac{k_f^2 R_m(0)}{\overline{W_s^2}}$$

Substituting this equation in Eq. 5-58, we have

$$\overline{W_f^2} = 2k_f^2 R_m(0)$$

$$= 2k_f^2 \overline{m^2(t)} \tag{5-62}$$

Note that, unlike for PM, the mean-square bandwidth for FM does not depend upon the frequency spectrum of the modulating signal. It depends upon the modulating signal only through its mean-square value. Compare Eq. 5-62 with Eq. 5-51 derived earlier heuristically. The difference between the two equations is as expected. Equation 5-51 gives the bandwidth which includes most of the significant frequency components of the FM carrier, whereas Eq. 5-62 yields the mean-square bandwidth of the carrier. The latter estimation of the bandwidth of the angle modulated signal proves more convenient in computing signal to noise ratios.

Power Spectrum of Angle-Modulated Signal by Series Expansion. For an angle-modulated wave

$$\phi(t) = A \cos[\omega_c t + \Theta + s(t)]$$

we have shown that

$$R_\phi(\tau) = \frac{A^2}{2} e^{[R_s(\tau) - R_s(0)]} \cos \omega_c \tau$$
$$= \frac{A^2}{2} e^{-R_s(0)} [e^{R_s(\tau)} \cos \omega_c \tau] \tag{5-63}$$

The power-density spectrum is a Fourier transform of $R_\phi(\tau)$ and cannot be obtained in a closed form in this case. One can, however, obtain a fairly good estimate of the power-density spectrum by expanding $R_\phi(\tau)$ as a Taylor series in $R_s(\tau)$ and finding the transform term by term. For convenience we let

$$R_s(0) = \overline{s^2(t)} = k$$

and

$$\frac{R_s(\tau)}{R_s(0)} = \frac{R_s(\tau)}{k} = a(\tau)$$

Also,

$$R_{\phi\ell}(\tau) = e^{[R_s(\tau) - R_s(0)]}$$
$$= e^{-k} e^{ka(\tau)}$$
$$= e^{-k} \left[1 + ka(\tau) + \frac{k^2}{2!} a^2(\tau) + \frac{k^3}{3!} a^3(\tau) + \cdots\right] \tag{5-64}$$

Taking the Fourier transform of both sides, we have

$$\mathcal{S}_{\phi\ell}(\omega) = e^{-k} \left[2\pi\delta(\omega) + kA(\omega) + \frac{k^2}{2!}\frac{1}{(2\pi)} A(\omega) * A(\omega) \right.$$
$$\left. + \frac{k^3}{3!}\frac{1}{(2\pi)^2} A(\omega) * A(\omega) * A(\omega) + \cdots\right] \tag{5-65}$$

where

$$a(\tau) \leftrightarrow A(\omega) = \frac{\mathcal{S}_s(\omega)}{k}$$

Note that $\mathcal{S}_\phi(\omega)$, the power-density spectrum of $\phi(t)$ is given by

$$\mathcal{S}_\phi(\omega) = \frac{A^2}{4} [\mathcal{S}_{\phi\ell}(\omega + \omega_c) + \mathcal{S}_{\phi\ell}(\omega - \omega_c)] \tag{5-66}$$

The power density $\mathcal{S}_{\phi\ell}(\omega)$ can be found from Eq. 5-65. The important question here is, how many terms in expansion 5-65 should be considered for a reasonably close approximation. The answer is not difficult. We observe that the power of $\phi_\ell(t)$ is given by $R_{\phi\ell}(0)$. From Eq. 5-64, we have

$$R_{\phi\ell}(0) = e^{-k} \sum_{n=0}^{\infty} \frac{k^n}{n!} = 1$$

If we truncate the expansion and consider only the first m terms, the power of the approximated signal will be

$$e^{-k} \sum_{n=0}^{m} \frac{k^n}{n!}$$

This is the fraction of power that will be included by the signal represented by the first m terms of the expansion. Observe that the first term represents the power concentrated at $\omega = 0$. This corresponds to the carrier-frequency power after the appropriate shift as indicated in Eq. 5-66. Hence the first term (e^{-k}) represents the power in the unmodulated carrier. The carrier signal itself does not carry any information. Hence it is desirable to reduce this fraction. Evidently increasing k achieves this result.

We shall now consider two extreme cases when k becomes very small (narrow-band angle modulation) and when k becomes very large. When k is very small, all but the first two terms of the expansion can be ignored. Hence

$$\mathcal{S}_{\phi\ell}(\omega) = e^{-k}[2\pi\delta(\omega) + kA(\omega)]$$

since

$$A(\omega) = \frac{\mathcal{S}_s(\omega)}{k}$$

$$\mathcal{S}_{\phi\ell}(\omega) = e^{-k}[2\pi\delta(\omega) + \mathcal{S}_s(\omega)]$$

It is obvious then that for small k the power-density spectrum of $\phi_\ell(t)$ has the same shape as that of the signal $s(t)$. The power spectrum of $\phi(t)$ will be the same but for translation by $\pm\,\omega_c$. This is similar to the DSB-AM case. Hence the bandwidth of $\phi(t)$ in this case is twice that of $s(t)$. From Eq. 5-58 it is evident that the bandwidth of $\phi(t)$ is proportional to k ($R_s(0)$). Hence for small values of k, the bandwidth of $\phi(t)$ becomes small. This case is referred to as *narrow-band angle modulation* because of the bandwidth of the carrier is small.

When k is large the bandwidth of $\phi(t)$ is also large and this case is referred to as *wide-band* modulation. We observe that the terms $k^n e^{-k}/n!$ are the terms of a Poisson distribution, and as k becomes large, the terms for which n is near k, dominate (see Fig. 2-19). The higher-order terms of expansion 5-65 represent multiple convolutions. As seen in Sec. 2-11, these terms approach gaussian distribution. Hence for very large k the terms for which n is near k dominate and each approaches the gaussian shape. It can be shown that

$$\lim_{k\to\infty} \mathcal{S}_{\phi\ell}(\omega) = \frac{\exp\left[-\omega^2/2\overline{W_\phi^2}\right]}{(2\pi)^{1/2}\sqrt{\overline{W_\phi^2}}}$$

Thus for large k (wide-band angle modulation), the power density of $\phi(t)$ becomes gaussian with variance equal to the mean square bandwidth.

Demodulation of Angle-Modulated Signals. The study of optimum demodulation systems for angle-modulated signals involves a deeper study of parameter estimation techniques and phase-locked loops (Ref. 5) and will not be discussed here. Instead we shall consider conventional demodulators. It can be shown that the performance of conventional demodulators approaches that of optimum demodulators for a large carrier-to-noise power ratio at the input of the demodulator.

1. *Frequency Modulation.* The conventional demodulator consists of a bandpass filter to suppress the out-of-band noise. The modulated signal bandwidth is $2(\Delta\omega + 2W)$. Hence the passband of this filter is $\omega_c \pm (\Delta\omega + 2W)$, where $\Delta\omega$ is the deviation in the carrier frequency and W is the bandwidth (rps) of $m(t)$. The output of this filter is fed to a limiter-discriminator. The limiter clips the amplitude of the signal to a constant value thus eliminating AM effects. The signal then passes through the discriminator, which yields an output proportional to the instantaneous frequency of the input signal. The conventional FM system is shown in Fig. 5-19.

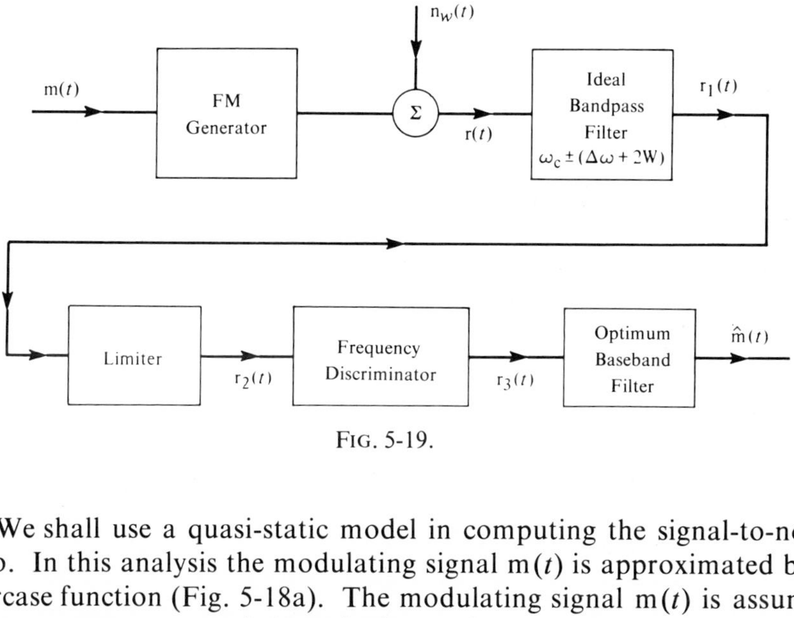

FIG. 5-19.

We shall use a quasi-static model in computing the signal-to-noise ratio. In this analysis the modulating signal $m(t)$ is approximated by a staircase function (Fig. 5-18a). The modulating signal $m(t)$ is assumed to be constant over each Nyquist interval. Over the kth interval, the signal is $m(t_k)$, the FM carrier frequency ω_i over this interval is given by $\omega_c + k_f m(t_k)$, and the carrier can be represented as

$$A \cos\{[\omega_c + k_f m(t_k)]t + \theta\}$$

Therefore, $r(t)$, the received signal (over the kth interval) at the receiver, is given by

$$r(t) = A \cos\{[\omega_c + k_f m(t_k)]t + \theta\} + n_w(t)$$

where $n_w(t)$ is the white gaussian channel noise. The bandpass filter suppresses the out-of-band noise component, and the output $r_1(t)$ is given by

$$r_1(t) = A\cos\{[\omega_c + k_f m(t_k)]t + \Theta\} + n(t)$$

where $n(t)$ is bandpass noise with constant power density $\mathfrak{N}/2$ over the band $\omega_c \pm (\Delta\omega + 2W)$. We shall represent $n(t)$ in terms of quadrature components with center frequency $\omega_c + k_f m(t_k)$. For convenience, let

$$\omega_c + k_f m(t_k) = \omega_k$$

Thus

$$\begin{aligned} r_1(t) &= A\cos(\omega_k t + \Theta) + n_c(t)\cos(\omega_k t + \Theta) + n_s(t)\sin(\omega_k t + \Theta) \\ &= \mathcal{E}(t)\cos[\omega_k t + \Theta + \psi(t)] \end{aligned} \tag{5-67}$$

where

$$\mathcal{E}(t) = \sqrt{[A + n_c(t)]^2 + n_s^2(t)} \tag{5-68a}$$

and

$$\psi(t) = -\tan^{-1}\left[\frac{n_s(t)}{A + n_c(t)}\right] \tag{5-68b}$$

As seen from Eq. 5-67, the envelope $\mathcal{E}(t)$ varies with time. The limiter clips the envelope to a constant magnitude A_L. Hence $r_2(t)$, the output of the limiter (neglecting the distortion due to clipping) is given by

$$r_2(t) = A_L\cos[\omega_k t + \Theta + \psi(t)]$$

The frequency discriminator yields an output proportional to the instantaneous frequency of $r_2(t)$:

$$r_3(t) = \alpha\omega_i$$

where $r_3(t)$ is the output of the frequency discriminator, and α is the constant of proportionality of the discriminator. For computing signal-to-noise ratio, the value of α is immaterial and so we shall assume $\alpha = 1$, for convenience. Therefore

$$\begin{aligned} r_3(t) &= \omega_i \\ &= \frac{d}{dt}[\omega_k t + \Theta + \psi(t)] \\ &= \omega_k + \dot{\psi}(t) \end{aligned}$$

But

$$\psi(t) = -\tan^{-1}\left[\frac{n_s(t)}{A + n_c(t)}\right]$$

We shall here consider only the small-noise case where

$$A >> n_c(t), \text{ and } n_s(t) \quad \text{for almost all } t$$

Hence

$$\psi(t) \simeq -\tan^{-1}\left[\frac{n_s(t)}{A}\right]$$

$$\simeq -\frac{n_s(t)}{A}$$

and

$$\dot{\psi}(t) \simeq \frac{-\dot{n}_s(t)}{A}$$

Therefore

$$r_3(t) = \omega_k - \frac{\dot{n}_s(t)}{A}$$

$$= \omega_c + k_f m(t_k) - \frac{\dot{n}_s(t)}{A} \tag{5-69}$$

The signal $r_3(t)$ contains the signal component $k_f m(t_k)$ and the noise component $\dot{n}_s(t)/A$. We use a final optimum baseband filter to filter the noise component as best we can. To find the optimum filter, we shall first find the power density of $\dot{n}_s(t)$. We can obtain $\dot{n}_s(t)$ by passing $n_s(t)$ through an ideal differentiator whose transfer function is $j\omega$. Hence the power density of $\dot{n}_s(t)$ is

$$|j\omega|^2 S_{n_s}(\omega) = \omega^2 S_{n_s}(\omega)$$

Obviously the power density of $-\dot{n}_s(t)/A$ is $(\omega^2/A^2)S_{n_s}(\omega)$. Thus the noise power density is $(\omega^2/A^2)S_{n_s}(\omega)$. But $S_{n_s}(\omega)$ is obtained by shifting $S_n(\omega)$ by $\pm\,\omega_k$ and discarding the spectrum centered at the double frequency $2\omega_k$ (see Eq. 5-3). This is shown in Fig. 5-20b. Note that the

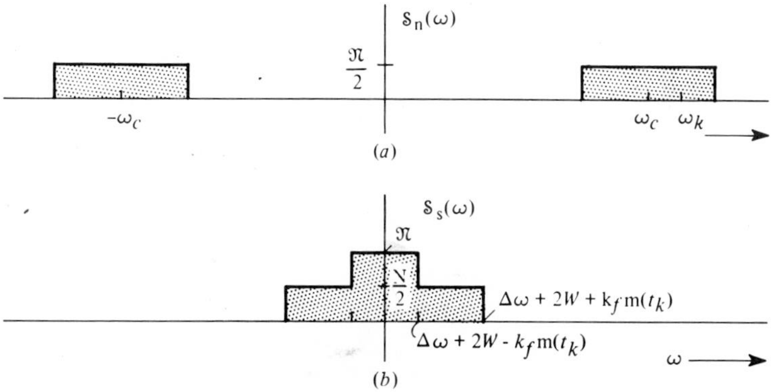

FIG. 5-20.

final baseband filter will suppress all the noise outside the range $\omega = W$. Hence over the frequency range of interest $S_{n_s}(\omega) = \mathfrak{N}$ (for $W < \Delta\omega + 2W - k_f m(t_k)$) and the power density of $-\dot{n}_s(t)/A$ is $(\mathfrak{N}\omega^2/A^2)$ over the band $(0, W)$ as shown in Fig. 5-21. Note that the power-density spectrum of the noise is independent of value $m(t_k)$ of the message signal.

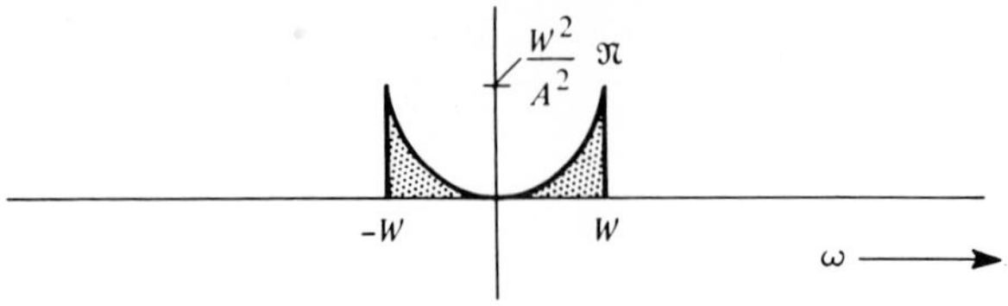

FIG. 5-21.

Hence this analysis holds over the entire time interval. We now have the desired signal k_f m(t) mixed with noise of power density $(\mathfrak{N}\omega^2/A^2)$. Hence the transfer function of the optimum filter is given by

$$H(\omega) = \frac{k_f^2 \mathcal{S}_m(\omega)}{k_f^2 \mathcal{S}_m(\omega) + \left(\frac{\mathfrak{N}\omega^2}{A^2}\right)}$$

$$= \frac{k_f^2 A^2 \mathcal{S}_m(\omega)}{k_f^2 A^2 \mathcal{S}_m(\omega) + \mathfrak{N}\omega^2} \tag{5-70}$$

Since we are considering the small-noise case (carrier power > > noise power), $k_f^2 A^2 \mathcal{S}_m(\omega) >> \mathfrak{N}\omega^2$ and

$$H(\omega) \simeq 1$$

Thus the optimum baseband filter approaches an ideal low-pass filter with cutoff frequency W rps (B Hz). In this case the noise-power output $\overline{n_0^2}$ is given by

$$\overline{n_0^2} = \frac{1}{\pi}\int_0^W \frac{\mathfrak{N}\omega^2}{A^2}\, d\omega$$

$$= \frac{\mathfrak{N}W^3}{3\pi A^2} = \frac{2}{3}\frac{\mathfrak{N}B}{A^2} W^2 \quad (W = 2\pi B)$$

The signal output is k_f m(t) (see Eq. 5-69). Hence the output signal power is $k_f^2 \overline{m^2(t)}$, and

$$\Lambda = \frac{k_f^2 \overline{m^2(t)}}{\overline{n_0^2}}$$

$$= 3\left(\frac{A^2}{2\mathfrak{N}B}\right)\frac{k_f^2 \overline{m^2(t)}}{W^2} \tag{5-71a}$$

As seen from Eq. 5-62, $2k_f^2 \overline{m^2(t)}$ is $\overline{W_f^2}$, the mean-square bandwidth of the FM carrier. Hence

$$\Lambda = \frac{3}{2}\frac{A^2}{2\mathfrak{N}B}\frac{\overline{W_f^2}}{W^2} \tag{5-71b}$$

When studying amplitude modulation, we expressed Λ as a function of the transmitted power. For a meaningful comparison of amplitude and angle modulation, we should express Λ as a function of transmitted power. For the angle-modulated signal, the carrier amplitude A is always constant. The instantaneous frequency and phase vary with time. Since the power of a sinusoid of amplitude A is $A^2/2$ regardless of its frequency and phase, one may guess that the power of angle-modulated carrier will be $A^2/2$ regardless of the value of k_f (or k_p) and $\overline{m^2(t)}$. This is indeed true. In our case we observe from Eq. 5-45 that the power of angle-modulated carrier $\phi(t)$ is given by

$$\overline{\phi^2(t)} = R_\phi(0) = \frac{A^2}{2}$$

We shall therefore study the relationship between Λ and $A^2/2$. Analogous to the AM case, we shall normalize the transmitted power $A^2/2$ by a noise power $\mathfrak{N}B$ in the baseband. Thus $A^2/2\mathfrak{N}B$ is the channel SNR within the modulating-signal bandwidth and plays the same role in angle modulation as does $\sigma_m^2/2\mathfrak{N}B$ in AM. The quantity $\overline{W_f^2}/W^2$ is the ratio

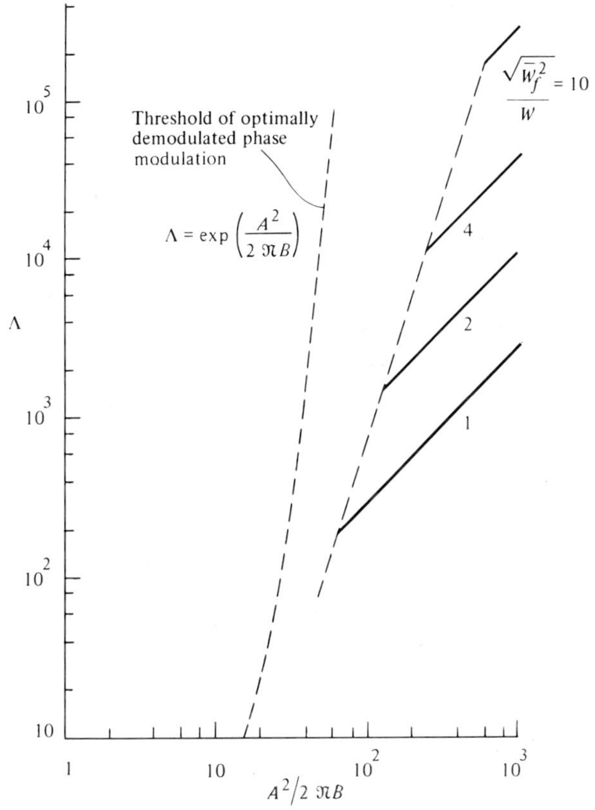

FIG. 5-22. Performance of conventional demodulator for frequency modulation (from Viterbi, Ref. 5, McGraw-Hill, with permission).

of the mean square bandwidth of transmission to the square of the bandwidth of the modulating signal. It is evident from Eq. 5-71b that for FM, Λ, the output signal to noise ratio varies as the square of the bandwidth (actually mean square bandwidth) of transmission. Fig. 5-22 shows Λ, the output signal-to-noise ratio as a function of channel signal-to-noise ratio $A^2/2\mathfrak{N}B$ for various values of bandwidth ratio $\sqrt{\overline{W_f^2}}/W$.

2. *Phase Modulation.* If we differentiate m(t) and use this derivative to frequency modulate a carrier, we obtain the phase-modulated carrier. At the receiver, one may use an FM receiver followed by an integrator to obtain the original signal. This is equivalent to the con-

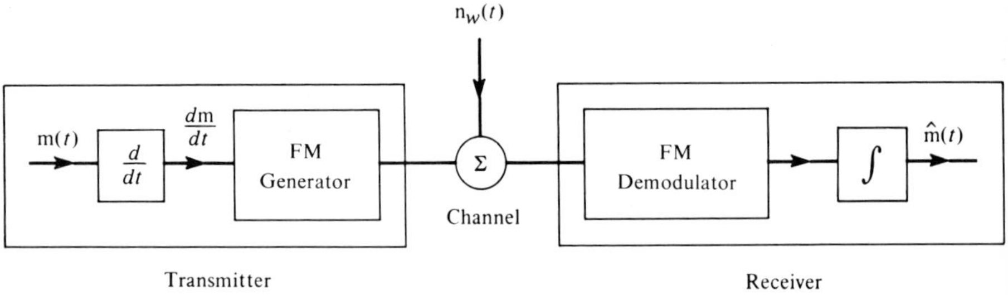

FIG. 5-23.

ventional PM system (Fig. 5-23). If the PM carrier is given by

$$\phi_{\text{pm}}(t) = A \cos [\omega_c t + \theta + k_p \text{m}(t)]$$

then it is evident that the final signal at the receiver output will be $k_p \text{m}(t)$ and the noise will be same as in FM case but integrated with respect to t. The transfer function of an integrator is $1/j\omega$. Hence the power density of the output noise in this case will be

$$\frac{1}{\omega^2}\left(\frac{\mathfrak{N}\omega^2}{A^2}\right) = \frac{\mathfrak{N}}{A^2}$$

This is of course band-limited to W rps. Hence the noise power at the output is given by

$$\overline{\text{n}_0^2} = \frac{2\mathfrak{N}B}{A^2} \tag{5-72}$$

and the output signal power is $k_p^2 \overline{\text{m}^2(t)}$. Therefore

$$\Lambda = \frac{k_p^2 \overline{\text{m}^2(t)}}{2\mathfrak{N}B/A^2}$$

$$= \left(\frac{A^2}{2\mathfrak{N}B}\right) k_p^2 \overline{\text{m}^2(t)} \tag{5-73a}$$

Use of Eq. 5-60 now yields

$$\Lambda = \frac{1}{2}\left(\frac{A^2}{2\mathfrak{N}B}\right)\frac{\overline{W_p^2}}{\overline{W_m^2}} \tag{5-73b}$$

where $\overline{W_p^2}$ is the mean-square bandwidth of PM carrier and $\overline{W_m^2}$ is the mean-square bandwidth of m(t). Note that the ratio $\overline{W_p^2}/\overline{W_m^2}$ has significance similar to the ratio $\overline{W_f^2}/W^2$ in the FM. It is evident from Eqs. 5-71b and 5-73b that Λ, the output signal-to-noise ratio exhibits similar behavior in both FM and PM. Figure 5-24 shows Λ as a function of the $A^2/2\mathfrak{N}B$ for various values of bandwidth ratios $\sqrt{\overline{W_p^2}/\overline{W_m^2}}$.

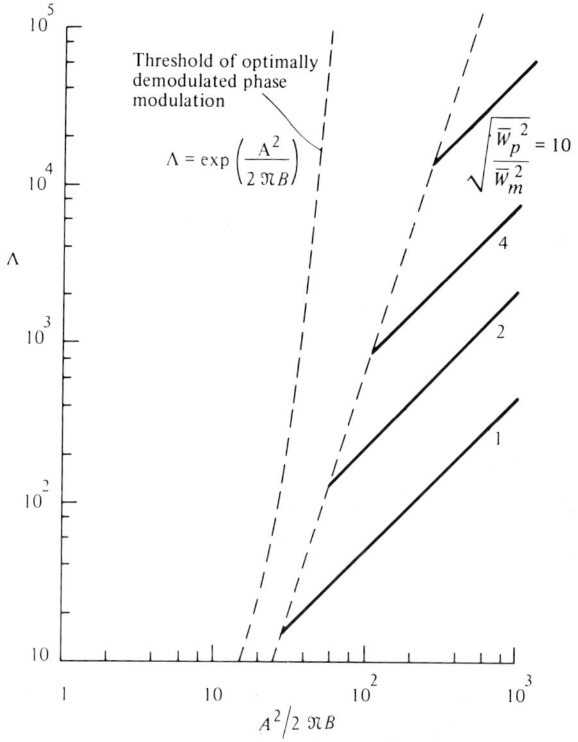

FIG. 5-24. Performance of conventional demodulator for phase modulation (from Viterbi, Ref. 5, Chapter 5, McGraw Hill, with permission).

Threshold in Angle Modulation. The derivation of Λ in Eqs. 5-71 and 5-73 is based on the small-noise assumption. The threshold occurs when the ratio of the carrier power to the noise power at the demodulator input approaches a ratio of 10 (see Ref. 7). We shall represent this ratio SNR by Λ_c.

The noise power at the discriminator input will be taken as the noise power within the rms bandwidth $\sqrt{\overline{W_f^2}}$. Thus for FM the noise power at the discriminator input is $\mathfrak{N}\sqrt{\overline{W_f^2}}$ and

$$\Lambda_c = \frac{A^2/2}{\mathfrak{N}\sqrt{\overline{W_f^2}}} \tag{5-74}$$

For PM

$$\Lambda_c = \frac{A^2/2}{\mathfrak{N}\sqrt{\overline{W_p^2}}} \tag{5-75}$$

At the threshold

$$\Lambda_c \simeq 10$$

If the ratio Λ_c falls below 10, the performance of the demodulator deteriorates rapidly. Equations 5-71 and 5-73 are valid for $\Lambda_c > 10$—that is, when the carrier-to-noise power at the discriminator input is greater than 10.

It can be shown that the performance of the optimum-demodulation systems (which use the phase-locked loop) is identical for small amounts of noise ($\Lambda_c > 10$). However, for the large-noise case, the optimum demodulator (which is a phase-locked loop) has a much superior performance (Ref. 5). The threshold of an optimum demodulator occurs well below $\Lambda_c = 10$. This is shown in Figs. 5-22 and 5-24 for the sake of comparison. In general for optimum angle-modulation systems, the threshold is extended by the order of 3 to 6 db beyond that of conventional demodulators.

Comments on Angle-Modulated Systems. The output signal-to-noise ratio Λ is directly proportional to the channel signal-to-noise ratio Λ_c $(=(A^2/2\mathfrak{N}B))$ and the mean-square bandwidth of transmission ($\overline{W_f^2}$ or $\overline{W_p^2}$). Hence Λ can be improved only by increasing the transmitted power and/or increasing the bandwidth of transmission. To maintain a required value of Λ, one can trade transmitted power with the bandwidth of transmission. Thus reduction in transmitted power should be offset by an increase in the mean-square bandwidth by the same factor (and vice versa). It is also obvious that Λ, the output SNR improves as the square of the bandwidth of transmission.

From Fig. 5-22, it is obvious that for a given mean-square bandwidth $\overline{W_f^2}$, Λ can be increased by increasing $A^2/2\mathfrak{N}B$ (increasing the channel SNR). On the other hand, for a given $A^2/2\mathfrak{N}B$ (given channel SNR) Λ can be increased by increasing the bandwidth of transmission ($\overline{W_f^2}$) but this cannot be done indefinitely. For higher bandwidth ratios the threshold appears sooner than for lower bandwidth ratios (see Fig. 5-22 or 5-24). The reason for this is obvious. As the bandwidth is increased the channel noise increases. This reduces Λ_c below the threshold level. Similar conclusions apply to phase modulation.

5-6. DIGITAL MODULATION TO TRANSMIT ANALOG DATA: PULSE-CODE MODULATION (PCM)

Earlier we have mentioned the possibility of transmitting analog data in digital form by quantization of the data. In this section we shall consider the performance that can be expected of this method.

A signal $m(t)$ band-limited to B Hz is completely specified by its samples taken at intervals $1/2B$ (or less) seconds apart. Therefore in order to transmit the information contained in $m(t)$ one could transmit these samples instead of the signal itself. Thus the problem of transmitting a continuous message reduces to that of transmitting a *finite* number of values (samples) per second. If the sample values were to assume one of the values in a given finite set, the problem would reduce to one of transmitting a sequence of finite number of known symbols (digital-data transmission). But unfortunately, the sample amplitudes can assume any value in a given continuous range. To overcome this problem we *quantize* the signal, so that each sample is approximated or "rounded off" to the nearest quantized level as shown in Fig. 5-25. Let us assume that the amplitudes of $m(t)$ lie in the range $-V$ to V volts. The entire range $(-V, V)$ is partitioned into L intervals, each of length[11] $2V/L$. Each sample amplitude is then approximated by the midpoint value of that interval (Fig. 5-25a). This procedure is known as *quantizing*. The quantizing obviously is a way of approximating $m(t)$ and will cause an error when the quantized samples are used to reconstruct the signal $m(t)$. The mean-square error between $m(t)$ and the signal reconstructed from the quantized samples is called the *quantizing error*. It can be shown that the quantizing error is a minimum when the quantizing is done so that each sample is assigned the midpoint value of the quantizing interval in which it lies (Ref. 8). The input-output characteristic of a quantizer is shown in Fig. 5-25b.

By using quantization, we have converted the analog data into digital data. The problem of transmitting $m(t)$ is now reduced to that of transmitting $2B$ values per second from a set of L known values or symbols. Some specific known waveforms may be assigned to each of

[11]The intervals need not be uniform. In fact, for optimum operation it can be shown that for smaller signal levels there should be more quantizing levels. Here we are going to assume amplitudes of $m(t)$ uniformly distributed in the range $(-V, V)$. In this case it can be shown that for optimum operation quantized steps should be uniform (Ref. 8).

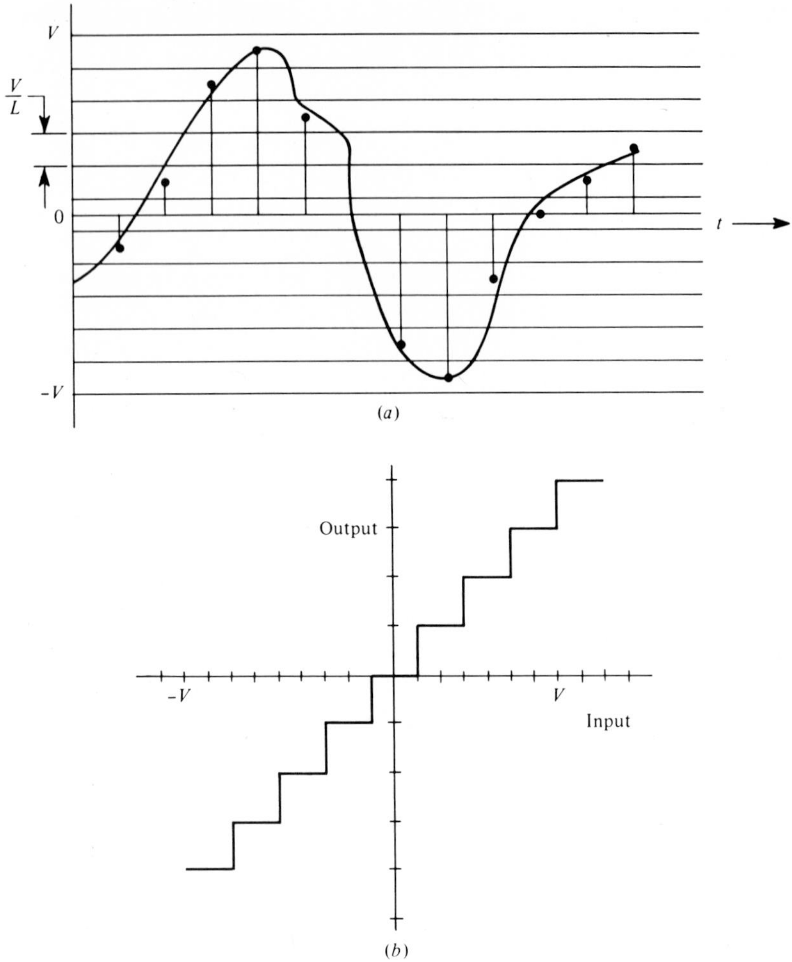

FIG. 5-25.

the L symbols. Thus transmitting m(t) reduces to that of transmitting a sequence of L known waveforms. The waveforms may be assigned in many ways. This problem is similar to that of devising a method to represent L numbers. We may use a binary, trinary, or M-ary number system to represent these numbers. We can use binary waveform (two waveforms) in the proper sequence to represent any number (or quantized sample value). In general, one can use M-ary waveforms. Here we shall consider only the binary case. (For a general M-ary case, see Ref. 5.) This mode of transmission—of converting analog data into digital data by quantizing—is referred to as pulse-code modulation (PCM). If the digital data is transmitted by M waveforms the scheme is referred to as M-ary PCM. We shall here consider only binary PCM.

In binary PCM each sample is represented by a binary number using 1's and 0's. Note that n binary digits can form 2^n distinct patterns. For example, two binary digits can form 4 patterns 00, 01, 10, 11. Hence each of the L numbers can be represented by a minimum of n binary digits where

$$2^n = L \tag{5-76a}$$

Digit	Binary Number
0	0000
1	0001
2	0010
3	0011
4	0100
5	0101
6	0110
7	0111
8	1000
9	1001
10	1010
11	1011
12	1100
13	1101
14	1110
15	1110

Waveform

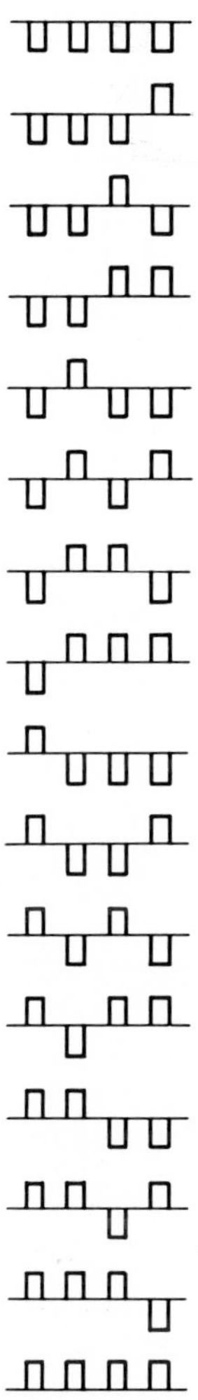

FIG. 5-26.

or

$$n = \log_2 L \tag{5-76b}$$

We now represent the number 1 by a pulse $p(t)$ and the number 0 by a pulse $-p(t)$. Figure 5-26 illustrates the case where $L = 16$ and $p(t)$ is a rectangular pulse of height $A/2$.

At the receiver, we use optimum detectors to process the incoming waveforms. The matched filter receiver discussed in Chapter 4 can be used for this purpose. It is shown in Chapter 6, that for white gaussian channel noise, the matched filter detector in Chapter 4 is indeed the optimum receiver as it has the lowest probability of error. We reconstruct the samples at the receiver and from these samples, the continuous signal $m(t)$ can be obtained as discussed in Chapter 1 (Sec. 1-11). The reconstructed signal at the receiver will be only an approximation of the original signal because of: (i) quantization error, and (ii) detection error (error in signal detection at the receiver). The latter error occurs because the optimum receiver will occasionally mistake the positive pulse $p(t)$ for the negative pulse $-p(t)$ (or vice versa) due to channel noise. The two kinds of error are independent. Hence the total mean-square error in sample values at the receiver is equal to the sum of the mean-square errors due to each kind of error. We shall now compute these errors.

1. *Quantizing Error.* At the transmitter, a sample is assigned the midpoint value of the interval in which it lies. It is therefore obvious (see Fig. 5-25a), that the quantizing error lies in the range $-V/L$, V/L. For computational convenience, we make a simplifying assumption that the amplitudes of $m(t)$ are uniformly distributed in the range $(-V, V)$. The results, however, can be extended to any other amplitude-limited source (Ref. 1). With this assumption, it is obvious that the quantizing error will also have a uniform probability density over the range $-V/L$, V/L. If e_q is the quantizing error, then e_q is a random variable (over the ensemble) and has a first-order probability $p(e_q)$ given by

$$p(e_q) = \begin{cases} \dfrac{L}{2V} & |e_q| < \dfrac{V}{L} \\ 0 & \text{Otherwise} \end{cases}$$

The mean-square quantizing error $\overline{e_q^2}$ is given by

$$\begin{aligned} \overline{e_q^2} &= \int_{-\infty}^{\infty} e_q^2 p(e_q)\, de_q \\ &= \int_{-V/L}^{V/L} \frac{L}{2V} e_q^2\, de_q \\ &= \frac{1}{3}\left(\frac{V}{L}\right)^2 \end{aligned} \tag{5-77}$$

2. *Detection Error.* To compute this error we shall use the results for error probabilities derived for the matched-filter in Chapter 4. The magnitude of the error probabilities in practical cases are of the order of 10^{-4} to 10^{-7}. Therefore it is extremely unlikely to have more than one error in a group of n consecutive pulses, even for a fairly large value of n. We shall therefore assume that the receiver will at most err in recognizing one of the digits (pulses) in any sample transmitted. Since each of the digits is independently transmitted, the probability that any one digit will be in error will be the same in all the samples. Let this probability be P_e. We also note that an error in a digit has a weight or significance depending on its positions in the sequence. As seen in Fig. 5-26, in the sequence 1011 (decimal 11) for example, if the first digit is in error, the result will be 0011 (decimal 3) and the error magnitude is 8 or 2^3. If on the other hand the second digit is in error, the result is 1111 (decimal 15) and the error magnitude is 4 or 2^2. Similarly for the third and the fourth digit error, the resulting error magnitudes are 2^1 and 2^0 respectively. In this case the full scale is 16 (2^4). If we define the relative error as the ratio of the actual error magnitude to the full scale, then the relative errors are 2^{-1}, 2^{-2}, 2^{-3} and 2^{-4} respectively. It can be seen that, in general, the relative error occurring due to ith digit in error is 2^{-i}. The actual error is the relative error multiplied by the full scale.

In the case under consideration we have a sequence of n digits and a full scale of $2V$. Hence the actual error due to the error in ith digit is $2^{-i}(2V) = 2^{-(i-1)}V$. The square of the error in the sample value, when the ith digit is in error is obviously $2^{-2(i-1)}V^2$. Let us now recapitulate and state the problem once again. The receiver makes a decision about a sequence of n digits representing a particular sample. There are $(n + 1)$ possibilities in the final reception: (1) only the first digit is in error, (2) only the second digit is in error . . . (n) only the nth digit is in error, ($n + 1$) no digit is in error. The probabilities of these events are P_e, $P_e, \ldots, P_e$, and $1 - nP_e$ respectively.[12] The square of the error magnitudes are V^2, $2^{-2}\ V^2, \ldots,\ 2^{-2(n-1)}\ V^2$, and 0 respectively. Obviously, the mean-square error $\overline{e_d^2}$ is given by

$$\overline{e_d^2} = P_eV^2 + P_e2^{-2}V^2 + \cdots + P_e2^{-2(n-1)}V^2 + (1 - nP_e)0$$

$$= P_eV^2 \sum_{i=1}^{n} 2^{-2(i-1)}$$

Summation of this geometric series yields

$$\overline{e_d^2} = \frac{4P_eV^2(L^2 - 1)}{3L^2} \tag{5-78}$$

[12]This is true under the assumption that there is only one digit in error in the whole sequence of k digits.

The total mean-square error $\overline{e_t^2}$ in the received sample is the sum of the quantizing mean-square error and the detection mean-square error.

$$\begin{aligned}\overline{e_t^2} &= \overline{e_q^2} + \overline{e_d^2} \\ &= \frac{1}{3}\frac{V^2}{L^2} + \frac{4P_eV^2(L^2-1)}{3L^2} \\ &= \frac{V^2}{3L^2}[1 + 4P_e(L^2-1)] \end{aligned} \tag{5-79}$$

Note that the process m(t) is assumed to be stationary (it is at least wide-sense stationary). Hence the mean-square value of any of its samples is the mean-square value $\overline{\mathrm{m}^2(t)}$. Similarly the mean-square error $\overline{e_t^2}$ is the quantity $\overline{[\mathrm{m}(t) - \hat{\mathrm{m}}(t)]^2}$. Hence

$$\overline{[\mathrm{m}(t) - \hat{\mathrm{m}}(t)]^2} = \frac{V^2}{3L^2}[1 + 4P_e(L^2-1)] \tag{5-80}$$

The signal mean-square value $\overline{\mathrm{m}^2(t)}$ is obtained by observing that each sample has a uniform probability distribution in the range $(-V, V)$. If m_j is the jth sample, then $p(m_j)$, the probability density of m_j is given by

$$p(m_j) = \begin{cases} 1/2V & |\mathrm{m}_j| < V \\ 0 & |\mathrm{m}_j| > V \end{cases}$$

and

$$\begin{aligned}\overline{\mathrm{m}_j^2} &= \int_{-\infty}^{\infty} m_j^2\, p(m_j)\, dm_j \\ &= \int_{-V}^{-V} \frac{1}{2V} m_j^2\, dm_j \\ &= \frac{V^2}{3}\end{aligned}$$

and

$$\overline{\mathrm{m}^2(t)} = \overline{m_j^2} = \frac{V^2}{3} \tag{5-81}$$

From Eqs. 5-80 and 5-81 we have

$$\Lambda = \frac{L^2}{1 + 4P_e(L^2-1)} \tag{5-82}$$

Here P_e is the probability of error in the binary system. We have shown in Chapter 4 (Sec. 4-20), that the binary system using two waveforms has an error probability given by Eq. 4-144b:

$$P_e = \mathrm{erfc}\left(\sqrt{\frac{2ST}{\mathfrak{N}}}\right) \tag{5-83}$$

where S is the average signal power and T is the width of the pulse. Note that there are n pulses per sample and there are $2B$ samples per second. Hence we need to transmit $2Bn$ pulses per second. Therefore

$$T = \frac{1}{2Bn} \tag{5-84}$$

and

$$P_e = \operatorname{erfc}\left(\sqrt{\frac{S}{\mathcal{N}Bn}}\right) \tag{5-85}$$

Also

$$L = 2^n$$

so

$$n = \log_2 L$$

Equation 5-82 can now be expressed as

$$\Lambda = \frac{2^{2n}}{1 + 4(2^{2n} - 1)\operatorname{erfc}\left(\sqrt{\frac{S}{\mathcal{N}Bn}}\right)} \tag{5-86}$$

Here again we encounter the term $S/\mathcal{N}B$ which represents the channel signal-to-noise ratio. This is the ratio of transmitted signal power S to $\mathcal{N}B$, the noise power in baseband. The variation of the final signal-to-noise ratio is shown in Fig. 5-27 for various values of n. Observe the threshold effect for this system of transmission. For a given value of n, Λ deteriorates rapidly for $S/\mathcal{N}B$ below a certain value.

From various equations derived so far and from Fig. 5-27, we can draw some useful conclusions. From Eq. 5-77 it is obvious that the quantizing error decreases as L, the number of levels is increased. This fact is also obvious intuitively. Increasing L reduces the quantizing interval and the approximation improves. On the other hand, increasing L increases the detection error as seen from Eq. 5-78. This means that there is an optimum value of L for which the total mean-square error is a minimum. Since $L = 2^n$, this implies that there exists an optimum value of n for which Λ is maximum for a given value of $S/\mathcal{N}B$. This fact is obvious from Fig. 5-27. This value of n may be obtained from $d\Lambda/dn = 0$ and using Eq. 5-86. Any other value of n yields a lower value of Λ for that particular $S/\mathcal{N}B$ ratio.

From Fig. 5-27, we also observe that for a given value of n (or L), there is a saturation effect for large values of $S/\mathcal{N}B$. Thus for a given n, increasing $S/\mathcal{N}B$ beyond some point does not appreciably increase Λ. This behavior can be explained by the fact that for a given n the quantizing error is constant and increasing $S/\mathcal{N}B$ increases the transmitted signal power. This means the pulse amplitudes in the transmitted signal

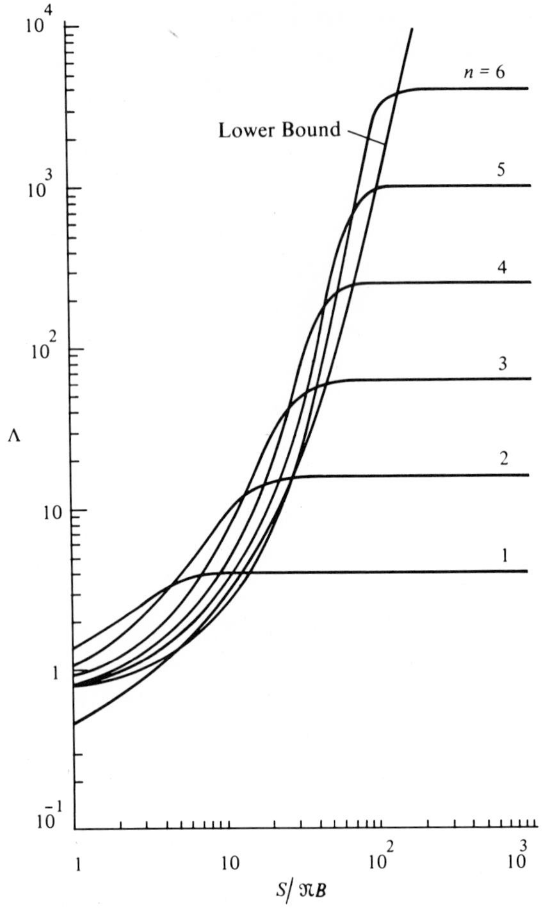

FIG. 5-27. Output signal-to-noise ratio for binary PCM (from Viterbi, Ref. 5, Chapter 5, McGraw-Hill, with permission).

are increased. This definitely reduces the error probability P_e and hence reduces the detection error. Increasing $S/\mathfrak{N}B$ may reduce the detection error to zero but it cannot remove the constant quantizing error given by $V^2/3L^2$. Hence Λ reaches a constant value. The error in this case is entirely due to the quantizing error. Hence Λ_s, the saturation value of Λ, is given by

$$\begin{aligned}\Lambda_s &= \frac{\overline{m^2(t)}}{\overline{e_q^2}} \\ &= \frac{V^2/3}{V^2/3L^2} \\ &= L^2 \\ &= 2^{2n} \qquad (5\text{-}87)\end{aligned}$$

Thus for a given n, nothing is gained by increasing the transmitted power beyond a certain point. If there is a large transmitter power available, then one should increase n in order to take advantage of it in improving Λ.

We also observe from Eq. 5-86 (or Fig. 5-27) that Λ can in general be increased by increasing the transmitted signal power and/or by increasing n. So far we have not considered how the bandwidth of the transmitted signal is affected. There are n digits per sample and there are $2B$ samples per second. Hence we need to transmit a total of $2nB$ independent pieces of information per second. Here we observe that a signal band-limited to B Hz can be specified by $2B$ independent pieces of information per second. Conversely, $2B$ independent pieces of information per second can be transmitted over a channel of bandwidth B Hz. This can be done by constructing a continuous signal band-limited to B Hz from these $2B$ independent pieces of information per second using Eq. 1-103. It is therefore obvious that $2nB$ independent pieces of information will need a bandwidth nB Hz. Hence B_{pcm}, the bandwidth of transmission, is given by

$$B_{pcm} = nB$$

or

$$\frac{B_{pcm}}{B} = n$$

Obviously, the bandwidth of transmission is multiplied by the factor n. We therefore conclude that, in general, the output SNR Λ can be increased by increasing the channel SNR—(that is, the transmitted power) and/or the transmission bandwidth. We have come to similar conclusion for angle modulation. For angle modulation, we observed that Λ was directly proportional to the square of the bandwidth of transmission (mean-square bandwidth). It is interesting to see how Λ is related to the bandwidth in the present case (binary PCM). From Eq. 5-82 we have

$$\Lambda = \frac{L^2}{1 + 4P_e(L^2 - 1)}$$

In practical cases P_e is of the order of 10^{-4} to 10^{-8}. Therefore as a first approximation $P_e(L^2 - 1)$ may be ignored, and

$$\begin{aligned} \Lambda &\simeq L^2 \\ &= 2^{2n} \\ &= 2^{2B_{\text{pcm}}/B} \end{aligned} \tag{5-88}$$

where B_{pcm} is the bandwidth of transmission. It therefore follows that Λ increases exponentially with B_{pcm}, the bandwidth of transmission. For angle modulation, it was shown that Λ increases linearly as the square of the bandwidth. It should be noted that the trade between SNR and bandwidth is not possible for AM because the bandwidth that is required is fixed and cannot be altered in AM systems. On the other hand, by properly changing the bandwidth of transmission, SNR and bandwidth can be exchanged in angle-modulated and PCM systems. Of the two (angle modulation and PCM), we have shown that exchange between the input SNR and the bandwidth of transmission is more effective in increasing Λ in PCM (Λ increases exponentially as the bandwidth) as compared to angle modulation (Λ increases as the square of bandwidth).

In Chapter 7 we shall study the ideal relationship between input and output SNR and the bandwidth of transmission in the light of Information theory and Shannon's theorem. This theorem predicts that output SNR (Λ) increases exponentially with the bandwidth of transmission. Hence the performance of PCM system approaches this ideal performance. It can, however, be shown that angle-modulation systems also asymptotically approach the ideal performance as the bandwidth of transmission approaches infinity (Ref. 5). This will be discussed in more detail in Chapter 7.

REFERENCES

1. Wozencraft, J. M., and I. M. Jacobs, *Principles of Communication Engineering*, Wiley, New York, 1965.
2. Middleton, D., *An Introduction to Statistical Communication Theory*, McGraw-Hill, New York, 1960.
3. Davenport, W. B., and W. L. Root, *Introduction to Random Signal and Noise*, McGraw-Hill, New York, 1958.
4. Abramson, N., "Bandwidth and Spectra of Phase- and Frequency-Modulated Waves," *IEEE Trans. on Communication Systems*, Vol. CS-11, No. 4 (December 1963), pp. 401–414.
5. Viterbi, A. J., *Principles of Coherent Communication*, McGraw-Hill, New York, 1966.
6. Viterbi, A. J., "Optimum Coherent Demodulation for Continuous Modulation Systems," *Proc. Nat. Electron. Conf.*, Vol. 18 (October 1962), pp. 498–508.
7. Rice, S. O., "Noise in FM Receivers," in M. Rosenblatt (ed.), *Time Series Analysis*, Wiley, New York, 1963.
8. Panter, P. F., and W. Dite, "Quantization Distortion in Pulse Code Modulation with Nonuniform Spacing of Levels," *Proc. IRE*, *39* (January 1951).
9. Van Trees, H. L., *Detection Estimation and Modulation Theory*, Vol. 1, Wiley, New York, 1967.
10. Lathi, B. P., *Communication Systems*, Wiley, New York, 1968.
11. Deutsch, R., *Nonlinear Transformation of Random Processes*, Prentice-Hall Englewood Cliffs, N.J., 1962.

PROBLEMS

5-1. A white noise of power density $\mathfrak{N}/2$ is transmitted through a bandpass filter whose characteristics are shown in Fig. P. 5-1. The output of the filter is a bandpass noise. Determine the power-density spectra and the mean-square values of $n_c(t)$ and $n_s(t)$.

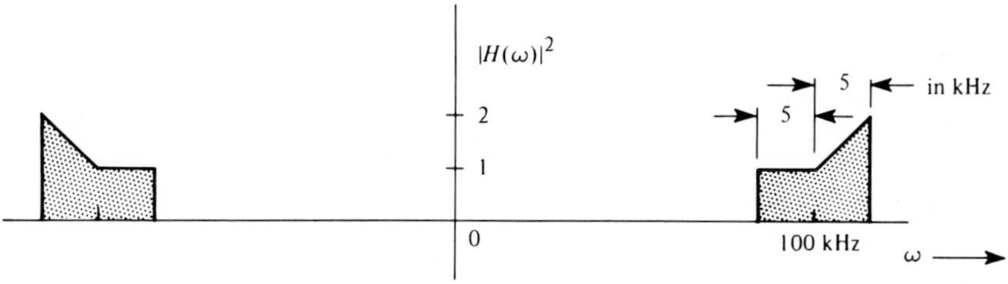

FIG. P.5-1.

5-2. A noise signal with a power-density spectrum $\mathcal{S}_n(\omega)$ as shown in Fig. P. 5-2a is transmitted through an ideal bandpass filter in Fig. P. 5-2b. Determine the power-density spectra and the mean-square values of $n_c(t)$ and $n_s(t)$ for the filter output.

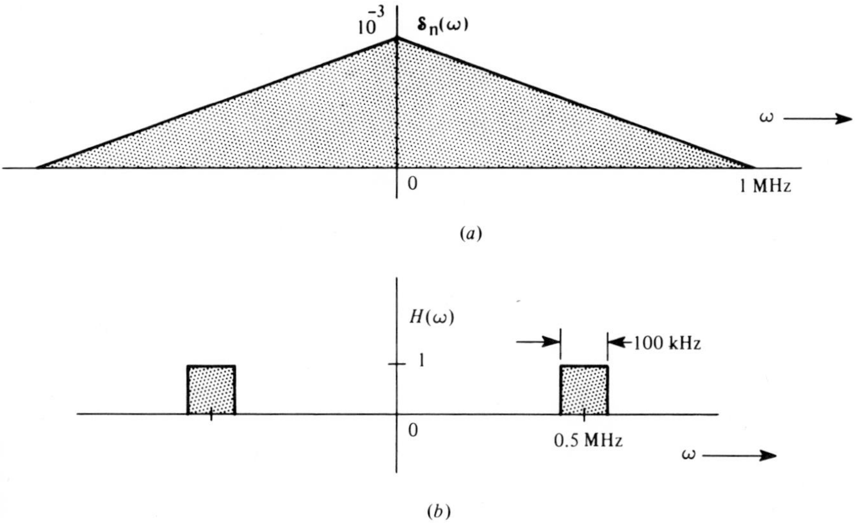

FIG. P.5-2.

5-3. A certain channel has a uniform-noise power-density spectrum $\mathcal{S}_n(\omega) = 10^{-3}$. Over this channel a DSB-SC signal is transmitted. The modulated signal $m(t)$ has a uniform power-density spectrum $\mathcal{S}_m(\omega) = 1$, and is band-limited to 5 kHz. The carrier frequency is 100 kHz. (a) Find the optimum demodulator (using Wiener filter). (b) Determine the channel SNR. (c) Find the output SNR.

5-4. Assume the data in Prob. 5-3. (a) Find the demodulator (using Wiener filter) for SSB-SC (USB) transmission. (b) Determine the output SNR for this case (assume the message signal power transmission same as in Prob. 5-3).

5-5. Given a channel disturbed by white noise of power density $\mathcal{S}_n(\omega) = 10^{-3}$ Over this channel a message m(t) is transmitted by DSB-SC. It is given that

$$S_m(\omega) = \frac{2\sigma_m^2}{W[1 + (\omega/W)^2]}$$

where

$$W = 2\pi(5 \times 10^3) \quad \text{rps}$$

The carrier frequency is 100 kHz. The demodulator shown in Fig. 5-12c is used. (a) Determine the transfer function of the optimum baseband filter (physically realizable, infinite lag Wiener filter). (b) Determine the output SNR.

5-6. Repeat Prob. 5-5 for the case of zero-lag physically realizable (Wiener) filter. (*Hint*: Use Eq. 4-87.) You may also need the definite integral

$$\frac{1}{2\pi}\int_{-\infty}^{\infty} \log\left[1 + \frac{a}{1 + (\omega/W)^2}\right] d\omega = W[(1 + a)^{1/2} - 1]$$

5-7. Given a channel disturbed by white noise of power density $\mathcal{S}_n(\omega) = 10^{-3}$. Over this channel, a message m(t) is transmitted by DSB-SC. It is given that

$$\mathcal{S}_m(\omega) = \frac{(\kappa^2/2B)\,\mathrm{Sa}(\pi/2n)}{1 + (\omega/2\pi B)^{2n}}$$

This spectrum has a 3-db cutoff at B Hz. The demodulator shown in Fig. 5-12c is used. (a) Determine the transfer function of the optimum baseband filter (physically realizable, infinite-lag Wiener filter). (b) Determine the output SNR.

5-8. Repeat Prob. 5-7 for the case of zero-lag physically realizable Wiener filter. (*Hint*: Use Eq. 4-87.) You may also need the definite integral

$$\int_{-\infty}^{\infty} \log\left[1 + \frac{a}{1 + (x/b)^{2n}}\right] dx = \frac{4nb}{\mathrm{Sa}(\pi/2n)}\,(1 + a)^{1/2n} - 1$$

5-9. (a) If $f_u(t)$ represents the SSB-SC (upper sideband) signal (Fig. 5-13b) corresponding to message signal m(t) (Fig. 5-13a), then show that

$$f_u(t) = m(t)\cos(\omega_c t + \Theta) - \check{m}(t)\sin(\omega_c t + \Theta)$$

where $\check{m}(t)$ is the Hilbert transform of m(t). (b) The corresponding equation for a lower sideband SSB-SC is given by

$$f_\ell(t) = m(t)\cos(\omega_c t + \Theta) + \check{m}(t)\sin(\omega_c t + \Theta)$$

(c) Show that m(t) can be recovered from $f_u(t)$ or $f_\ell(t)$ by synchronous detection, that is, m(t) can be recovered from $f_u(t)\cos(\omega_c t + \Theta)$ or $f_\ell(t)\cos(\omega_c t + \Theta)$. [*Hint for part a*: Find $R_{f_u}(\tau)$.]

5-10. Show that the quadrature components of the bandpass process form a Hilbert transform pair, that is, in Eq. 5-2, $x_s(t)\sin\omega_c t$ is the Hilbert transform of $x_c(t)\cos\omega_c t$.

5-11. In a wideband FM system, the carrier amplitude is 100 volts and the carrier frequency is 100 mHz. The power-density spectrum of the modulating signal is given by

$$\mathcal{S}_m(\omega) = \frac{2\sigma_m^2}{W[1 + (\omega/W)^2]}$$

where $W = 2\pi(5 \times 10^3)$. The demodulator shown in Fig. 5-19 is used. (a) Find the transfer function of the final filter (infinite-lag Wiener filter). (b) Find the output SNR.

5-12. The modulating signal $m(t)$ is band-limited to 5 kHz and has a uniform amplitude distribution in the range -2 to 2. Also $\overline{m^2(t)} = 4/3$. This signal is sampled at a rate 10,000 samples per second. The samples are quantized and coded by using binary pulses. The quantizing levels are 1/32 volt apart. The channel noise is white with power-density spectrum $\mathcal{S}_n(\omega) = 10^{-3}$. The binary pulses are represented by $p(t)$ and $-p(t)$ where $p(t)$ is a pulse of width 16.67 μs, and having an energy 0.05 joules. The average power is obviously 3 kw. (a) Find Λ, the output SNR. (b) If the transmitted signal power is reduced to 300 watts, find the output SNR. (c) If the signal power available for transmission is 300 watts, is it possible to attain higher SNR than that in (b) by changing the quantizing level separation? What is the maximum value of Λ that can be attained?

6

Digital Data Communication: Signal Detection

6-1. INTRODUCTION

In Chapter 5 we studied communication systems in which the transmitted and received messages were continuous functions of time. These systems are called *continuous communication* (or analog communication) *systems*. In these systems the chief objective is fidelity of reproduction of waveforms. Hence the signal-to-noise power ratio was used as a criterion for system performance. The choice of this figure of merit stems from the fact that the signal-to-noise ratio bears a relationship to the ability of the listener to interpret a message. In contrast to this, we have *discrete communication* (or digital communication) *systems* in which the transmitter input is chosen from a finite set of possible messages. These messages can assume only a finite number of different specified forms. The set of discrete messages may, for example, consist of the twenty-six letters of the English alphabet, or it may be the mark and space in binary pulse-code modulation. To recapitulate: In general, in a discrete communication system, the information source consists of a finite number of discrete messages which are coded into sequence of waveforms or symbols. The transmitter, therefore, transmits a sequence of waveforms chosen from a specified and finite set. This contrasts with the continuous communication system where the resulting set of waveforms is infinite. In a discrete communication system the objective at the receiver is *not* to reproduce the waveform with fidelity because the possible waveforms are known exactly and the details of the waveform really do not carry any information. Our goal is to determine, from the received signal masked by channel noise, which of the finite waveforms has been transmitted. It is evident that the appropriate figure of merit in a discrete communication system is not the signal-to-noise ratio but the probability of error in making the decisions at the receiver.

We have already touched upon this subject in Chapter 4 in connection with the matched filter. However, our scope there was limited. In the first place, we had restricted ourselves to the binary case (mark and space). Secondly, the system was constrained to be linear. In addition, we made the arbitrary assumption that increasing the signal-to-noise ratio at some instant will increase the reliability of observation (reduce the probability of error in the decision). Qualitatively this assumption is sound, but we have not proved anywhere that this is the method that will yield the least probability of error. There is no general theorem which states that maximum output signal-to-noise ratio ensures maximum gain of information.

In this chapter, we shall study the problem of discrete communication (signal detection of known waveforms) from a more fundamental point of view. We shall consider the case of M waveform (M-ary signaling). The figure of merit of the receiver will be the probability of error in making decisions. There will be no constraints on the systems and we shall not presume that maximum signal-to-noise ratio at the output yields minimum-error probability. Rather we shall try to answer the question: What receiver will yield the minimum-error probability? The only restriction we shall accept is that the channel noise be gaussian.

The comprehension of the signal-detection problem is greatly facilitated by geometrical representation of signals. In a signal space we can represent a signal by a fixed point (or a vector). A random process can be represented by a random point (or a random vector). The region in which the random point may lie will be shown shaded, with the shading intensity proportional to the probability of observing the signal in that region. Let there be M symbols or messages $m_1, m_2, \ldots, m_M$ to be transmitted. Each of these symbols is represented by a specified waveform. Let the corresponding waveforms be $s_1(t), s_2(t), \ldots, s_M(t)$. Thus the symbol (or message) m_k is sent by transmitting the waveform $s_k(t)$. These waveforms are corrupted by a gaussian channel noise

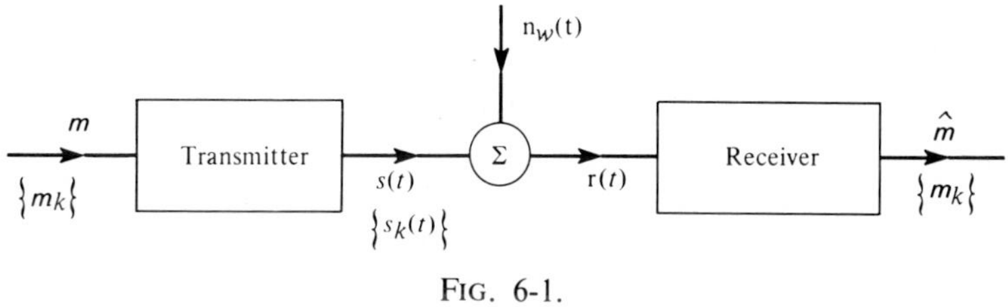

FIG. 6-1.

in additive manner as shown in Fig. 6-1. By assumption the channel noise $n_w(t)$ is a white gaussian noise with power-density spectrum $\mathcal{S}_n(\omega)$ given by

$$\mathcal{S}_n(\omega) = \frac{\mathfrak{N}}{2} \tag{6-1}$$

At the receiver the received signal r(t) consists of one of the M-message waveform $s_k(t)$ plus the channel noise. For the sake of generality, the message waveform will be denoted by $s(t)$, where it is understood that $s(t)$ is one of the m waveforms $s_k(t)$:

$$\mathrm{r}(t) = s(t) + \mathrm{n}_w(t) \tag{6-2}$$

Let us consider the message m_k. Corresponding to this message, the received signal is

$$\mathrm{r}(t) = s_k(t) + \mathrm{n}_w(t)$$

We can represent r(t) in a signal space, by letting $\mathbf{r}$, $\boldsymbol{s}_k$, and $\mathbf{n}_w$ be the points (or vectors) representing signals r(t), $s_k(t)$, and n$_w(t)$ respectively. Then it is evident that

$$\mathbf{r} = \boldsymbol{s}_k + \mathbf{n}_w \tag{6-3}$$

The vector $\boldsymbol{s}_k$ is a fixed vector, since the waveform $s_k(t)$ is nonrandom. The vector $\mathbf{n}_w$ (or the point $\mathbf{n}_w$) is random. Hence the vector $\mathbf{r}$ is also random. Since n$_w(t)$ is a gaussian white noise, the probability distribution of $\mathbf{n}_w$ has spherical symmetry in signal space (see Sec. 3-16). Hence the distribution of $\mathbf{r}$ has a spherical distribution centered at a fixed point $\boldsymbol{s}_k$, as shown in Fig. 6-2. Whenever the message m_k is

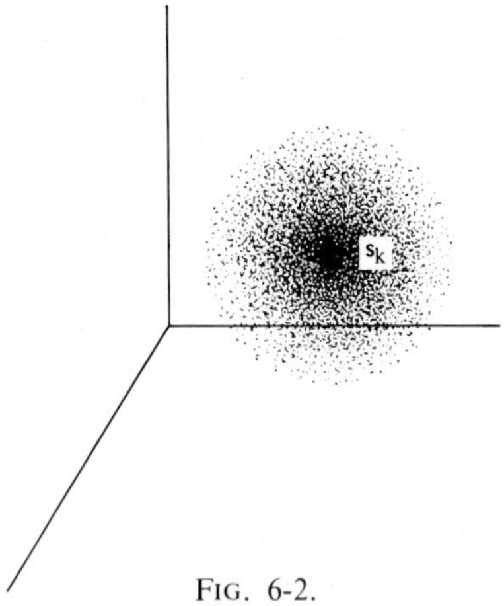

FIG. 6-2.

transmitted, the probability of observing the received signal r(t) in a given region is indicated by the intensity of the shading in Fig. 6-2. Actually, because the noise is white, the space has an infinite number dimensions (see Sec. 1-15). However, for simplicity, we have shown the space to be three-dimensional. This will suffice to indicate our line of reasoning. We can draw similar regions for various messages $\boldsymbol{s}_1, \boldsymbol{s}_2$, $\ldots, \boldsymbol{s}_M$. Figure 6-3 shows the regions for two messages m_j and m_k wher

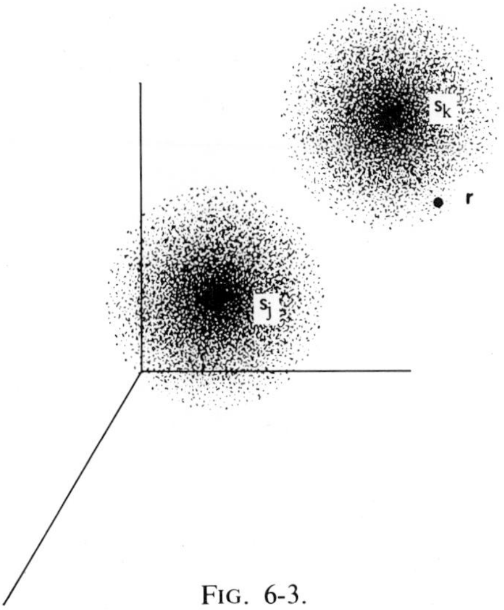

FIG. 6-3.

$\boldsymbol{s}_j$ and $\boldsymbol{s}_k$ are widely separated in signal space. In this case there is virtually no overlap in the two regions. If either m_j or m_k is transmitted, the received signal will lie in one of the two regions. From the position of the received signal, one can decide with a very small probability of error whether m_j or m_k was transmitted. Note that theoretically each region extends to infinity although its probability diminishes rapidly as one moves away from the center. Hence there will always be an overlap in the two regions and this will result in a nonzero error probability. In Fig. 6-3, the received signal $\boldsymbol{r}$ is much closer to $\boldsymbol{s}_k$ than $\boldsymbol{s}_j$. It is therefore more likely that m_k was transmitted.

Figure 6-4 illustrates the case when the points $\boldsymbol{s}_j$ and $\boldsymbol{s}_k$ are spaced

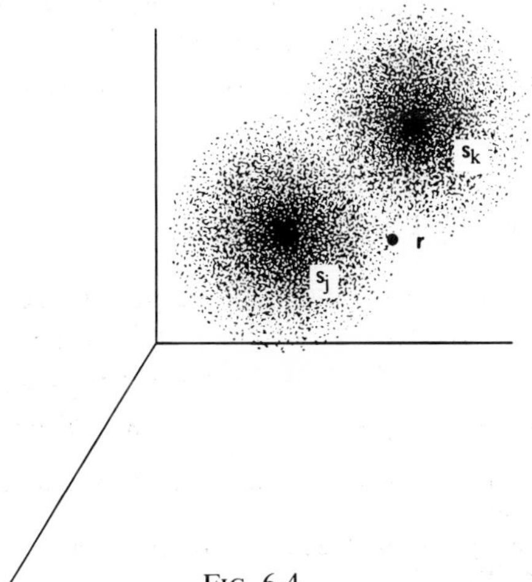

FIG. 6-4.

closely together. In this case there is a considerable overlap in the two regions. In Fig. 6-4, the received signal $\boldsymbol{r}$ is closer to $\boldsymbol{s}_j$ than $\boldsymbol{s}_k$. Hence it is more likely that m_j was transmitted. But in this case there is also a considerable probability that m_k may have been transmitted. Hence in this situation there will be a much higher probability of error in any decision scheme.

The optimum receiver must decide which message has been transmitted from a knowledge of $\boldsymbol{r}$. The signal space must be divided into M nonoverlapping or disjoint regions $R_1, R_2, \ldots, R_M$ corresponding to M messages $m_1, m_2, \ldots, m_M$. If $\boldsymbol{r}$ falls in the region R_k, the decision is m_k. The problem of designing the receiver, then reduces to choosing the boundaries of these regions $R_1, R_2, \ldots, R_M$ such that the probability of error in decision making is minimum.[1] The next few sections deal with this problem of designing an optimum receiver.

A Word about Notation. We shall briefly discuss the notations used here to avoid confusion later. As before, we use roman type to denote a random variable or a random process. Thus x or $\mathrm{x}(t)$ represent a random variable or a random process. A particular value assumed by the random variable in a certain trial is denoted by italic type. Thus x represents the value assumed by x. Similarly, $x(t)$ represents a particular sample function of the random process $\mathrm{x}(t)$. In the case of random vectors, we follow the same convention; a random vector is denoted by roman boldface type and a particular value assumed by the vector in a certain trial is represented by boldface italic. Thus $\mathbf{r}$ denotes a random vector representing $\mathrm{r}(t)$, but $\boldsymbol{r}$ is a particular value of $\mathbf{r}$ and represents a particular received waveform $r(t)$ in some trial. Note that the roman types represent random entities and italic types represent particular values (which are, of course, nonrandom). The signals $s_1(t), s_2(t), \ldots, s_M(t)$ being deterministic signals, are represented by boldface italic, $\boldsymbol{s}_1, \boldsymbol{s}_2, \ldots, \boldsymbol{s}_M$.

6-2. THE OPTIMUM RECEIVER[2]

Before proceeding any further it is helpful to recapitulate the problem: A transmitter transmits a sequence of messages from a set of M messages $m_1, m_2, \ldots, m_M$. These messages are represented by waveforms $s_1(t), s_2(t), \ldots, s_M(t)$ respectively. Each waveform has a duration

[1]Note that here we are implicitly assigning equal costs to all kinds of errors (false dismissal and false alarm).

[2]This approach closely follows that of Wozencraft and Jacobs (Ref. 2). See also E. Arthurs and H. Dym (Ref. 3).

of T seconds and has finite energy. One waveform is transmitted every T seconds. We assume that the receiver is time synchronized with the transmitter. The waveforms are corrupted during transmission by a white gaussian channel noise of power density $\mathfrak{N}/2$. The disturbance is assumed to be additive, that is the noise adds on to the signal waveforms. Knowing the received waveform, the receiver must make a decision as to which waveform was transmitted. The merit criterion of the receiver is the minimum probability of error in making this decision.[3]

Let us now discuss the dimensionality of the signal space in our problem. If there was no noise, we would be dealing with only M waveforms $s_1(t), s_2(t), \ldots, s_M(t)$. In this case a signal space of at most M dimensions would suffice. This is because the dimensionality of a signal space is always equal to or less than the number of independent signals in this space (see Sec. 1-15). For the sake of generality we shall assume the space to have the largest possible number of dimensions, viz. M. The white gaussian noise has infinite bandwidth and consequently has infinite number of dimensions. Let $g_1(t), g_2(t), \ldots$ be a complete set of orthonormal basis signals. This implies that

$$\int_{-\infty}^{\infty} g_j(t)\, g_k(t)\, dt = \begin{cases} 1 & j = k \\ 0 & j \neq k \end{cases} \tag{6-4}$$

The channel noise $\mathrm{n}_w(t)$ can be represented (Sec. 3-16) by a random vector $\mathbf{n}_w(\mathrm{n}_1, \mathrm{n}_2, \ldots)$ where

$$n_w(t) = \mathrm{n}_1 g_1(t) + \mathrm{n}_2 g_2(t) + \cdots$$

$$= \sum_{j=1}^{\infty} \mathrm{n}_j g_j(t) \tag{6-5a}$$

$$\mathrm{n}_j = \int_{-\infty}^{\infty} \mathrm{n}_w(t)\, g_j(t)\, dt \tag{6-5b}$$

We shall now show that for a white noise, the components $\mathrm{n}_1, \mathrm{n}_2, \ldots$ are uncorrelated, and each has a mean square value $\mathfrak{N}/2$, that is,

$$\overline{\mathrm{n}_i \mathrm{n}_j} = \begin{cases} \mathfrak{N}/2 & i = j \\ 0 & i \neq j \end{cases} \tag{6-6}$$

To prove this we have from Eq. 6-5b

$$\overline{\mathrm{n}_i \mathrm{n}_j} = \overline{\int_{-\infty}^{\infty} \mathrm{n}_w(\alpha)\, g_i(\alpha)\, d\alpha \int_{-\infty}^{\infty} \mathrm{n}_w(\beta)\, g_j(\beta)\, d\beta}$$

$$= \int_{-\infty}^{\infty} \int_{-\infty}^{\infty} \overline{\mathrm{n}_w(\alpha)\, \mathrm{n}_w(\beta)}\, g_i(\alpha)\, g_j(\beta)\, d\alpha\, d\beta$$

$$= \int_{-\infty}^{\infty} \int_{-\infty}^{\infty} R_{\mathrm{n}_w}(\beta - \alpha)\, g_i(\alpha)\, g_j(\beta)\, d\alpha\, d\beta$$

[3]The other merit criteria will be discussed later.

For a white noise with power density spectrum $\mathfrak{N}/2$, the autocorrelation function $R_{n_w}(\tau)$ is given by

$$R_{n_w}(\tau) = \mathfrak{F}^{-1}\left[\frac{\mathfrak{N}}{2}\right] = \frac{\mathfrak{N}}{2}\,\delta(\tau)$$

Hence

$$\begin{aligned}\overline{n_i n_j} &= \frac{\mathfrak{N}}{2}\int_{-\infty}^{\infty}\int_{-\infty}^{\infty}\delta(\beta-\alpha)g_i(\alpha)g_j(\beta)\,d\alpha\,d\beta \\ &= \frac{\mathfrak{N}}{2}\int_{-\infty}^{\infty}\left(\int_{-\infty}^{\infty}\delta(\beta-\alpha)g_i(\alpha)\,d\alpha\right)g_j(\beta)\,d\beta \\ &= \frac{\mathfrak{N}}{2}\int_{-\infty}^{\infty}g_i(\beta)g_j(\beta)\,d\beta\end{aligned}$$

Using Eq. 6-5, we obtain

$$\overline{n_i n_j} = \begin{cases}\mathfrak{N}/2 & i = j \\ 0 & i \neq j\end{cases}$$

This proves that the components of the noise vector $\mathbf{n}_w$ are uncorrelated and each has a mean square value of $\mathfrak{N}/2$. Since we are assuming the process to be gaussian, this implies that all the components are independent with variance (mean square value in this case since the mean is zero) of $\mathfrak{N}/2$. This is a very useful result. Note that the above result is independent of the basis signals used. It is valid for any complete set of orthonormal signals.

Earlier, we had seen that the message signals can be represented in M-dimensional hyperspace. Let $\phi_1(t), \phi_2(t), \ldots, \phi_M(t)$ be the orthonormal basis set for this space. Such a set can always be constructed by using Gram-Schmidt procedure discussed in Section 1-16. We can then represent the signal waveform $s_k(t)$ as

$$\begin{aligned}s_k(t) &= s_{k1}\phi_1(t) + s_{k2}\phi_2(t) + \cdots + s_{kM}\phi_M(t) \\ &= \sum_{j=1}^{M} s_{kj}\phi_j(t)\end{aligned} \tag{6-7a}$$

where

$$s_{kj} = \int_{-\infty}^{\infty} s_k(t)\,\phi_j(t)\,dt \tag{6-7b}$$

Now consider the white gaussian channel noise $n_w(t)$. This signal has an infinite bandwidth ($B = \infty$). It has an infinite number of dimensions and obviously cannot be represented in the M-dimensional signal space discussed above. We can, however, split $n_w(t)$ into two components: (i) the projection of $n_w(t)$ on the M-dimensional signal space, and

(ii) the remaining component which will be orthogonal to the M-dimensional signal space. Let us denote the two components by $\mathrm{n}(t)$ and $\mathrm{n}_0(t)$. Thus

$$\mathrm{n}_w(t) = \mathrm{n}(t) + \mathrm{n}_0(t) \tag{6-8}$$

where

$$\mathrm{n}(t) = \sum_{k=1}^{M} \mathrm{n}_j \phi_j(t) \tag{6-9a}$$

and

$$\mathrm{n}_j = \int_{-\infty}^{\infty} \mathrm{n}(t)\,\phi_j(t)\,dt \tag{6-9b}$$

Since $\mathrm{n}_0(t)$ is orthogonal to M-dimensional space, it is orthogonal to every signal in this space. Hence

$$\int_{-\infty}^{\infty} \mathrm{n}_0(t)\,\phi_j(t)\,dt = 0 \qquad \text{for } j = 1, 2, \ldots, M \tag{6-10}$$

From Eqs. 6-5, 6-7, and 6-8 it follows that

$$\begin{aligned} \mathrm{n}_j &= \int_{-\infty}^{\infty} [\mathrm{n}(t) + \mathrm{n}_0(t)]\,\phi_j(t)\,dt \\ &= \int_{-\infty}^{\infty} \mathrm{n}_w(t)\,\phi_j(t)\,dt \end{aligned} \tag{6-11}$$

From Eqs. 6-11 and 6-9a it is evident that we can filter out the component $\mathrm{n}_0(t)$ from $\mathrm{n}_w(t)$. This can be seen from the fact that $\mathrm{r}(t)$, the received signal can be expressed as

$$\begin{aligned} \mathrm{r}(t) &= s_k(t) + \mathrm{n}_w(t). \\ &= s_k(t) + \mathrm{n}(t) + \mathrm{n}_0(t) \\ &= \mathrm{q}(t) + \mathrm{n}_0(t) \end{aligned} \tag{6-12}$$

where $\mathrm{q}(t)$ is the projection of $\mathrm{r}(t)$ on the M-dimensional space. Thus

$$\mathrm{q}(t) = s_k(t) + \mathrm{n}(t) \tag{6-13}$$

We can obtain the projection $\mathrm{q}(t)$ from $\mathrm{r}(t)$ by observing that (see Eqs. 6-7 and 6-9a)

$$\mathrm{q}(t) = \sum_{j=1}^{M} (s_{kj} + \mathrm{n}_j)\,\phi_j(t) \tag{6-14}$$

From Eqs. 6-7b, 6-11, and 6-14 it is obvious that if we feed the received signal $\mathrm{r}(t)$ to a system shown in Fig. 6-5, the resultant outcome will be $\mathrm{q}(t)$. Thus the orthogonal noise component can be filtered out without disturbing the message signal. The question here is: Would such a filtering help in our decision making? We can easily show that it cannot hurt us. The noise $\mathrm{n}_w(t)$ is independent of the signal waveform $s_k(t)$.

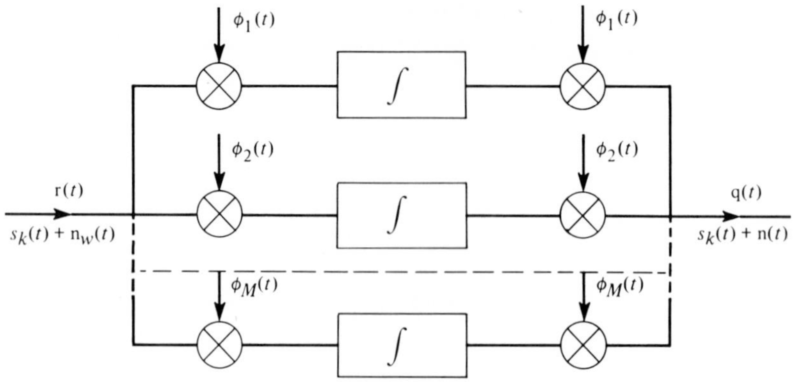

FIG. 6-5.

Therefore its component $n_0(t)$ is also independent of $s_k(t)$. Thus $n_0(t)$ contains no information about the transmitted signal. It therefore appears that discarding such a component from the received signal $r(t)$ will not cause any loss of information regarding the signal waveform $s_k(t)$. This, however, is not enough. We must also make sure that the noise being discarded $[n_0(t)]$ is not in any way related to the remaining noise component $n(t)$. If $n_0(t)$ and $n(t)$ are related in any way, it will be possible to obtain some information about $n(t)$ from $n_0(t)$ and thus enable us to detect that signal with less error probability. This topic has already been discussed in Sec. 5-3. If the components $n_0(t)$ and $n(t)$ are independent random processes, the component $n_0(t)$ does not carry any information about $n(t)$ and can be discarded. Under these conditions, $n_0(t)$ is irrelevant to the decision making at the receiver. We have already shown that the components of white gaussian noise are independent (Eq. 6-6). The process $n(t)$ is represented by M components $n_1, n_2, \ldots, n_M$ along $\phi_1(t), \phi_2(t), \ldots, \phi_M(t)$, and $n_0(t)$ is represented by the remaining components (infinite number) along the remaining basis signals in the complete set, $\{\phi_k(t)\}$. From Eq. 6-6, we observe that all the components are independent. Hence the components representing $n_0(t)$ are independent of components representing $n(t)$. Consequently $n_0(t)$ is independent of $n(t)$ and is irrelevant data. We can, therefore, consider only $n(t)$ the component of $n_w(t)$ in the signal space.

The received signal $r(t)$ is now reduced to signal $q(t)$ which contains the desired signal waveform and the projection of the channel noise on the M-dimensional signal space. Thus the signal $q(t)$ can be completely represented in signal space. Let the vectors representing $n(t)$ and $q(t)$ be denoted by $\mathbf{n}$ and $\mathbf{q}$. Thus

$$\mathbf{q} = \boldsymbol{s} + \mathbf{n}$$

where $\boldsymbol{s}$ may be any one of vectors $\boldsymbol{s}_1, \boldsymbol{s}_2, \ldots, \boldsymbol{s}_M$.

The random vector $\mathbf{n}(n_1, n_2, \ldots, n_M)$ is represented by M independent gaussian variables, each with zero mean and variance $\sigma^2 = \mathfrak{N}/2$. The joint probability-density function of vector $\mathbf{n}$ in such a case has a spherical symmetry as shown in Eq. 3-108:

$$p_{\mathbf{n}}(n) = \frac{1}{(2\pi\sigma^2)^{M/2}} e^{-|n|^2/2\sigma^2}$$

$$= \frac{1}{(\pi\mathfrak{N})^{M/2}} e^{-|n|^2/\mathfrak{N}} \tag{6-15a}$$

Note that this actually represents

$$p_{n_1, n_2, \ldots, n_M}(n_1, n_2, \ldots, n_M) = \frac{1}{(\pi\mathfrak{N})^{M/2}} e^{-(n_1^2 + n_2^2 + \cdots + n_M^2)/\mathfrak{N}} \tag{6-15b}$$

The Decision Procedure. Our problem is now considerably simplified. The irrelevant noise component has been filtered out. The residual signal $q(t)$ can be represented in M-dimensional signal space. We proceed to determine the M decision regions $R_1, R_2, \ldots, R_M$ in this space. The regions must be so chosen that the probability of error in making the decision is minimized.

Suppose the received vector $\mathbf{q} = q$. Then if the receiver decides $\hat{m} = m_k$, the conditional probability of making the correct decision, given that $\mathbf{q} = q$ is

$$P(C \mid \mathbf{q} = q) = P(m_k \mid \mathbf{q} = q) \tag{6-16}$$

where $P(C \mid \mathbf{q} = q)$ is the conditional probability of making the correct decision given $\mathbf{q} = q$ and $P(m_k \mid \mathbf{q} = q)$ is the conditional probability that m_k was transmitted given $\mathbf{q} = q$. The unconditional probability $P(C)$ is given by

$$P(C) = \int_q P(C \mid \mathbf{q} = q) p_{\mathbf{q}}(q)\, dq \tag{6-17}$$

where the integration is performed over the entire region occupied by $\mathbf{q}$. Note that this is an M-fold integration with respect to the variables $(q_1, q_2, \ldots, q_M)$ over the range $-\infty$ to ∞. Also, since $p_{\mathbf{q}}(q) \geq 0$, this integral is maximum when $P(C \mid q = q)$ is maximum. From Eq. 6-16 it now follows that if a decision $\hat{m} = m_k$ is made, the error probability is minimized if

$$P(m_k \mid \mathbf{q} = q)$$

is maximized. The probability $P(m_k \mid \mathbf{q} = q)$ is called the *a posteriori probability* of m_k. This is because it represents the probability that m_k was transmitted when q is received.

The decision procedure is now clear. Once we receive $\mathbf{q} = q$, we evaluate all the M *a posteriori* probabilities. Then we make the decision in favor of that message for which the *a posteriori* probability is highest—that is, the receiver decides that $\hat{m} = m_k$ if

$$P(m_k \mid \mathbf{q} = q) > P(m_j \mid \mathbf{q} = q) \qquad \text{for all } j \neq k \tag{6-18}$$

We can use Bayes' mixed rule (Eq. 2-43c) to determine the a posteriori probabilities. We have

$$P(m_k \mid \mathbf{q} = q) = \frac{P(m_k)\,p_{\mathbf{q}}(q \mid m_k)}{p_{\mathbf{q}}(q)} \tag{6-19}$$

Hence the receiver decides $\hat{m} = m_k$ if the decision function

$$\frac{P(m_i)\,p_{\mathbf{q}}(q \mid m_i)}{p_{\mathbf{q}}(q)} \qquad i = 1, 2, \ldots, M$$

is maximum for $i = k$.

Note that the denominator $p_{\mathbf{q}}(q)$ is common to all decision functions and hence may be ignored. Thus the receiver sets $\hat{m} = m_k$ if the decision function

$$P(m_i)\,p_{\mathbf{q}}(q \mid m_i) \qquad i = 1, 2, \ldots, M \tag{6-20}$$

is maximum for $i = k$.

Thus, once q is obtained, we compute the decision function 6-20 for all messages $m_1, m_2, \ldots, m_M$ and decide that message for which the function is maximum is the one most likely to have been sent.

We now turn our attention to computing the decision functions. The a priori probability $P(m_i)$ represents the probability that the message m_i will be transmitted. These probabilities must be known if the above criterion is used.[4] The term $p_{\mathbf{q}}(q \mid m_i)$ represent the probability-density function of $\mathbf{q}$ when $s(t) = s_i(t)$. Under this condition,

$$\mathbf{q} = s_i + \mathbf{n}$$

and

$$\mathbf{n} = \mathbf{q} - s_i$$

The point s_i is constant and $\mathbf{n}$ is a random point. Obviously $\mathbf{q}$ is a random point with the same distribution as $\mathbf{n}$ but centered at the points s_i. Hence from Eq. 6-15a, we have

$$p_{\mathbf{q}}(q \mid m_i) = p_{\mathbf{n}}(q - s_i) = \frac{1}{(\pi \mathfrak{N})^{M/2}}\, e^{|q - s_i|^2/\mathfrak{N}} \tag{6-21}$$

The decision function in Eq. 6-20 now becomes

$$\frac{P(m_i)}{(\pi \mathfrak{N})^{M/2}}\, e^{-|q - s_i|^2/\mathfrak{N}} \tag{6-22}$$

[4] In case these probabilities are unknown, one must use other merit criteria such as maximum likelihood or minimax, as discussed in later sections.

Note that the decision function is always nonnegative for all values of i. Hence comparing these functions is equivalent to comparing their logarithms, since the logarithm is a monotonic function for positive argument. Hence for convenience, the decision function will be chosen as the logarithm of Eq. 6-22. In addition the factor $(\pi\mathfrak{N})^{M/2}$ being common for all i, can be left out. Hence the decision function is

$$\ln P(m_i) - \frac{1}{\mathfrak{N}} \ln |\boldsymbol{q} - \boldsymbol{s}_i|^2 \tag{6-23}$$

Note that $|\boldsymbol{q} - \boldsymbol{s}_i|^2$ is the square of the length of the vector $\boldsymbol{q} - \boldsymbol{s}_i$. Hence

$$\begin{aligned} |\boldsymbol{q} - \boldsymbol{s}_i|^2 &= \sum_{k=1}^{M} (q_k - s_{ik})^2 \\ &= \sum_{k=1}^{M} q_k^2 + \sum_{k=1}^{M} s_{ik}^2 - 2\sum_{k=1}^{M} q_k s_{ik} \\ &= |\boldsymbol{q}|^2 + |\boldsymbol{s}_i|^2 - 2\boldsymbol{q}\cdot\boldsymbol{s}_i \end{aligned} \tag{6-24}$$

Hence the decision function in Eq. 6-23 becomes (after multiplying throughout by $\mathfrak{N}/2$)

$$\frac{\mathfrak{N}}{2} \ln P(m_i) - \frac{1}{2}[|\boldsymbol{q}|^2 + |\boldsymbol{s}_i|^2 - 2\boldsymbol{q}\cdot\boldsymbol{s}_i] \tag{6-25}$$

Note that the term $|\boldsymbol{s}_i|^2$ is the square of the length of $\boldsymbol{s}_i$ and represents E_i, the energy of signal $s_i(t)$ (see Eq. 1-135). The terms $\mathfrak{N} \ln P(m_i)$ and E_i are constants in the decision function. Let

$$a_i = \tfrac{1}{2}[\mathfrak{N} \ln P(m_i) - E_i] \tag{6-26}$$

Now the decision function in Eq. 6-25 becomes

$$a_i + \boldsymbol{q}\cdot\boldsymbol{s}_i - \tfrac{1}{2}|\boldsymbol{q}|^2$$

The term $(1/2)|\boldsymbol{q}|^2$ is common to all the M decision functions, and can be omitted for purposes of comparison. Thus the new decision function b_i is

$$b_i = a_i + \boldsymbol{q}\cdot\boldsymbol{s}_i \tag{6-27}$$

We compute this function b_i for $i = 1, 2, \ldots, M$ and the receiver decides that $\hat{m} = m_k$ if this function is the largest for $i = k$.

By assumption all signals are of finite duration. The signals exist over the interval $(0, T)$ and are zero outside this interval. We have (see Eq. 1-134)

$$\boldsymbol{q}\cdot\boldsymbol{s}_i = \int_{-\infty}^{\infty} q(t)\, s_i(t)\, dt \tag{6-28}$$

If the signal $q(t)$ is applied at the input terminals of a system whose impulse response is $h(t)$, the output is given by

$$\int_{-\infty}^{\infty} q(\lambda)h(T-\lambda)\,d\lambda$$

If we choose $h(t) = s_i(T-t)$,

$$h(T-\lambda) = s_i(\lambda)$$

and the output is

$$\int_{-\infty}^{\infty} q(\lambda)s_i(\lambda)\,d\lambda$$

Note that since $s_i(t)$ exists over the interval $(0, T)$ only, the limits of integration in the above integral may be replaced by 0 to T. Hence $q \cdot s_i$ is the output at $t = T$ of a system with impulse response $s_i(T-t)$ when the signal $q(t)$ is fed to it. This is obviously the filter matched to $s_i(t)$.

Actually we do not have $q(t)$. The incoming signal $r(t)$ is given by

$$\begin{aligned} r(t) &= s_i(t) + n_w(t) \\ &= \underbrace{s_i(t) + n(t)}_{q(t)} + \underbrace{n_0(t)}_{\text{irrelevant noise}} \end{aligned}$$

where $n_0(t)$ is the (irrelevant) component of $n_w(t)$ orthogonal to the M-dimensional signal space. Since $n_0(t)$ is orthogonal to this space, it is orthogonal to every signal in this space. Hence it is orthogonal to the signal $s_i(t)$, and

$$\int_{-\infty}^{\infty} n_0(t)s_i(t)\,dt = 0$$

Equation 6-28 therefore becomes

$$\begin{aligned} q \cdot s_i &= \int_{-\infty}^{\infty} q(t)s_i(t)\,dt + \int_{-\infty}^{\infty} n_0(t)s_i(t)\,dt \\ &= \int_{-\infty}^{\infty} [q(t) + n_0(t)]\,s_i(t)\,dt \\ &= \int_{-\infty}^{\infty} r(t)s_i(t)\,dt \end{aligned}$$

Hence it is immaterial whether we use $q(t)$ or $r(t)$ at the input. We thus apply the incoming signal $r(t)$ to a parallel bank of matched filters and the output of the filters is sampled at $t = T$. To this a constant a_i is added to the ith filter and the output of all the filters are compared. The decision is made in favor of the signal for which this output is largest. The receiver implementation for this decision procedure is shown in Fig. 6-6a. As shown in Chapter 4, a matched filter is equivalent to a correlator. One may therefore use correlators instead of matched filters. Such an arrangement is shown in Fig. 6-6b.

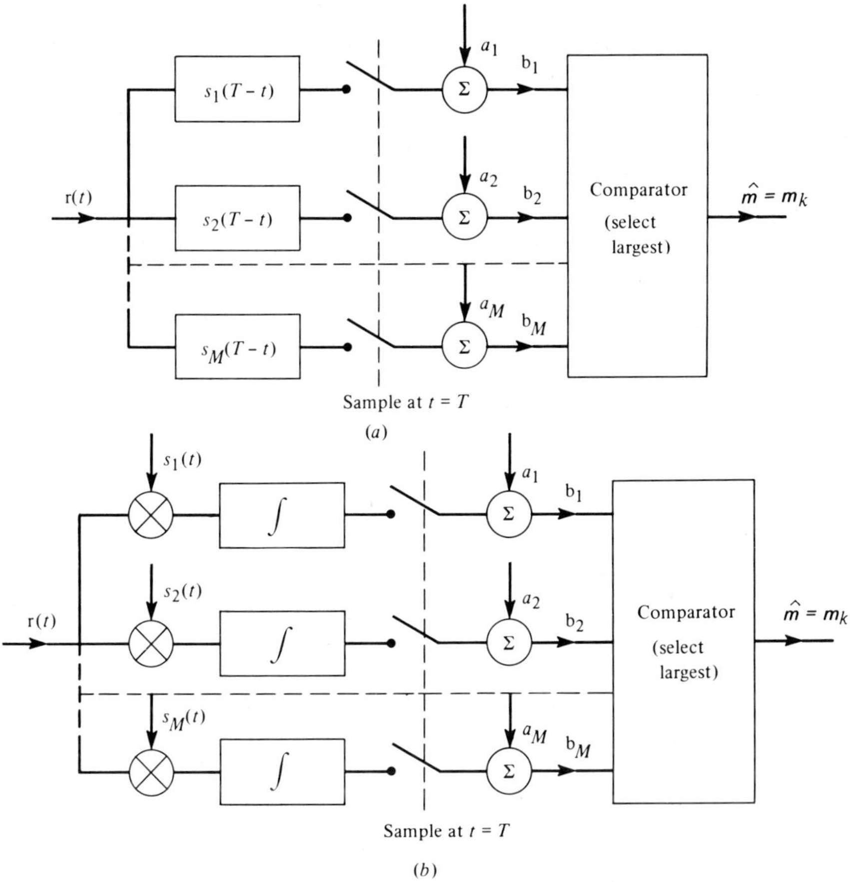

FIG. 6-6.

We have shown that in the presence of additive white gaussian noise, the matched-filter receiver is the optimum receiver when the merit criterion is the minimum error probability. Note that the system is linear although it was not constrained to be so. It is therefore obvious that for white gaussian noise the optimum receiver happens to be linear.

The matched filter obtained in Chapter 4 and the decision procedure is identical to that derived here (for $M = 2$) when the false-alarm cost

(C_a) and the false-dismissal cost (C_d) are identical, and when the energies of the two signals are equal (see Fig. 4-28).

In Chapter 4, we arbitrarily assumed the structure of the optimum filter to be the one which maximizes signal-to-noise power ratio at some instant. Here we have shown that when the noise is gaussian such a structure is indeed optimum.

6-3. DECISION REGIONS

We shall now consider the problem of determining the decision regions for the optimum receiver.

As mentioned earlier, the signal space is divided into M nonoverlapping or disjoint decision regions $R_1, R_2, \ldots, R_M$ corresponding to M messages. If $\boldsymbol{q}$ falls in the region R_k, the decision is that m_k was transmitted. The decision regions are so chosen that the probability of error of the receiver is minimum. In the light of this geometrical representation, we shall now try to interpret how the optimum receiver sets these decision regions.

The decision function is given by Eq. 6-23. The optimum receiver sets $\hat{m} = m_k$ if the decision function

$$\mathfrak{N} \ln P(m_i) - |\boldsymbol{q} - \boldsymbol{s}_i|^2$$

is maximum for $i = k$. This equation defines the decision regions.

For simplicity, let us first consider the case of equiprobable messages —that is, $P(m_i) = 1/M$ for all i. In this case the first term in the decision function is the same for all i and hence can be dropped. Thus the receiver decides that $\hat{m} = m_k$ if the term $-|\boldsymbol{q} - \boldsymbol{s}_i|^2$ has its largest (numerically the smallest) value for $i = k$. Alternatively this may also be stated as follows: the receiver decides that $\hat{m} = m_k$ if the decision function $|\boldsymbol{q} - \boldsymbol{s}_i|^2$ is minimum for $i = k$. Note that $\boldsymbol{q} - \boldsymbol{s}_i$ is the distance of point $\boldsymbol{q}$ from point $\boldsymbol{s}_i$. Thus the decision procedure in this case has a very simple interpretation in geometrical space. We take the projection of the received signal $r(t)$ in the signal space. This is represented by point $\boldsymbol{q}$. Then the decision is made in favor of that signal which is closest to $\boldsymbol{q}$. This result is as expected on qualitative grounds for gaussian noise. This is because the gaussian noise has a spherical symmetry. If, however, the messages are not equiprobable, we cannot go too far on purely qualitative grounds. Nevertheless, we can draw certain broad conclusions. If a particular message m_i is more likely than the others, one will be safer in deciding more often in favor of m_i than other messages. Hence in such a case the decision regions will be biased or weighted in favor of m_i. This fact is shown by the term $\ln P(m_i)$ appear-

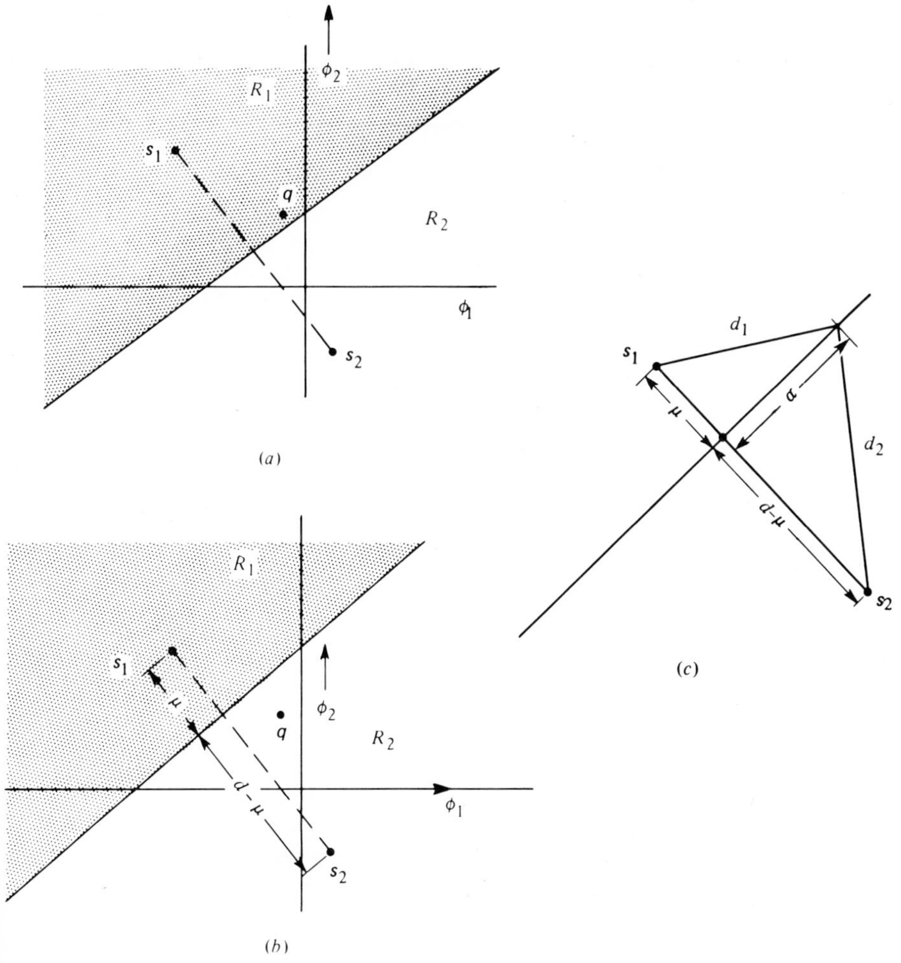

FIG. 6-7.

ing in the decision function. To understand this point a little better, let us consider a two-dimensional signal space and two signals $\boldsymbol{s}_1$ and $\boldsymbol{s}_2$ as shown in Fig. 6-7a. In this figure, the decision regions R_1 and R_2 are shown for equiprobable messages; $P(m_1) = P(m_2) = 0.5$. It is obvious from Fig. 6-7a that the decision is made in favor of that message which is closest to $\boldsymbol{q}$. The boundary of the decision region is the perpendicular bisector of the line joining points $\boldsymbol{s}_1$ and $\boldsymbol{s}_2$. Note that any point on the boundary is equidistant from $\boldsymbol{s}_1$ and $\boldsymbol{s}_2$. If $\boldsymbol{q}$ happens to fall on the boundary, we just flip a coin and decide whether to select m_1 or m_2. Figure 6-7b shows the case when the two messages are not equiprobable. To delineate the boundary of the decision regions, we use Eq. 6-23. The decision is m_1 if

$$|\mathbf{q} - \mathbf{s}_i|^2 - \mathfrak{N} \ln P(m_1) < |\mathbf{q} - \mathbf{s}_2|^2 - \mathfrak{N} \ln P(m_2)$$

Otherwise the decision is m_2.

Note that $|\mathbf{q} - \mathbf{s}_1|$ and $|\mathbf{q} - \mathbf{s}_2|$ represent d_1 and d_2, the distance of $\mathbf{q}$ from $\mathbf{s}_1$ and $\mathbf{s}_2$ respectively. Thus the decision is m_1 if

$$d_1^2 - d_2^2 < \mathfrak{N} \ln \frac{P(m_1)}{P(m_2)} \tag{6-29}$$

The right-hand side of the above inequality is a constant c:

$$c = \mathfrak{N} \ln \frac{P(m_1)}{P(m_2)} \tag{6-30}$$

Thus the decision is m_1 if

$$d_1^2 - d_2^2 < c \tag{6-31}$$

The decision is m_2 if

$$d_1^2 - d_2^2 > c$$

On the boundary of the decision regions,

$$d_1^2 - d_2^2 = c$$

We can easily show that such a boundary is given by a straight line as shown in Fig. 6-7b. This line is perpendicular to line $\mathbf{s}_1\mathbf{s}_2$ and passes through $\mathbf{s}_1\mathbf{s}_2$ at a distance μ from $\mathbf{s}_1$, where

$$\mu = \frac{c + d^2}{2d} = \frac{\mathfrak{N}}{2d} \ln \left[\frac{P(m_1)}{p(m_2)}\right] + \frac{d}{2} \tag{6-32}$$

d being the distance between $\mathbf{s}_1$ and $\mathbf{s}_2$. To prove this we redraw the pertinent part of Fig. 6-7b in Fig. 6-7c. It is evident from this figure that

$$d_1^2 = \alpha^2 + \mu^2$$

$$d_2^2 = \alpha^2 + (d - \mu)^2$$

Hence

$$d_1^2 - d_2^2 = 2d\mu - d^2 = c$$

Therefore

$$\mu = \frac{c + d^2}{2d}$$

This is the desired result. Thus along the decision boundary $d_1^2 - d_2^2$ is constant and equal to c. The boundaries of the decision regions for $M > 2$ may be determined along similar lines. The decision regions for the case of three equiprobable, 2-dimensional signals are shown in Fig. 6-8. The boundaries of the decision regions are perpendicular bisectors of the lines joining the original transmitted signals. If the signals are not equiprobable, then the boundaries will be shifted away from the signals with larger probabilities of occurrence.

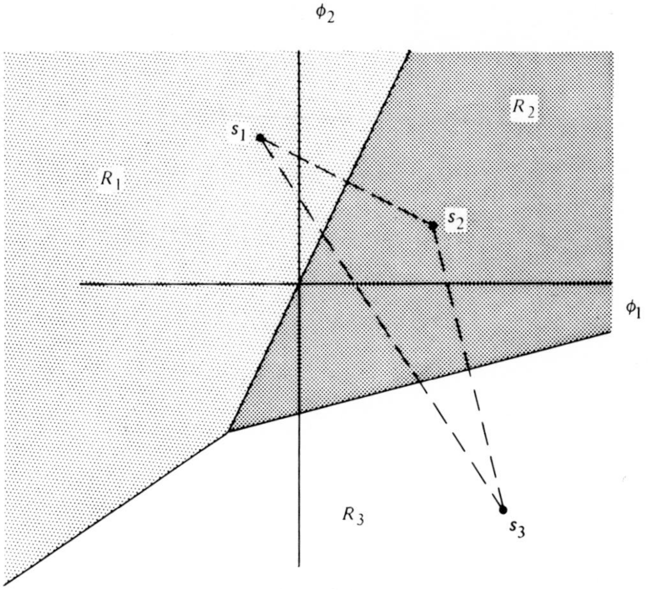

FIG. 6-8.

For signals in M-dimensional space, the decision regions will be M-dimensional hypercones.

6-4. PROBABILITY OF ERROR

If there are M messages $m_1, m_2, \ldots, m_M$ with decision regions R_1, $R_2, \ldots, R_M$, then $P(C \mid m_i)$, the probability of correct decision when m_i is transmitted, is given by

$$P(C \mid m_i) = \text{probability that } \boldsymbol{q} \text{ lies in } R_i$$

and $P(C)$, the probability of a correct decision is given by

$$P(C) = \sum_{i=1}^{n} P(m_i)\, P(C \mid m_i) \tag{6-33}$$

and $P(\epsilon)$, the probability of error is given by

$$P(\epsilon) = 1 - P(C)$$

Let us consider one of the conditional probabilities $P(C \mid m_i)$. Figure 6-9a displays the signal $\boldsymbol{s}_i$ and the hypothetical decision region R_i (shown shaded). The conditional probability $P(C \mid m_i)$ is the probability that $\boldsymbol{q}(\boldsymbol{q} = \boldsymbol{n} + \boldsymbol{s}_i)$ lies inside the region R_i. This is also the probability that the noise vector drawn from the point $\boldsymbol{s}_i$ lies within R_i. Note that this probability does not depend upon the origin of the coordinate system. We may translate the coordinate system any way we wish. This

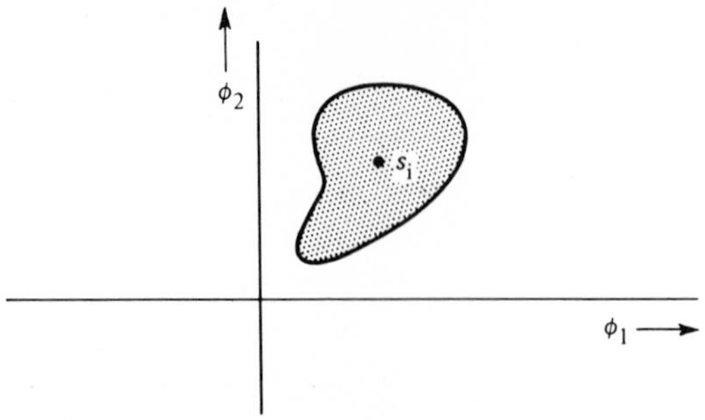

(a)

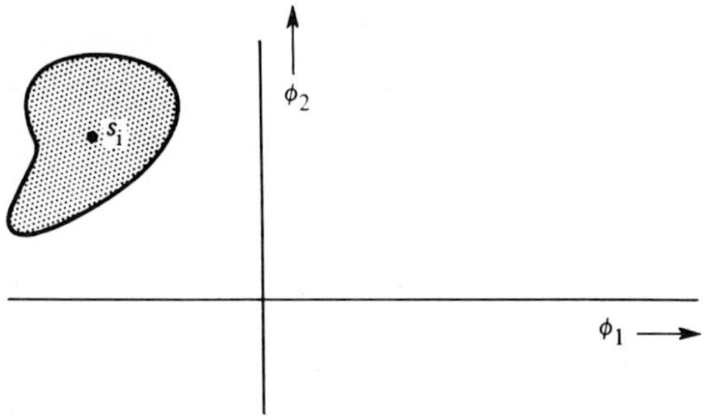

(b)

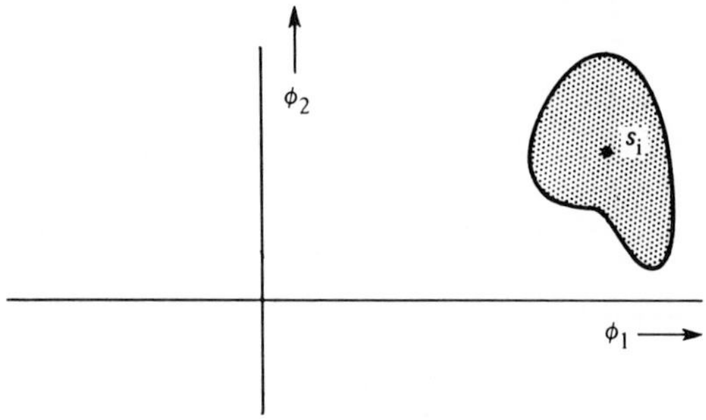

(c)

FIG. 6-9.

is equivalent to translating the signal and the corresponding decision region. Thus the $P(C \mid m_i)$ for the translated system shown in Fig. 6-9b is identical to that of the system in Fig. 6-9a. We therefore conclude that the translation of the entire signal system and the corresponding decision regions do not affect the error probabilities.

In the case of gaussian noise we make another important observation. The rotation of the coordinate system does not affect the error probability because the noise-vector probability density has spherical symmetry. To show this we shall consider Fig. 6-9c where the system in Fig. 6-9a is shown translated and rotated. Note that a rotation of the coordinate system is equivalent to a rotation of the signal system in the opposite sense. Here for convenience we rotate the signal system instead of the coordinate system. It can be seen that the probability that noise vector **n** drawn from $\boldsymbol{s}_i$ should lie in R_i is the same in Fig. 6-9a and 6-9c. This is because this probability is given by the integral of the noise probability density $p_{\mathbf{n}}(\boldsymbol{n})$ over the region R_i. Since $p_{\mathbf{n}}(\boldsymbol{n})$ has a spherical symmetry for Gaussian noise, the probability will remain unaffected by rotation of the region R_i.

We therefore conclude that for additive channel noise a translation of the coordinate system (or translation of the signal system) does not affect the error probability. If in addition the noise is gaussian, the error probability is also unaffected by rotation of the coordinate system. Note that when we rotate or translate a set of signals, the resulting set represents an entirely different set of signals. Yet the error probabilities of the two sets are identical. Such sets are called *equivalent sets*. Two sets are said to be equivalent if their error probabilities are equal. Thus translation and rotation of a signal set corrupted by gaussian noise yields an equivalent set. Note that rotation is allowed only in the case of gaussian noise.

It often proves advantageous to translate and rotate a system in order to compute the error probability. As an example, consider the case of the signal set $\boldsymbol{s}_1$ and $\boldsymbol{s}_2$ in Fig. 6-7b. This figure is reproduced in Fig. 6-10a. The received signal (projection) is represented by point $\boldsymbol{q}$. Here

$$P(C \mid m_1) = \text{probability that } \boldsymbol{q} \text{ lies in } R_1 \text{ when } m_1 \text{ is transmitted}$$

$$= \int_{R_1} p_{\mathbf{q}}(\boldsymbol{q} \mid m_1)\, d\boldsymbol{q} \tag{6-34}$$

Figure 6-10a shows the case when m_1 was transmitted, but $\boldsymbol{q}$ lies in R_2. In this case, the decision of the receiver will be in error. From this figure it is evident that the event that $\boldsymbol{q}$ lies in R_1 when m_1 is transmitted is identical to the event that the noise vector $\boldsymbol{n}$ drawn from $\boldsymbol{s}_1$ should lie entirely in R_1. Evaluation of integral 6-34 is not very convenient. If, however, we translate and rotate the signal set as shown in Fig. 6-10b, we form the equivalent set and the corresponding integral is much easier to evaluate. In Fig. 6-10b it is evident that $P(C \mid m_1)$ is the probability that the noise vector $\boldsymbol{n}$ drawn from $\boldsymbol{s}_1$ should lie in R_1. This is equivalent to the probability that n_1 (component of $\boldsymbol{n}$ along ϕ_1) be less than μ. We

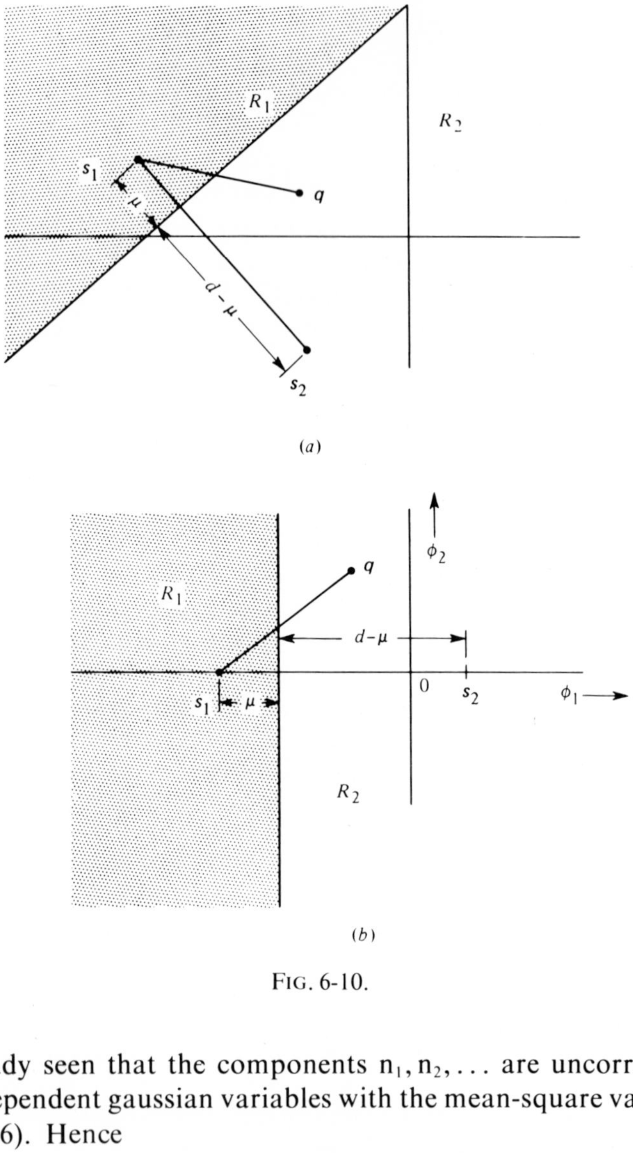

FIG. 6-10.

have already seen that the components $n_1, n_2, \ldots$ are uncorrelated and hence independent gaussian variables with the mean-square value of $\mathfrak{N}/2$ (see Eq. 6-6). Hence

$$p_{n_1}(n_1) = \frac{1}{\sqrt{\pi \mathfrak{N}}} e^{-n_1^2/\mathfrak{N}}$$

and

$$P(C \mid m_1) = \text{probability } (n_1 < \mu)$$

$$= \frac{1}{\sqrt{\pi \mathfrak{N}}} \int_{-\infty}^{\mu} e^{-n_1^2/\mathfrak{N}} \, dn_1$$

and

$$P(\epsilon \mid m_1) = 1 - P(C \mid m_1)$$
$$= \frac{1}{\sqrt{\pi \mathfrak{N}}} \int_{\mu}^{\infty} e^{n_1^2/\mathfrak{N}} \, dn_1$$
$$= \operatorname{erfc}\left(\frac{\mu}{\sqrt{\mathfrak{N}/2}}\right)$$

Similarly,[5]

$$P(\epsilon \mid m_2) = \text{probability } [\mathrm{n}_1 < -(d - \mu)]$$
$$= \frac{1}{\sqrt{\pi \mathfrak{N}}} \int_{-\infty}^{-(d-\mu)} e^{-n_1^2/\mathfrak{N}} \, dn_1$$
$$= \operatorname{erfc}\left(\frac{d - \mu}{\sqrt{\mathfrak{N}/2}}\right)$$

and

$$P(\epsilon) = P(m_1)\, P(\epsilon \mid m_1) + P(m_2)\, P(\epsilon \mid m_2)$$
$$= P(m_1) \operatorname{erfc}\left(\frac{\mu}{\sqrt{\mathfrak{N}/2}}\right) + P(m_2) \operatorname{erfc}\left(\frac{d - \mu}{\sqrt{\mathfrak{N}/2}}\right) \qquad (6\text{-}35)$$

where μ is given by Eq. 6-32:

$$\mu = \frac{\mathfrak{N}}{2d} \ln \left[\frac{P(m_1)}{P(m_2)}\right] + \frac{d}{2}$$

If the two signals are equiprobable, then

$$P(m_1) = P(m_2) = 0.5 \qquad \text{and} \qquad \mu = \frac{d}{2}$$

Therefore

$$P(\epsilon) = \operatorname{erfc}\left(\frac{d}{\sqrt{2\mathfrak{N}}}\right) \qquad (6\text{-}36)$$

This example clearly illustrates the advantages of translating and rotating the coordinates (or the decision regions). If this were not done, the computation of $P(\epsilon)$ would be much more difficult. As another example consider the 3-ary case where

$$s_1(t) = \sqrt{\frac{2}{T}} \sin\left(\frac{2\pi t}{T}\right)$$
$$s_2(t) = \sqrt{\frac{2}{T}} \sin\left(\frac{4\pi t}{T}\right) \qquad 0 \leq t < T$$
$$s_3(t) = \sqrt{\frac{1}{2T}} \left[\sin\left(\frac{2\pi t}{T}\right) + \sin\left(\frac{4\pi t}{T}\right)\right]$$

[5] For discussion on erf (x) and erfc (x) see p. 297.

The signals are zero for $t \geq T$. Compute the probability of error assuming all the messages to be equiprobable.

We observe that $s_1(t)$ and $s_2(t)$ are orthonormal basis signals. It follows that

$$s_1(t) = \phi_1(t), \quad s_2(t) = \phi_2(t)$$

and

$$s_3(t) = \tfrac{1}{2}[\phi_1(t) + \phi_2(t)]$$

The geometrical representation of the three signals is shown in Fig. 6-11a. Figure 6-11b shows a set equivalent to the set in Fig. 6-11a.

Since the signals are equiprobable, the decision region R_1, R_2, and R_3 are as shown in Fig. 6-11b. From this figure, it is clear that

$$P(C \mid m_1) = \text{probability}\left(n_1 < \frac{1}{2\sqrt{2}}\right)$$

$$P(C \mid m_2) = \text{probability}\left(n_1 > -\frac{1}{2\sqrt{2}}\right)$$

$$P(C \mid m_3) = \text{probability}\left(|n_1| < \frac{1}{2\sqrt{2}}\right)$$

Note that $P(C \mid m_1) = P(C \mid m_2)$

$$P(C \mid m_1) = P(C \mid m_2) = \frac{1}{\sqrt{\pi \mathfrak{N}}} \int_{-\infty}^{1/2\sqrt{2}} e^{-n_1^2/\mathfrak{N}}\, dn_1$$

$$= \frac{1}{\sqrt{2\pi}} \int_{-\infty}^{1/2\sqrt{\mathfrak{N}}} e^{-x^2/2}\, dx$$

$$= \operatorname{erf}\left(\frac{1}{2\sqrt{\mathfrak{N}}}\right)$$

Similarly

$$P(C \mid m_3) = 2 \operatorname{erf}\left(\frac{1}{2\sqrt{\phantom{\mathfrak{N}}}}\right) - 1$$

and

$$\begin{aligned} P(\epsilon) &= 1 - P(C) \\ &= 1 - P(m_1)P(C \mid m_1) + P(m_2)P(C \mid m_2) + P(m_3)P(C \mid m_3) \\ &= 1 - \frac{1}{3}\left[2 \operatorname{erf}\left(\frac{1}{2\sqrt{\mathfrak{N}}}\right) + 2 \operatorname{erf}\left(\frac{1}{2\sqrt{\mathfrak{N}}}\right) - 1\right] \\ &= 1 - \frac{1}{3}\left[4 \operatorname{erf}\left(\frac{1}{2\sqrt{\mathfrak{N}}}\right) - 1\right] \end{aligned}$$

General Method of Computing Error Probability When Decisio[n] Regions Are Known. So far, we have considered rather simple cases i[n] computing error probabilities. In these problems rotation and translation of coordinates simplified the computations. In more complex problems, however, such techniques (rotation and translation of axes) has limited utility. For this reason we shall now demonstrate a general method of computing error probability when the decision regions are known. Error probability is $1 - P(C)$ where $P(C)$ is the probability of correct decision. It can be seen from Eq. 6-33 that $P(C)$ can be determined from the knowledge of a priori probabilities $\{P(m_i)\}$ and the conditional probabilities $\{P(C \mid m_i)\}$. The conditional probability $P(C \mid m_i)$ is the probability that the received signal $\boldsymbol{q}$ lies in R_i when m_i is transmitted and is given by

$$P(C \mid m_i) = \int_{R_i} p_{\mathbf{q}}(\boldsymbol{q} \mid m_i)\, d\boldsymbol{q} \tag{6-37a}$$

We may also express $P(C \mid m_i)$ in alternate way. If m_i is transmitted, the received signal is $\mathbf{q} = \mathbf{n} + s_i$. If the noise $\mathbf{n}$ is such that $\mathbf{n} + s_i$ remains

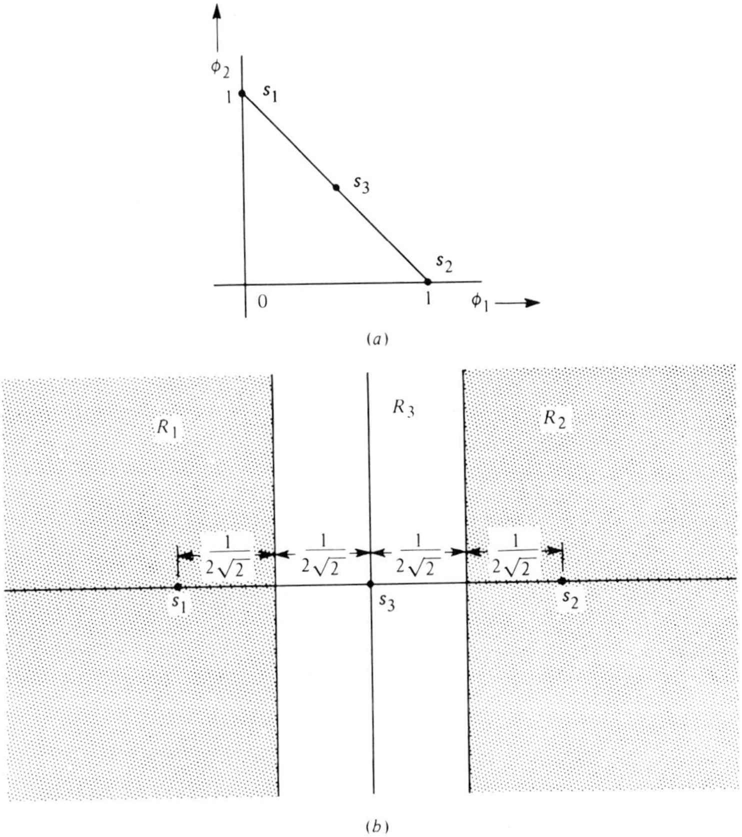

FIG. 6-11.

inside R_i, the correct decision will be made. Hence $P(C \mid m_i)$ is the probability that the noise **n** centered at s_i lie in R_i. This is given by the integral over R_i of probability density of the noise centered at s_i. Note that $\mathbf{q} - s_i$ represents the noise centered at s_i. Hence

$$P(C \mid m_i) = \int_{R_i} p_{\mathbf{n}}(q - s_i)\, dq \tag{6-37b}$$

This expression is also evident from Eq. 6-37a because

$$p_{\mathbf{q}}(q \mid m_i) = p_{\mathbf{n}}(q - s_i)$$

For a gaussian signal

$$p_{\mathbf{n}}(q - s_i) = \frac{1}{(\pi\mathfrak{N})^{M/2}} e^{-|q - s_i|^2/\mathfrak{N}}$$

$$= \frac{1}{(\pi\mathfrak{N})^{M/2}} \exp\left[-\sum_{k=1}^{M} (q_k - s_{ik})^2/\mathfrak{N}\right]$$

and

$$P(C \mid m_i) = \frac{1}{(\pi\mathfrak{N})^{M/2}} \int\int \cdots \int_{R_i} \exp\left[-\sum_{k=1}^{M} (q_k - s_{ik})^2/\mathfrak{N}\right] dq_1 dq_2 \ldots dq_M \tag{6-37c}$$

where the limits of integration of variables $q_1, q_2 \ldots q_M$ will be determined by the region R_i.

As a simple example, consider a 2-dimensional case with four equiprobable signals s_1, s_2, s_3 and s_4 and the corresponding decision regions R_1, R_2, R_3, and R_4 as shown in Fig. 6-12.

This problem can easily be solved by rotating the axes by $\pi/2$ radians. However, to demonstrate the general method, we shall compute the error probability without any rotation of axes. Let us compute $P(C \mid m_1)$.

$$P(C \mid m_1) = \int_{R_1} p_{\mathbf{n}}(q - n)\, \mathrm{d}q$$

$$= \frac{1}{\pi\mathfrak{N}} \int_{q_1} \int_{q_2} \exp\left[\frac{-(q_1 - s_{11})^2 - (q_2 - s_{12})^2}{\mathfrak{N}}\right] dq_1 dq_2$$

Note that in this problem

$$s_1 = \phi_1$$

Hence

$$s_{11} = 1 \qquad \text{and} \qquad s_{12} = 0$$

and

$$P(C \mid m_1) = \frac{1}{\pi\mathfrak{N}} \int_{q_1} \int_{q_2} \exp\left[\frac{-(q_1 - 1)^2 - q_2^2}{\mathfrak{N}}\right] dq_1 dq_2$$

$$= \frac{1}{\pi\mathfrak{N}} \int_{q_1} \left(\int_{q_2} e^{-q_2^2/\mathfrak{N}}\, dq_2 \right) e^{-(q_1 - 1)^2/\mathfrak{N}}\, dq_1$$

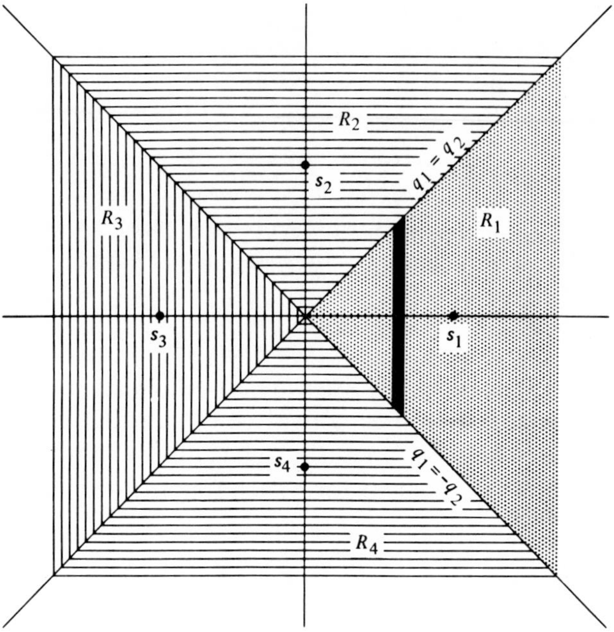

FIG. 6-12.

To integrate over R_1, we first integrate over the shaded strip shown in Fig. 6-12. Note that along the upper boundaries of the R_1, $q_1 = q_2$ and on the lower boundary $q_1 = -q_2$. Hence,

$$P(C \mid m_1) = \frac{1}{\pi \mathfrak{N}} \int_0^\infty \left(\int_{-q_1}^{q_1} e^{-q_2^2/\mathfrak{N}} dq_2 \right) e^{-(q_1-1)^2/\mathfrak{N}} dq_1$$

$$= \frac{1}{\pi \mathfrak{N}} \int_0^\infty \left[1 - 2 \operatorname{erfc}\left(\frac{q_1}{\sqrt{\mathfrak{N}/2}} \right) \right] e^{-(q_1-1)^2/\mathfrak{N}} dq_1$$

This integral can not be evaluated in a closed form and must be computed.

Minimum-Energy Signal Set. A glance at Eq. 6-35 (and Eq. 6-36) shows that the probability of error is a function of d the distance between $\mathbf{s}_1$ and $\mathbf{s}_2$. The error probability is not influenced by the actual location of $\mathbf{s}_1$ and $\mathbf{s}_2$ as long as their distance of separation d is maintained constant. Therefore there is an infinite number of possible arrangements of signals which will attain a given error probability. The energy of a signal, however, depends upon its location and is equal to the square of the distance of the signal point from the origin. Thus among the infinitely many arrangements which achieve a given error probability, one should choose the one in which the signals are closest to the origin in order to minimize the signal energy.

This conclusion is valid for any general case. We have observed that the translation and the rotation of coordinates (or the decision regions) does not influence the error probability. One can, therefore, always find an arrangement of signals which will have a minimum energy required to attain a given error probability by properly translating and rotating the signal set. Let $m_1, m_2, \ldots, m_M$ be the M messages with waveforms $s_1(t), s_2(t), \ldots, s_M(t)$ represented by points $\boldsymbol{s}_1, \boldsymbol{s}_2, \ldots, \boldsymbol{s}_M$ in the signal space. The mean energy of these signals is $\overline{E}$ given by

$$\overline{E} = \sum_{i=1}^{M} P(m_i) \, | \boldsymbol{s}_i |^2$$

The translation of these signals by some amount will not affect the error probability. The translation is equivalent to subtracting some vector $\boldsymbol{a}$ from each signal. Let this operation yield the minimum mean-energy set. We now wish to find the vector $\boldsymbol{a}$ such that the new mean energy

$$\overline{E} = \sum_{i=1}^{M} P(m_i) \, | \boldsymbol{s}_i - \boldsymbol{a} |^2 \tag{6-38a}$$

is minimum. We can easily show that $\boldsymbol{a}$ must be the center of gravity of M points located at $\boldsymbol{s}_1, \boldsymbol{s}_2, \ldots, \boldsymbol{s}_M$ with masses $P(m_1)$, $P(m_2)$, . . . , $P(m_M)$ respectively:

$$\boldsymbol{a} = \sum_{i=1}^{M} P(m_i) \boldsymbol{s}_i = \overline{\boldsymbol{s}_i} \tag{6-38b}$$

This result follows from the fact that the expression for mean energy in Eq. 6-38a is equivalent to the moment of inertia about point $\boldsymbol{a}$ of M particles of masses $P(m_1), P(m_2), \ldots, P(m_M)$ respectively. It is known that the moment of inertia is minimum about the centroid or the center of gravity of the particle system. This can also be proved easily. Let the mean energy be minimum for some translation $\boldsymbol{b}$. Then

$$\begin{aligned}\overline{E} &= \sum_{i=1}^{M} P(m_i) \, | \boldsymbol{s}_i - \boldsymbol{b} |^2 \\ &= \sum_{i=1}^{M} P(m_i) \, | (\boldsymbol{s}_i - \boldsymbol{a}) + (\boldsymbol{a} - \boldsymbol{b}) |^2 \\ &= \sum_{i=1}^{M} P(m_i) \, | \boldsymbol{s}_i - \boldsymbol{a} |^2 + 2(\boldsymbol{a} - \boldsymbol{b}) \sum_{i=1}^{M} P(m_i)(\boldsymbol{s}_i - \boldsymbol{a}) \\ &\qquad + \sum_{i=1}^{M} P(m_i) \, | \boldsymbol{a} - \boldsymbol{b} |^2\end{aligned}$$

Observe that the second term in the above expression vanishes because of relationship 6-38b. Hence

$$E = \sum_{i=1}^{M} P(m_i) \, | \boldsymbol{s}_i - \boldsymbol{a} |^2 + \sum_{i=1}^{M} P(m_i) \, | \boldsymbol{a} - \boldsymbol{b} |^2$$

This is obviously minimum, when $\boldsymbol{b} = \boldsymbol{a}$. Note that rotation of coordinates does not change the energy and hence there is no need to rotate the signals to minimize the energy.

As an example, consider the case of binary orthogonal signals shown in Fig. 6-13a. Let the orthonormal basis signals $\phi_1(t)$ and $\phi_2(t)$ be given by

$$\begin{aligned} \phi_1(t) &= \sqrt{\frac{2}{T}} \sin\left(\frac{2\pi}{T} t\right) \\ \phi_2(t) &= \sqrt{\frac{2}{T}} \sin\left(\frac{4\pi}{T} t\right) \end{aligned} \qquad 0 \leq t < T$$

Let us choose

$$\begin{aligned} s_1(t) &= \frac{d}{\sqrt{2}} \phi_1(t) = \frac{d}{\sqrt{T}} \sin\left(\frac{2\pi}{T} t\right) \\ s_2(t) &= \frac{d}{\sqrt{2}} \phi_2(t) = \frac{d}{\sqrt{T}} \sin\left(\frac{4\pi}{T} t\right) \end{aligned} \qquad 0 \leq t < T$$

The signals $s_1(t)$ and $s_2(t)$ are shown in Fig. 6-13b and the geometrical representation is shown in Fig. 6-13a. Both signals are located at a

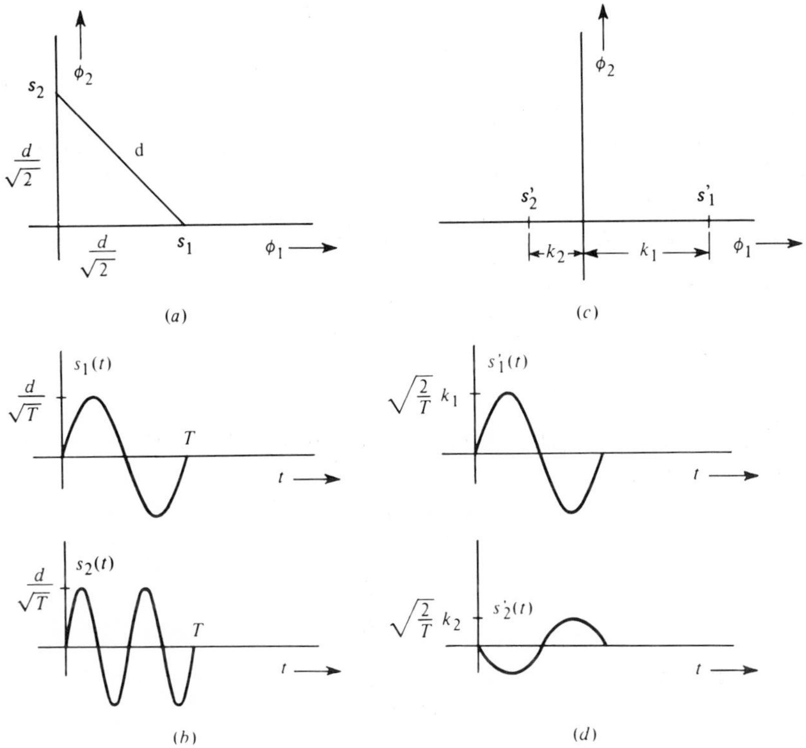

FIG. 6-13.

distance $d/\sqrt{2}$ from the origin and the distance between the signals is d. The error probability in this case is given by Eq. 6-35.

The minimum-energy set for this case is shown in Fig. 6-13c. The origin lies at the center of gravity of the signals. We have also rotated the signals for convenience. The distances k_1 and k_2 must be such that

$$k_1 + k_2 = d$$

and

$$k_1 P(m_1) = k_2 P(m_2)$$

That is,

$$\frac{k_1}{k_2} = \frac{P(m_2)}{P(m_1)}$$

Hence

$$k_1 = P(m_2)\, d$$

and

$$k_2 = P(m_1)\, d$$

The new signals $s_1'(t)$ and $s_2'(t)$ for this set are given by

$$\left.\begin{aligned} s_1'(t) &= \sqrt{\frac{2}{T}}\, P(m_2)\, d \sin\left(\frac{2\pi}{T}\, t\right) \\ s_2'(t) &= -\sqrt{\frac{2}{T}}\, P(m_1)\, d \sin\left(\frac{2\pi}{T}\, t\right) \end{aligned}\right\} \quad 0 \leq t < T$$

These signals are sketched in Fig. 6-13d.

Both the sets have the same error probability, but the latter has a smaller mean energy. If $\overline{E}$ and $\overline{E'}$ are the respective mean energies of the two sets, then

$$\overline{E} = P(m_1)\frac{d^2}{2} + P(m_2)\frac{d^2}{2} = \frac{d^2}{2}$$

and

$$\begin{aligned} \overline{E'} &= P(m_1)\, k_1^2 + P(m_2)\, k_2^2 \\ &= P(m_1)\, P^2(m_2)\, d^2 + P(m_2)\, P^2(m_1)\, d^2 \\ &= P(m_1)\, P(m_2)\, d^2 \end{aligned}$$

Note that the product $P(m_1)\,P(m_2)$ is maximum when $P(m_1) = P(m_2) = 1/2$, and in this case

$$P(m_1)\, P(m_2) = 1/4$$

Hence

$$\overline{E'} \leq \frac{d^2}{4}$$

It is obvious that

$$\overline{E'} \leq \frac{\overline{E}}{2}$$

For the case of equiprobable signals,

$$\overline{E'} = \frac{\overline{E}}{2}$$

In this case

$$k_1 = k_2 = \frac{d}{2}$$

$$\overline{E} = \frac{d^2}{2} \qquad \text{and} \qquad \overline{E'} = \frac{d^2}{4}$$

The signals in Fig. 6-13c are called antipodal signals when $k_1 = k_2$.

The probability of error given by Eq. 6-36 can be expressed in terms of mean energy of the set:

$$P(\epsilon) = \text{erfc}\left(\frac{d}{\sqrt{2\mathfrak{N}}}\right)$$

For orthogonal signals both signals have the same energy $E = d^2/2$. Hence

$$P(\epsilon) = \text{erfc}\left(\sqrt{\frac{E}{\mathfrak{N}}}\right) \tag{6-39a}$$

For antipodal signals, $k_1 = k_2 = d/2$. Both signals have the same energy $E = d^2/4$. Hence

$$P(\epsilon) = \text{erfc}\left(\sqrt{\frac{2E}{\mathfrak{N}}}\right) \tag{6-39b}$$

For a given mean energy the error probability $P(\epsilon)$ for orthogonal set is plotted in Fig. 6-16 ($M = 2$) from Eqs. 3-39a and 3-39b, it is obvious that for a given error probability, the energy requirements of the antipodal set is half as much as that for orthogonal signals.

From the above discussion one significant fact emerges. The error probability does not depend upon the specific waveforms $s_k(t)$ but is related to these signals only through their energies. This is a rather surprising result. As long as two signal sets have the same geometrical configuration (or equivalent as derived through translation and rotation), their error probabilities are identical. The actual waveforms of the signals will depend upon the orthogonal basis signals used. Thus even though two signal sets are geometrically identical (or equivalent), their waveforms in general will be entirely different depending upon the basis signal set chosen. Two cases of orthogonal signals are shown in Fig. 6-14. Both sets have identical geometrical representations. Hence they have the same error probability.

General Expression for Error Probability. So far, we have considered, rather simple schemes where decision regions can be easily found.

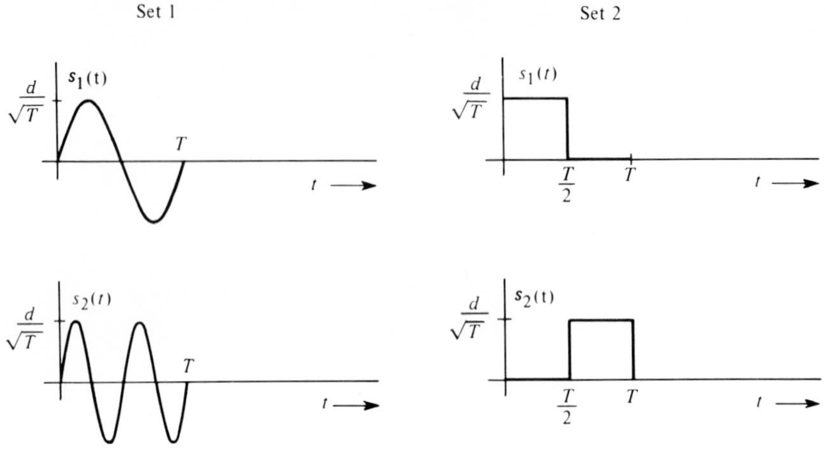

FIG. 6-14.

The method of computing error probabilities from the knowledge of decision regions has also been discussed. When the dimensions of signal space increase, it becomes difficult to visualize the decision regions graphically and as a result the method loses its power. We shall now develop the analytical expression for computing probability of error for a general M-ary scheme.

From the structure of the optimum receiver in Fig. 6-6, we observe that if m_1 is transmitted, then the correct decision will be made only if

$$\mathrm{b}_1 > \mathrm{b}_2, \mathrm{b}_3, \ldots, \mathrm{b}_M$$

In other words

$$P(C \mid m_1) = \text{probability}\,(\mathrm{b}_1 > \mathrm{b}_2, \mathrm{b}_3, \ldots, \mathrm{b}_M \mid m_1) \tag{6-40}$$

If m_1 is transmitted, then (Fig. 6-6)

$$\mathrm{b}_k = \int_0^T [s_1(t) + \mathrm{n}(t)]\, s_k(t)\, dt + a_k \tag{6-41}$$

Let

$$\rho_{ij} = \int_0^T s_i(t)\, s_j(t)\, dt \qquad i,j = 1, 2, \ldots, M \tag{6-42}$$

ρ_{ij} are known as *crosscorrelation coefficients* and is a measure of similarity of two signals. Substituting Eq. 6-42 in Eq. 6-41, we obtain (if m_1 is transmitted)

$$\mathrm{b}_k = \rho_{1k} + \int_0^T \mathrm{n}(t)\, s_k(t)\, dt + a_k \tag{6-43}$$

Using Eq. 6-7a in Eq. 6-43a, we obtain

$$\mathrm{b}_k = \rho_{1k} + a_k + \int_0^T \mathrm{n}(t) \sum_{j=1}^{M} s_{kj}\phi_j(t)\ dt$$

$$= \rho_{1k} + a_k + \sum_{j=1}^{M} s_{kj} \int_0^T \mathrm{n}(t)\,\phi_j(t)\,dt$$

From Eq. 6-9b, we have

$$\mathrm{b}_k = \rho_{1k} + a_k + \sum_{j=1}^{M} s_{kj}\mathrm{n}_j \qquad \text{(6-43b)}$$

where n_j is the component of $\mathrm{n}(t)$ along $\phi_j(t)$. Note that $\rho_{1k} + a_k$ is a constant and $\mathrm{n}_j(j = 1, 2, \ldots, M)$ are independent jointly gaussian variables; each with zero mean and a variance of /2. Thus variables b_k are a linear combination of jointly gaussian variables. It follows that (Sec. 2-13) the variables $\mathrm{b}_1, \mathrm{b}_2, \ldots, \mathrm{b}_M$ are also jointly gaussian. The joint probability density function of $\mathrm{b}_1, \mathrm{b}_2, \ldots, \mathrm{b}_M$ can be obtained by using the method discussion in Sec. 2-7. In Sec. 2-13, we have discussed this problem of computing joint probability density function of variables which are linear combination of jointly gaussian variables. Once the joint probability density of variables $\mathrm{b}_1, \mathrm{b}_2, \ldots, \mathrm{b}_M$ is computed, the probability of making correction decision when m_1 is transmitted can be computed from Eq. 6-40.

$$P(C \mid m_1) = \text{probability}\ (\mathrm{b}_1 > \mathrm{b}_2, \mathrm{b}_3, \ldots, \mathrm{b}_M) \qquad \text{(6-44a)}$$

Note that b_1 can lie anywhere in the range $-\infty$, ∞. More precisely if $p(b_1, b_2, \ldots, b_M \mid m_1)$ is the joint probability density function of b_1, $\mathrm{b}_2, \ldots, \mathrm{b}_M$, then Eq. 6-44a can be expressed as

$$P(C \mid m_1) = \int_{-\infty}^{\infty} \int_{-\infty}^{b_1} \cdots \int_{-\infty}^{b_1} p(b_1, b_2, \ldots, b_M \mid m_1)\, db_1, db_2, \ldots, db_M \qquad \text{(6-44b)}$$

where the limits of integration of b_1 are $(-\infty, \infty)$ and for the remaining variables the limits are $-\infty$, b_1. Thus

$$P(C \mid m_1) = \int_{-\infty}^{\infty} db_1 \int_{-\infty}^{b_1} db_2 \cdots \int_{-\infty}^{b_1} p(b_1, b_2, \ldots, b_M \mid m_1)\, db_M \qquad \text{(6-44c)}$$

Similarly $P(C \mid m_2), \ldots, P(C \mid m_M)$ can be computed and

$$P(C) = \sum_{j=1}^{M} P(C \mid m_j)\, P(m_j)$$

and

$$P(\epsilon) = 1 - P(C)$$

Orthogonal Signal Set. In an orthogonal signal set all the signals are mutually orthogonal. The two-dimensional orthogonal case has

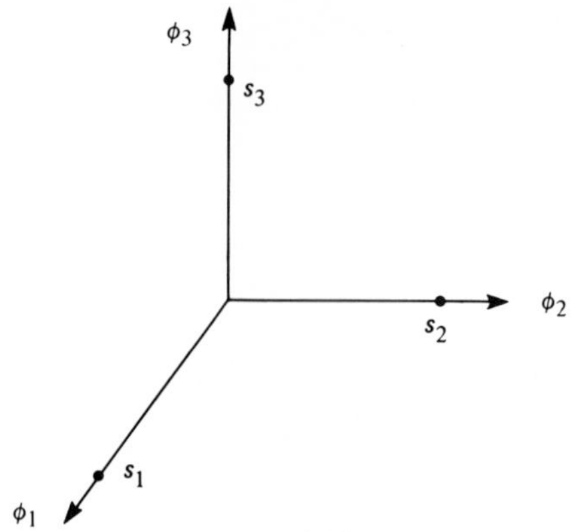

FIG. 6-15.

already been discussed in a previous section (Fig. 6-13a). The arrangement in three dimensional signal space is shown in Fig. 6-15.

The orthogonal set $\{s_k(t)\}$ is characterized by the fact that

$$\mathbf{s}_j \cdot \mathbf{s}_k = \begin{cases} 0 & j \neq k \\ E & j \neq k \end{cases}$$

Hence

$$\rho_{ij} = \mathbf{s}_i \cdot \mathbf{s}_j = \begin{cases} 0 & i \neq j \\ E & i = j \end{cases} \tag{6-45a}$$

Further we shall assume all signals to be equiprobable. This yields

$$a_k = \frac{1}{2}\left[\mathfrak{N} \ln\left(\frac{1}{M}\right) - E_k\right]$$

$$= -\frac{1}{2}\left[\mathfrak{N} \ln M + E\right]$$

Where E is the energy of each signal. Note that a_k is the same for all values of k. Let this constant be a.

$$a_k = a \qquad k = 1, 2, \ldots, M \tag{6-45b}$$

Note that for orthogonal set

$$s_k(t) = \sqrt{E}\,\phi_k(t)$$

Hence

$$s_{kj} = \begin{cases} \sqrt{E} & k = j \\ 0 & k \neq j \end{cases} \tag{6-45c}$$

Hence from Eq. 6-43b, 6-45a, 6-45b, and 6-45c, we have (when m_1 is transmitted)

$$\mathrm{b}_k = \begin{cases} E + a + \sqrt{E}\,\mathrm{n}_1 & k = 1 \\ a + \sqrt{E}\,\mathrm{n}_k & k = 2, 3, \ldots, M \end{cases} \tag{6-46}$$

Note that $n_1, n_2, \ldots, n_M$ are independent gaussian variables each with zero mean and variance $\mathfrak{N}/2$. Obviously b_k's which are of the form $(\alpha n_k + \beta)$ are also independent gaussian variables. From Eq. 6-46, it is evident that the variable b_1 has a mean $E + a$ and variance $(\sqrt{E})^2(\mathfrak{N}/2) = \mathfrak{N}E/2$. Hence,

$$p_{b_1}(b_1) = \frac{1}{\sqrt{\pi \mathfrak{N} E}} e^{-(b_1 - E - a)^2/\mathfrak{N}E}$$

$$p_{b_k}(b_k) = \frac{1}{\sqrt{\pi \mathfrak{N} E}} e^{-(b_k - a)^2/\mathfrak{N}E}$$

Since $b_1, b_2, \ldots, b_M$ are independent, the joint probability density is the product of individual densities

$$p(b_1, b_2, \ldots, b_M \mid m_1) = \frac{1}{\sqrt{\pi \mathfrak{N} E}} e^{-(b_1 - E - a)^2/\mathfrak{N}E} \prod_{k=2}^{M} \left[\frac{1}{\sqrt{\pi \mathfrak{N} E}} e^{-(b_k - a)^2/\mathfrak{N}E}\right]$$

and

$$P(C \mid m_1) = \frac{1}{\sqrt{\pi \mathfrak{N} E}} \int_{-\infty}^{\infty} db_1 \left[e^{-(b_1 - E - a)^2/\mathfrak{N}E}\right] \prod_{k=2}^{M} \cdot \left[\int_{-\infty}^{b_1} \frac{1}{\sqrt{\pi \mathfrak{N} E}} e^{-(b_k - a)^2/\mathfrak{N}E} \, db_k\right]$$

$$= \frac{1}{\sqrt{\pi \mathfrak{N} E}} \int_{-\infty}^{\infty} db_1 \left[e^{-(b_1 - E - a)^2/\mathfrak{N}E}\right] \cdot \left[\int_{-\infty}^{b_1} \frac{1}{\sqrt{\pi \mathfrak{N} E}} e^{-(x - a)^2/\mathfrak{N}E} \, dx\right]^{M-1}$$

$$= \frac{1}{\sqrt{\pi \mathfrak{N} E}} \int_{-\infty}^{\infty} \left[\operatorname{erf}\left(\frac{b_1 - a}{\sqrt{\mathfrak{N}E/2}}\right)\right]^{M-1} e^{-(b_1 - E - a)^2/\mathfrak{N}E} \, db_1 \quad (6\text{-}47)$$

changing the variable, so that

$$\frac{b_1 - E - a}{\sqrt{\mathfrak{N}E/2}} = y$$

the integral becomes

$$P(C \mid m_1) = \frac{1}{\sqrt{2\pi}} \int_{-\infty}^{\infty} \left[\operatorname{erf}\left(y + \sqrt{\frac{2E}{\mathfrak{N}}}\right)\right]^{M-1} e^{-y^2/2} \, dy \qquad (6\text{-}48)$$

Note that since this signal set is geometrically symmetrical,

$$P(C \mid m_1) = P(C \mid m_2) = \cdots = P(C \mid m_M)$$

Hence

$$P(C) = P(C \mid m_1)$$

and

$$P(\epsilon) = 1 - P(C)$$

The integral in Eq. 6-48 cannot be evaluated in a closed form. It has been tabulated for various values of M. Figure 6-16 shows the plot of $P(\epsilon)$ as a function of $E/\mathfrak{N}$ for various values of M.

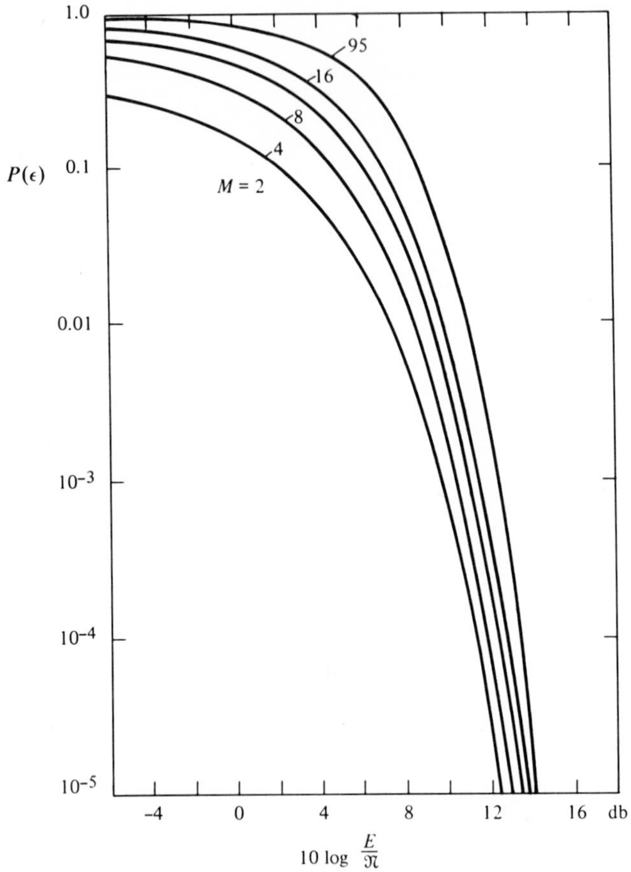

FIG. 6-16.

Transorthogonal (Simplex) Signals. A *transorthogonal* signal set is a minimum-energy equivalent set of the equiprobable orthogonal signal set. It can easily be seen that the orthogonal set is not a minimum-energy set. In general we must shift the origin to the center of gravity of the signal points in order to obtain a minimum-energy set. This is shown in Fig. 6-17 for the case of two- and three-dimensional signal space. Note that energy minimization actually reduces the number of dimension of the signal space by one. For a two-dimensional orthogonal

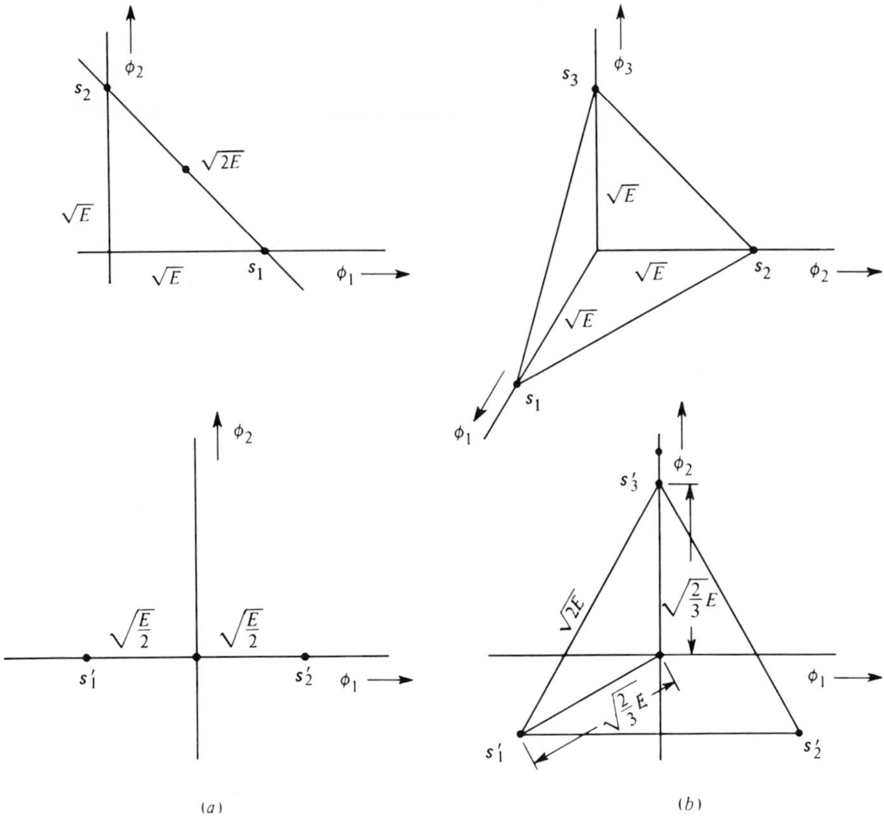

FIG. 6-17.

case, the corresponding transorthogonal signals lie on the same straight line, that is in one dimension (Fig. 6-17a). For a three-dimensional case, the plane formed by the three signals is constructed. The center of gravity of the signals on this plane is now the new origin. It is obvious that in this case the transorthogonal signals are reduced to two-dimensions (Fig. 6-17b). Similarly, for a four-dimensional case the minimum-energy set lies at the vertices of a tetrahedron, reducing it to a three-dimensional space.

The transorthogonal set is a minimum-energy set formed from the orthogonal set. Obviously the two sets are equivalent. The transorthogonal set is also known as *simplex* (or *regular simplex*) set. It can be shown that for a given mean energy the simplex signal set is the optimum (minimum-error probability) for the case of equiprobable signals embedded in white Gaussian noise (Refs. 4 and 5).

We can calculate the mean energy of the simplex set by noting that it is obtained by translating the orthogonal signal by the vector $\boldsymbol{a}$ given by Eq. 6-38b:

$$\boldsymbol{a} = \frac{1}{M}\sum_{i=1}^{M} \boldsymbol{s}_i$$

For orthogonal signals

$$\boldsymbol{s}_i = \sqrt{E}\Phi_i$$

Therefore

$$\boldsymbol{a} = \frac{\sqrt{E}}{M}\sum_{i=1}^{M} \Phi_i$$

where E is the energy of each signal in the orthogonal set and Φ_i is the unit vector along ith coordinate axis. The signals in the simplex set are given by

$$\begin{aligned} \boldsymbol{s}'_k &= \boldsymbol{s}_k - \boldsymbol{a} \\ &= \sqrt{E}\Phi_k - \frac{\sqrt{E}}{M}\sum_{i=1}^{M} \Phi_i \end{aligned} \tag{6-49}$$

The energy E' of signal $\boldsymbol{s}'_k$ is given by $|\boldsymbol{s}'_k|^2$.

$$E' = \boldsymbol{s}'_k \cdot \boldsymbol{s}'_k \tag{6-50}$$

Substituting Eq. 6-49 in Eq. 6-50 and observing that the set Φ_i is orthonormal, we have

$$\begin{aligned} E' &= E - \frac{E}{M} \\ &= E\left(1 - \frac{1}{M}\right) \end{aligned} \tag{6-51}$$

Hence for the same performance (error probability), the mean energy of the simplex signal set is $(1 - 1/M)$ times that of the orthogonal signal set. It can be seen that for large values of M, the performance of orthogonal signals approaches that of the simplex set.

In the next chapter we will show that in the limit as $M \to \infty$, the orthogonal set (or the simplex set) attains the upper bound of performance that is predicted by Shannon's theorem when the constraint of bandwidth is removed.

6-5. BANDWIDTH REQUIREMENT FOR TRANSMISSION OF M-ARY SIGNALS

Associated with every M-ary signal set, is the bandwidth required for transmission and the signal-to-noise ratio required to attain a given error probability. So far we have studied the relationship between signal-to-noise ratio and the error probability. We shall see in the next chapter, that the rate of information transmission over a channel is a function of signal-to-noise ratio and the bandwidth of transmission. Hence complete evaluation of any channel and a signal set involves not only the relationship between signal-to-noise ratio and the error probability, but also the bandwidth of transmission. We shall now demonstrate that the minimum bandwidth required for the transmission of M-ary signals each of time duration T seconds each is $M/2T$ Hz.

We have observed in Sec. 1-15 that a signal, time limited to T seconds and band-limited to B Hz, has a minimum of $2BT$ dimensions in a signal space. Each signal in any M-ary set has a duration of T seconds and the number of dimensions fewer than or equal to M. Hence it follows that the bandwidth required to transmit these M-ary signals is at most $M/2T$ Hz. If an M-ary set can be represented in N-dimensional space, where $N \leq M$, then the bandwidth of transmission is given by $N/2T$ Hz. For more discussion on this topic, the reader may consult ref. 5-5. It should be noted that reducing the dimensionality of the M-ary set reduces the bandwidth of transmission but in general increases the error probability unless the signal-to-noise ratio is increased. Thus the reduced bandwidth must be compensated for by an increased signal-to-noise ratio to obtain the same performance. The exact relationship between such exchange will be studied in the next chapter.

6-6. OTHER USEFUL PERFORMANCE CRITERIA

The purpose of the optimum receiver is to use that decision strategy which makes the best possible use of the observed data and any a priori information available. The strategy will also depend upon the weights assigned to various kinds of errors. In this chapter we have so far assumed that all errors have equal weight (or equal cost). This assumption may not be justified in some cases, and so, we may have to alter the decision rule.

Generalized Bayes Receiver. If we are given a priori probabilities and the cost functions of various kind of errors, the receiver which minimizes the average cost of decision is called the *Bayes receiver* and the decision rule is *Bayes' decision rule*. Note that the receiver that has been discussed so far is Bayes receiver under the condition that all errors have equal costs (equal weight). Thus Eq. 6-20 represents Bayes' rule when all errors are weighted equally. We shall now generalize this rule for the case where different errors have different costs. Let

$$C_{kj} = \text{cost of deciding that } \hat{m} = m_k \text{ when } m_j \text{ was transmitted} \qquad (6\text{-}52)$$

and as usual

$$P(m_i \mid \boldsymbol{q}) = \text{conditional probability that } m_i \text{ was transmitted when } \boldsymbol{q} \text{ is received} \tag{6-53}$$

If $\boldsymbol{q}$ is received and the receiver decides that $\hat{m} = m_k$, then the probability that m_j was transmitted is $P(m_j \mid \boldsymbol{q})$ for all $j = 1, 2, \ldots, M$. Hence the average cost of this decision $\hat{m} = m_k$ is σ_k, given by

$$\sigma_k = C_{k1}P(m_1 \mid \boldsymbol{q}) + C_{k2}P(m_2 \mid \boldsymbol{q}) + \cdots + C_{kM}(m_M \mid \boldsymbol{q}) \tag{6-54}$$

$$= \sum_{j=1}^{M} C_{kj}\, P(m_j \mid \boldsymbol{q}) \tag{6-55}$$

Similarly, if $\boldsymbol{q}$ is received and the receiver sets $\hat{m} = m_i$, the average cost σ_i is given by

$$\sigma_i = \sum_{j=1}^{M} C_{ij}\, P(m_j \mid \boldsymbol{q}) \tag{6-56}$$

Obviously if $\boldsymbol{q}$ is received, the optimum receiver decides that $\hat{m} = m_k$ if

$$\sigma_k < \sigma_i \qquad \text{for all } i \neq k$$

or

$$\sum_{j=1}^{M} C_{kj}\, P(m_j \mid \boldsymbol{q}) < \sum_{j=1}^{M} C_{ij}\, P(m_j \mid \boldsymbol{q}) \tag{6-57}$$

for all $i \neq k$.

Use of the mixed Bayes' rule in Eq. 6-57 yields

$$\sum_{j=1}^{M} C_{kj}\, P(m_j)\, p_{\mathbf{q}}(\boldsymbol{q} \mid m_j) < \sum_{j=1}^{M} C_{ij}\, P(m_j)\, p_{\mathbf{q}}(\boldsymbol{q} \mid m_j) \tag{6-58}$$

for all $i \neq k$. Note that C_{kk} is the cost of setting $\hat{m} = m_k$ when m_k is transmitted. This cost is generally zero. If we assign equal weight to all other errors, then

$$C_{kj} = \begin{cases} 0 & k = j \\ 1 & k \neq j \end{cases}$$

and the decision rule in Eq. 6-58 reduces to the rule in Eq. 6-20 as expected. The generalized Bayes' receiver for $M = 2$, assuming $C_{11} = C_{22} = 0$, sets $\hat{m} = m_1$ if

$$C_{12}P(m_2)\, p_{\mathbf{q}}(\boldsymbol{q} \mid m_2) < C_{21}P(m_1)\, p_{\mathbf{q}}(\boldsymbol{q} \mid m_1) \tag{6-59}$$

Otherwise the receiver decides that $\hat{m} = m_2$. The implementation of this receiver has been studied in Chapter 4.

Maximum-Likelihood Receiver. The strategy used in the Bayes receiver discussed above is very general except that it can be implemented only when the a priori probabilities $P(m_1), P(m_2), \ldots, P(m_M)$ are known. Frequently this information is not available. Under these conditions various possibilities exist depending upon the assumptions made. When, for example, there is no reason to expect any one signal to be more likely than the other, we may assign equal probabilities to all the messages

$$P(m_1) = P(m_2) = \cdots = P(m_M) = \frac{1}{M}$$

The Bayes' rule (Eq. 6-20) in this case becomes; set $\hat{m} = m_k$ if

$$p_{\mathbf{q}}(\boldsymbol{q} \mid m_k) > p_{\mathbf{q}}(\boldsymbol{q} \mid m_i) \qquad \text{for all } i \neq k \tag{6-60}$$

Observe that $p_{\mathbf{q}}(\boldsymbol{q} \mid m_k)$ represents the probability of observing $\boldsymbol{q}$ when m_k is transmitted. Thus the receiver chooses that signal which when transmitted will maximize the likelihood (probability) of observing the received $\boldsymbol{q}$. Hence this receiver is called the *maximum-likelihood receiver.* Note that the maximum-likelihood receiver is Bayes receiver under the condition that the a priori message probabilities are equal. In terms of geometrical concepts, the maximum-likelihood receiver decides in favor of that message signal which is closest to the received data $\boldsymbol{q}$. The practical implementation of the maximum-likelihood receiver is the same as Bayes receiver (Figs. 6-6a and 6-6b) under the condition that all a priori probabilities are equal to $1/M$.

If the signal set is geometrically symmetrical, and if all a priori probabilities are equal (maximum-likelihood receiver), then the decision regions for various signals are congruent. In this case because of symmetry, the conditional probability of a correct decision is the same no matter which signal is transmitted—that is,

$$P(C \mid m_i) = \text{const.} \qquad \text{for all } i$$

Since

$$P(C) = \sum_{i=1}^{M} P(m_i)\,P(C \mid m_i)$$

In this case

$$P(C) = P(C \mid m_i) \tag{6-61}$$

Thus the error probability of the maximum-likelihood receiver is independent of the actual source statistics $P(m_i)$ for the case of symmetric signal sets. It should, however, be realized that if the actual source statistic were known beforehand, one can design a better receiver using Bayes' decision rule.

It is apparent that if the source statistics is not known, the maximum-likelihood receiver proves very attractive for a symmetrical signal set. In such receiver one can specify the error probability independently of the actual source statistics.

Minimax Receiver. Designing a receiver with a certain decision rule completely specifies the conditional probabilities $P(C \mid m_i)$. The probability of error is given by

$$P(\epsilon) = 1 - P(C)$$
$$= 1 - \sum_{i=1}^{M} P(m_i) P(C \mid m_i)$$

Thus, in general, for a given receiver (with some specified decision rule) the error probability depends upon the source statistics $\{P(m_i)\}$. The error probability is the largest for some source statistics. The error probability in this worst possible case is $[P(\epsilon)]_{\max}$ and represents the upper bound or the error probability of the given receiver. This upper bound $[P(\epsilon)]_{\max}$ serves as an indication of the quality of the receiver. Each receiver (with a certain decision rule) will have a certain $[P(\epsilon)]_{\max}$. The receiver which has the smallest upper bound on the error probability, that is the minimum $[P(\epsilon)]_{\max}$ is called the *minimax receiver.*

We shall illustrate the concept of minimax receiver for a binary case ($M = 2$). For this case we design a Bayes receiver using some threshold a. [See discussion in Sec. 4-18 (Fig. 4-23).] Figure 4-23 is reproduced

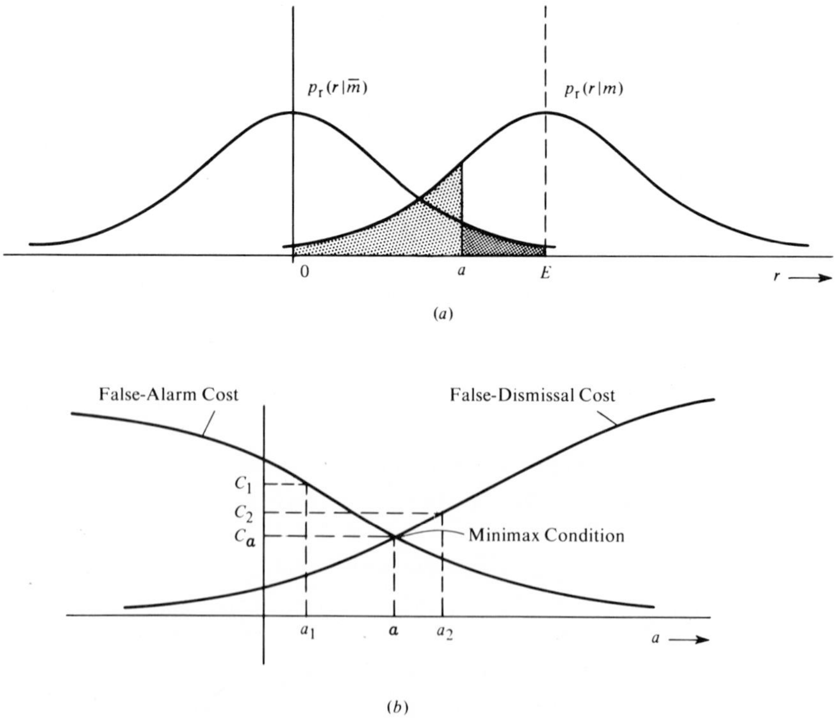

FIG. 6-18.

here in Fig. 6-18a. If we choose the hypothesis "signal present" (m), the shaded area to the right of a is proportional to the false-alarm cost. If we choose the hypothesis "signal absent" ($\overline{m}$), the shaded area to the left of a is proportional to false-dismissal cost.[6] It is obvious that if a is increased, the false-dismissal cost increases whereas the false-alarm cost decreases as shown in Fig. 6-18b.

We shall now find the minimax condition for this receiver. For the minimax receiver, we consider all possible receivers (all possible values of a in this case) and find the maximum error probability (or cost) which occurs under the worst possible a priori probability distribution. Let us choose $a = a_1$, as shown in Fig. 6-18b. In this case the worst possible case occurs when $P(\overline{m}) = 0$ and $P(m) = 1$, that is when the signal $s(t)$ is always present. The type of error in this case is false alarm. These errors have a cost C_1. On the other hand if we choose $a = a_2$, the worst possible case occurs when $P(\overline{m}) = 1$ and $P(m) = 0$, that is when the signal is always absent, causing only false-dismissal type of errors. These errors have a cost C_2. It is evident that for the setting $a = \alpha$, the cost of false alarm and false dismissal are equal, viz., C_α. Hence for all possible source statistics the cost is C_α. Since $C_\alpha < C_1, C_2$ it is obvious that this cost is the *minimum* of the maximum possible cost (since worst cases are considered) that accrues for all values of a. Hence $a = \alpha$ represents the minimax setting.

It is obvious from this discussion that the minimax receiver is rather conservative. It is designed under the pessimistic assumption that worst possible source statistics exist. The maximum-likelihood receiver, on the other hand, is designed on the assumption that all messages are equally likely. It can, however, be shown that for a symmetric signal set, the maximum-likelihood receiver is in fact the minimax receiver. This can easily be proved by observing that for a symmetric set, the probability of error of a maximum-likelihood receiver (equal a priori probabilities) is independent of the source statistics (see Eq. 6-61). Hence for a symmetric set, the error probability $P(\epsilon) = \alpha$ of a maximum-likelihood receiver is also equal to its $[P(\epsilon)]_{\max}$. We now show that no other receiver exists whose $[P(\epsilon)]_{\max}$ is less than α of a maximum-likelihood receiver for a symmetrical signal set. This is easily seen from the fact that for equiprobable messages, the maximum-likelihood receiver is optimum by definition. All other receivers must have $P(\epsilon) > \alpha$ for equiprobable messages. Obviously $[P(\epsilon)]_{\max}$ for these receivers can never be less than α. This proves that the maximum-likelihood receiver is indeed the minimax receiver for a symmetrical signal set.

[6]Here for convenience, we assume the cost of both errors to be equal. If this is not true, then the areas sould be multiplied by the respective cost factors.

REFERENCES

1. Woodward, P. M., *Probability and Information Theory with Applications to Radar*, Pergamon Press, London, 1953.

2. Wozencraft, J. M., and I. M. Jacobs, *Principles of Communication Engineering*, Wiley, New York, 1965.

3. Arthurs, E., and H. Dym, "On Optimum Detection of Digital Signals in the Presence of White Gaussian Noise—A Geometric Interpretation and a Study of Three Basis Data Transmission Systems," *IRE Trans. Commun. Systems*, CS-10 (December 1962), pp. 336–372.

4. Balakrishnan, A. V., "Contribution to the Sphere-packing Problem of Communication Theory," *J. Math Analysis and Appls*. vol. 3, no. 3 (December 1961), pp. 485–506.

5. Landau, H. J., and D. Slepian, *On the Optimality of the Regular Simplex Code*, B.S.T.J. 45, 1966, pp. 1247–1272.

ADDITIONAL REFERENCES

6. Helstrom, C. W., *Statistical Theory of Signal Detection*, Pergamon Press, London, 1960.

7. Hancock, J., and P. Wintz, *Signal Detection Theory*, McGraw-Hill, New York, 1966.

8. Schwarz, M., W. Bennett, and S. Stein, *Communication Systems and Techniques*, McGraw-Hill, New York, 1966.

9. Van Trees, H. L., Detection, Estimation, and Modulation Theory, vol. 1, Wiley, New York, 1968.

PROBLEMS

6-1. A source emits M equiprobable messages which are assigned waveforms $s_1, s_2, \ldots, s_M$ shown in Fig. P. 6-1. Determine the error probability of the optimum receiver when the channel is disturbed by an additive white gaussian noise of power density $\mathfrak{N}/2$.

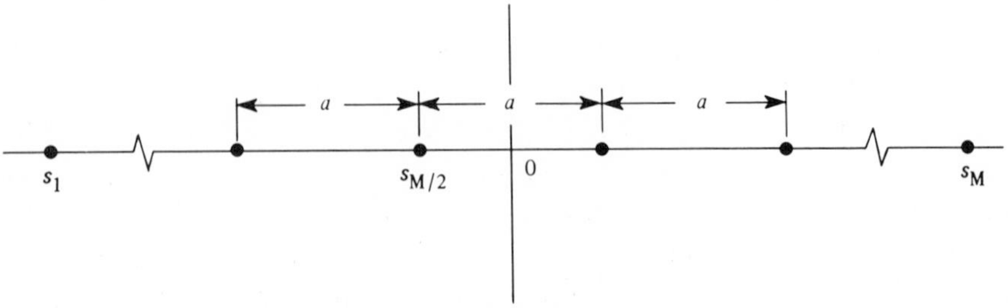

FIG. P. 6-1.

6-2. A source emits eight equiprobable messages which are assigned waveforms $s_1, s_2, \ldots, s_8$ as shown in Fig. P. 6-2. Determine the decision regions and the error probability of the optimum receiver, assuming additive white gaussian channel noise with power density $\mathfrak{N}/2$.

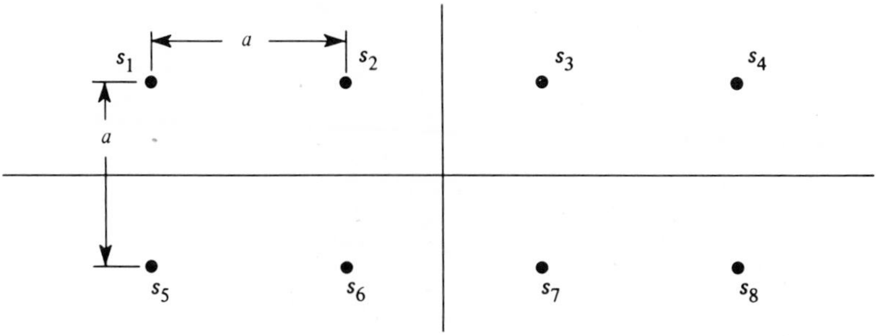

FIG. P. 6-2.

6-3. A source emits five messages which are assigned waveforms $s_1, s_2, \ldots, s_5$ shown in Fig. P. 6-3. The messages are equiprobable. Find the optimum decision regions. Determine the conditional error probability at the receiver when s_1 is transmitted. Compute similar probabilities for each of the remaining four messages. Assume the channel to be disturbed by additive white gaussian noise of power density $\mathfrak{N}/2$.

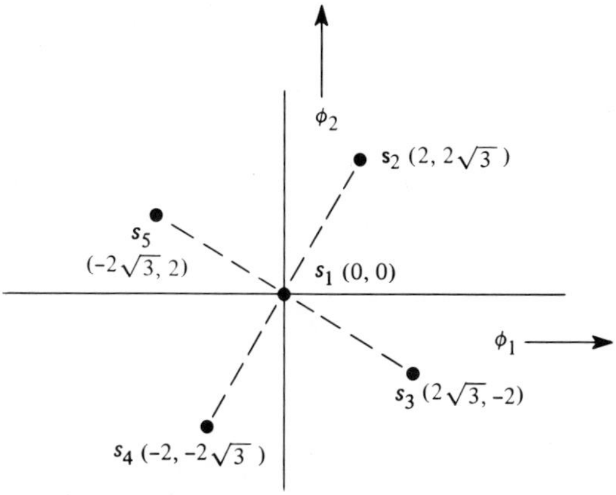

FIG. P. 6-3.

6-4. It is given that the signals $\phi_1(t)$ and $\phi_2(t)$ are mutually orthogonal and each having a unit energy and duration T seconds. A discrete communication system has four equiprobable messages m_1, m_2, m_3 and m_4. These messages are assigned waveforms $s_1(t)$, $s_2(t)$, $s_3(t)$ and $s_4(t)$ respectively where

$$\begin{aligned} s_1(t) &= (\sqrt{3} - 1)\,\phi_1(t) \\ s_2(t) &= -2\phi_1(t) + (\sqrt{3} - 1)\,\phi_2(t) \\ s_3(t) &= -(\sqrt{3} + 1)\,\phi_1(t) - 2\phi_2(t) \\ s_4(t) &= -(\sqrt{3} + 1)\,\phi_2(t) \end{aligned}$$

The channel is disturbed by additive white gaussian noise with power density $\mathfrak{N}/2 = 0.20$ (a) Find the optimum receiver. (b) Represent the message signals geometrically. (c) Determine the decision regions.

6-5. Determine the error probability of the system in Prob. 6-4. *Hint*: Determine first the minimum-energy equivalent set.

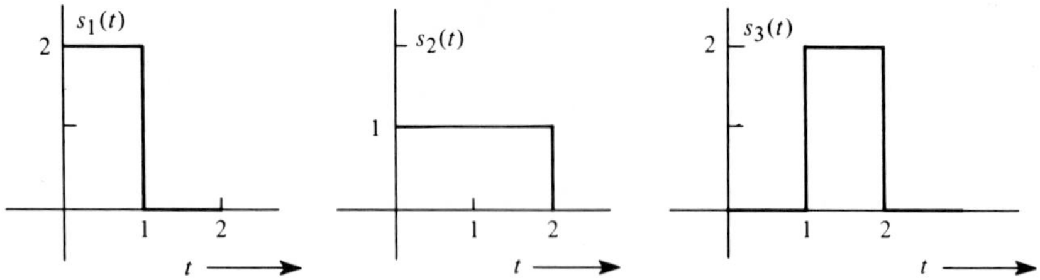

FIG. P. 6-6.

6-6. A source emits three equally likely messages which are assigned waveforms $s_1(t)$, $s_2(t)$ and $s_3(t)$ respectively as shown in Fig. P. 6-6. The channel is disturbed additively by a white gaussian noise of power density $\mathfrak{N}/2 = 0.01$. Determine the error probability of the optimum receiver.

6-7. A source generates three equiprobable messages m_1, m_2, and m_3. These messages are transmitted by waveforms $s_1(t)$, $s_2(t)$ and $s_3(t)$ shown in Fig. P. 6-7. The channel is disturbed by a white gaussian noise with power-density spectrum $\mathfrak{N}/2 = 0.1$. (a) Design the optimum receiver. (b) Find the probability of error.

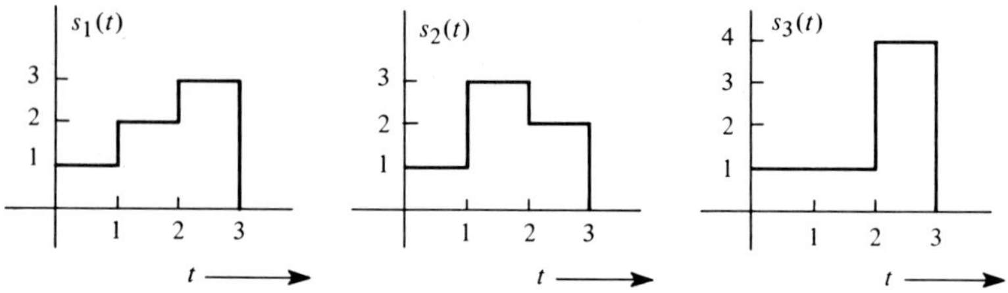

FIG. P. 6-7.

6-8. Three messages m_1, m_2, and m_3 with probabilities 0.3, 0.4, and 0.3 are assigned signals $\boldsymbol{s}_1$, $\boldsymbol{s}_2$, $\boldsymbol{s}_3$ shown in Fig. P. 6-8. The channel is disturbed by additive white gaussian noise with power density $\mathfrak{N}/2 = 0.05$. Determine the optimum receiver (to minimize the error probability) and the corresponding error probability.

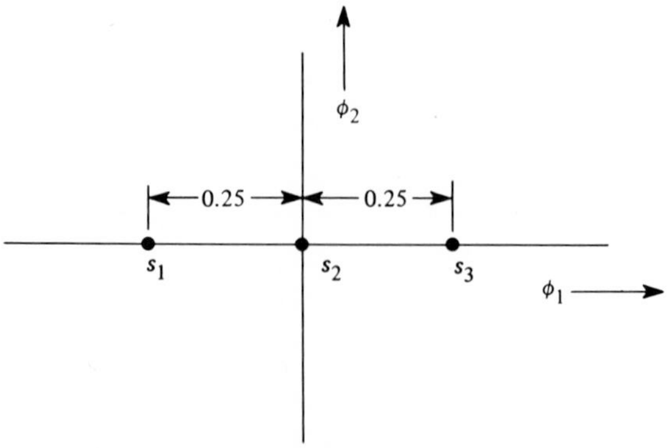

FIG. P. 6-8.

6-9. Compare the performance of the signal set in Prob. 6-8, with the orthogonal signal set (three-dimensional) when all the three signals are equiprobable. Assume the mean energy of each signal set to be unity and the white noise power density $\mathfrak{N}/2 = 0.05$.

6-10. Compare the performance of the signal set in Prob. 6-8, with the corresponding simplex (transorthogonal) set when all the three signals are equiprobable. Assume the mean energy of each signal set to be unity and the white noise power density $\mathfrak{N}/2 = 0.05$.

6-11. A certain signal set $\{s_i(t)\}$ has an orthogonal basis signals given by

$$\{\phi_i(t)\}$$

The signals $\{s_i(t)\}$ and $\{\phi_i(t)\}$ exist over the interval $(0, T)$ only and are zero outside this interval. Give an alternative arrangement for an optimum receiver which uses the filter with transfer function $\phi_i(T - t)$ instead $s_i(T - t)$.

6-12. A signal set consists of 2^N signals $s_1(t), s_2(t), \ldots, s_{2N}(t)$ given by

$$s_k(t) = \frac{d}{2} \sum_{j=1}^{N} a_{kj}\, \phi_j(t)$$

where $\phi_1(t), \phi_2(t), \ldots, \phi_N(t)$ are the N orthonormal basis signals, and a_{kj} is either 1 or -1. Note that all the 2^N signals have equal energies $\left(E = \frac{Nd^2}{4}\right)$. For $N = 2$, there are 4 signals in the set at the vertices of a square and for $N = 3$ there are 8 signals at the vertices of a cube. In general, such a set is called vertices of hypercube. Assuming all signals to be equiprobable, derive the error probability of the optimum receiver.

6-13. For orthogonal signal set

$$s_k(t) = \sqrt{E}\, \phi_k(t)$$

A biorthogonal signal set is formed from orthogonal set by augmenting it with negative of each signal. Thus we add to orthogonal set, another set

$$s_{-k}(t) = -\sqrt{E} \phi_k(t)$$

This gives $2N$ signals in N dimensional space. Assuming all signals to be equiprobable obtain the error probability of the optimum receiver.

7

Introduction to Information Theory

7-1. INTRODUCTION

In Chapter 5 we observed that the output signal-to-noise ratio is increased by increasing the input signal-to-noise ratio. The increased signal-to-noise ratio at the output makes the received message more reliable. We also observed in Chapter 6, that increasing signal-to-noise ratio at the input reduces the error probability in detection. The obvious conclusion is that an increased signal-to-noise ratio at the transmitter increases the reliability of communication. The result is hardly surprising as it is intuitively appealing. In the present chapter we shall derive quantitatively the signal-to-noise ratio required for a given rate of information transmission over a channel. We shall show that for a given signal-to-noise power ratio, one can transmit messages with arbitrarily small error probability provided the rate of information transmission is maintained below a certain value, depending upon the average signal-to-noise power ratio. This implies that the presence of random disturbance in a channel does not, by itself, set any limit to the transmission accuracy. Instead, it sets a limit on the transmission rate for a specific transmission accuracy.

We have been using the phrase "the rate of information transmission" as if an information can be measured. This is indeed so. We shall first discuss the nature of information content in a message and then define the unit of information. This is essential for a quantitative discussion on transmission of information.

7-2. MEASURE OF INFORMATION

The purpose of communication is to convey information. We shall discuss the nature of the information contained in a message both from intuitive and from engineering point of view. Surprisingly, both the viewpoints lead to the same quantitative definition of the unit of information.

The Intuitive Concept of Information. All messages convey information but some messages convey more information than others. A close scrutiny shows that the probability of occurrence of an event is closely related to the amount of information. If someone tells us of the occurrence of a highly probable event he conveys less information than to that which will be conveyed if the event were less probable. The element of surprise or uncertainty in the occurrence of an event appears to be proportional to the amount of information. If someone says that the sun rises in the east, it conveys practically no information because everyone knows that it is true. There is no uncertainty in the event that the sun rises in the east every day. In other words, the probability of the event that the sun rises in the east every day is unity. On the other hand, if on some day in January a weatherman from a national broadcasting service said that the temperature in Minneapolis reached 150°F, that day the statement would convey a tremendous amount of information. This is because this event is totally unexpected and its probability of occurrence is very small ($P \to 0$). In other words, because this event is very uncertain its description conveys a lot of information. Again, consider a hypothetical news statement: "The United States invades Cuba." The statement comes as a surprise and definitely conveys a large amount of information simply because the event has such a small probability of occurrence. The surprise however is not nearly as great as it would be if the statement read "Cuba invades the United States." This is because the probability of the latter event occurring is even smaller than that of the former. The surprise of course comes as a result of uncertainty or unexpectedness. The more unexpected an event is, the greater surprise and hence the more information. The probability of occurrence of an event is a measure of its unexpectedness and hence is related to the information content of the event. Thus on an intuitive basis the amount of information received from a knowledge of the occurrence of a certain event is directly related to the uncertainty, or inversely related to its probability of its occurrence. What must be the nature of this relationship? It is obvious that if the event is certain (probability 1), it conveys no information. On the other hand, if the event is impossible (probability zero), then its occurrence conveys an infinite amount of information. This suggests that the amount of information should be a logarithmic function of the reciprocal of the event probability:

$$\text{Information } I \sim \log \frac{1}{P} \tag{7-1}$$

where P is the probability of occurrence of an event and I is the amount of information conveyed by a knowledge of the occurrence of the event.

The Engineering Point of View of Information. We shall now show from the engineering point of view that the information about an event is identical to that obtained on intuitive basis (Eq. 7-1). What do we mean by an engineering point of view? An engineer is concerned with efficiently communicating information bearing messages. From his point of view, an amount of information in a message is proportional to the time required to transmit the message. We shall now see that this concept of information also leads to Eq. 7-1. This implies that it takes a shorter time to transmit a message about an event with a higher certainty (or higher probability) as compared to an event with lower probability. This fact may be easily verified from the example of transmission of alphabetic symbols in the English language using the Morse code. This code is made up of various combinations of two symbols (such as a mark and a space or pulses of height a and $-a$ volts). Each letter is represented by a certain combination of these symbols and has a certain length. Obviously for efficient transmission, shorter code words should be assigned to letters e, t, a, and o which occur more frequently. The longer code words are assigned to letters x, k, q, and z which occur less frequently. Each letter may be considered as a message. It is obvious that the letters which occur more frequently (with higher probability of occurrence) need a shorter time to transmit (shorter code words) compared to those with smaller probability of occurrence. We shall now show that on average, the time required to transmit a symbol (or a message) with probability of occurrence P is indeed proportional to log $(1/P)$ as was seen intuitively.

For the sake of simplicity, let us assume that we are required to transmit sequences of the two messages a_1 and a_2 which are equally likely to occur. We may use binary digits to encode these messages. Messages a_1 and a_2 may be represented by digits 0 and 1 respectively. We can assign two distinct signals (binary signals) to the digits 0 and 1 for the purpose of transmission. It is obvious that we must have a minimum of one binary digit (which can assume 2 values) to represent each of the two equally likely messages. Next consider the case of four equiprobable messages a_1, a_2, a_3, and a_4. If these messages are encoded in binary form we need a minimum of two binary digits per message. Each binary digit can assume two values. Hence a combination of two binary digits can form four code words 00, 01, 10, 11 which can be assigned to the four equiprobable messages a_1, a_2, a_3, and a_4 respectively. The binary digits are transmitted by two distinct signals (binary signals). It is obvious that each of these four messages takes twice as much transmission time as that required by each of the two equiprobable messages, and hence contains twice as much information. Similarly, we can encode any one of the eight equiprobable messages by a minimum of three binary digits. This is because three binary digits form eight distinct code words which can be assigned to each of the eight messages. It can be easily seen that, in general, we need $\log_2 n$ binary digits to encode each of the n equiprobable messages.[1] Since all the messages are equiprobable, P, the probability of any one message occurring is $1/n$. Hence each message (with probability P) needs $\log_2 (1/P)$ binary digits for encoding. From the engineering point of view it is then obvious that the information I contained in a message with probability of occurrence P is proportional to $\log_2 (1/P)$.

$$I = k \log_2 \frac{1}{P} \tag{7-2}$$

Once again we come to the conclusion (from the engineering point of view) that the information content of a message (or event) is proportional to the logarithm of the reciprocal of the probability of the event.

We shall now define information conveyed by a message (or an event) according to Eq. 7-2. The constant of proportionality is taken as unity for convenience, and the information is then in terms of binary units, abbreviated *bit* (*bi*nary uni*t*).

$$I = \log_2 \frac{1}{P} \quad \text{bits} \tag{7-3}$$

According to this definition, the information I in a message can be interpreted as the minimum number of binary digits required to encode the message. This is given by $\log_2 (1/P)$, where P is the probability of occurrence of the message. Although we have shown this result here for a highly special case of equiprobable messages, we shall show in a later section that this is true for nonequiprobable messages also.

Next we shall consider the case of r-ary digits instead of binary digits for encoding. Each of the r-ary digits can assume r values $(0, 1, 2, \ldots, r)$. To each of these digits we then assign one of r distinct signals (r-ary signals). Each of the n messages (encoded by r-ary digits) can then be transmitted by a particular sequence of r-ary signals.

Since each r-ary digit can assume r values, k r-ary digits can form a maximum of r^k distinct codewords. Hence to encode each of the n equiprobable messages, we need a minimum of $\log_r n$, r-ary digits.[2] But $n = 1/P$, where P is the probability of occurrence of each message.

[1]Here we are assuming that the number n is such that $\log_2 n$ is an integer. Later on we shall observe that this restriction is not necessary.

[2]Here again, we are assuming that n is such that $\log_r n$ is an integer. As we shall see later, this restriction is not necessary.

Obviously we need a minimum of $\log_r n$, r-ary digits. The information I per message can be considered as $\log_r (1/P)$ r-ary units:

$$I = \log_r \frac{1}{P} \qquad r\text{-ary units} \tag{7-4}$$

Equation 7-4 is a general definition of information and Eq. 7-3 is a special case when $r = 2$. According to definition 7-4, the information I (in r-ary units) of a message is equal to the minimum number of r-ary digits required to encode the message. This is given by $\log_r (1/P)$, where P is the probability of the message. Although we have shown this result here for a highly special case of equiprobable messages, we shall prove it later for any arbitrary message probabilities.

From Eqs. 7-3 and 7-4, it is evident that

$$I = \log_2 \frac{1}{P} \quad \text{bits} = \log_r \frac{1}{P} \quad r\text{-ary units}$$

Hence

$$1\ r\text{-ary unit} = \log_2 r \quad \text{bits} \tag{7-5}$$

In general,

$$1\ r\text{-ary unit} = \log_s r \quad s\text{-ary units} \tag{7-6}$$

The 10-ary unit of information is called the Hartley in honor of R. V. L. Hartley (Ref. 1) who was one of the pioneers [along with Nyquist (Ref. 2) and Carson] in the area of information transmission in the twenties. The rigorous mathematical foundations of information theory, however, were established by C. E. Shannon (Ref. 3) in 1948:

$$\begin{aligned} 1 \text{ hartley} &= \log_2 10 \quad \text{bits} \\ &= 3.32 \quad \text{bits} \end{aligned}$$

Sometimes the unit "nat" is used:

$$\begin{aligned} 1 \text{ nat} &= \log_2 e \quad \text{bits} \\ &= 1.44 \quad \text{bits} \end{aligned}$$

A Comment on the Information Conveyed by an *r*-ary Digit. From Eqs. 7-5 and 7-6 it is evident that as r is increased, the information content per digit also increases logarithmically. This may seem odd in view of the fact that one r-ary digit is transmitted as one r-ary signal. Why should such a signal carry more information if r is increased? It appears as if we can transmit a higher amount of information without paying any price. A closer examination, however, shows that an increase in r demands an increased bandwidth. We have seen in Chapter 6 that for M-ary signals, the bandwidth of transmission is proportional to M. Thus increasing r increases the required bandwidth of transmission i proportion to r. Furthermore, over a noisy channel, to attain a given error probability, the signal energy must be increased with r, as seen from Fig. 6-16.

A Note on the Unit of Information. From the earlier discussion, it follows that a general unit of information is the r-ary unit. So we should use r as the base of logarithm everywhere. However, the binary unit bit ($r = 2$) is very commonly used in the literature. There is, of course, no loss of generality in using $r = 2$. These units can always be converted into any other units by using Eq. 7-6. Henceforth, we shall always use the binary unit (bit) for information unless otherwise stated. The bases of the logarithm functions will be omitted but will be understood to be 2.

7-3. AVERAGE INFORMATION PER MESSAGE: ENTROPY OF A SOURCE

Consider a source m emitting messages $m_1, m_2, \ldots, m_n$ with probabilities $P_1, P_2, \ldots, P_n$ respectively ($P_1 + P_2 + \cdots + P_n = 1$). The source emits the n messages randomly with given probabilities. By the definition in Eq. 7-3 (or Eq. 7-4), the information content of message m_i is I_i given by

$$I_i = \log \frac{1}{P_i} \quad \text{bits}$$

The probability of the occurrence of m_i is P_i. Hence the mean or average information per message emitted by the source is given by $\sum_{i=1}^{n} P_i I_i$ bits. The average information of a source m is called its entropy, denoted by $H(\mathrm{m})$. Hence[3]

$$\begin{aligned} H(\mathrm{m}) &= \sum_{i=1}^{n} P_i I_i \quad \text{bits} \\ &= \sum_{i=1}^{n} P_i \log \frac{1}{P_i} \quad \text{bits} \end{aligned} \tag{7-7a}$$

[3]The base of the logarithm term is understood to be 2. If we take the base of logarithm as r, then the entropy $H_r(\mathrm{m})$ of the source is in r-ary units:

$$\begin{aligned} H_r(\mathrm{m}) &= \sum_{i=1}^{n} P_i \log_r \frac{1}{P_i} \quad r\text{-ary units} \\ &= \frac{H(\mathrm{m})}{\log r} \quad r\text{-ary units} \end{aligned}$$

$$= -\sum_{i=1}^{n} P_i \log P_i \quad \text{bits} \tag{7-7b}$$

It is evident that the entropy of a source is a function of the message probabilities. It is interesting to find that probability distribution of messages which yields the maximum entropy. Since the entropy is a measure of uncertainty, the probability distribution which generates the maximum uncertainty will have maximum entropy. On qualitative grounds one expects entropy to be maximum when all the messages are equiprobable. We shall now show that this is indeed true.

To show this, we use the result in Appendix 7-1 (at the end of this chapter), which states that if

$$\sum_{i=1}^{n} P_i = \sum_{i=1}^{n} Q_i = 1$$

then

$$\sum_{i=1}^{n} P_i \log \frac{1}{P_i} \leq \sum_{i=1}^{n} P_i \log \frac{1}{Q_i}$$

with equality if and only if

$$P_i = Q_i \qquad \text{for all } i$$

Let

$$Q_1 = Q_2 = \cdots = Q_n = \frac{1}{n}$$

Then

$$H(\mathrm{m}) = \sum_{i=1}^{n} P_i \log \frac{1}{P_i} \leq \sum_{i=1}^{n} P_i \log \frac{1}{Q_i}$$

$$\leq \sum_{i=1}^{n} P_i \log n$$

$$\leq \log n \sum_{i=1}^{n} P_i$$

Hence

$$H(\mathrm{m}) \leq \log n$$

with equality if and only if

$$P_i = Q_i = \frac{1}{n}$$

for all i. Hence the entropy $H(m)$ is maximum if all messages are equiprobable

$$H_{\max}(\mathrm{m}) = \log n \quad \text{bits}$$

The Intuitive and Engineering Interpretations of Entropy. Earlier we observed that both the intuitive and the engineering viewpoints lead to the same definition of the information associated with an event (or a message). The conceptual bases, however, are entirely different in the two points of view. Consequently, we have two physical interpretations of information. According to the engineering point of view, the information content of any message is equal to the minimum number of digits required to encode the message, and therefore entropy $H(\mathrm{m})$ is equal to the minimum number of digits per message required on average for encoding. From the intuitive standpoint, on the other hand, information was thought of as being synonymous with the amount of surprise or uncertainty associated with the event (or message). A smaller probability of occurrence implies more uncertainty about the event. Uncertainty is, of course, associated with surprise. Hence intuitively the information associated with an event is a measure of the uncertainty (unexpectedness) of the event. Therefore, $\log(1/P_i)$ is a measure of the uncertainty of the event (message) m_i and $\sum_{i=1}^{n} P_i \log(1/P_i)$ is the average uncertainty (per message) of the source which generates messages $m_1, m_2, \ldots, m_n$ with probability $P_1, P_2, \ldots, P_n$. Both these interpretations prove very useful in qualitative understanding of the mathematical definitions and results in information theory. Entropy may also be viewed as a function associated with a random variable m which assumes values $m_1, m_2, \ldots, m_n$ with probabilities $P(m_1), P(m_2), \ldots, P(m_n)$:

$$H(\mathrm{m}) = \sum_{i=1}^{n} P(m_i) \log \frac{1}{P(m_i)}$$

$$= \sum_{i=1}^{n} P_i \log \frac{1}{P_i}$$

Thus we can associate an entropy with every discrete random variable.

7-4. CODING THEOREM FOR NOISELESS CHANNELS

If all the messages of a source are equiprobable, then the minimum number of digits required to encode a message was shown to be equal to the information ($\log 1/P$) carried by the message. We shall now generalize this result to the case of nonequiprobable message. We shall show that the average number of binary digits per message required for encoding is given by $H(\mathrm{m})$ (in bits) for an arbitrary probability distribution of the messages. This implies that on an average, a message with probability P_i will need a minimum of $\log(1/P_i)$ binary digits for encoding. This also shows that the information content (from the engineering standpoint) of a message (or event) with probability P_i is $\log(1/P_i)$.

Consider a sequence of N messages from the source m. We shall make N very large. Let k_i be the number of times message m_i occurs in N trials. Then according to relative frequency interpretation

$$\lim_{N\to\infty} \frac{k_i}{N} = P_i$$

This result is also justified on the basis of the *law of large numbers* in the axiomatic approach to probability. Thus the message m_i occurs NP_i times in the whole sequence of N messages (provided $N \to \infty$). Therefore in a typical sequence of N messages, m_1 will occur NP_i times, m_2 will occur NP_2 times, ... m_n will occur NP_n times. All other compositions are extremely unlikely to occur ($P \to 0$). Thus any typical sequence (where $N \to \infty$) has the same proportion of the n messages, although in general the order may be different. We shall assume a zero memory source that is, the message is emitted from the source independently of the previous messages. Consider now a typical sequence S_N of N messages from the source. Since the n types of messages (of probability $P_1, P_2, \ldots, P_n$) occur $NP_1, NP_2, \ldots, NP_n$ times respectively and since each message is independent, the probability of occurrence of a typical sequence S_N is given by

$$P(S_N) = (P_1)^{NP_1}(P_2)^{NP_2}\cdots(P_n)^{NP_n} \tag{7-8}$$

Since all possible sequences of N messages from this source have the same composition, all the sequences (of N messages) are equiprobable with probability $P(S_N)$. We can consider these sequences as new messages (which are now equiprobable). To encode one such sequence we need L_N binary digits where

$$L_N = \log\left(\frac{1}{P(S_N)}\right) \qquad \text{binary digits}$$

The binary digit will be abbreviated to *binit* for convenience (note the difference between bit and binit).[4] Thus

$$L_N = \log\left(\frac{1}{P(S_N)}\right) \quad \text{binits} \tag{7-9}$$

[4]The distinction between bit and binit should be clearly understood. The bit (binary unit) represents the unit of information based upon a certain definition whereas the binit is an abbreviation of "binary digit." As we have indicated here, the minimum number of binits required to encode a message is numerically equal to the information content (in bits) of the message. The number of binits required to encode a message depends upon a method of coding where as the information content (in bits) of a message is independent of how it is encoded.

Substituting Eq. 7-8 in Eq. 7-9, we obtain

$$L_N = N \sum_{i=1}^{n} P_i \log \frac{1}{P_i}$$
$$= NH(\mathrm{m}) \quad \text{binits}$$

Note that L_N is the length (number of binary digits) of the codeword required to encode N messages in sequence. Obviously L, the average number of binits required per message is L_N/N and is given by

$$L = \frac{L_N}{N} = H(\mathrm{m})$$

This is the desired result which states that it is possible to encode the messages emitted by a source using on an average $H(\mathrm{m})$ number of binits (binary digits) per message, where $H(\mathrm{m})$ is the entropy of the source (in bits). Although it does not prove that on the average this is the minimum number of digits required, one can show that $H(\mathrm{m})$ is indeed the minimum. It is not possible to find any uniquely decodable code whose average length is less than $H(\mathrm{m})$. For a proof of this statement, see Refs. 4, 5, and 6.

If the channel is noiseless, then all the information transmitted will be recovered at the receiver. Hence one can transmit messages from a source with entropy $H(\mathrm{m})$ bits, by transmitting on an average a minimum of $H(\mathrm{m})$ binary digits (binits) per message. This is the substance of the results in this section. This result is known as the *coding theorem for noiseless (discrete) channels.* (This theorem is also known as *Shannon's first theorem.*) Later on we shall prove Shannon's second theorem (theorem for noisy channels), which states that for a reliable (error-free) communication, we need to transmit on an average $H(\mathrm{m})/C$ binits (binary digits) per message where C is the channel capacity per binit ($C < 1$).

In all our discussion so far we have used the bit as the unit of information and the binary digit (binit) as the unit used in coding. We can use r-ary digits instead. Following the same lines, the reader can prove that it is possible, on an average to encode a source with $H_r(\mathrm{m})$ number of r-ary digits per message where $H_r(\mathrm{m})$ is the entropy of the source in r-ary units. The coding theorem for noiseless channels may be stated as follows: on a noiseless channel, it is possible to transmit the information emitted by a source with entropy $H_r(\mathrm{m})$ (in r-ary units) using on an average a minimum of $H_r(\mathrm{m})$ r-ary digits per message.

7-5. COMPACT CODES

If a source m can send n messages $m_1, m_2, \ldots, m_n$ with probabilities $P_1, P_2, \ldots, P_n$, we have shown that on an average, a minimum of

H(m) binits per message are needed for encoding. The number of binary digits in a code word will be called the *length* of the code word. The minimum average length L of the code word is H(m)

$$L = H(\mathrm{m}) = \sum_{i=1}^{n} P_i \log \frac{1}{P_i} \tag{7-10}$$

From the coding theorem for noiseless channels, we know that there is no other possible code (that is uniquely decodable) with average length less than H(m). It should be recalled that in order to attain this minimum length, we have to encode sequences of N messages ($N \to \infty$). If we wish to encode each message directly without using longer sequences, then in general, the average length of the code word per message will be greater than H(m). In practice, it is not very desirable to use long sequences (also called long extensions) for coding. Hence we prefer to encode messages directly and not in sequences even if we have to pay for it with an increased average word length. In general, however, this is a small price to pay. The advantages of coding long sequences are only marginal; that is, the improvement achieved by using longer sequences is, in general, only marginal. The next logical question is: how shall we construct the optimum code which will yield the minimum average length when the messages are encoded directly (one at a time)? Following is a procedure given without proof, for finding the optimum code (called the Huffman code). The proof that this code is optimum can be found in Refs. 4–7.

We shall illustrate the procedure with an example using the binary code. We first arrange the messages in order of descending probability as shown in Table 7-1. Here we have six messages with probabilities

TABLE 7-1A

Original Source		Reduced Sources			
Messages	Probabilities	S_1	S_2	S_3	S_4
m_1	.30	.30	.30	.43	.57
m_2	.25	.25	.27	.30	.43
m_3	.15	.18	.25	.27	
m_4	.12	.15	.18		
m_5	.10	.12			
m_6	.08				

.30, .25, .15, .12, .10, and .08 respectively. We now combine the last two messages into one message with probability $P_5 + P_6 = 0.18$. We now have five messages with probabilities .30, .25, .15, .12, and .18. These messages are now rearranged in the second column in order of descending probability. We repeat this procedure by combining the last two messages in the second column and rearranging them in order of descending probability. This is done until the number of messages is reduced to 2. These two (reduced) messages are now assigned 0 and 1 as their first digits in the code sequence. We now go back and assign the numbers 0 and 1 to the second digit for the two messages which were combined in the previous step. We keep regressing this way until the first column is reached. The code finally obtained (for the first column) can be shown to be the optimum (Ref. 7). The complete procedure is shown in Tables 7-1A and 7-1B.

TABLE 7-1B

	Original Source		Reduced Sources			
Messages	Probabilities	Code	S_1	S_2	S_3	S_4
m_1	.30	00	.30 00	.30 00	.43 1	.57 0
m_2	.25	10	.25 10	.27 01	.30 00	.43 1
m_3	.15	010	.18 11	.25 10	.27 01	
m_4	.12	011	.15 010	.18 11		
m_5	.10	110	.12 011			
m_6	.08	111				

The optimum (Huffman) code obtained this way is also called a *compact code*. The average length of the compact code in the present case is given by

$$L = \sum_{i=1}^{n} P_i L_i = .60 + .50 + .45 + .36 + .30 + .24$$

$$= 2.45 \quad \text{binary digits}$$

The entropy $H(\text{m})$ of the source is given by

$$H(\text{m}) = \sum_{i=1}^{n} P_i \log_2 \frac{1}{P_i}$$

$$= 2.418 \quad \text{bits}$$

Obviously the minimum possible length (attained by an infinitely long sequence of messages) is equal to 2.418 binary digits. Using direct coding (Huffman code), it is possible to attain an average length of 2.45 bits in the example given. This is a very close approximation to the optimum performance attainable Thus very little is gained by complex coding, using long sequences of messages. It can be shown that the Huffman code is uniquely decodable, that is, a sequence of coded messages can be decoded unambiguously.

The goodness of any code is measured by its average length in comparison to $H(\mathrm{m})$ (the average minimum possible length). We define *code efficiency* η as

$$\eta = \frac{H(\mathrm{m})}{L}$$

where L is the average length of the code. In our present example,

$$\eta = \frac{2.418}{2.45} = 0.976$$

The *redundancy* γ is defined as

$$\gamma = 1 - \eta = 0.024$$

A similar procedure is used to find a compact r-ary code. In this case we arrange the messages in descending order of probability and combine the last r messages into one message and rearrange the new set (reduced set) in order of descending probability. We repeat the procedure until the final set reduces to r messages. Each of these messages is now assigned one of the r numbers $0, 1, 2, \ldots, r$. We now regress in exactly the same way as in binary case until each of the original message is assigned a code.

For an r-ary code, we shall have exactly r messages left in the last reduced set if and only if the total number of original messages is equal to $r + k(r - 1)$ where k is an integer. This is obvious since each reduction decreases the number of messages by $r - 1$. Hence if there is a total of k reductions, the total number of original messages must be $r + k(r - 1)$. In case the original messages do not satisfy this condition, we must add some dummy messages with zero probability of occurrence until this condition is fulfilled. As an example, if $r = 4$, and the number of messages n is 6, then we must add one dummy message with zero probability of occurrence to make the total messages 7, that is, $[4 + 1(4 - 1)]$, and proceed as usual. The procedure is illustrated in Table 7-2.

The length L of this code is

$$L = .30(1) + .25(1) + .15(1) + .12(2) + .1(2) + .08(2) = 1.3 \quad \text{4-ary digits}$$

$$H_4(\mathrm{m}) = \sum_{i=1}^{6} P_i \log_4 \frac{1}{P_i} = 1.209 \quad \text{4-ary units}$$

TABLE 7-2

Original Source			Reduced Sources	
Messages	Probabilities	Code		
m_1	.30	0	.30	0
m_2	.25	2	.30	1
m_3	.15	3	.25	2
m_4	.12	10	.15	3
m_5	.10	11		
m_6	.08	12		
m_7	.0	13		

Code efficiency η is given by

$$\eta = \frac{1.209}{1.3} = 0.93$$

Redundancy $\gamma = 1 - \eta = 0.07$.

In some special cases the length of the compact code is equal to $H(\mathrm{m})$. This occurs when the probabilities $P_1, P_2, \ldots, P_n$ are of the form[5]

$$P_i = 2^{-L_i} \quad (L_i \text{ an integer})$$

In such case it can be shown that the compact code attains the minimum possible length $H(\mathrm{m})$. As an example consider the probabilities of 6 messages $m_1, m_2, \ldots, m_6$ as 1/2, 1/4, 1/8, 1/16, 1/32, and 1/32. These probabilities are of the form 2^{-L_i}. The Huffman code obtained by the usual procedure is 0, 10, 110, 1110, 11110, 11111. The average length L of this code is given by

$$L = \sum_{i=1}^{6} P_i L_i$$

$$= \tfrac{1}{2} + 2(\tfrac{1}{4}) + 3(\tfrac{1}{8}) + 4(\tfrac{1}{16}) + 5(\tfrac{1}{32}) + 5(\tfrac{1}{32})$$

$$= 1\tfrac{15}{16} \quad \text{binits}$$

The entropy $H(\mathrm{m})$ of this source is

$$H(\mathrm{m}) = \sum_{i=1}^{6} P_i \log_2 \frac{1}{P_i}$$

$$= 1\tfrac{15}{16} \quad \text{bits}$$

Hence

$$L = H(\mathrm{m})$$

[5]For an r-ary code this condition is

$$P_i = r^{-L_i}$$

7-6. TRANSMISSION OF INFORMATION OVER A CHANNEL: MUTUAL INFORMATION

Let us consider a source whose messages are encoded into certain symbols $x_1, x_2, \ldots, x_r$ (these may be just r-ary digits or some other symbols). Each symbol is in turn transmitted over the channel by one of the r-ary signals assigned to it. The receiver receives the symbols y_1, $y_2, \ldots, y_s$. The set of symbols $\{y_k\}$ may or may not be identical to the set $\{x_k\}$, depending upon the nature of the receiver. If we use the types of receivers discussed in Chapter 6, obviously the set of received symbols will be the same as the set transmitted. This is because the optimum receiver, upon receiving a signal, decides which of the r symbols $x_1, x_2, \ldots, x_r$ has been transmitted. Here we shall be more general and shall not constrain the set $\{y_k\}$ to be identical to the set $\{x_k\}$.

If the channel is noiseless, then the reception of some symbol y_j uniquely determines the message transmitted. However because of noise, there is a certain amount of uncertainty regarding the transmitted symbol, when y_j is received. If $P(x_i \mid y_j)$ represents the conditional probabilities that x_i was transmitted when y_j is received, then there is an uncertainty of $\log [1/P(x_i \mid y_j)]$ about x_i when y_j is received. If the channel were noiseless, the uncertainty about x_i when y_j is received is zero. Obviously the uncertainty $\log [1/P(x_i \mid y_j)]$ is caused by channel noise. This evidently is the loss of information due to channel noise. The average loss of information when y_j is received is therefore

$$\sum_{\mathrm{x}} P(x_i \mid y_j) \log \frac{1}{P(x_i \mid y_j)} \quad \text{bits/symbol}$$

Note that this is the average uncertainty about the transmitted symbol when y_j is received, and is denoted by $H(\mathrm{x} \mid y_j)$:

$$H(\mathrm{x} \mid y_j) = \sum_{\mathrm{x}} P(x_i \mid y_j) \log \frac{1}{P(x_i \mid y_j)} \quad \text{bits/symbol} \tag{7-11}$$

Next we average the information loss over all the received symbols. This is denoted by $H(\mathrm{x} \mid \mathrm{y})$:

$$H(\mathrm{x} \mid \mathrm{y}) = \sum_{\mathrm{y}} P(y_j) H(\mathrm{x} \mid y_j) \quad \text{bits/symbol} \tag{7-12}$$

$$= \sum_{\mathrm{x}} \sum_{\mathrm{y}} P(y_j) P(x_i \mid y_j) \log \frac{1}{P(x_i \mid y_j)} \quad \text{bits/symbol} \tag{7-13}$$

$$= \sum_{\mathrm{x}} \sum_{\mathrm{y}} P(x_i, y_j) \log \frac{1}{P(x_i \mid y_j)} \quad \text{bits/symbol} \tag{7-14}$$

Thus $H(\mathrm{x} \mid \mathrm{y})$ represents the average uncertainty about the transmitted symbol, averaged over all the received symbols. Evidently $H(\mathrm{x} \mid \mathrm{y})$ is the average loss of information per received symbol. If the channel is noiseless, then $H(\mathrm{x} \mid \mathrm{y})$ should be zero. This can easily be verified by observing that for a noiseless channel, the received symbol uniquely determines the transmitted symbol. Hence $P(x_i \mid y_j)$ is either 0 or 1. Therefore, $H(\mathrm{x} \mid y_j)$ as given by Eq. 7-11 must be zero. Hence $H(\mathrm{x} \mid \mathrm{y})$ in Eq. 7-12 is also zero. $H(\mathrm{x} \mid \mathrm{y})$ is called the *equivocation* of x with respect to y.

Note that $P(y_j \mid x_i)$ represents the probability that y_j is received when x_i is transmitted. This is obviously a characteristic of the channel and receiver. Thus a given channel (with its receiver) is specified by the *channel matrix*:

$$\text{Inputs}\quad \begin{array}{c} \\ x_1 \\ x_2 \\ \\ x_r \end{array} \overset{\text{Outputs}}{\begin{array}{cccc} y_1 & y_2 & \cdots & y_s \\ \left[\begin{array}{c} P(y_1 \mid x_1) \\ P(y_1 \mid x_2) \\ \cdots \\ P(y_1 \mid x_r) \end{array}\right. & \begin{array}{c} P(y_2 \mid x_1) \\ P(y_2 \mid x_2) \\ \cdots \\ P(y_2 \mid x_r) \end{array} & \begin{array}{c} \cdots \\ \cdots \\ \cdots \\ \cdots \end{array} & \left.\begin{array}{c} P(y_s \mid x_1) \\ P(y_s \mid x_2) \\ \cdots \\ P(y_s \mid x_r) \end{array}\right] \end{array}}$$

We can obtain the reverse conditional probabilities $P(x_i \mid y_j)$ using Bayes' rule:

$$P(x_i \mid y_j) = \frac{P(y_j \mid x_i)\,P(x_i)}{P(y_j)} \tag{7-15a}$$

$$= \frac{P(y_j \mid x_i)\,P(x_i)}{\sum_{\mathrm{x}} P(x_i, y_j)} \tag{7-15b}$$

$$= \frac{P(y_j \mid x_i)\,P(x_i)}{\sum_{\mathrm{x}} P(x_i)\,P(y_j \mid x_i)} \tag{7-15c}$$

Thus if the input symbol probabilities $P(x_i)$ and the channel matrix is known, the reverse conditional probabilities can be computed from Eq. 7-15. The reverse conditional probability $P(x_i \mid y_j)$ is the probability that x_i is transmitted when y_j is received.

If the channel were noise-free, the average amount of information received is $H(\mathrm{x})$ bits (entropy of the source) per received symbol. Note that $H(\mathrm{x})$ is the average information transmitted over the channel per symbol. Because of channel noise, we lose an average of $H(\mathrm{x} \mid \mathrm{y})$ bits of information per symbol. Obviously in this transaction the receiver receives on average amount of information of $I(\mathrm{x};\mathrm{y})$ bits per received symbol where

$$I(\mathrm{x};\mathrm{y}) = H(\mathrm{x}) - H(\mathrm{x} \mid \mathrm{y}) \quad \text{bits/symbol} \tag{7-16}$$

$I(\mathrm{x};\mathrm{y})$ is called the *mutual information* (of x and y). We have

$$H(\mathrm{x}) = \sum_{\mathrm{x}} P(x_i) \log \frac{1}{P(x_i)} \quad \text{bits} \tag{7-17}$$

where $P(x_i)$ is the probability of symbol x_i being transmitted. Substitution of Eqs. 7-17 and 7-14 in Eq. 7-16 yields

$$I(\mathrm{x};\mathrm{y}) = \sum_{\mathrm{x}} P(x_i) \log \frac{1}{P(x_i)} - \sum_{\mathrm{x}} \sum_{\mathrm{y}} P(x_i, y_j) \log \frac{1}{P(x_i \mid y_j)}$$

since

$$\sum_{\mathrm{y}} P(x_i, y_j) = P(x_i)$$

$$I(\mathrm{x};\mathrm{y}) = \sum_{\mathrm{x}} \sum_{\mathrm{y}} P(x_i, y_j) \log \frac{1}{P(x_i)} - \sum_{\mathrm{x}} \sum_{\mathrm{y}} P(x_i, y_j) \log \frac{1}{P(x_i \mid y_j)}$$

$$= \sum_{\mathrm{x}} \sum_{\mathrm{y}} P(x_i, y_j) \log \frac{P(x_i \mid y_j)}{P(x_i)} \tag{7-18}$$

$$= \sum_{\mathrm{x}} \sum_{\mathrm{y}} P(x_i, y_j) \log \frac{P(x_i, y_j)}{P(x_i)\, P(y_j)} \tag{7-19}$$

Alternatively using Bayes' rule in Eq. 7-18 $I(\mathrm{x};\mathrm{y})$ may be expressed as

$$I(\mathrm{x};\mathrm{y}) = \sum_{\mathrm{x}} \sum_{\mathrm{y}} P(x_i, y_j) \log \frac{P(y_j \mid x_i)}{P(y_j)}$$

or we may substitute Eq. 7-15c in Eq. 7-18:

$$I(\mathrm{x};\mathrm{y}) = \sum_{\mathrm{x}} \sum_{\mathrm{y}} P(x_i)\, P(y_j \mid x_i) \log \frac{P(y_j \mid x_i)}{\sum_{\mathrm{x}} P(x_i)\, P(y_j \mid x_i)} \tag{7-20}$$

Equation 7-20 expresses $I(\mathrm{x};\mathrm{y})$ in terms of input symbol probabilities and the channel matrix.

The units of $I(\mathrm{x};\mathrm{y})$ should be carefully noted. $I(\mathrm{x};\mathrm{y})$ is the average amount of information received per symbol transmitted (or symbol received). Hence its units are bits/symbol. If we use binary digits at the input, then the symbol is a binary digit and the units of $I(\mathrm{x};\mathrm{y})$ are bits per binit. If the input symbols are in r-ary digits, then the units are bits per r-ary digit. If we use s as the base of logarithms, then the units are s-ary units per symbol.

Some Properties of Mutual Information. From Eq. 7-19, we have

$$I(\mathrm{x};\mathrm{y}) = \sum_{\mathrm{x},\mathrm{y}} P(x_i, y_j) \log \frac{P(x_i, y_j)}{P(x_i)\,P(y_j)}$$

$$= \sum_{\mathrm{x},\mathrm{y}} P(x_i, y_j) \log P(x_i, y_j) - \sum_{\mathrm{x},\mathrm{y}} P(x_i, y_j) \log P(x_i)\,P(y_j) \tag{7-21}$$

Observe that

$$\sum_{\mathrm{x},\mathrm{y}} P(x_i, y_j) = \sum_{\mathrm{x},\mathrm{y}} P(x_i)\,P(y_j) = 1 \tag{7-22}$$

Therefore direct application of inequality A7-1 in Appendix 7 to Eq. 7-21 yields

$$I(\mathrm{x};\mathrm{y}) \geq 0 \tag{7-23}$$

with equality if and only if

$$P(x_i, y_j) = P(x_i)\,P(y_j) \tag{7-24}$$

This is to be expected. The mutual information $I(\mathrm{x};\mathrm{y})$ represents the average amount of information received through a channel. Obviously the transmitted information can never be negative, so, on an average, we can never lose information, by observing the output of a channel. Moreover, as Eq. 7-22 implies, the received information is zero (on average) if and only if Eq. 7-24 is satisfied. This is also logical, since Eq. 7-24 implies that the received symbol is statistically independent of the transmitted symbol. In this case there is clearly no information being received through the channel.

The second important property of mutual information follows from the symmetry in x and y of Eq. 7-19. It therefore follows that

$$I(\mathrm{x};\mathrm{y}) = I(\mathrm{y};\mathrm{x}) \tag{7-25}$$

Hence

$$I(\mathrm{x};\mathrm{y}) = H(\mathrm{y}) - H(\mathrm{y} \mid \mathrm{x}) \tag{7-26}$$

where

$$H(\mathrm{y}) = \sum_{\mathrm{y}} P(y_j) \log \frac{1}{P(y_j)} \tag{7-27}$$

and

$$H(\mathrm{y} \mid \mathrm{x}) = \sum_{\mathrm{x}} \sum_{\mathrm{y}} P(x_i, y_j) \log \frac{1}{P(y_j \mid x_i)} \tag{7-28}$$

The quantity $H(\mathrm{y} \mid \mathrm{x})$ is called *equivocation* of y with respect to x. It is obvious that $H(\mathrm{y} \mid \mathrm{x})$ represents the average uncertainty about the received symbol when the transmitted symbol is known. Note that $P(y_j \mid x_i)$ is the probability that y_j will be received when x_i is transmitted. We can rewrite Eq. 7-25 as

$$H(\mathrm{x}) - H(\mathrm{x} \mid \mathrm{y}) = H(\mathrm{y}) - H(\mathrm{y} \mid \mathrm{x}) \tag{7-29}$$

Example 7-1. Find the mutual information $I(x; y)$ of a channel where the messages are transmitted by binary digits (binary channel). The channel is symmetric with respect to 0 and 1. The probability of error and the probability of correct reception are given by p and $\bar{p}$ respectively. This means that the probability of receiving 1 when 0 is transmitted is the same as the probability of receiving 0 when 1 is transmitted, viz., p. Similarly the probability of receiving 1 when 1 is transmitted is the same as probability of receiving 0 when 0 is transmitted, viz. $\bar{p} = 1 - p$. Such a channel is called a *binary symmetric channel* (*BSC*).

The various probabilities can conveniently be illustrated by the channel diagram, Fig. 7-1.

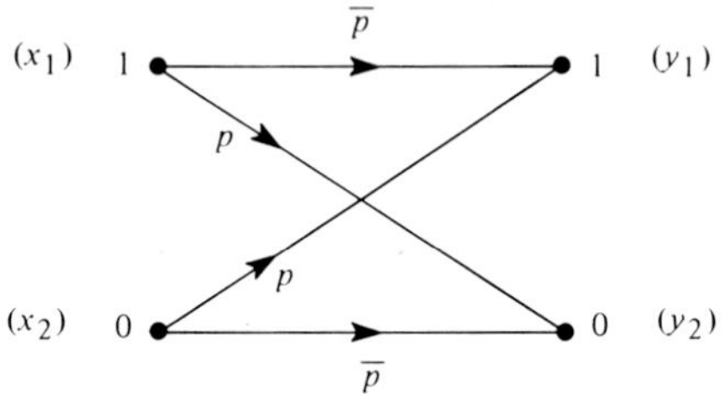

FIG. 7-1.

In this problem the symbols x_1 and x_2 are represented by 0 and 1 respectively. Similarly, symbols y_1 and y_2 are represented by 0 and 1 respectively. Let

$$P(x_1) = \alpha; \quad P(x_2) = 1 - \alpha = \bar{\alpha} \tag{7-30a}$$

Also,

$$\begin{aligned} P(y_1 \mid x_2) &= P(y_2 \mid x_1) = p \\ P(y_1 \mid x_1) &= P(y_2 \mid x_2) = \bar{p} = 1 - p \end{aligned} \tag{7-30b}$$

Since we know the input symbol probabilities (Eq. 7-30a) and the channel matrix (Eq. 7-30b), we can compute $I(x; y)$ at once from Eq. 7-20:

$$I(x; y) = \sum_{x} \sum_{y} P(x_i) P(y_j \mid x_i) \log \frac{P(y_j \mid x_i)}{\sum_{x} P(x_i) P(y_j \mid x_i)} \quad \text{bits/binit}$$

$$= \alpha\bar{p} \log \frac{\bar{p}}{\alpha\bar{p} + \bar{\alpha}p} + \alpha p \log \frac{p}{\alpha p + \bar{\alpha}\bar{p}}$$

$$+ \bar{\alpha}p \log \frac{p}{\alpha\bar{p} + \bar{\alpha}p} + \bar{\alpha}\bar{p} \log \frac{\bar{p}}{\alpha p + \bar{\alpha}\bar{p}} \tag{7-31}$$

$$= \left[(\alpha p + \bar{\alpha}\bar{p}) \log \frac{1}{(\alpha p + \bar{\alpha}\bar{p})} + (\alpha\bar{p} + \bar{\alpha}p) \log \frac{1}{(\alpha\bar{p} + \bar{\alpha}p)} - \left(p \log \frac{1}{p} + \bar{p} \log \frac{1}{\bar{p}}\right)\right] \quad \text{bits/binit} \tag{7-32}$$

If we define a function $\Omega(z)$ as

$$\Omega(z) = z \log \frac{1}{z} + \bar{z} \log \frac{1}{\bar{z}}$$

where

$$\bar{z} = 1 - z$$

then Eq. 7-31 can be expressed as

$$I(\mathrm{x};\mathrm{y}) = \Omega(\alpha p + \bar{\alpha}\bar{p}) - \Omega(p) \quad \text{bits/binit} \tag{7-33}$$

The function $\Omega(z)$ as a function of z is shown in Fig. 7-2. This function reaches a maximum value of 1 when $z = 1/2$. [Note that we are interested in $\Omega(z)$ only over the interval $0 < z < 1$.]

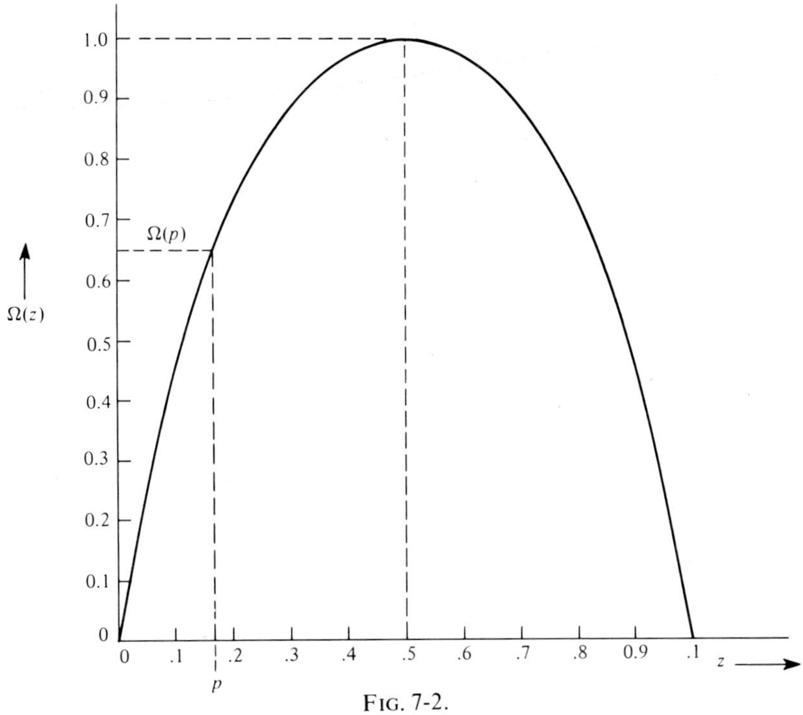

FIG. 7-2.

The mutual information $I(\mathrm{x};\mathrm{y})$ represents the average information received over the channel (per binary digit received). It is interesting to find the maximum information that can be transmitted over the channel per digit. If we are given some value of p, then $\Omega(p)$ is fixed (Fig. 7-2). Then $I(\mathrm{x};\mathrm{y})$ can be maximized by maximizing $\Omega(\alpha p + \bar{\alpha}\bar{p})$. The Ω function is maximum (unity) when

$$\alpha p + \bar{\alpha}\bar{p} = \tfrac{1}{2}$$

or

$$\alpha p + (1 - \alpha)(1 - p) = \tfrac{1}{2}$$

This relation is satisfied when

$$\alpha = \tfrac{1}{2}$$

and for $\alpha = 1/2$,

$$\Omega(\alpha p + \bar{\alpha}\bar{p}) = 1$$

Hence

$$I_{\max}(\mathrm{x};\mathrm{y}) = 1 - \Omega(p) \quad \text{bits/binit}$$
$$= 1 - \left(p \log \frac{1}{p} + \bar{p} \log \frac{1}{\bar{p}}\right) \quad \text{bits/binit} \tag{7-34}$$

Therefore for a given value of p, the maximum information can be transmitted over the channel when

$$\alpha = P(x_1) = P(x_2) = \tfrac{1}{2}$$

that is, when equiprobable symbols are used. The maximum information transmitted in bits per binary digit (binit) is given by Eq. 7-34.

7-7. CHANNEL CAPACITY

We have seen in Ex. 7-1, that for a given channel (that is, a given p), $I(\mathrm{x};\mathrm{y})$, the amount of information transmitted over the channel, was a function of α (the input-symbol probabilities). This is also obvious from Eq. 7-20. It can be seen from Eq. 7-20, that $I(\mathrm{x};\mathrm{y})$ is a function of channel matrix $[P(y_j \mid x_i)]$ and the input-symbol probabilities $P(x_i)$ only. Hence for a given channel (that is a given channel matrix), $I(\mathrm{x};\mathrm{y})$, the information transmitted over a channel per symbol is a function of the input-symbol probabilities only. For one set of input-symbol probabilities $I(\mathrm{x};\mathrm{y})$ reaches a maximum. This represents the maximum amount of information that can be transmitted over the channel per symbol (or per digit). The maximum value of $I(\mathrm{x};\mathrm{y})$ obtained for a certain set of input-symbol probabilities is called the *channel capacity per symbol* (or per digit) transmitted. The channel capacity is denoted by C_s.

$$C_s = \max_{P(x_i)} I(\mathrm{x};\mathrm{y}) \quad \text{units/symbol} \tag{7-35}$$

For a binary symmetric channel as in Ex. 7-1,

$$C_s = 1 - \Omega(p) \quad \text{bits/binit}$$
$$= 1 - \left(p \log_2 \frac{1}{p} + \bar{p} \log_2 \frac{1}{\bar{p}}\right) \quad \text{bits/binit} \tag{7-36}$$

We have seen that the maximum rate C_s is attained when the two input symbols are equiprobable.

$$P(x_1) = P(x_2) = \tfrac{1}{2}$$

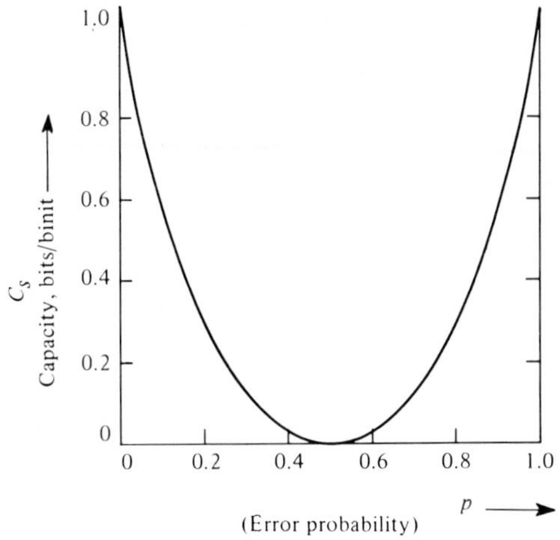

FIG. 7-3.

The channel capacity C_s of a BSC as a function of error probability p is shown in Fig. 7-3. From this figure we see that the maximum value of C_s is unity. This means we can transmit at most 1 bit of information per binary digit. This is the expected result, since one binary digit can convey one of the two equiprobable messages. The information content of one of the two equiprobable messages is $\log_2 2 = 1$ bit. Secondly, we observe that C_s is maximum when the error probability $p = 0$ or $p = 1$. When error probability $p = 0$, the channel is noiseless and we expect C_s to be maximum. But surprisingly, C_s is maximum when $p = 1$. This is easy to explain, because a channel which consistently and with certainty makes errors is as good as a noiseless channel. All we have to do is to reverse the decision that is made, and we have error-free reception, that is, if 0 is received, we decide that 1 was actually sent, and vice versa. The channel capacity C_s is zero (minimum) when $p = 1/2$. If the error probability is $1/2$, then the transmitted symbols and received symbols are statistically independent. If we received 0, for example, either 1 or 0 is equally likely to have been transmitted and it does not matter what decision is made at the receiver.

Channel Capacity per Second. The channel capacity C_s in Eq. 7-35 gives the maximum possible rate of information transmission when one symbol (digit) is transmitted. If there are K symbols being transmitted per second, then obviously the maximum rate of transmission of information per second is KC_s. This is the channel capacity in information units per seconds, and will be denoted by C (bits/second):

$$C = KC_s \quad \text{bits/second}$$

A Comment on Channel Capacity. The beginning student often thinks of channel capacity as a property of a particular physical channel over which the information is transmitted. This is true, provided the term channel is correctly interpreted. A channel means not only the physical path of propagation of messages, but also includes the specifications of the kind of signals (binary, r-ary, etc., or orthogonal, trans-orthogonal, etc.), and the kind of receiver used (the receiver determines error probability). All these specifications are included in the channel matrix. A channel matrix completely specifies a channel. If we decide to use, for example, 4-ary digits instead of binary digits over the same physical channel, the channel matrix changes (it becomes a 4×4 matrix) as does the channel capacity. Similarly, change in receiver or signal power or noise power will change the channel matrix and hence the channel capacity.

Magnitude of Channel Capacity. Since C_s, the channel capacity is the maximum value of $H(\mathrm{x}) - H(\mathrm{x} \mid \mathrm{y})$, it is evident that $C_s \leq H(\mathrm{x})$, (since $H(\mathrm{x} \mid \mathrm{y}) \geq 0$). But $H(\mathrm{x})$ is the average information per input symbol. Hence C_s is always less than (or equal to) the average information per input symbol. If we use binary symbols at the input, the maximum value of $H(\mathrm{x})$ is one bit (per binit) occurring when $P(x_1) = P(x_2) = 1/2$. Hence for a binary channel, $C_s \leq 1$ bit per binit. If we use r-ary symbols, the maximum value of $H_r(\mathrm{x})$ is $\log_r r = 1$ r-ary unit. Hence $C_s \leq 1$ r-ary unit per symbol.

Example 7-2. Find the channel capacity of a channel over which the messages are transmitted by binary digits. The digits 0 and 1 are transmitted as antipodal signals $s(t)$ and $-s(t)$ each of energy E. The channel-noise power density is $\mathfrak{N}/2$.

The channel capacity is the value of $I(\mathrm{x};\mathrm{y})$ when the input symbols are equiprobable for the binary case. The error probability for equiprobable antipodal signals is given by Eq. 6-39b:

$$p = P(\epsilon) = \operatorname{erfc}\left(\sqrt{\frac{2E}{\mathfrak{N}}}\right)$$

and

$$C_s = 1 - \left[p \log \frac{1}{p} + (1 - p) \log \frac{1}{1 - p}\right] \text{ bits/binit} \tag{7-37}$$

7-8. ERROR-FREE TRANSMISSION OVER A NOISY CHANNEL: SHANNON'S SECOND THEOREM

In our discussion so far, we have accepted the fact that because of channel noise, there will always be an error in the information received over a noisy channel. It appears plausible that we can reduce the error by increasing the signal-to-noise ratio and/or reducing the rate of information transmission. This is indeed true, because increasing SNR and reducing the rate of transmission reduce the error probability. However, no matter how much we increase SNR or reduce the rate of transmission, superficially it appears that the error can never be reduced to zero. As we shall see in this section, this conclusion is not true. We shall show that for any given channel information can be transmitted without error, provided the rate of information transmission per second is maintained below C_s bits per binit or C bits per second.

This is an extremely important theorem due to Shannon (called *Shannon's coding theorem for noisy channels*). For many years communication engineers assumed that by increasing the signal power and/or reducing the transmission rate, communication could be made more reliable (but never error-free). Shannon's theorem states that the presence of random disturbance (noise) over a channel does not, by itself, set any limit to the transmission accuracy; instead, it sets a limit to the transmission rate at which any degree of transmission accuracy can be achieved. This is the essence of Shannon's coding theorem for noisy channels.

Here we shall give a proof for Shannon's theorem for a special case of the binary symmetric channel. Consider a binary symmetric channel with error probability $p = 10^{-2}$. We shall now consider the techniques of reducing error probability by reducing the rate of transmission. This can be achieved by repeating the same digit several times instead of transmitting it just once. If we wish to transmit 0 for example, we may repeat it, say three times instead of transmitting it only once. Similarly, the digit 1 may be transmitted by repeating 1 thrice. Thus we transmit the digits 0 and 1 by 000 and 111. At the receiver, some of the digits will be in error and we expect in general to receive any of the eight possible patterns formed by three binary digits. This is shown in Table 7-3.

TABLE 7-3

Transmitted Messages	Received Messages	Decision
000	000	
	001	0
	010	
	100	
	011	
	101	1
	110	
111	111	

At the receiver, the problem is to decide which of the two messages 000, 111 was transmitted, when any one of the eight messages are received.

One possible decision procedure is to use the majority rule. The received signal is decided in favor of 0 or 1 depending on whether it has a majority of 0's or 1's. In this case any one of the top four received messages (Table 7-3) will be interpreted as 0 and the bottom four as 1. What is the error probability in this scheme? An error can occur if any of the two or all three digits are in error. The probability that all three digits are in error is p^3. Two simultaneous errors can occur in three different ways (2 digits in error and one digit correct). Hence the probability of this event is $3p^2\bar{p}$. The new error probability p_1 is therefore given by

$$p_1 = 3p^2\bar{p} + p^3$$

for

$$p = 10^{-2} \qquad \bar{p} = 1 - 10^{-2} \simeq 1$$

and

$$p_1 \simeq 3 \times 10^{-4}$$

In conclusion, we observe that by reducing the rate of transmission by a factor of 3, the error probability is reduced from 10^{-2} to 3×10^{-4}. Proceeding along similar lines, we can show that by reducing the rate of transmission by a factor of 5 (5 repetitions), the probability of error can be reduced from 10^{-2} to 10^{-5}. When we have 5 repetitions there will be $2^5 = 32$ possible messages at the receiver. We again use the majority rule to decide whether 0 or 1 has been transmitted.

At this point it becomes very convenient to talk in terms of what is called the *Hamming distance* between binary sequences. If two binary sequences of the same length differ in j places (j digits), then the Hamming distance between the sequences is said to be j. Thus the Hamming distance between 000, 010 is 1, and that between 0011 and 1100 is 4. In future discussion, the Hamming distance will be called the *distance between two sequences.* The majority rule of decision is really a rule which decides in favor of that message which is nearest (in the sense of Hamming-distance) to the received sequence. Thus in the case of 3 repetitions, the sequences 000, 001, 010, 100 are less than or equal to one unit of distance from 000 but are at least 2 units away from 111. Similarly, the sequences 011, 101, 110, 111 are less than or equal to one unit of distance from 111 but at least 2 units away from 000.

In the technique discussed here, the probability of error is still finite. We wish to reduce the error probability to zero. How can this be done? We shall first use a heuristic argument to show that it is possible to have a zero-error probability. Consider sequences of N binary digits.

If we transmit one such sequence, the received sequence will be different from the transmitted sequence because of channel noise (causing error in detection). The channel noise therefore creates a separation (of a certain distance) between the received and the transmitted sequences. If N becomes very large, then by the law of large numbers there will be a distance of Np between the received sequence and the transmitted sequence. If the transmitted sequences are spaced widely apart at a distance exceeding Np, then the decisions at the receiver can be made without any error.

We may then draw Hamming spheres of radii Np, centered at each of the transmitted sequences.[6] If the transmitted sequences are separated by distances greater than Np, then all the spheres will be nonoverlapping. All the received messages must then necessarily lie inside one of these spheres (if $N \to \infty$). We decide in favor of that sequence in whose sphere the received sequence lies. There is no error in this decision (as $N \to \infty$). This is the qualitative argument to show the plausibility of Shannon's theorem. The reader may wonder here why if $N \to \infty$, this does not cause the rate of information transmission to go to zero. This would be true if M, the number of transmitted sequences, were fixed while N was increased to ∞. But as N increases, we can transmit more and more sequences. This is because there are in all 2^N possible sequences of N binary digits and as N increases, we can pack in more and more nonoverlapping spheres of radius Np in Hamming space. If there are M sequences (equally likely) transmitted, then the information per sequence is log M units. Hence the information transmitted is $\log M/N$ units per symbol. We shall now show that if N is increased, M also increases so that $\log M/N$ approaches a constant C_s (the channel capacity per symbol).

To show this we consider a source with M equiprobable messages. (If the messages are not equiprobable, they can effectively be made equiprobable by considering long extensions as discussed in Sec. 7-4.) The information per message is log M bits. We transmit these messages by using sequences of N binary digits. There are in all 2^N possible sequences of N binary digits. Out of these 2^N sequences we choose only M sequences and assign them to M messages. Around these M sequences, we construct Hamming spheres each of radius Np where p is the probability of error. The decision rule is, if a received sequence falls inside a sphere belonging to message m_1, then we decide in favor of m_1.

[6]Note that a Hamming sphere is not a geometrical hypersphere because the Hamming distance is not the true geometrical distance. The sequences 001, 010, 100 lie on a Hamming sphere of radius 2 and centered at 111.

As $N \to \infty$, all received sequences must fall in one of the M spheres because by the law of large numbers all received sequences are exactly at a distance Np from the transmitted sequence. The decision regarding the transmitted sequence can be made without any error (we have assumed that the spheres are nonoverlapping).

We shall now find how many of all the possible 2^N sequences lie in a sphere of radius Np centered at some message m_1. Among all sequences of length N, the total number of sequences that differ from a given sequence in exactly one digit is N. Thus there are N sequence within a Hamming sphere of unit radius. Similarly there are $N(N-1)/2$ number of sequences which differ from a given sequence in exactly 2 digits. In general, the number of sequences which differ from a given sequence is exactly k digits is given by the binomial coefficient $\binom{N}{k}$:

$$\binom{N}{k} = \frac{N!}{k!(N-k)!} \tag{7-38}$$

Hence the total number of sequences which differ from a given sequence by k or less number of digits is given by $\alpha(k)$:

$$\alpha(k) = \binom{N}{1} + \binom{N}{2} + \cdots + \binom{N}{k}$$

Note that $\alpha(k)$ is the number of sequences falling inside a Hamming sphere of radius k around a sequence. Hence the number of sequences falling inside a Hamming sphere of radius Np is $\alpha(Np)$, given by

$$\alpha(Np) = \sum_{k=0}^{Np} \binom{N}{k} \tag{7-39}$$

Next of all the possible 2^N sequences we shall choose M sequences to be assigned to M messages. How shall we choose these sequences? From the decision procedure, it is clear that if we choose one sequence, then any sequence lying within a sphere of radius Np cannot be assigned to the other message. Thus we choose one sequence for message m_1. The corresponding $\alpha(Np)$ sequences are now ineligible for consideration. From the remaining $2^N - \alpha(Np)$ we choose another sequence for m_2 and eliminate corresponding $\alpha(Np)$ sequences lying within the sphere of radius Np surrounding the message m_2. We proceed this way until all the 2^N sequences are exhausted. This procedure, however, is very tedious. Let us see what happens if we choose the required M sequences randomly from 2^N sequences. In this procedure there is a danger that we may select two sequences which lie within a distance of Np units. If, however, M is sufficiently small, the probability of making such a choice is extremely small. The probability of choosing any particular sequence s_1, for this set of M sequences out of 2^N sequences, is $M/2^N$. Remembering that $\alpha(Np)$ sequences lie within a distance Np digits from s_1, what is the probability that we may also choose any other sequence s_2 which is within a distance Np digits from s_1? If this probability be P_M, then we have

$$P_M = M \frac{\alpha(Np)}{2^N}$$

$$= \frac{M}{2^N} \sum_{k=0}^{Np} \binom{N}{k} \tag{7-40}$$

Here we use an inequality often used in information theory (Refs. 8, 9).

$$\sum_{k=0}^{Np} \binom{N}{k} \leq 2^{N\Omega(p)} \qquad \text{for } p < \frac{1}{2}$$

where

$$\Omega(p) = p \log_2 \frac{1}{p} + \bar{p} \log_2 \frac{1}{\bar{p}}$$

Hence

$$P_M \leq \frac{M2^{N\Omega(p)}}{2^N} \qquad \text{for } p < \frac{1}{2}$$

$$\leq M2^{-N[1-\Omega(p)]} \tag{7-41}$$

But for a binary symmetric channel (as we have assumed), the channel capacity C_s is given by

$$C_s = 1 - \Omega(p) \quad \text{bits/binary digit}$$

Hence

$$P_M \leq M2^{-NC_s} \tag{7-42}$$

If we choose

$$M = 2^{N(C_s - \epsilon)} \tag{7-43}$$

where $\epsilon \to 0$, then

$$P_M = 2^{-N\epsilon}$$

As $N \to \infty$, $P_M \to 0$ no matter how small ϵ may be. When ϵ is made very small (Eq. 7-43),

$$M \to 2^{NC_s}$$

Equation 7-43 may be expressed as

$$\log_2 M = N(C_s - \epsilon) \qquad \epsilon \to 0 \tag{7-44}$$

Hence the probability of choosing two sequences randomly within a distance Np digits approaches zero as $N \to \infty$ provided $M < 2^{NC_s}$.

For M equiprobable messages, the information per message is

$\log_2 M$ bits. Since we are transmitting these messages as sequences of N binary digits, the rate R of transmission per binary digit is $\log_2 M/N$ bits/binit. From Eq. 7-44 we now have

$$R = \frac{\log_2 M}{N} = (C_s - \epsilon) \quad \text{bits/binary digit} \quad (\epsilon \to 0) \tag{7-45}$$

Thus if the information is transmitted over the channel at a rate approaching channel capacity C_s, it is possible to attain error-free communication. There is a converse of this theorem which will not be proved here. It states that it is impossible to transmit error-free information at a rate equal to or greater than C_s bits/symbol.

Comment on Maximum Rate of Transmission. The implications of Shannon's theorem on noisy channels are often not clear to beginning student. An attempt is therefore made here to clarify the substance of this theorem. This discussion refers to the binary channel (channel using two symbols for communication) but can be extended to the *r*-ary channel in general.

If a channel were noiseless, we need exactly one binary digit to transmit one bit of information. If, however, the channel is noisy, we have shown that the amount of information transmitted per binary digit is reduced to C_s bit (the channel capacity in bits per binary digit) under the most favorable conditions. So even if we are feeding in the information at the source at a rate of 1 bit per binary digit, only C_s bit per binary digit transmitted are recovered at the receiver. The remaining $1 - C_s$ bit of information is lost. We know that one binary digit can carry at most one bit of information. If there is some redundancy in coding, the information carried by one binary digit is less than one bit. The presence of redundancy in a code is, however, helpful in reducing the error probability. We have seen that repeating a symbol (equivalent to adding redundancy) reduces the error probability but also reduces the rate of information transmission. Thus for reliable communication, redundancy is essential. Shannon's theorem essentially points out the minimum redundancy required for error-free communication. The theorem says that for an error-free communication the minimum redundancy required is $1 - C_s$. We must encode the messages so that the information transmitted per binary digit does not exceed C_s bit. If the information per binary digit exceeds C_s bit, then any information in excess over C_s bit will be lost in the process of transmission.

Let us consider a source m with entropy $H(\mathrm{m})$ bits (per message of course). According to Shannon's first theorem (coding theorem for noiseless channels), to transmit these messages error-free over a noiseless channel, we need to transmit an average of $H(\mathrm{m})$ binary digits per message. Shannon's second theorem (coding theorem for noisy channels) deals with error-free transmission over noisy channels. It states that these messages can be transmitted error-free over a noisy channel if the information is transmitted at a rate approaching C_s bit per binary digit transmitted. Since information per message is $H(\mathrm{m})$ bits on average, this implies that for error-free transmission we need a minimum of $H(\mathrm{m})/C_s$ binary digits per message on average. The implications are that the presence of random disturbance (noise) in a channel does not, by itself, set any limit to the transmission accuracy. Instead it sets a limit to the transmission rate for which any required degree of transmission accuracy can be achieved. We have proved the channel capacity theorem here for a special case of a binary symmetric channel. The theorem, however, is valid for any channel (see Refs. 4, 5, and 6).

It should be noted that to attain the maximum error-free rate of communication, we must let $N \to \infty$. This implies that $M \to \infty$. Hence if a source has a finite number of messages, we must consider the long sequences of input messages and transmit these long (ideally of infinite length) strings. This necessarily involves a large amount of delay. The encoder must wait until a large sequence of input messages is delivered. This will take a long time. This is, in general, the characteristic of any channel which allows error-free communication at a rate near to channel capacity. It is obviously not worth the cost to operate a channel near its capacity. The great value of Shannon's second theorem lies in the fact that it essentially defines the upper bound on the rate of error-free transmission.

7-9. ANALOGY BETWEEN INFORMATION FLOW AND FLUID FLOW

Analogies often prove helpful in understanding abstract phenomena. To some extent there is an analogy between information flow over an information channel and a fluid flow through a fluid channel. The analogy is by no means perfect, but is helpful in imparting an understanding of Shannon's second theorem (coding theorem for noisy channels).

Consider the transmission of a fluid from one location to another via a pipe of a certain cross section. This pipe can transmit fluid at a certain maximum rate σ, depending on the cross section of the pipe and the pressure of the fluid which will be assumed to have a constant value. Note that the pipe can transmit fluid at a rate of at most σ, and this occurs under the most favorable conditions. For example, if at the source (transmitting end), the fluid is flowing in at a rate less than σ, the rate of transmission will obviously be lower than σ, or, if the pipe inlet and the storage-tank outlet are not properly matched, then the transmission rate may be less than σ, even if the rate of outflow from the storage tank is greater than σ. Thus because of mismatch (which may occur due to nonalignment or unequal cross sections of pipe inlet and storage-tank outlet) fluid will be spilled and only part of it will enter the pipe. Thus the fluid channel can transmit at a rate σ only under proper conditions. This is analogous to an information channel which can at most transmit at a rate C_s under proper conditions (by proper encoding). In a fluid channel, if the storage tank tries to dispense fluid into the pipe at a rate greater than σ, then the pipe will reject the fluid at a rate equal to the excess over σ, even if there is a proper match existing at the input terminal. Similarly in an information channel, information transmitted at a rate greater than C_s will be lost [the lost information is $H(\mathrm{x} \mid \mathrm{y})$] even if the conditions are most favorabe. In the case of fluid transmission, if we wish to transmit fluid at a maximum possible rate without any spillage, the rate of inflow of the fluid must be exactly σ, and there must be a proper match between the outlet of the storage tank and the transmitting-pipe inlet. Likewise for information channels: for the maximum rate of information transmission (without any loss), the information must be fed in to the channel exactly at the rate C_s (actually slightly less than C_s) and it must be fed in properly (by proper encoding of long sequences, etc.).

The analogy can be carried further for transmission channels in tandem (or in parallel). If the fluid is transmitted over two (or more) pipes in tandem, then the rate at which the fluid is received at the destination is, at most, equal to the rate of transmission of fluid through the first pipe. This is because there is a possibility of spillage at the junction of the first and the second pipe, either due to mismatch or because the capacity of the second pipe may be smaller than that of the first. One can find similar relationships for information channels (see Ref. 4).

7-10. THE CONTINUOUS CHANNEL

So far we have considered the case of discrete messages (digital data). For a discrete random variable x, assuming values $x_1, x_2, \ldots, x_n$, we defined a meaningful quantity which we call the entropy $H(\mathrm{x})$ as

$$H(\mathrm{x}) = \sum_{i=1}^{n} P(x_i) \log \frac{1}{P(x_i)} \tag{7-46}$$

For analog data, we have to deal with continuous random variables. It is therefore desirable to extend the definition of entropy to continuous random variables. One is tempted to state that $H(\mathrm{x})$ for continuous random variables is obtained by using the integral instead of discrete summation in Eq. 7-46:

$$H(\mathrm{x}) = \int_{-\infty}^{\infty} p(x) \log \frac{1}{p(x)}\, dx \tag{7-47}$$

We shall see, that Eq. 7-47 is indeed the meaningful definition of entropy for a continuous random variable. However, we cannot blindly accept this definition unless we show that it is indeed a reasonable and meaningful definition of $H(\mathrm{x})$ and has the meaningful interpretation of entropy. To verify this we shall consider the continuous random variable x as a limiting form of a discrete random variable, which assumes discrete values $0, \pm\Delta x, \pm 2\Delta x, \ldots,$ etc. Let $k\Delta x = x_k$. Obviously $\Delta x_k = \Delta x$. The continuous random variable x can thus be approximated by a discrete random variable. The random variable x assumes a value in the range $(x_k,\ x_k + \Delta x_k)$ with probability $p(x_k)\,\Delta x_k$ in the limit as $\Delta x_k \to 0$. The error in the approximation will vanish in the limit as $\Delta x_k \to 0$. Hence $H(\mathrm{x})$, the entropy of a continuous random variable x is given by

$$\begin{aligned}
H(\mathrm{x}) &= \lim_{\Delta x_k \to 0} \sum_k p(x_k)\,\Delta x_k \log \frac{1}{p(x_k)\,\Delta x_k} \\
&= \lim_{\Delta x_k \to 0} \left[\sum_k p(x_k)\,\Delta x_k \log \frac{1}{p(x_k)} - \sum_k p(x_k)\,\Delta x_k \log \Delta x_k \right] \\
&= \int_{-\infty}^{\infty} p(x) \log \frac{1}{p(x)}\, dx - \lim_{\Delta x_k \to 0} \log \Delta x_k \int_{-\infty}^{\infty} p(x)\, dx \\
&= \int_{-\infty}^{\infty} p(x) \log \frac{1}{p(x)}\, dx - \lim_{\Delta x \to 0} \log \Delta x
\end{aligned} \tag{7-48}$$

In the limit as $\Delta x \to 0$, $\log \Delta x \to -\infty$. It therefore appears that the entropy of a continuous random variable is infinite. This is quite true. The magnitude of uncertainty associated with a continuous random variable is infinite. This fact is also apparent intuitively. A continuous random variable assumes a nonenumerably infinite number of values and hence the uncertainty is of the order of infinity. Does this mean that there is no meaningful definition of entropy for a continuous random variable? On the contrary, we shall see that the first term in Eq. 7-48 serves as a meaningful measure of the entropy (average information) of a continuous random variable x. This may be argued as follows. We can consider $\int p(x) \log [1/p(x)]\, dx$ as a relative entropy with

$-\log \Delta x$ serving as a datum or reference. The information transmitted over a channel is actually the difference between the two quantities of information $H(\mathrm{x})$ and $H(\mathrm{x} \mid \mathrm{y})$. Obviously, if we have a common datum for both $H(\mathrm{x})$ and $H(\mathrm{x} \mid \mathrm{y})$, the difference $H(\mathrm{x}) - H(\mathrm{x} \mid \mathrm{y})$ will be the same as the difference between their relative entropies. We are therefore justified in considering the first term in Eq. 7-48 as the entropy of x. We must, however, always remember that this is a relative entropy and not the absolute entropy. Failure to realize this subtle point generates many apparent fallacies, one of which is given following the next paragraph.

Based on the above argument, we define $H(\mathrm{x})$, the entropy of a continuous random variable x as

$$H(\mathrm{x}) = \int_{-\infty}^{\infty} p(x) \log \frac{1}{p(x)}\, dx \quad \text{bits} \tag{7-49a}$$

$$= -\int_{-\infty}^{\infty} p(x) \log p(x)\, dx \quad \text{bits} \tag{7-49b}$$

We shall now demonstrate an apparent fallacy which may arise if one loses sight of the fact that Eq. 7-49 defines the relative entropy and not the absolute entropy of x. Consider a random variable x with uniform probability distribution over the interval $(-1, 1)$. Thus

$$p(x) = \begin{cases} \tfrac{1}{2} & -1 < x < 1 \\ 0 & \text{Otherwise} \end{cases}$$

The entropy $H(\mathrm{x})$ is given by

$$H(\mathrm{x}) = \int_{-1}^{1} \tfrac{1}{2} \log 2\, dx$$

$$= 1 \quad \text{bit}$$

The random variable x may be represented by a voltage of amplitude x. Let us amplify this voltage by an amplifier of constant gain 2. The output of the amplifier is a continuous random variable y given by

$$\mathrm{y} = 2\mathrm{x}$$

It follows that

$$p(y) = \begin{cases} \tfrac{1}{4} & -2 < \mathrm{y} < 2 \\ 0 & \text{Otherwise} \end{cases}$$

The entropy $H(\mathrm{y})$ is given by

$$H(\mathrm{y}) = \int_{-2}^{2} \tfrac{1}{4} \log_2 4\, dx$$

$$= 2 \quad \text{bits}$$

The entropy of the random variable y is twice that of x. This result may come as a surprise, since a knowledge of x uniquely determines y, and vice versa because y = 2x. Hence the average uncertainty of x and y should be identical. Amplification itself can neither add nor subtract information. Why then is H(y) twice as large as H(x)? This becomes clear when we remember that H(x) and H(y) are relative entropies and they will be equal if and only if their datum (or reference) entropies are equal. The reference entropy R_1 for x is $-\log \Delta x$ and the reference entropy R_2 for y is $-\log \Delta y$ (in the limit as $\Delta x, \Delta y \rightarrow 0$).

$$R_1 = \lim_{\Delta x \rightarrow 0} -\log \Delta x$$

$$R_2 = \lim_{\Delta y \rightarrow 0} -\log \Delta y$$

and

$$\begin{aligned} R_1 - R_2 &= \lim_{\Delta x, \Delta y \rightarrow 0} \log \left(\frac{\Delta y}{\Delta x}\right) \\ &= \log \frac{dy}{dx} \\ &= \log 2 = 1 \quad \text{bit} \end{aligned}$$

It is evident that R_1, the reference entropy of x, is higher than the reference entropy R_2 for y. Obviously if x and y have equal absolute entropies, their relative entropies must differ by 1 bit:

$$\text{Absolute entropy of x} = R_1 + H(\text{x}) = R_1 + 1$$

$$\text{Absolute entropy of y} = R_2 + H(\text{y}) = R_2 + 2$$

Since $R_1 = R_2 + 1$, we have the absolute entropy of x = the absolute entropy of y. Hence information is neither gained nor lost in the process of amplification. Indeed, this conclusion is true for any reversible operation. If we perform a reversible operation on random variable x to obtain a random variable y, then the absolute entropies of x and y must be identical. The relative entropies, however, will in general be different.

Maximum Entropy for a Given Mean-Square Value of x. For discrete random variables, we observed that the entropy is a maximum when all the outcomes (messages) were equally likely (uniform probability distribution). For continuous random variables, there also exists a probability distribution $p(x)$ which maximizes H(x) in Eq. 7-49. In the case of a continuous distribution, however, we may have additional constraints on x. Either the maximum value of x or the mean-square value of x may be given. We shall here find the distribution $p(x)$ which will yield maximum entropy when $\overline{x^2}$ is given to be a constant σ^2. The problem then is to maximize $H(x)$:

$$H(x) = \int_{-\infty}^{\infty} p(x) \log \frac{1}{p(x)}\, dx \tag{7-50a}$$

with the constraints

$$\int_{-\infty}^{\infty} p(x)\, dx = 1 \tag{7-50b}$$

$$\int_{-\infty}^{\infty} x^2 p(x)\, dx = \sigma^2 \tag{7-50c}$$

To solve this problem, we use a theorem from the calculus of variation, which concerns itself with finding p as a function of x, which maximizes the integral I,

$$I = \int_a^b F(x, p)\, dx \tag{7-51}$$

subject to the following constraints:

$$\begin{aligned} \int_a^b \phi_1(x, p)\, dx &= \lambda_1 \\ \int_a^b \phi_2(x, p)\, dx &= \lambda_2 \\ &\vdots \\ \int_a^b \phi_k(x, p)\, dx &= \lambda_k \end{aligned} \tag{7-52}$$

where $\lambda_1, \lambda_2, \ldots, \lambda_k$ are given constants. The result from the calculus of variation, states that the form of $p(x)$ which maximizes I in Eq. 7-51 with constraints in 7-52 is found from the solution of the equation

$$\frac{\partial F}{\partial p} + \alpha_1 \frac{\partial \phi_1}{\partial p} + \alpha_2 \frac{\partial \phi_2}{\partial p} + \cdots + \alpha_k \frac{\partial \phi_k}{\partial p} = 0 \tag{7-53}$$

The quantities $\alpha_1, \alpha_2, \ldots, \alpha_k$ are adjustable constants (called *undetermined multipliers*), which can be found by substituting the solution of $p(x)$ (obtained from Eq. 7-53) in Eqs. 7-52. In the present case,

$$F(p, x) = p \log \frac{1}{p}$$

$$\phi_1(x, p) = p$$

$$\phi_2(x, p) = x^2 p$$

Hence the solution for p is given by

$$\frac{\partial}{\partial p}\left(p \log \frac{1}{p}\right) + \alpha_1 + \alpha_2 \frac{\partial}{\partial p} x^2 p = 0$$

or

$$-(1 + \log p) + \alpha_1 + \alpha_2 x^2 = 0$$

Solving for p, we have

$$p = e^{(\alpha_1 - 1)} e^{\alpha_2 x^2} \tag{7-54}$$

Substituting Eq. 7-54 in Eq. 7-50b, we have

$$\begin{aligned} 1 &= \int_{-\infty}^{\infty} e^{\alpha_1 - 1} e^{\alpha_2 x^2} \, dx \\ &= 2e^{\alpha_1 - 1} \int_0^{\infty} e^{\alpha_2 x^2} \, dx \\ &= 2e^{\alpha_1 - 1} \left(\frac{1}{2} \sqrt{\frac{\pi}{-\alpha_2}} \right) \end{aligned}$$

provided α_2 is negative or

$$e^{\alpha_1 - 1} = \sqrt{\frac{-\alpha_2}{\pi}} \tag{7-55}$$

Next we substitute Eqs. 7-54 and 7-55 in Eq. 7-50c:

$$\begin{aligned} \sigma^2 &= \int_{-\infty}^{\infty} x^2 \sqrt{\frac{-\alpha_2}{\pi}} \, e^{\alpha_2 x^2} \, dx \\ &= 2 \sqrt{\frac{-\alpha_2}{\pi}} \int_0^{\infty} x^2 e^{\alpha_2 x^2} \, dx \\ &= -\frac{1}{2\alpha_2} \end{aligned}$$

or

$$\alpha_2 = -\frac{1}{2\sigma^2} \tag{7-56a}$$

and

$$e^{\alpha_1 - 1} = \sqrt{\frac{1}{2\pi\sigma^2}} \tag{7-56b}$$

Substituting Eqs. 7-56 in Eq. 7-54, we have

$$p(x) = \frac{1}{\sigma\sqrt{2\pi}} e^{-x^2/2\sigma^2} \tag{7-57}$$

We therefore conclude that for a given mean-square value, the maximum entropy (or maximum uncertainty) is obtained when the distribution of x is gaussian. This maximum entropy or uncertainty is given by

$$H(\mathrm{x}) = \int_{-\infty}^{\infty} p(x) \log_2 \frac{1}{p(x)} \, dx$$

Note that

$$\log \frac{1}{p(x)} = \log\left(\sqrt{2\pi\sigma^2}e^{x^2/2\sigma^2}\right)$$
$$= \frac{1}{2}\log(2\pi\sigma^2) + \frac{x^2}{2\sigma^2}\log e$$

Hence

$$H(x) = \int_{-\infty}^{\infty} p(x)\left[\frac{1}{2}\log(2\pi\sigma^2) + \frac{x^2}{2\sigma^2}\log e\right]dx$$
$$= \frac{1}{2}\log(2\pi\sigma^2)\int_{-\infty}^{\infty} p(x)\,dx + \frac{\log e}{2\sigma^2}\int_{-\infty}^{\infty} x^2 p(x)\,dx$$
$$= \frac{1}{2}\log(2\pi\sigma^2) + \frac{\log e}{2\sigma^2}\sigma^2$$
$$= \frac{1}{2}\log(2\pi e\sigma^2) \qquad (7\text{-}58a)$$
$$= \frac{1}{2}\log(17.1\sigma^2) \qquad (7\text{-}58b)$$

To reiterate, for a given mean-square value $\overline{x^2}$, the entropy is maximum for gaussian distribution and the corresponding entropy is 1/2 log $(2\pi e\sigma^2)$.

The reader can similarly show that if x is constrained to some peak value M $(-M < x < M)$, then the entropy is maximum when x is uniformly distributed:

$$p(x) = \begin{cases} 1/2M & -M < x < M \\ 0 & \text{Otherwise} \end{cases} \qquad (7\text{-}59)$$

Mutual Information $I(x; y)$. The ultimate test of any concept is its usefulness. We shall now show that the relative entropy defined in Eq. 7-49 does lead to meaningful results when we consider $I(x; y)$, the mutual information of continuous random variables x and y. We wish to transmit a random variable x over a channel. Each value of x in a given continuous range is now a message which may be transmitted, for example, as a pulse of height x. The message recovered by the receiver will be a continuous random variable y. If the channel were noise-free, the received value y will uniquely determine the transmitted value x. But channel noise introduces a certain uncertainty about the true value of x. Consider the event that at the transmitter a value of x in the range x, $x + \Delta x$ has been transmitted $(\Delta x \to 0)$. The probability of this event is $p(x)\,\Delta x$ in the limit $\Delta x \to 0$. Hence the amount of information transmitted is $\log [1/p(x)\,\Delta x]$. Let the value of y at the receiver be y and $p(x \mid y)$ be the conditional probability density of x when y = y. Then $p(x \mid y)\,\Delta x$ is the probability that x will lie in the interval x, $x + \Delta x$ when y = y (provided $\Delta x \to 0$). Obviously there is an uncertainty about the event that x lies in the interval $(x, x + \Delta x)$. This uncertainty is $\log [1/p(x \mid y)\,\Delta x]$. This uncertainty arises because of channel noise and therefore represents a loss of information. Since $\log [1/p(x)\,\Delta x]$ is the information transmitted and $\log [1/p(x \mid y)\,\Delta x]$ is the information lost over the channel, the net information received is $I(x; y)$ given by

$$I(x; y) = \log \frac{p(x \mid y)}{p(x)} \tag{7-60}$$

Note that this relation is true in the limit $\Delta x \to 0$. $I(x; y)$, therefore, represents the information transmitted over a channel when we receive y(y = y) when x is transmitted (x = x). We are interested in finding the average information transmitted over a channel when some x is transmitted and a certain y is received. We must therefore average $I(x; y)$ over all values of x and y. The average information transmitted will be denoted by $I(\mathrm{x}; \mathrm{y})$, where

$$I(\mathrm{x}; \mathrm{y}) = \int_{-\infty}^{\infty} \int_{-\infty}^{\infty} p(x, y)\, I(x; y)\, dx\, dy \tag{7-61}$$

$$= \int_{-\infty}^{\infty} \int_{-\infty}^{\infty} p(x, y) \log \frac{p(x \mid y)}{p(x)}\, dx\, dy \tag{7-62}$$

$$= \int_{-\infty}^{\infty} \int_{-\infty}^{\infty} p(x, y) \log \frac{1}{p(x)}\, dx\, dy + \int_{-\infty}^{\infty} \int_{-\infty}^{\infty} p(x, y) \log p(x \mid y)\, dx\, dy$$

$$= \int_{-\infty}^{\infty} \int_{-\infty}^{\infty} p(x) p(y \mid x) \log \frac{1}{p(x)}\, dx\, dy + \int_{-\infty}^{\infty} \int_{-\infty}^{\infty} p(x, y) \log p(x \mid y)\, dx\, dy$$

$$= \int_{-\infty}^{\infty} p(x) \log \frac{1}{p(x)}\, dx \int_{-\infty}^{\infty} p(y \mid x)\, dy + \int_{-\infty}^{\infty} \int_{-\infty}^{\infty} p(x, y) \log p(x \mid y)\, dx\, dy$$

Note that

$$\int_{-\infty}^{\infty} p(y \mid x)\, dy = 1$$

and

$$\int_{-\infty}^{\infty} p(x) \log \frac{1}{p(x)}\, dx = H(\mathrm{x})$$

Hence

$$I(\mathrm{x};\mathrm{y}) = H(\mathrm{x}) + \int_{-\infty}^{\infty}\int_{-\infty}^{\infty} p(x,y) \log p(x \mid y)\, dx\, dy \tag{7-63}$$

$$= H(\mathrm{x}) - \int_{-\infty}^{\infty}\int_{-\infty}^{\infty} p(x,y) \log \frac{1}{p(x \mid y)}\, dx\, dy \tag{7-64}$$

The integral on the right-hand side is the average over x and y of $\log [1/p(x \mid y)]$. But $\log [1/p(x \mid y)]$ represents the uncertainty about x when y is received. This, as we have seen, is the information lost over the channel. The average of $\log [1/p(x \mid y)]$ is the average loss of information when some x is transmitted and some y is received. This, by definition is $H(\mathrm{x} \mid \mathrm{y})$, the equivocation of x with respect to y.

$$H(\mathrm{x} \mid \mathrm{y}) = \int_{-\infty}^{\infty}\int_{-\infty}^{\infty} p(x,y) \log \frac{1}{p(x \mid y)}\, dx\, dy \tag{7-65}$$

Hence

$$I(\mathrm{x};\mathrm{y}) = H(\mathrm{x}) - H(\mathrm{x} \mid \mathrm{y}) \tag{7-66}$$

Thus when some value of x is transmitted and some value of y is received, the average information transmitted over the channel is $I(\mathrm{x};\mathrm{y})$, given by Eq. 7-66. We can define the channel capacity C_s as the maximum amount of information that can be transmitted on average per sample or per value transmitted:

$$C_s = \max I(\mathrm{x};\mathrm{y}) \tag{7-67}$$

For a given channel, $I(\mathrm{x};\mathrm{y})$ is a function of the input probability density $p(x)$ alone. This can be seen from Eq. 7-62. We have

$$p(x,y) = p(x)p(y \mid x) \tag{7-68}$$

$$\frac{p(x \mid y)}{p(x)} = \frac{p(y \mid x)}{p(y)} = \frac{p(y \mid x)}{\int_{-\infty}^{\infty} p(x,y)\, dx} = \frac{p(y \mid x)}{\int_{-\infty}^{\infty} p(x)p(y \mid x)\, dx} \tag{7-69}$$

Substituting Eqs. 7-68 and 7-69 in Eq. 7-62, we obtain

$$I(\mathrm{x};\mathrm{y}) = \int_{-\infty}^{\infty}\int_{-\infty}^{\infty} p(x)p(y \mid x) \log \left(\frac{p(y \mid x)}{\int_{-\infty}^{\infty} p(x)p(y \mid x)\, dx} \right) dx\, dy \tag{7-70}$$

The conditional probability density $p(y \mid x)$ is characteristic of a given channel. Hence for a given channel $I(\mathrm{x};\mathrm{y})$ is a function of the input probability density $p(x)$ alone. Thus

$$C_s = \max_{p(x)} I(\mathrm{x};\mathrm{y})$$

If the channel allows transmission of K values per second, then C, the channel capacity per second is given by

$$C = KC_s \quad \text{bits/second} \tag{7-71}$$

It should be noted that just as in the case of discrete variables, $I(\mathrm{x};\mathrm{y})$ is symmetrical with respect to x and y for continuous random variables also. This can easily be seen from Eq. 7-62. Using Bayes' rule, we can express this equation as

$$I(\mathrm{x};\mathrm{y}) = \int_{-\infty}^{\infty}\int_{-\infty}^{\infty} p(x,y)\log\frac{p(x,y)}{p(x)p(y)}\,dx\,dy \tag{7-72}$$

It is evident from this equation that $I(\mathrm{x};\mathrm{y})$ is symmetrical with respect to x and y. Hence

$$I(\mathrm{x};\mathrm{y}) = I(\mathrm{y};\mathrm{x})$$

From Eq. 7-66, it now follows that

$$I(\mathrm{x};\mathrm{y}) = H(\mathrm{x}) - H(\mathrm{x} \mid \mathrm{y}) = H(\mathrm{y}) - H(\mathrm{y} \mid \mathrm{x}) \tag{7-73}$$

7-11. ENTROPY OF A WHITE GAUSSIAN BAND-LIMITED NOISE

Let us consider a white gaussian noise with a power density $\mathfrak{N}/2$ band-limited to B Hz. We have seen in Chapter 3 that this process can be completely specified by $2B$ uniform samples per second. Furthermore, all these samples are independent, gaussian random variables with mean-square value $\mathfrak{N}B$. We have already shown that the entropy of a gaussian random variable with mean-square value $\mathfrak{N}B$ is given by (see Eq. 7-58)

$$H(\mathrm{x}) = \tfrac{1}{2}\log_2(2\pi e\mathfrak{N}B) \tag{7-74}$$

Thus entropy per sample is $(1/2)\log(2\pi e\mathfrak{N}B)$ bits. The signal is completely specified by $2B$ samples per second. Can we say that the entropy of the signal per second is the sum of entropies of $2B$ samples? This will be true provided the samples are independent of each other. If the samples are dependent, then the knowledge of one sample gives us some information about other samples and if we add the entropies (information content) of all the $2B$ samples, then we shall be duplicating certain portions of information. Hence the total entropy (information content) of the signal per second will in general be less than or equal to the sum of the entropies of the $2B$ samples. However, when all the samples are independent (as in the case of white gaussian noise), the entropy of the signal per second is the sum of entropies of all the samples ($2B$ per second). If we denote the entropy of white gaussian noise (band-limited to B Hz) by $H(\mathrm{n})$, then

$$\begin{aligned} H(\mathrm{n}) &= 2B[\tfrac{1}{2}\log(2\pi e\mathfrak{N}B)] \\ &= B\log(2\pi e\mathfrak{N}B) \qquad (7\text{-}75) \end{aligned}$$

From the results derived so far, we can draw one significant conclusion. Among all signals band-limited to B Hz and constrained to have a certain mean-square value σ^2, the white gaussian band-limited signal has the largest entropy per second. The reason for this lies in the fact that for a given mean-square value, gaussian samples have the largest entropy; moreover, all the $2B$ samples of a gaussian band-limited process are independent. Hence the entropy per second is the sum of the entropies of all the $2B$ samples. In processes which are not white, the Nyquist samples are correlated and hence the entropy per second is less than the sum of entropies of $2B$ samples. If the signal is not gaussian, then its samples are not gaussian and hence the entropy per sample is also less than the maximum possible entropy for a given mean-square value. To reiterate, *for a class of band-limited signals contrained to a certain mean-square value, the white gaussian signal has the largest entropy per second or the largest amount of uncertainty.* This is also the reason why white gaussian noise is the worst possible noise in terms of interference with signal transmission.

7-12. CAPACITY OF A BAND-LIMITED CHANNEL DISTURBED BY WHITE GAUSSIAN NOISE

The channel capacity C is by definition, the maximum rate of information transmission over the channel. The mutual information $I(\mathrm{x};\mathrm{y})$ is given by Eq. 7-73:

$$I(\mathrm{x};\mathrm{y}) = H(\mathrm{y}) - H(\mathrm{y} \mid \mathrm{x})$$

The channel capacity C is the maximum value of mutual information $I(\mathrm{x};\mathrm{y})$ per second. Let us first find the maximum value of $I(\mathrm{x};\mathrm{y})$ per sample. We shall here find the capacity of a channel band-limited to B Hz and disturbed by a white gaussian noise of power density $\mathfrak{N}/2$. In addition, we shall constrain the signal power (or its mean-square value) to S. The disturbance is assumed to be additive—that is, the received signal $\mathrm{y}(t)$ is given by

$$\mathrm{y}(t) = \mathrm{x}(t) + \mathrm{n}(t)$$

Since the channel is band-limited, both the signal $\mathrm{x}(t)$ and the noise $\mathrm{n}(t)$ are band-limited to B Hz. Obviously $\mathrm{y}(t)$ is also band-limited to B Hz. All these signals can therefore be completely specified by samples taken at the uniform rate of $2B$ samples per second. Let us find the maximum information that can be transmitted per sample. Let x, n, and y represent samples of $\mathrm{x}(t)$, $\mathrm{n}(t)$ and $\mathrm{y}(t)$ respectively. The information $I(\mathrm{x};\mathrm{y})$ transmitted per sample is given by Eq. 7-73:

$$I(\mathrm{x};\mathrm{y}) = H(\mathrm{y}) - H(\mathrm{y} \mid \mathrm{x})$$

We shall now find $H(\mathrm{y} \mid \mathrm{x})$. By definition (Eq. 7-65),

$$\begin{aligned} H(\mathrm{y} \mid \mathrm{x}) &= \int_{-\infty}^{\infty} \int_{-\infty}^{\infty} p(x, y) \log \frac{1}{p(y \mid x)}\, dx\, dy \\ &= \int_{-\infty}^{\infty} p(x)\, dx \int_{-\infty}^{\infty} p(y \mid x) \log \frac{1}{p(y \mid x)}\, dy \end{aligned}$$

Remember that

$$p(y \mid x) = p_{\mathrm{y}}(y \mid \mathrm{x} = x)$$

Since

$$y = x + n$$

for a given x, y is equal to n plus a constant (x). Hence the distribution of y, when x has a given value, is identical to that of n except for a translation of x. If $p_{\mathrm{n}}(\cdot)$ represents the probability density of noise sample n, then obviously

$$p(y \mid x) = p_{\mathrm{n}}(y - x)$$

$$\int_{-\infty}^{\infty} p(y \mid x) \log \frac{1}{p(y \mid x)}\, dy = \int_{-\infty}^{\infty} p_{\mathrm{n}}(y - x) \log \frac{1}{p_{\mathrm{n}}(y - x)}\, dy$$

Letting $y - x = z$, we have

$$\int_{-\infty}^{\infty} p(y \mid x) \log \frac{1}{p(y \mid x)}\, dy = \int_{-\infty}^{\infty} p_{\mathrm{n}}(z) \log \frac{1}{p_{\mathrm{n}}(z)}\, dz$$

The right-hand side is obviously the entropy $H(\mathrm{n})$ of the noise sample n. Hence

$$\begin{aligned} H(\mathrm{y} \mid \mathrm{x}) &= H(\mathrm{n}) \int_{-\infty}^{\infty} p(x)\, dx \\ &= H(\mathrm{n}) \end{aligned} \tag{7-76}$$

In deriving Eq. 7-76, we made no assumptions about the noise. Hence Eq. 7-76 is very general and applies to all types of noise. The only condition is that the noise disturbs the channel in an additive fashion. Thus

$$I(\mathrm{x};\mathrm{y}) = H(\mathrm{y}) - H(\mathrm{n}) \quad \text{bits/sample} \tag{7-77}$$

We have assumed that the mean-square value of the signal $x(t)$ is constrained to have a value S, and the mean-square value of the noise is N. We shall also assume that the signal $x(t)$ and noise $n(t)$ are independent. In such a case, the mean square value of y will be the sum of the mean square values of x and n. Hence

$$\overline{y^2} = S + N$$

For a given noise [given $H(n)$], $I(x; y)$ is maximum when $H(y)$ is maximum. We have seen that for a given mean-square value of y ($\overline{y^2} = S + N$), $H(y)$ will be maximum if y is gaussian, and the maximum entropy $H_{max}(y)$ is then given by

$$H_{max}(y) = \tfrac{1}{2} \log [2\pi e(S + N)] \tag{7-78}$$

Since

$$y = x + n$$

and n is gaussian, y will be gaussian only if x is gaussian. As the mean-square value of x is S, this implies that

$$p(x) = \frac{1}{\sqrt{2\pi S}} e^{-x^2/2S}$$

and

$$\begin{aligned} I_{max}(x; y) &= H_{max}(y) - H(n) \\ &= \tfrac{1}{2} \log [2\pi e(S + N)] - H(n) \end{aligned}$$

For a white gaussian noise with mean-square value N,

$$H(n) = \tfrac{1}{2} \log 2\pi e N \qquad (N = \mathfrak{N} B)$$

and

$$I_{max}(x; y) = \frac{1}{2} \log \left(\frac{S + N}{N}\right) \tag{7-79a}$$

$$= \frac{1}{2} \log \left(1 + \frac{S}{N}\right) \tag{7-79b}$$

The channel capacity per second will be the maximum information that can be transmitted per second. Equation 7-79 represents the maximum information transmitted per sample. If all the samples are statistically independent, the total information transmitted per second will be $2B$ times $I_{max}(x; y)$. If the samples are not independent, then the total information will be less than $2BI_{max}(x; y)$. Since channel capacity C represents the maximum possible information transmitted per second,

$$C = 2B \left[\frac{1}{2} \log \left(1 + \frac{S}{N}\right)\right]$$

$$= B \log \left(1 + \frac{S}{N}\right) \quad \text{bits/second} \tag{7-80a}$$

$$= B \log (1 + \Lambda) \quad \text{bits/second} \tag{7-80b}$$

where Λ is the signal-to-noise power ratio S/N.

The samples of a band-limited gaussian signal are independent if and only if the signal power density is uniform over the band (see Sec. 3-15). Obviously, to transmit information at the maximum rate (Eq. 7-80), the power density of signal $y(t)$ must be uniform. The power density of y is given by

$$\mathcal{S}_y(\omega) = \mathcal{S}_x(\omega) + \mathcal{S}_n(\omega)$$

Since $\mathcal{S}_n(\omega) = \mathfrak{N}/2$, the power density of $x(t)$ must also be uniform. Thus the maximum rate of transmission (C bits per second) is attained when $x(t)$ is also white and gaussian signal.

To recapitulate, *when the channel noise is additive and is white gaussian, with mean-square value N ($N = \mathfrak{N}B$), the channel capacity C of a band-limited channel under the constraint of a given signal power S is given by*

$$C = B \log \left(1 + \frac{S}{N}\right) \quad \text{bits/second}$$

where B is the channel bandwidth in Hertz. The maximum rate of transmission (C bits/second) can be realized only if the input signal is a white gaussian signal.

Capacity of a Channel of Infinite Bandwidth. Superficially, Eq. 7-80 seems to indicate that the channel capacity goes to ∞ as the channel's bandwidth B goes to ∞. This, however, is not true. For white noise, the noise power $N = \mathfrak{N}B$. Hence as B increases, N also increases. It can be shown that in the limit as $B \to \infty$, C approaches a limit

$$C = B \log \left(1 + \frac{S}{N}\right)$$

$$= B \log \left(1 + \frac{S}{\mathfrak{N}B}\right)$$

$$\lim_{B \to \infty} C = \lim_{B \to \infty} B \log \left(1 + \frac{S}{\mathfrak{N}B}\right)$$

$$= \lim_{B \to \infty} \frac{S}{\mathfrak{N}} \left[\frac{\mathfrak{N}B}{S} \log \left(1 + \frac{S}{\mathfrak{N}B}\right)\right]$$

This limit can be found by noting that

$$\lim_{x \to \infty} x \log_2 \left(1 + \frac{1}{x}\right) = \log_2 e = 1.44$$

Hence

$$\lim_{B \to \infty} C = 1.44 \frac{S}{\mathfrak{N}} \quad \text{bits/second} \tag{7-81}$$

Thus for a white gaussian channel noise, the channel capacity C approaches a limit 1.44 $S/\mathfrak{N}$ as $B \to \infty$. The variation of C with B is shown in Fig. 7-4. It is evident that capacity can be made infinite only by increasing the signal power S to infinity. For finite signal and noise powers, the channel capacity always remains finite.

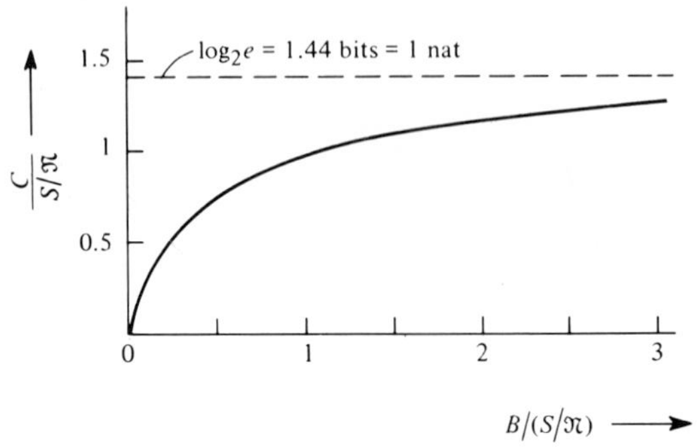

FIG. 7-4.

7-13. EFFICIENCY OF ORTHOGONAL SIGNALS

When a channel is disturbed by white gaussian noise, we have shown that under the constraint of a given signal power, the channel capacity C is given by

$$C = B \log\left(1 + \frac{S}{N}\right) \quad \text{bits/second}$$

where B is the bandwidth of transmission. This implies that it is possible to attain an error-free transmission if the rate is maintained below C. We shall now consider M-ary orthogonal signals (discussed in Chapter 6) and show that when optimum reception conditions exist, the performance of M-ary orthogonal signals approaches the upper bound predicted by Shannon's theorem in the limit as $M \to \infty$. This implies that M-ary orthogonal signals can transmit information error-free, at a rate approaching C bits per second in the limit as $M \to \infty$.

Let us consider M-ary signals each of energy E and duration T seconds. Assuming that all the M signals are equiprobable, the information per signal is $\log M$ bits. Since there are $1/T$ signals transmitted per second, R, the rate of information transmission is $\log M/T$ bits per second.

$$R = \frac{\log M}{T} \quad \text{bits/second}$$

The average signal power S is given by

$$S = \frac{E}{T}$$

Therefore

$$R = \frac{S \log M}{E} \quad \text{bits/second} \tag{7-82}$$

The probability of making a correct decision, using the optimum receiver is $P(C)$, given by Eq. 6-48:

$$P(C) = \frac{1}{\sqrt{2\pi}} \int_{-\infty}^{\infty} e^{-y^2/2} \left[\text{erf}\left(\sqrt{\frac{2E}{\mathfrak{N}}} + y\right)\right]^{M-1} dy \tag{7-83}$$

Substituting Eq. 7-82 in Eq. 7-83, we have

$$P(C) = \frac{1}{\sqrt{2\pi}} \int_{-\infty}^{\infty} e^{-y^2/2} \left[\text{erf}\left(\sqrt{\frac{2S \log M}{\mathfrak{N}R}} + y\right)\right]^{M-1} dy$$

Letting $\dfrac{S}{\mathfrak{N}R} = \alpha$,

$$P(C) = \frac{1}{\sqrt{2\pi}} \int_{-\infty}^{\infty} e^{-y^2/2} [\text{erf}(\sqrt{2\alpha \log M} + y)]^{M-1} dy \tag{7-84}$$

We shall now show that[7]

$$P(C) \longrightarrow 1 \text{ as } M \longrightarrow \infty \text{ provided } R < 1.44\,(S/\mathfrak{N}) \text{ bits/second}$$

Consider the logarithm of the bracketed term in Eq. 7-84.

$$\lim_{M\to\infty} \ln [\text{erf}(\sqrt{2\alpha \log M} + y)]^{M-1} = \lim_{M\to\infty} \frac{\ln [\text{erf}(\sqrt{2\alpha \log M} + y)]}{1/(M-1)}$$

As $M \to \infty$, both the numerator and the denominator of the right-hand side approach 0. We shall therefore use l'Hospital's rule while treating M as a continuous variable. The right-hand side is therefore given by

$$= \lim_{M\to\infty} \frac{[\text{erf}\sqrt{2\alpha \log M} + y)]^{-1} \dfrac{1}{\sqrt{2\pi}} \exp \times \left[-\dfrac{1}{2}(2\alpha \log M + y^2 + 2y\sqrt{2\alpha \log M})\right]\left[\dfrac{1}{2M}\sqrt{\dfrac{2\alpha}{\ln 2 \ln M}}\right]}{-1/(M-1)^2}$$

$$= \lim_{M\to\infty} \left[\frac{-(M-1)^2}{M^{1+\alpha/\ln 2}}\right]\left[\sqrt{\frac{\alpha}{4\pi \ln 2 \ln M}}\right]\left[\frac{\exp[-y^2/2 - y\sqrt{2\alpha \log M}]}{\text{erf}(\sqrt{2\alpha \log M} + y)}\right]$$

$$= \begin{cases} -\infty & \text{if } \alpha < \ln 2 \\ 0 & \text{if } \alpha > \ln 2 \end{cases}$$

[7]The derivation follows Viterbi (Ref. 10).

Thus

$$\lim_{M\to\infty} \ln\,[\text{erf}\,(\sqrt{2\alpha \log M} + y)]^{M-1} = \begin{cases} -\infty & \text{if } \alpha < \ln 2 \\ 0 & \text{if } \alpha > \ln 2 \end{cases}$$

Hence

$$\lim_{M\to\infty} \text{erf}\,(\sqrt{2\alpha \log M} + y) = \begin{cases} 0 & \text{if } \alpha < \ln 2 \\ 1 & \text{if } \alpha > \ln 2 \end{cases} \tag{7-85}$$

It is obvious from Eq. 7-85, that the quantity inside the bracket in Eq. 7-84 becomes unity for $\alpha > \ln 2$, and

$$P(C) = \frac{1}{\sqrt{2\pi}} \int_{-\infty}^{\infty} e^{-y^2/2}\,dy = 1$$

Therefore the error probability $P(\epsilon) \to 0$:

$$P(\epsilon) = 1 - P(C) = 0 \qquad \text{for } \alpha > \ln 2$$

Since $\alpha = (S/\mathfrak{N}R)$, we have error-free transmission if

$$\alpha = \frac{S}{\mathfrak{N}R} > \ln 2$$

or

$$R < \frac{S}{\mathfrak{N}\ln 2} = 1.44\,\frac{S}{\mathfrak{N}} \quad \text{bits/second} \tag{7-86}$$

This result can be easily seen from Fig. 7-5. This figure represents the probability of error $P(\epsilon)$ for orthogonal set for various values of M. Note that this figure is the same as the Fig. 6-16 except that $P(\epsilon)$ is shown as a function of $S/\mathfrak{N}R$ instead of $E/\mathfrak{N}$. It is obvious that as $M \to \infty$, $P\,(\epsilon) \to 0$ provided that $S/\mathfrak{N}R < 0.695$ or $R < 1.44\,S/\mathfrak{N}$.

Let us now find the maximum rate predicted by the channel capacity formula. In Sec. 6-5 we have shown that the channel bandwidth required for transmission of M-ary signals is $(M/2T)$ Hz. Obviously as $M \to \infty$, this bandwidth becomes infinite. The channel capacity (for white gaussian channel noise) in this case was found to be (Eq. 7-81) $1.44\,S/\mathfrak{N}$. Hence the maximum rate of error-free transmission predicted by Shannon's second theorem approaches $1.44\,S/\mathfrak{N}$. We have shown that M-ary signals attains this maximum error-free rate asymptotically as $M \to \infty$. This behavior is also exhibited by transorthogonal (simplex) signals. Thus, with infinite bandwidth, the orthogonal, and transorthogonal sets all attain the upper bound of performance predicted by Shannon's theorem.

Note, however, that the optimum receiver for M-ary signals consists of M correlators or M matched filters. Hence as $M \to \infty$, the receiver tends to become extremely unwieldy. Also as $M \to \infty$, the information

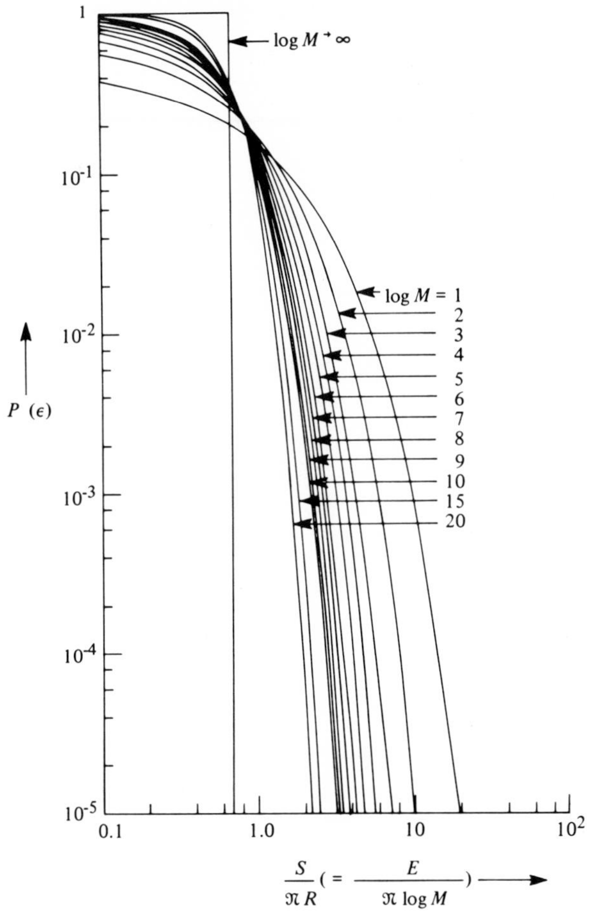

FIG. 7-5. (From Viterbi, Ref. 5-5, with permission.)

contained in each M-ary signal ($\log M$ bits) approaches infinity. Since the rate of transmission is finite, this implies that the duration T of each signal approaches infinity. It is evident that there is infinite delay associated with transmission. The encoder must wait for a long time (ideally infinite) before it receives enough information to be transmitted as one M-ary signal. This is of course the characteristic of any channel used for error-free transmission at a rate approaching its capacity.

7-14. EXCHANGE OF BANDWIDTH AND SIGNAL-TO-NOISE RATIO

From Eq. 7-80 it is apparent that for a given channel capacity, the bandwidth and the signal-to-noise ratio can be exchanged. To maintain a given rate of transmission, we may reduce the SNR provided the bandwidth is increased correspondingly (as given in Eq. 7-80). On the other hand, a given amount of information may be squeezed in a smaller bandwidth provided the SNR is increased appropriately.

Let us consider two channels of identical capacity when the channel noise is white gaussian. Let the bandwidths of these channels be B_1 and B_2 and SNR be Λ_1 and Λ_2 respectively. Then

$$B_1 \log(1 + \Lambda_1) = B_2 \log(1 + \Lambda_2)$$

or

$$1 + \Lambda_2 = (1 + \Lambda_1)^{B_1/B_2} \tag{7-87}$$

Equation 7-87 gives the rule of exchange between SNR and bandwidth under ideal conditions. In general, Λ_1 and $\Lambda_2 >> 1$ and Eq. 7-87 can be approximated as

$$\Lambda_2 = (\Lambda_1)^{B_1/B_2} \tag{7-88}$$

Equation 7-88 shows that under ideal conditions, the SNR improves exponentially with bandwidth. Let us see how wide-band systems used to transmit analog data fare in the light of Eq. 7-88. For FM and PM, we have shown that for conventional demodulation methods, the output SNR varies as the square of the bandwidth of transmission. This is obviously inferior to the ideal performance (Eq. 7-88), where the SNR increases exponentially with bandwidth of transmission. It can be shown, however, that using optimum demodulators (phase-locked loop), the performance of a phase-modulated system approaches the upper bound predicted by Shannon's theorem in the limit as the bandwidth of transmission is made infinite (see Ref. 5, Chapter 5). The optimum-frequency demodulator behaves similarly.

If the output SNR of a demodulator is Λ, and the signal bandwidth is B Hz, then ideally such a signal will need a channel of capacity

$$C = B \log(1 + \Lambda)$$

If to transmit this signal, if we use a channel of infinite bandwidth and a signal of power S, then the capacity of the channel is (Eq. 7-81):

$$C = 1.44 \frac{S}{\mathfrak{N}} = \log e \frac{S}{\mathfrak{N}}$$

Hence

$$(\log e) \frac{S}{\mathfrak{N}} = B \log(1 + \Lambda)$$

and

$$\frac{S}{\mathfrak{N}} = B \ln(1 + \Lambda)$$

$$1 + \Lambda = \exp\left(\frac{S}{\mathfrak{N}B}\right) \tag{7-89}$$

for $\Lambda >> 1$,

$$\Lambda \simeq \exp\left(\frac{S}{\mathfrak{N}B}\right) \tag{7-90}$$

Note that $(S/\mathfrak{N}B)$ represents the channel SNR (the ratio of signal power over the channel to the noise power within the bandwidth of the baseband signal). Therefore theoretically under ideal conditions when the channel bandwidth is made infinite, the output SNR increases exponentially with the channel SNR. This behavior is exhibited by the optimum phase-demodulation system when the input signal has a uniform power-density spectrum (see Ref. 5, Chapter 5). This is shown in Figs. 5-22 and 5-24.

Next we come to digital systems used to transmit analog data. We have seen that in such a system (PCM), the output SNR increases exponentially with bandwidth. However, the performance of the binary PCM system (discussed in Chapter 5), is still somewhat inferior to the ideal performance by 10 to 20 db.

Information theory in general and the channel capacity formula (Eq. 7-80) in particular have many other applications. The reader may consult Refs. 11 and 12 of this chapter for some interesting applications.

APPENDIX 7-1.

Given

$$\sum_{i=1}^{n} P_i = \sum_{i=1}^{n} Q_i = 1 \tag{A7-1}$$

we shall prove that

$$\sum_{i=1}^{n} P_i \log \frac{1}{P_i} \leq \sum_{i=1}^{n} P_i \log \frac{1}{Q_i} \tag{A7-2}$$

with equality if and only if

$$P_i = Q_i \qquad \text{for all } i \tag{A7-3}$$

To prove this result, we first show that

$$\ln x \leq x - 1 \tag{A7-4}$$

with equality if and only if $x = 1$. The functions $\ln x$ and $x - 1$ are sketched in Fig. A7-1. It is evident from this figure that

$$\ln x < x - 1$$

with equality if and only if $x = 1$.

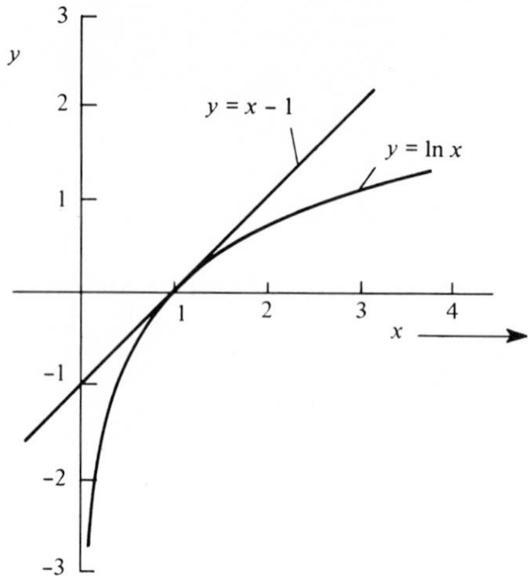

FIG. A7-1.

We have

$$\sum_{i=1}^{n} P_i \log \frac{Q_i}{P_i} = \frac{1}{\ln 2} \sum_{i=1}^{n} P_i \ln \frac{Q_i}{P_i} \tag{A7-5}$$

Applying the inequality in Eq. A7-4 to each right-hand side term in Eq. A7-5, we have

$$\begin{aligned} \sum_{i=1}^{n} P_i \log \frac{Q_i}{P_i} &\leq \frac{1}{\ln 2} \sum_{i=1}^{n} P_i \left(\frac{Q_i}{P_i} - 1 \right) \\ &\leq \frac{1}{\ln 2} \sum_{i=1}^{n} Q_i - \sum_{i=1}^{n} P_i \\ &\leq 0 \end{aligned}$$

or

$$\sum_{i=1}^{n} P_i \log \frac{1}{P_i} \leq \sum_{i=1}^{n} P_i \log \frac{1}{Q_i}$$

with equality if and only if

$$P_i = Q_i \qquad \text{for all } i.$$

REFERENCES

1. Hartley, R. V. L., "Transmission of Information," *Bell System Tech. J.*, vol. 7, no. 3 (July 1928), pp. 535–563.

2. Nyquist, H., "Certain Factors Affecting Telegraph Speed," *Bell System Tech. J.*, vol. 3, no. 2 (April 1924), pp. 324–346.

3. Shannon, C. E., and W. Weaver, *The Mathematical Theory of Communication*, U. of Illinois Press, Urbana, Ill., 1949.

4. Abramson, N., *Information Theory and Coding*, McGraw-Hill, New York, 1963.
5. Reza, F. M., *An Introduction to Information Theory*, McGraw-Hill, New York, 1961.
6. Ash, R., *Information Theory*, Wiley, New York, 1965.
7. Huffman, D. A., "A Method for Construction of Minimum Redundancy Codes," *Proc. IRE*, vol. 40, no. 10 (1952), pp. 1098–1101.
8. Peterson, W. W., *Error-Correcting Codes*, Wiley, New York, 1961.
9. Wozencraft, I. M., and B. Reiffen, *Sequential Decoding*, Wiley, New York, 1961.
10. Viterbi, A., "On Coded Phase-Coherent Communication," *IRE Trans. Space Electron. Telemetry*, vol. SET–7 (March 1961), pp. 3–12.
11. Pierce, J. R., *Symbols, Signals, and Noise*, Harper & Brothers, New York, 1961.
12. Beckman, Petr., *Probability in Communication Engineering*, Harcourt, Brace & World, Inc., New York, 1967.

PROBLEMS

7-1. A television picture is composed of approximately 300,000 basic picture elements (about 600 picture elements in a horizontal line and 500 horizontal lines per frame). Each of these elements can assume 10 distinguishable brightness levels (such as black and shades of gray). A picture element can assume these 10 levels of brightness with equal probability. Find the information content of a television picture frame.

7-2. A radio announcer describes a television picture orally in 1,000 words out of his vocabulary of 10,000 words. Assume that each of the 10,000 words in his vocabulary is equally likely to occur in the description of this picture (a crude approximation, but good enough to give the idea). Determine the amount of information broadcast by the announcer in describing the picture. Would you say the announcer can do justice to the picture in 1,000 words? Is the old adage "a picture is worth a thousand words" an exaggeration or underrating of the reality?

7-3. A source emits six messages with probabilities 1/2, 1/4, 1/8, 1/16, 1/32, and 1/32 respectively. Find the entropy of the source. Obtain the compact binary code and find the average length of the code word. Determine the efficiency and the redundancy of the code.

7-4. A source emits five messages with probabilities 1/3, 1/3, 1/9, 1/9 and 1/9 respectively. Find the entropy of the source. Obtain the compact 3-ary code and find the average length of the code word. Determine the efficiency and the redundancy of the code.

7-5. For the messages in Prob. 7-3, obtain the compact 3-ary code and find the average length of the code word. Determine the efficiency and the redundancy of this code.

7-6. For the messages in Prob. 7-4, obtain the compact binary code and find the average length of the code word. Determine the efficiency and the redundancy of this code.

7-7. A binary channel matrix is given by

$$\begin{bmatrix} \frac{2}{3} & \frac{1}{3} \\ \frac{1}{10} & \frac{9}{10} \end{bmatrix}$$

The probabilities of the two symbols being transmitted are 1/3 and 2/3 respectively. (a) Determine the probabilities of the two symbols received at the destination. (b) Determine $H(\mathrm{x})$, $H(\mathrm{x} \mid \mathrm{y})$, and $I(\mathrm{x}; \mathrm{y})$.

7-8. A cascade of two channels is shown in Fig. P. 7-8. The symbols at the source, at the output of the first channel and at the output of the second channel will be denoted

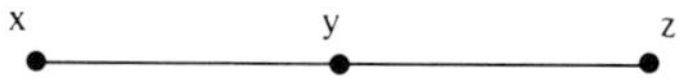

FIG. P.7-8.

by $\{\mathrm{x}\}$, $\{\mathrm{y}\}$, and $\{\mathrm{z}\}$. Show that

$$H(\mathrm{x} \mid \mathrm{z}) \geq H(\mathrm{x} \mid \mathrm{y})$$

and

$$I(\mathrm{x}; \mathrm{y}) \geq I(\mathrm{x}; \mathrm{z})$$

This shows that the information that can be transmitted over a cascaded channel can be no greater than that transmitted over one link. In effect, information channels tend to leak information. *Hint*: For cascaded channel, observe that

$$P(\mathrm{z}_k \mid \mathrm{y}_j, \mathrm{x}_i) = P(\mathrm{z}_k \mid \mathrm{y}_j)$$

Hence by Bayes' rule,

$$P(\mathrm{x}_i \mid \mathrm{y}_j, \mathrm{z}_k) = P(\mathrm{x}_i \mid \mathrm{y}_j)$$

7-9. For a continuous random variable x constrained to a peak magnitude $M(-M < x < M)$ show that the entropy is maximum when x is uniformly distributed in the range $(-M, M)$ and has zero probability density outside this range. Show that the maximum entropy is given by $\log 2M$.

7-10. For a continuous random variable x constrained to only positive values $0 < x < \infty$ and a mean value A, show that the entropy is maximum for the probability-density function.

$$p_{\mathrm{x}}(x) = \frac{1}{A} e^{-x/A} u(x)$$

The corresponding entropy is given by

$$H(\mathrm{x}) = \log eA$$

7-11. Use the data in Prob. 7-1 to estimate the minimum channel bandwidth required to transmit TV signals. Assume channel noise to be white gaussian and a signal-to-noise power ratio of 30 db.

7-12. Show that the channel capacity of a band-limited channel disturbed by a colored gaussian noise under the constraint of a given signal power is

$$C = B \log [\mathcal{S}_s(\omega) + \mathcal{S}_n(\omega)] - \int_{f_1}^{f_2} \log \mathcal{S}_n(\omega)\, df$$

where B is the channel bandwidth (in Hz) over the frequency range f_1, $f_2(f_2 - f_1 = B)$. $\mathcal{S}_s(\omega)$ and $\mathcal{S}_n(\omega)$ are the signal and the noise power densities respectively.

Show that this maximum rate of information transmission is attained if the desired signal is gaussian and its power-density spectrum satisfies the condition.

$$\mathcal{S}_s(\omega) + \mathcal{S}_n(\omega) = \alpha \text{ (a constant)}$$

(*Hint*: Consider a narrow-band Δf in the range (f_1, f_2). The maximum rate of transmission over this band is given by

$$\Delta f \log \left[\frac{\mathcal{S}_s(\omega)\,\Delta f + \mathcal{S}_n(\omega)\,\Delta f}{\mathcal{S}_n(\omega)\,\Delta f}\right] = \Delta f \log \left[\frac{\mathcal{S}_s(\omega) + \mathcal{S}_n(\omega)}{\mathcal{S}_n(\omega)}\right]$$

provided the signal over this band is gaussian. The rate of transmission over the entire band is given by

$$\int_{f_1}^{f_2} \log \left[\frac{\mathcal{S}_s(\omega) + \mathcal{S}_n(\omega)}{\mathcal{S}_n(\omega)}\right] df$$

Now maximize this under the constraint

$$\int_{f_1}^{f_2} \mathcal{S}_s(\omega)\, df = S)$$

7-13. Using the results in Prob. 7-12, show that the worst kind of gaussian noise is a white gaussian noise when it is constrained to a given mean-square value. (*Hint*: Use the expression for channel capacity in Prob. 7-12. The first term in this expression is constant. Now show that the second term attains a maximum value when $\mathcal{S}_n(\omega)$ is constant under the constraint

$$\int_{f_1}^{f_2} \mathcal{S}_n(\omega)\, df = N)$$

Index